Scientific Writing and Communication

Scientific Writing and Communication

PAPERS, PROPOSALS, AND PRESENTATIONS

Angelika H. Hofmann, Ph.D.
Yale University

New York   Oxford
OXFORD UNIVERSITY PRESS
2010

Oxford University Press, Inc., publishes works that further Oxford University's
objective of excellence in research, scholarship, and education.

Oxford   New York
Auckland   Cape Town   Dar es Salaam   Hong Kong   Karachi
Kuala Lumpur   Madrid   Melbourne   Mexico City   Nairobi
New Delhi   Shanghai   Taipei   Toronto

With offices in
Argentina   Austria   Brazil   Chile   Czech Republic   France   Greece
Guatemala   Hungary   Italy   Japan   Poland   Portugal   Singapore
South Korea   Switzerland   Thailand   Turkey   Ukraine   Vietnam

Published by Oxford University Press, Inc.,
198 Madison Avenue, New York, New York 10016
http://www.oup-usa.org

Oxford is a registered trademark of Oxford University Press

Library of Congress Cataloging-in-Publication Data
Hofmann, Angelika H., 1965-
Scientific writing and communication : papers, proposals,
   and presentations / Angelika H. Hofmann.
   p. cm.
ISBN 978-0-19-539005-6 (pbk. : alk. paper)   1. Communication in science.
2. Scientific literature.   3. Technical writing.   I. Title.
Q223.H63 2010
808'.0665—dc22
2009041549

9 8 7 6 5 4
Printed in the United States of America
on acid-free paper

CONTENTS

v

FOREWORD

He who writes carelessly makes first and foremost the confession that he himself does not place any great value on his thoughts. For the enthusiasm which inspires the unflagging endurance necessary for discovering the clearest, most forceful and most attractive form of expressing our thoughts is begotten only by the conviction of their weightiness and truth—just as we employ silver or golden caskets only for sacred things or priceless works of art.

—Arthur Schopenhauer

It is customary for those charged with editorial duties to make a distinction between the substance of the work and its expression, between its content and its form. Never is it a simple matter, though, of a sharp idea being obscured by cloudy syntax, or a valuable contribution being dismantled by misplaced commas. Clear thought, by necessity, begs clear communication, and the burden of its delivery is always on the author, not the reader. No greater error exists for a writer than to rely upon the readership and demand that it interpret the text by guessing at the author's intentions. In doing so, authors invite mistranslations and lose the opportunity to engage their readers in a fair and true dialogue with the work.

One of the most frequent pieces of advice given to writers is based on the recognition that clarity (of thought) begets clarity (of expression). "Say what you are thinking" is an exercise that can be, however, absolutely futile if the writer has not considered what precisely those thoughts are. Read through a freshman composition essay, and you can easily locate on the page where the author felt most confident. In those areas, the grammar tends to be much improved if not perfect, and the flow of the sentences is easy and even elegant. Conversely, as a student wrestles with a difficult theory, the punctuation wanders, and the coherence suffers. It is not by accident that our greatest thinkers are often the most quoted in the canon.

Thus, we have at heart two issues to consider: how to think clearly and how to articulate our thoughts. With education and practice, the two come into an equilibrium with one another, and the writing just flows. By education I mean that one must first learn the basic tools to be employed. The grammatical elements of the written language are akin to the paintbrushes and paints of the artist, or the notes and instruments of the musician. Likewise, as artists and musicians seclude themselves in their studios and practice their techniques every day before delivering the masterpiece or the recital, so must writers pour over their craft and gain the necessary familiarity with the concepts and the language they will use to express it. Although any of us can close our eyes and envision a spectacular landscape, few can sit in front of an easel and unfold with a full palette of oil paints what we have seen in our minds. Why, then, should we expect that without learning the basics and without practice, any of us can simply produce, in final form, a text that communicates with no error and no distraction our original thoughts and discoveries?

It takes more than education and practice, however, to produce the kind of technical text that secures major grant funding or impresses an audience of one's peers. Scientific writers typically are highly trained in their field, but when asked to compose and present their ideas, they often lack the ability to recognize and acknowledge how a reader receives and processes their work. The language of their discipline is a specialized one, its alphabet unintelligible to most, and unless they learn how to write *for the reader* and not just *to the reader*, they will not move beyond producing mediocre works that detract from their scientific contributions.

A cursory look at the table of contents will convince even the most skeptical of academics that this book will remain a standard on your desk for the duration of your professional career. I cannot here do justice to the myriad subjects covered, as I would have to commit a run-on sentence that would be especially egregious in a *Foreword* such as this. Suffice it to say that Dr. Hofmann has an enviable talent for communicating communication, for teaching those of all levels how to write papers and proposals and how to deliver effective and arresting presentations. This invaluable book will illustrate for the reader how to translate clarified thought into a crisp and engaging text, one that is a pleasure to read and ponder, and one that invites further engagement, either through grant support, intellectual exchange, or, hopefully, both.

Tahia Thaddeus Reynaga, Ph.D.
Yale University

PREFACE

Communication plays a fundamental role in the sciences. It is the engine that propels virtually all scientific progress. Without good communication skills, scientists stand little chance of publishing their work, obtaining funds, receiving award, or attracting a wide audience when giving a talk. Even the most promising discovery means little if it cannot be communicated successfully. In fact, it is more often the case that advancements are limited not by their technical merit but by ineffective articulation. Thus, clear communication is a requirement, not an option, for a good scientist. Yet, most researchers are not formally trained in scientific writing, and thus have only a skeletal knowledge of basic scientific writing principles at best.

***Scientific Writing and Communication: Papers, Proposals, and Presentations* serves as a comprehensive "one-stop" reference guide to scientific writing and communication for researchers in various scientific fields.** This book can be used as a textbook, a self-guided handbook, or as a general reference guide. It covers all areas of scientific communication a scientist needs to know and to master in order to successfully promote his or her research. The level of presentation is geared for those looking to improve their writing without having to read many different books on the subject. The handbook includes practical advice for writing coherent and succinct research papers, review articles, and grant proposals, as well as for organizing academic presentations and posters. Its discussion of the basic scientific writing principles also applies to theses and annual reports. This book shows you how to write clearly as a scientific author and how to recognize shortcomings in your own writing. It does so not only by providing crucial knowledge about the structure and delivery of written material but also by explaining how readers go about reading. In addition, it presents easy-to-understand and easy-to-remember principles and guidelines. Potential problem areas in written and oral presentations are pointed out, and many guidelines and practical examples are provided for wording certain sections in research papers, grant proposals, or scientific talks. Examples range from student drafts to text passages of published works, including those of Nobel Prize papers edited by the author. The text contains numerous hallmark features, including:

Practical and conceptual organization. As a result of extensive class-testing, the text has been revised repeatedly to reflect the interests, concerns, and problems undergraduate and graduate students, postdoctoral fellows, and faculty encounter when communicating in their disciplines. The table of contents is divided into six distinctive parts. The first three parts of the text follow a logical progression from the basics of scientific writing style and composition, to planning and organizing the foundational precursor of a manuscript, to the specifics of how to write each major section of a research paper and review article. The last three parts focus on grant proposals, posters, and presentations, and job applications.

Broad-based appeal. This text is written for an audience ranging from upper-level undergraduate students to graduate students, from postdoctoral fellows and junior faculty to fully-fledged researchers. Although *Scientific Writing and Communication* can be used as a textbook, it is structured such that it is equally self-explanatory, allowing readers to understand how to write English publications or proposals and to present scientific talks without having to take a class.

Numerous real-world, relevant, and multi-disciplinary examples. The in-chapter examples are derived directly from real scientific documents and cover a broad range of disciplines, serving to accommodate the interest and need of scientists in the physical and biomedical fields, including medicine, molecular biology, ecology, chemistry, engineering, physics, and more.

Extensive exercise sets and end-of-chapter summaries. Chapter summaries reference the most important concepts in an easy-to-understand bulleted format. The end-of-chapter exercises and problems review style and composition principles and encourage readers to apply the presented principles and guidelines to their own writing. Answers to the exercises are provided in a separate appendix.

Writing guidelines and checklists for revisions. Straightforward writing principles and guidelines presented in the book provide the basis for writing scientific articles, proposals, and job applications and for creating clear posters and oral presentations. Explanations of these writing principles and guidelines are followed by common pitfall examples, as well as by suggestions and advice to revise one's work successfully. Furthermore, annotated examples of text passages bring to life the guidelines and principles presented throughout the chapter. Checklists at the end of each chapter aid readers in remembering and applying these rules when writing or revising a document.

Sample wording for scientific documents and presentations. Beginning scientific writer, non-native speaker, or those struggling with writer's block, will especially value the many tables with sample sentences that apply to different sections of a scientific paper, review article, or grant proposal. Anyone presenting data at meetings and conferences will also find the sample phrases and advice on creating and delivering a talk or poster highly useful.

Special features for ESL students and researchers. *Scientific Writing and Communication* is written in an easy-to-follow style, appealing to both native and nonnative English speakers. In addition, given the broad international authorship (and readership) in science, special consideration is given to the unique problems international authors face. If you are an author whose native language is not English, you should, however, have a solid, basic knowledge of the language to maximize use of the book.

Scientific Writing and Communication: Papers, Proposals, and Presentations will not teach you how to write in the English language, *i.e.*, it is not a grammar book. You, the writer, must *practice* writing and thinking within this structure, and learn by example from the writings of others. Familiarity with the nuances of these elements will be enhanced as you read scientific literature and pay attention to how professional scientists write about their work. You will see improvement in your own writing skills by repeatedly practicing reading, writing, and critiquing of others' work.

Writing a clear research paper or grant proposal, and presenting an articulate talk, can be difficult for any scientist, but this difficulty is by no means insurmountable. Ultimately, with guidance and practice, any scientist should be able to write a paper or proposal that sparkles with clarity and to deliver an engaging presentation. As you write your own papers or prepare your talks, you will recognize that every project has its unique challenges and that you will need practice and good judgment to apply all the writing and communication principles presented herein. In giving due attention to composition, style, and impact, your communication skills will improve significantly, and this book will have accomplished its purpose.

ACKNOWLEDGMENTS

Without the contributions and advice of others, a book such as this would not have been possible. Thus, I would like to acknowledge my students, friends, and colleagues from the Max-Planck Institute, Indiana University, Fritz-Haber Institute, Humboldt University, University of Carabobo, University of Massachusetts at Worcester, and Yale University who have shared information and ideas across the sciences. I am particularly thankful to all those who were courageous enough to allow me to use draft sentences, paragraphs, or sections as examples or problems in this book as well as to those providing me with extensive and very specific samples: Kan Biao, Irene Bosch, Mark Bradford, Jaclyn Brown, Stephane Budel, Neeta Connally, Cindy Crusto, Alexey Federov, Alison Galvani, Roland Geerken, Daniela Grunow, Moshe Herzberg, Robert Homer, Jun Korenaga, Andres Franceschi Larrea, Stefanie Leacock, Patty Lee, Jin-Yu Lu, Rudolf Lurz, Richard Phillips, Hanna Richter, Michael Robek, Robert J. Schneider, Klaus von Schwarzenberg, Hong Tang, Jeffrey Townsend, and Jimin Wang. Without these samples the book would not be nearly as effective in exemplifying clear writing.

Furthermore, I am grateful to all the reviewers who have edited and commented on various draft chapters, including:

Allison Abbott	Marquette University
Stephen Amato	Northeastern University
Gavin E. Arteel	University of Louisville
Brian Avery	Westminster College of Salt Lake City
James Bednarz	Arkansas State University
Kathy Bernard	State University of New York—Buffalo North Campus
J. Harrison Carpenter	University of Colorado—Boulder
Mark Clarke	University of Houston
Carleton DeTar	University of Utah
Jeffrey A. Donnell	Georgia Institute of Technology
Carlton Erickson	UT Austin
Joseph W. Francis	The Masters College
Christine Freeman	Ohio University—Main Campus
Cathryn Frere	West Virginia University—Medical
Todd Hurd	Shippensburg University
Marilyn James-Kracke	University of Missouri—Columbia
Katherine Kantardjieff	California State University—Fullerton
Thomas Kolb	Northern Arizona University
Michael Laughter	Georgia Institute of Technology
Theo Light	Shippensburg University of Pennsylvania
William Matter	University of Arizona
Curtis Meadow	University of Maine
Anthony A. Miller	Shenandoah University
Gene Ness	University of South Florida, College of Medicine
Steve Nizielski	Grand Valley State University
Daniel O'Connor	University of Houston
Katherine Palacio	University of Illinois—Urbana Champ
David Penetar	Harvard University
Florence Petrofes	University of Texas at El Paso
Roger A. Powell	North Carolina State
Irving Rothman	University of Houston
Tony Schountz	University of Northern Colorado
Bernard G. Schreurs	West Virginia University—Medical
John Thomlinson	California State University—Dominquez Hills
Patricia Weis-Taylor	University of Colorado, Boulder
Linda Werling	George Washington University
Dan Williams	University of Utah

Anne Windham	Brown University
Timothy Wright	New Mexico State University
Sarah Wyatt	Ohio University—Main Campus

I would especially like to thank Betty Liu, Francois Franceschi, Bettina Holzheimer, Tahia Thaddeus Reynaga, Lisa DeCrosta, Tracy Plumley, Anita and Peter Todd, James Hagen, Roopashree Narasimhaiah, Gail Emilsson, Zandra Ruiz, Paola Crucitti, Riccardo Missich, Francisco Triana, and Douglas Barnes for their encouragement as well as their critical comments and the many, many, helpful discussions over the years.

Finally, I would like to express appreciation to everyone at Oxford University Press: Jason Noe, Senior Editor; Melissa Rubes, Editorial Assistant; Patrick Lynch, Editorial Director; John Challice, Publisher and Vice President; Adam Glazer, Director of Marketing; Preeti Parasharami, Product Manager; Steven Cestaro, Production Director; Jennifer Bossert, Production Editor; Paula Schlosser, Art Director; Dan Niver and Binbin Li, Designers.

CHAPTER 1

Prelude

1.1 IMPORTANCE OF WRITING IN SCIENCE

Success in the professional scientific world hinges not only on good data but also on good communication. Without good communication, scientists stand little chance of publishing their work or moving up in their career path. Most colleges and universities, however, do not formally train their students, postdoctoral fellows, or faculty in technical writing but rather expect their students and staff to rely on undergraduate English composition classes and to pick up good composition and style by reading the publications of others. If you fall into one of these categories, you know that these approaches are usually not sufficient to provide the tools for authors in professional scientific fields. Not only the lack of these tools hinders your writing; most of the time you are also too close to your own writing to recognize any potential problems readers might have when trying to interpret what you have written.

Many readers think that science is generally hard to read because of the complexity of scientific concepts and topics. However, this complexity does not need to result in difficult communication. It is important that readers accurately perceive what you as the author had in mind. To become an effective and successful author, you can and should strive to communicate clearly without oversimplifying scientific issues.

To understand how best to write clearly, you have to understand better how readers go about reading. Expectations and perceptions of readers have been widely studied in the fields of rhetoric, linguistics, and cognitive psychology. In this book, I provide an overview of these expectations and perceptions and apply them to a broad range of scientific fields, thus giving scientists the tools needed to become better writers.

1

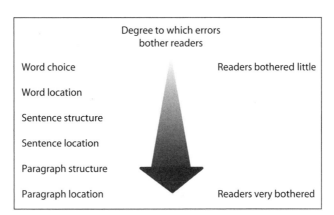

Figure 1.1 **Degree to which errors in writing bother readers. Based on the perception of readers, it is more important to logically organize and present one's ideas than to worry about perfect grammatical form or word choice.**

1.2   ABOUT READERS

Readers do not simply read—they interpret. Any sentence can be interpreted in many different ways, none of which may coincide with the interpretation intended by the author. Your goal as a writer should be to communicate the intended meaning of your writing to as many readers as possible.

How do readers interpret what is written? There is no single answer to this question. When reading scientific papers, readers are affected not only by the content and format of a paper but also by its composition and style. Their interpretations are based not only on words, sentences, and paragraphs but above all on the structural location of these elements. Thus, readers are bothered much more when a sentence is misplaced than when a word is imprecise. They are bothered much more when a paragraph that clearly belongs in the Materials and Methods section appears in the Introduction of a manuscript than when a word is misspelled. In other words, the logical and structural organization of your document is much more important than using perfect grammatical form (see Figure 1.1).

As an author, you need to be conscious of these elements when you write. Understanding the correlation of structure and function in a sentence, paragraph, or section is what underlies the science of scientific writing. The adage "form fits function" applies as much to writing as it does to the shape of a bird's wing or to the conformation of a protein.

1.3   ABOUT WRITERS

GUIDELINE:
Writing principles are the tools of professional writing.

Most of us can identify unclear writing by others, but we have a harder time recognizing our own. Scientists who write unclearly rarely think they do, much less intend to. Similarly, your own writing may appear

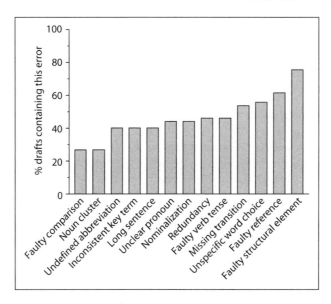

Figure 1.2 Commonness of writing errors among senior graduate students and postdoctoral fellows.

Explanation of elements:

Faulty comparison	Incomplete and ambiguous comparisons—comparing apples and oranges (Our findings are similar to Frater et al.)
Noun cluster	Two or more nouns in a row (water bath temperature variability results)
Undefined abbreviation	A nonstandard abbreviation is not defined in the text (CMMT)
Inconsistent key term	Different terms with the same meaning are used without being linked (nematode—worm—*C. elegans*)
Long sentence	Use of too many long sentences in document
Nominalization	Use of noun instead of more active and more interesting verb (measurement instead of to measure)
Redundancy	The same thing is said repeatedly in different ways, or unnecessarily long phrases are used instead of shorter versions ("the majority of" instead of "most")
Faulty verb tense	Present tense is used instead of past tense ("In our study, X crosses Y." instead of "In our study, X crossed Y.")
Unspecific word choice	Words are imprecise or unclear (The sample was incubated for several hours.)
Faulty reference	Reference is faultily placed in text, wrongly cited, inaccurate, or missing (It was reported (5) that x is the only element.)
Faulty structural element	Necessary component of a paper is missing, misplaced, or obscured (example: purpose of experiment is not stated, data are not interpreted, or conclusion is missing)

clear to you, and it will come as a surprise when your readers say that it is not. The reason for this insensitivity is simple: Anyone who writes about something and understands its content is more likely to think the passage is clearly written than someone who knows less. However, we all know

how difficult it is to understand "legalese" for anybody outside the legal field or "bureaucratese" for anyone other than a bureaucrat. In the same way that legalese and bureaucratese sound funny to most people outside these fields, "academese" sounds funny to most people outside the field of science.

To illustrate the commonness of errors in style and composition, I have randomly selected and analyzed 100 prepublication drafts, which were composed by senior graduate students and postdoctoral fellows from several U.S. universities for a range of biological and biomedical topics. Half the authors were native speakers, the other half were not—a ratio that reflects approximately the ratio of native to nonnative speakers among postdoctoral fellows in the United States today.

All drafts contained errors in style and composition, but each type of error was scored only once. For the most common grammatical, structural, and stylistic problems, the percentage of drafts containing these errors was calculated and is shown in Figure 1.2.

Interestingly, most drafts from these students and postdoctoral fellows contained faulty structural locations (76%). Faulty or missing references were the second most common error encountered (62%) followed by unspecific word choice (56%) and missing transitions (54%). Other errors, such as faulty verb tense, use of redundancies, and nominalization, as well as unclear pronouns, also occurred relatively often (40%–46%). These results demonstrate the variety as well as the amazingly high frequency of writing problems scientists appear to have and show the importance and need for good scientific writing guidelines and instructions.

Why do these problems exist in scientific writing? Many writers simply want to unburden themselves of all the information they have collected and are happy to get their data onto paper. Others learn to imitate a particular style to fit into a field. Imitation is employed particularly by authors new in a field or unfamiliar with a topic. Some plump up their writing to impress readers, convinced that dense style sounds sophisticated and reflects deep thinking. The main reason for bad style and composition, however, is ignorance and lack of training. Scientific writers are unaware of how to identify words, sentences, or paragraphs that may give readers a problem. Most are aware of certain "rules" taught to them in high school and undergraduate English composition classes but not of the writing principles that would benefit them as professionals.

To be successful scientists and professional writers, basic English composition is usually not enough. Particularly because you are often too close to your own writing to judge it, you need a set of writing principles to make you aware of how readers will interpret what has been written. The main goal of this book is to provide you with these principles to ensure that your work has the highest possible impact and is clearly understood by the majority of readers, be they editors, students, or fellow scientists.

1.4   ABOUT THIS BOOK

This book covers all elements of style and composition needed for writing a scientific document and for presenting a scientific talk. It explains choice of words, word location, sentence construction, and paragraph structure as well as the arrangement of these elements into the larger structures of a research paper, review article, or grant proposal. At the same time, the book emphasizes how the reader interprets what has been written. Special sections of the book have also been devoted to oral presentations and posters as well as to job applications.

Scientific Writing and Communication is both a practical text with teaching exercises and a ready reference guide, which can be used long after class is over. In writing this book, my goal was to target as broad of a scientific audience as possible, ranging from both undergraduate and graduate students to postdoctoral fellows, junior faculty, and other scientists inside and outside the academic field. Considering the growing globalization, especially in science, I have also added suggestions for writers and readers for whom English is a second language.

Within each chapter, potential problem areas in scientific writing are pointed out in well-explained examples, and thorough instructions on how to avoid these problems are provided. It is important for readers and students to understand the writing principles presented and to learn how to apply them. To make examples and exercises directly applicable and more readily understandable to scientific writers, most examples were taken from prepublication drafts written by scientists in various fields and represent typical mistakes made when writing about science. I have collected most of the sample sentences and paragraphs over the years and also added a few of my own. Quite a few examples come from writers whose native language is not English and thus illustrate typical mistakes made by nonnative English writers. Other examples and exercises also present writings commonly encountered by scientific editors.

Aside from the examples, the book provides ample exercises to practice the discussed writing principles. I encourage you to work through these problems. The way you will learn and improve is by struggling with the problems, trying to apply the relevant writing principles yourself even if you miss the point in your answers. It is important to practice on paper because the only way to learn how to write is by writing. So give yourself the opportunity to make your own mistakes and achieve your own successes. Know that some of the problems are challenging and may be frustrating. Try not to get stuck in the scientific details, but rather try to think about the writing and to spend the time needed to understand these problems and their revisions. In this respect, although examples and exercises from many different areas of science are used in the book, it is not important from what scientific discipline examples or exercises arise; what is important is that you recognize mistakes of style and composition. You may disagree with some choices of answers. In fact, many exercises may have more than one possible answer, as there may be more than one way to express an idea and to improve written passages. Some choices are better, some worse, and some just a matter of personal style. The goal is to improve the original statements such that they are clearer and more easily understandable by the reader.

1.5   DESIGN OF THIS BOOK

This book consists of six parts:

I. Scientific Writing Principles: Style and Composition
II. Planning and Laying the Foundation
III. Manuscripts: Research Papers and Review Articles
IV. Grant Proposals
V. Posters and Presentations
VI. Job Applications

Part I presents 30 basic scientific writing principles of technical style and composition that every scientific writer should know. These principles range from principles on word choice to principles on sentence structure, sentence location, and paragraph construction. The importance of word location and details of grammar/technical style are particularly emphasized as is the interpretation by the reader. Many examples and their revised versions are used to illustrate the writing principles, and each chapter is followed by a variety of problems to practice the writing principles.

Part II deals with key elements necessary for preparing a manuscript. This part explains how to get started writing a manuscript and discusses authorship. It also explains how to collect, manage, and use references and how to avoid plagiarism. In addition, it provides the most important illustration principles for scientific authors in the biological, biomedical, and other scientific fields. It shows not only different types of figures but also explains the difference between figures constructed for a poster, an oral presentation, or a manuscript. Part II contains examples of good and bad graphs and reinforces guidelines and principles with practical problems.

In Part III authors learn to apply the writing principles of Part I to writing and revising individual sections of a scientific research paper or review article. Authors are introduced to structural guidelines important for writing each section of an article and are given many examples of well-written sections as well as examples of sections that would benefit from revisions. Here again, authors are encouraged to practice using the discussed guidelines on provided problems. In addition, Part III provides an overview of revising a manuscript and submitting it to a journal. It also explains the review process and gives advice on how to write a cover letter and how to respond to the editor and reviewer comments.

As garnering funding is an important part of the life of a scientist, a section on grant writing is included in Part IV. These chapters give information on federal and private funders, writing letters of inquiry, and the different sections of a grant proposal. Here, writing principles are applied to putting together proposals. These chapters also contain sample sections of successful proposals and sample wordings for less frequently encountered proposal sections. In addition, Part IV also provides instructions on how to submit a proposal and how to communicate with the funder.

The ability to communicate findings orally is an important part of the scientific process. Part V instructs scientists on preparing and presenting an oral presentation or poster. In these chapters, the book provides not only real examples of slides and posters but also gives advice on the mechanism of speaking, combating stage fright, and fielding questions. It also presents basic guidelines to instruct speakers on visual aids and the content of a talk or poster.

To complete the series on scientific writing skills all scientists should have at their disposal, Part VI of this book provides information on job applications. This part includes advice on how to compose a CV and how to ask for or write a letter of recommendation. Examples of well-written research and teaching statements are also included in this chapter.

PART ONE

Scientific Writing Principles

STYLE AND COMPOSITION

CHAPTER 2

Individual Words

Word choice in scientific research papers is one of the primary concerns of scientists and editors alike. A few basic principles can provide a good guideline for the choice of words in scientific papers. These principles are the focus of this chapter.

Other parts of this chapter teach you to distinguish between words whose meanings are similar but not exactly the same. English is a particularly rich language. It encompasses about half a million words and has many synonyms and near synonyms. Over time, the meaning of words may change, making it even more difficult to distinguish between words in English. Authors need to be aware of the exact meaning of words to convey their messages clearly to as many readers as possible.

2.1   THE CENTRAL PRINCIPLE

WRITING PRINCIPLE 1:

Write with the reader in mind.

In the professional world, success in writing is determined by whether your readers understand what you are trying to say. In the scientific fields, these readers may be reviewers of a paper or proposals, editors, students, Nobel laureates, scientists from a different discipline, or readers whose native language is not English; in fact, probably most of them will be non-native speakers. Because of this diversity in readership, the burden of clarity rests on you, the author. You need to write clearly so that readers can follow your thinking and so that you achieve the highest possible impact. In other words, you need to write with the reader in mind.

To "write with the reader in mind" is the central principle of this book, and all other principles follow from it. Many scientists think that the primary goal in science is to obtain great results, but good science alone will not bring you success. Your collection of data cannot speak for itself—it needs to be communicated and communicated well. Good science does not excuse poor writing. Authors have an obligation to their readers to ensure that science is communicated well.

2.2   WORD CHOICE

Preciseness

WRITING PRINCIPLE 2:

Use precise words.

The problem of many sentences in science is not grammar but word choice. Consider the following three examples:

Example 2-1   a   The current remained increased for <u>several</u> hours.

b   Nests were observed <u>frequently</u> for signs of predation.

c   The carbonate layer was prepared <u>with</u> sodium carbonate.

Although the words underlined in these examples can be found frequently in research papers, these word choices are problematic and disliked by editors and reviewers. In each of the three sentences of the above example, the underlined words violate the same writing principle: These words are not precise.

You can improve these sample sentences by revising the word choices.

Revised	a	The current remained increased for **6 hours**.
Example 2-1	b	Nests were observed **every 12 hours** for signs of predation.
	c	The carbonate layer was prepared **using** sodium carbonate. or The carbonate layer was prepared **in the presence of** sodium carbonate.

Why are the revised sentences better than those in Examples 2-1a to 2-1c? The revised sentences convey more precisely what the writer is describing. "Enhanced" is imprecise as well as the wrong word choice. "Increased" is the correct quantitative term for concentration. But how much was the increase? Writers should give a quantitative value such as "10%."

"Frequently" is also imprecise. How often is frequently? Use a quantitative term such as "every 12 hours," or "at 6 am and at 6 pm." Science is quantitative. A quantitative detail such as "every 12 hours" is much clearer than a qualitative term such as "frequently." Overall then, you should use precise terms and state the mean or a range when applicable.

Let us look at Example 2-1c more closely. The vague term underlined in this example is "with." "With" is one of the vaguest and most ambiguous terms in English. Because "with" can mean so many things, it is clearer to use a precise term whenever possible. If you do not use precise terms, the reader has to guess what you mean. Note that "with" does have legitimate uses such as "in the company of" as in "I went to school with Brian." Another standard meaning is "by the means of" as in "We washed the dishes with soap." "With" can also be used as an attribute as in "patients with diabetes." Furthermore, some verbs are followed by "with" such as "compared with." However, scientific writers often use "with" instead of a more precise term and thus confuse readers. In the preceding example, it is much more accurate to write "using" or "in the presence of" instead of "with."

Level of Sophistication

WRITING PRINCIPLE 3:

Use simple words.

Words in science should not only be precise, but they should also be as simple as possible. Consider the next examples:

| **Example 2-2** | a | Fractions of 0.8 ml were collected, <u>reduced to dryness</u>, and dissolved in 3.75 % methanol (v/v) prior to being sequenced. |
| | b | Our results <u>reflect deviations</u> from thermal equilibrium during desorption. |

These sentences are written in a style that appears heavy and dense to the reader. Admittedly, scientific writing has many technical terms. Therefore, to keep your writing from being too heavy, choose simple words for the rest of the sentence. "Reduced to dryness" can be expressed much simpler by writing "dried" and "reflect deviations" by "deviate."

 Revised   a   Fractions of 0.8 ml were collected, **dried**, and dissolved
Example 2-2   in 3.75 % methanol (v/v) prior to being sequenced.

               b   Our results **deviate** from thermal equilibrium during
                 desorption.

The revised sentences are more easily understood by readers because their word choice is much simpler.

Here is another example of pompous words that just cries out to be simplified:

Example 2-3   There is a large body of experimental evidence that clearly shows that members of the genus *Crotalus* congregate simultaneously in cases of prolonged decreased temperature conditions in the later part of the year.

 Revised   **Rattlesnakes come together when it gets cold in the fall.**
Example 2-3

ESL advice

Many English as a Second Language (ESL) authors convert vocabulary of their native language for use in English writing. In some cultures, pompous words are extensively used, and statements tend to be indirect. In Western cultures, however, statements are rather direct. Thus, these ESL authors need to pay special attention not to overuse pompous words and phrases. These authors also need to ensure that they do not use terms in English that they would use in their language.

Regardless of your native language, remember that most of your readers are probably nonnative English speakers. You have to ensure that these readers can understand what has been written. Use simple words. That is, aside from the technical terms, choose a level of words that you would use when talking about your work to a friend; choose "use" rather than "utilize," for example (see also http://www.userlab.com/Downloads/SE.pdf for more details on using simplified English for an international audience, last accessed October, 2009).

2.3   WORD CHOICE—SPECIAL CASES

Misused Words

GUIDELINE:

Watch out for misused words.

Words are not always what they seem. Quite a few words and expressions in science are commonly misused and confused, especially by ESL authors. Some of the words are used incorrectly so often that they sound right even when they are not. Watch out for these misused and confused scientific terms. Consult a dictionary when you write so that people do not need to have one on hand when they read what you have written.

Commonly misused words fall into several categories including words with suffixes, verbs, adverbs and adjectives, and links.

ESL advice

Suffixes

-ability Be aware of *-ability* words. Often the sentence should be rewritten using a stronger verb preceded by the verb *can*.

Example 2-4	a	<u>Changeability</u> of *X* occurs when *Y* is added.
Revised Example 2-4	a	*X* **can change** when *Y* is added.

-ization Challenge *-ization* nouns. Many writers tend to invent nouns by adding the ending *-ation* or *-ization* onto the verb.

Example 2-4	b	<u>Metabolization</u> of phosphates was different than expected.
Revised Example 2-4	b	Phosphates were **metabolized** differently than expected.

-ize Often nouns or adjectives are wrongly changed to verbs by adding *-ize* to a word.

Example 2-4	c	Older patients were <u>prioritized</u>.
Revised Example 2-4	c	Older patients were **given priority**.

-ized/-izing You should also challenge *-ized* or *-izing* adjectives and search for simpler substitutions.

👎 **Example 2-4**   d   <u>Individualized</u> doses were calculated.
Nanoscience has a <u>transformatizing</u> impact on various technologies.

👍 **Revised**   d   **Individual** doses were calculated.
Example 2-4   Nanoscience has a **transformative** impact on various technologies.

or even better:

Nanoscience transforms various technologies.

-ology   This ending means the study of something and is jargon when used in sentences such as these:

👎 **Example 2-4**   e   No <u>pathology</u> was found.
<u>Cytology</u> was normal.
<u>Symptomology</u> was severe.
<u>Serology</u> was negative.

👍 **Revised**   e   No **pathologic condition** was found.
Example 2-4   **Cytologic findings** were normal.
Symptoms were severe.
Serologic findings were normal.

Verbs

make   Like "to do," "to make" is often overused by ESL writers. Be sure to use the correct terms in context instead of simply substituting "to make" for any unknown term. If you are not sure about the correct terminology, consult an English textbook, journal, or scientist who is a native speaker.

ESL advice

👎 **Example 2-5**   a   A picture was made.
A gel was <u>made</u>.
We <u>made</u> a graph.
We <u>made</u> the following experiments.

👍 **Revised**   a   A picture was **taken**.
Example 2-5   A gel was **run**.
We **graphed** the data. OR
We **constructed** a graph.
We **performed** the following experiments.

affect, effect   "Affect" is usually used as a verb and means to act on or to influence.

👍 **Example 2-5**   b   The addition of KI-3 to MZ1 cells **affected** their growth rate (i.e., it could have increased or decreased or induced something else.)

More rare, "affect" can also be a noun with a specialized meaning in medicine and psychology: an emotion.

 Example 2-5   c   People can experience a positive or negative **affect** as a result of their thoughts.

"Effect" is usually used as a noun meaning a result or resultant condition.

 Example 2-5   d   We examined the **effect** of KI-3 on MZ1 cells.

When used as a verb (rarely), "effect" means to cause or bring about.

 Example 2-5   e   The addition of KI-3 to MZ1 cells **effected** a change in their growth rate (i.e., it caused or brought about change).

Adverbs and Adjectives

overnext   This word does not exist in English. What you probably mean is "The slide *after next*."

 Example 2-6   In the <u>overnext</u> slide, we will see...

 Revised Example 2-6   In the **slide after next**, we will see...

significant(ly)   Use only when you are talking about statistical significance, and give a *P* value. Otherwise, use important, substantial, markedly, meaningful, or notable.

Links

since, because   Use "since" only in its temporal sense, not as a substitute for "because." If you want to indicate causality, use "because."

Example 2-7   a   δ stopped increasing **since** being dampened by ψ.
The reaction rate decreased **because** temperature dropped.

which, that   Sometimes these words can be used interchangeably. More often, they cannot. Use "which" with commas for nondefining (nonessential) sentences.

ESL advice

Example 2-7 b   Dogs, which have been domesticated for millennia, recovered.

Use "that" without commas for essential sentences. A phrase or clause introduced by "that" cannot be omitted without changing the meaning of the sentence. Such essential material should not be set off with commas.

 Example 2-7 c   Dogs that were treated with the antidote recovered.

Be especially careful about words that are easily confused by writers and about words that look similar but mean different things. We have seen some examples of such word pairs already: "affect" and "effect," "since" and "because," and "which" and "that." More commonly confused words, including *as/like, while/whereas, principle/principal,* and *quantitate/quantify,* are listed in **Appendix** together with their corresponding meanings.

Handling Language Sensitively

GUIDELINE:
Avoid sexism.

In recent years, people have become much more aware of the ways in which language affects our thinking. To avoid being accused of chauvinism and insensitivity, carefully consider what you write. If readers get offended, they are likely to stop reading.

Sexism includes any verbal or visual reference that presents men and women as unequal or excludes one group in favor of another. Although it may be unconscious and unintentional, sexism in writing can take many forms. Some forms are so subtle that authors might not even notice them unless they are pointed out. Consider this example:

Example 2-8 a   <u>Man</u> is not the only host for this parasite.

The easiest solution to avoid sexism is to use "unisex" terms.

Revised   a   **Humans** are not the only host for this parasite.
Example 2-8

Writing gets more complicated when we have to consider which pronoun to use for singular nouns that do not indicate gender such as faculty, staff, teacher, scientist, student, and doctor. Although more formal language requires a singular pronoun (its, his, her), it raises the problem of biased language. The change in the English language is toward using a plural "they" and plural verbs for these cases as shown in the next example:

 Example 2-8   b   A nurse should double-check her IV settings.

 Revised Example 2-8   b   **Nurses** should double-check **their** IV settings.

2.4   REDUNDANCIES AND JARGON

WRITING PRINCIPLE 4:

Omit unnecessary words and phrases.

Avoid any verbosity and wordiness by omitting unnecessary words and phrases and jargon.

Redundancies

Redundant words or phrases unnecessarily qualify other words and phrases. Many sentences in science appear complex because they contain redundancies. Writers should be as brief as possible consistent with clarity. However, if it takes more words to be clear, use more words.

Here are three examples of unnecessarily complex sentences:

 Example 2-9   a   The sample size was not <u>quite sufficiently large</u> enough.

b   High pH values <u>have been observed</u> to occur in areas that <u>have been determined</u> to have few pine trees.

c   Most galaxies with unusually luminous cores are highly asymmetric <u>in shape.</u>

Look at what can be cut out in the revision after unnecessary words and redundancies have been removed:

 Revised Example 2-9   a   The sample size was not **large** enough. OR: The sample size was too small.

b   High pH values occur in areas with few pine trees.

c   Most galaxies with unusually luminous cores are highly asymmetric.

Unnecessary words

The following individual words can and should be omitted because they add nothing to a text.

actually	basically	essentially	fairly	much	really
practically	quite	rather	several	very	virtually

Other examples of redundancies

In the next list, all the words in parentheses are redundant and can be omitted:

(already) existing	at (the) present (time)
(basic) fundamentals	blue (in color)
cold (temperature)	(completely) eliminate
(currently) underway	each (individual)
each and every [choose one]	(end) result
estimated (roughly) at	(final) outcome
first (and foremost)	(future) plans
(main) essentials	never (before)
period (of time)	reason is (because)
(still) persists	(true) facts

Other warning words

Sometimes words that are perfectly good on their own can still indicate potential trouble. The following "warning words" often indicate that your thought has to be sharpened and your writing needs to be tightened. Most of the time, these warning words can be omitted:

area	character	conditions	field
level	nature	problem	process
situation	structure	system	

Unnecessary phrases

Many unnecessary words and phrases are used by both native and nonnative English speakers. We have already looked at commonly misused words. Let us now look at commonly misused phrases. Avoiding these phrases is a simple way to make your writing shorter and clearer.

Certain phrases are often unnecessarily used to introduce previous studies or results. These phrases can almost always be deleted so that the facts are succinctly stated.

Example 2-10 <u>It is well known that</u> there are three subtypes of the KL-2 virus.

Revised Example 2-10 There are three subtypes of the KL-2 virus.

👎	**Example 2-11**	<u>In a previous study, it was demonstrated that</u> "nanowire" devices with excellent sensing characteristics can be defined by TMAH etching.
👍	**Revised Example 2-11**	"Nanowire" devices with excellent sensing characteristics can be defined by TMAH etching.

👎	**Example 2-12**	Eddies <u>have been shown to vary</u> depending on the time of year.
👍	**Revised Example 2-12**	Eddies **vary** depending on the time of year.

Other commonly used unnecessary phrases that can usually be deleted include the following:

there are many papers stating...	it is speculated that...
it was shown to...	it has been found that...
it was observed that...	it has been demonstrated...
it is reasonable to assume that...	it has been reported that...
evidence has been presented that shows that...	it has long been known that...

Phases that can be shortened include:

Avoid	Better	Avoid	Better
A considerable number of	many	in some cases	sometimes
an adequate amount of	enough	in the absence of	without
an example of this is the fact that	for example	in the event that	if
		in view of the fact that	because, since
as a consequence of	because	it is of interest to note that	note that
at no time	never	it is often the case that	often
based on the fact that	because	majority of	most
by means of	by	no later than	by
considerable amount of	much	number of	many
despite the fact that	although	on the basis of	by
due to the fact that	due to	prior to	before
during the time that	while, when	referred to as	called
first of all	first	regardless of the fact that	even though
for the purpose of	to	so as to	to
has the capability of	can, is able	utilization	use
in light of the fact that	because	with reference/regard to	about (or omit)
in many cases	often	with respect to	about
in order to	to	with the exception of	except

A more extensive list of redundancies can be found in Day, 1998, and O'Connor, 1975.

Jargon

Jargon is the use of terms specific to a technical or professional group. Jargon can also be pompous, and it is often not comprehensible for "outsiders." In science, jargon often includes "laboratory slang" as in the following examples:

Examples of jargon that should be avoided:

Southern blotted	This is laboratory jargon. The correct use is "… analyzed by Southern blot …"
Western blotted	Similar to "Southern blotted," "Western blotted" is laboratory slang. The correct use is "… subjected to western blot analysis" or "… analyzed by western blot."
electrophorized	the correct usage is "analyzed by or subjected to electrophoresis"
bugs	meaning bacteria, never used in scientific writing
lab	use "laboratory"
prep	use "prepare"
vet	the correct term to use is "veterinarian"
evidenced	use the noun "evidence" instead
vortexed	"vortex" exists only as a noun; use "was mixed by vortex" instead.

2.5   ABBREVIATIONS

WRITING PRINCIPLE 5:

Avoid too many abbreviations.

A special type of word choice to consider is the use of abbreviations. Too many abbreviations can be confusing to the reader and should therefore be kept to a minimum.

Similarly, nonstandard abbreviations need to be limited or the reader will get lost. Use International System (SI) units when you use standard abbreviations such as kg or m. Standard abbreviations are widely accepted. Check also that you have not used too many abbreviations, even those approved by your target journal. You can legitimately use abbreviations to replace lengthy terms that appear more than about 10 times in a 10-page manuscript or that appear several times in quick succession, but do not use more than four or five such abbreviations in a single paper. Additionally, avoid making sentences indigestible by using too many abbreviations in a short space:

Example 2-13 a   MPTP is converted by MAOB to MPP, which reaches SNpc nerve cells via DA uptake systems.

b   We assessed non-AGN galaxies, contained in the MUSYC survey in the ECDFS using HST for a target source.

The preceding examples may be perfectly intelligible to expert colleagues but will be unintelligible to most readers.

Define essential abbreviations at their first appearance, in a footnote at the beginning of the paper, or in both places, according to the journal's requirements. Once you have defined an abbreviation, use it whenever you need it—do not switch back to using the full term unless many pages have elapsed since its previous appearance—then you may remind the reader, once, what the abbreviation means. If you use—and define—an abbreviation in the title of a paper (although this is not recommended), redefine it in the text. Do the same for abbreviations used (and defined) in the abstract. If you are using many abbreviations in a long scientific document, consider adding a list of abbreviations with definitions to go with the document.

Special Abbreviations

Certain Latin-derived abbreviations are used often in science. Note that although the following are Latin derivatives, they are often used without italics:

e.g. = *exempli gratia*—for example
et al. = *et alia*—and others
i.e. = *id est*—that is

2.6   NOMENCLATURE AND TERMINOLOGY

WRITING PRINCIPLE 6:

Use correct nomenclature and terminology.

In science, it is important to use correct vocabulary, nomenclature, and terminology to avoid being misunderstood and to avoid confusing the reader. If you are not sure about a term, do not guess. Rather, take the time to look it up in a dictionary, thesaurus, or other reference book. Dictionaries for the biological, medical, and other scientific fields as well as online dictionaries are listed in section 2-7.

Common Terminology

Species and all Latin derivates are in *italics (in vivo, Physcomitrella patens,* etc.)

Human genes: all caps and *italics (ADH3, HBA1)*

Mouse genes: first letter cap, the rest lowercase, *italics (Sta, Shh, Glra1)*

Human proteins: caps, no italics (ADH3, HBA1)

Mouse proteins: like genes, but no italics (Sta, Shh, Glra1)

To distinguish the species of origin for homologous genes with the same gene symbol, an abbreviation of the species name is added as a prefix to the gene symbol. For example, human loci, (HSA)*G6PD*; homologous mouse loci, (MMU)*G6pd,* in which HSA = Homo sapiens, MMU = *Mus musculus.*

Restriction enzymes: a combination of *italics* and nonitalics (e.g., *Bam* HI). Check supplier.

2.7   DICTIONARIES

Dictionaries—Biological and Medical Sciences

Biological Sciences (general)

Jeffrey, C. (1992). *Biological nomenclature.* New York: Cambridge University Press.

Martin, E. (2004). *A dictionary of biology.* New York: Oxford University Press.

McGraw-Hill. (2003). *Dictionary of bioscience* (2nd ed.). New York: McGraw-Hill Book Company and Sybil P. Parker.

McGraw-Hill. (1993). *Dictionary of scientific and technical terms* (5th ed.). New York: Sybil P. Parker.

Morris, C. G. (Ed). (1992). *Academic press dictionary of science and technology.* New York: Academic.

Walker, J. M., & Cox, M. E. (1995). *The language of biotechnology: A dictionary of terms.* Washington, DC: American Chemical Society.

Walker, P. M. (1990). *Cambridge dictionary of science and technology.* New York: Cambridge University Press.

Biochemistry

Academic Press. (1984). *Enzyme nomenclature: Recommendations (1984) of the Nomenclature Committee of the International Union of Biochemistry.* Orlando, FL: Academic.

Cammack, R., Atwood, T., Campell, P., Parish, H., Smith, T., Vella, F., et al. (2006). *The Oxford dictionary of biochemistry and molecular biology.* New York: Oxford University Press.

International Union of Biochemistry. (1978). *Biochemical nomenclature and related documents.* London: Biochemical Society.

Biotechnology

Bains, W. (2004). *Biotechnology from A to Z*. New York: Oxford University Press.

Cell Biology

Lackie, J. M., & Dow, J. A. T. (1995). *The dictionary of cell biology*. New York: Academic.

Genetics

King, R. C., & Stansfield, W. D. (2002). *Dictionary of genetics*. New York: Oxford University Press.

Immunology

Herbert, W. J., Wilkinson, P. C., & Stott, D. I. (1995). *Dictionary of immunology*. New York: Academic.

Playfair, J. H. L., & Chain, B. M. (2000). *Immunology at a glance*. Oxford, England: Blackwell.

Rosen, F. S., Gamble, J. L., Steiner, L. A., & Unanue, E. R. (1989). *Macmillan dictionary of immunology*. New York: Macmillan.

Medical Sciences

International Anatomical Nomenclature Committee Subcommittees. (1989). *Nomina anatomica* (6th ed.). Edinburgh, England: Churchill Livingstone.

International Anatomical Nomenclature Committee Subcommittees. (1989). *Nomina embryologica* (3rd ed). Edinburgh, England: Churchill Livingstone.

International Anatomical Nomenclature Committee Subcommittees. (1989). *Nomina histologica* (3rd ed.). Edinburgh, England: Churchill Livingstone.

Miller, B. F., Keanne, C. B. & O'Toole, M. T. (Eds.). (2005). *Miller–Keanne encyclopedia and dictionary of medicine, nursing and allied health*. Philadelphia: Saunders.

Mosby. (2005). *Mosby's dictionary of medicine, nursing and health professions*. St. Louis, MO: C.V. Mosby.

Stedman's word books series. (2001–2004). Philadelphia: Lippincott.

Microbiology

Garrity, G. M. (2005). *Bergey's manual of systematic bacteriology, Vol. 2 (Parts A, B & C; Three-Volume Set)*. New York: Springer.

Garrity, G. M., & Boone, D. R. (2001). *Bergey's manual of systematic bacteriology: Volume 1. The archaea and the deeply branching and phototrophic bacteria*. New York: Springer.

Gillespie, S. H., & Bamford, K. B. (2003). *Medical microbiology and infection at a glance.* Malden, MA: Blackwell.

Singleton, P., & Sainsbury, D. (2002). *Dictionary of microbiology and molecular biology.* New York: Wiley.

Skerman, V. B. D., McGowan, V., & Sneath, P. H. A. (Eds.). (1980). Approved lists of bacterial names. *International Journal of Systematic Bacteriology, 30,* 225–240.

Molecular Biology

Cammack, R., Atwood, T., Campell, P., Parish, H., Smith, T., Vella, F., et al. (2006). *The Oxford dictionary of biochemistry and molecular biology.* New York: Oxford University Press.

Singleton, P., & Sainsbury, D. (2002). *Dictionary of microbiology and molecular biology.* New York: Wiley.

Plant Biology

Mabberley, D. J. (1987). *The plant-book: A portable dictionary of the higher plants.* New York: Cambridge University Press.

Macura, P. (2002). *Elsevier's dictionary of botany.* New York: Elsevier Science.

Virology

Hull, R., Brown, F., & Payne, C. (1989). *Virology: Directory & dictionary of animal, bacterial and plant viruses.* London: Macmillan.

Mahy, B. W. J. (2001). *A dictionary of virology.* New York: Academic.

Dictionaries—Other Scientific Fields

General

American Heritage Dictionary (Eds.). (2005). *American Heritage science dictionary.* Boston: Houghton Mifflin.

Chemistry

Conelly, N. G., Hartshorn, R. M., Damhus, T., & Hutton, A. T. (Eds.). (2005). *Nomenclature of inorganic chemistry: Recommendations.* London: Royal Society of Chemistry.

Daintith, J. (2004). *A dictionary of chemistry.* New York: Oxford University Press.

Dictionary of chemistry. (2003). New York: McGraw-Hill.

Hellwinkel, D. (2001). *Systematic nomenclature of organic chemistry: A directory to comprehension and application of its basic principles.* New York: Springer.

International Union of Pure and Applied Chemistry. (1971).

Nomenclature of inorganic chemistry (2nd ed.). Oxford, England: Pergamon.

International Union of Pure and Applied Chemistry. (1979). *Nomenclature of organic chemistry.* Oxford, England: Pergamon.

Marler, E. E. J. compiler. (1985). *Pharmacological and chemical synonyms: A collection of names of drugs, pesticides and other compounds drawn from the medical literature of the world* (8th ed.). Amsterdam: Elsevier.

The Merck index: an encyclopedia of chemicals, drugs, and biologicals (14th ed.). (2006). Merck, Whitehouse Station, NY.

Geology

Bates, R. L., & Jackson, I. A. (2005). *Glossary of geology.* Alexandria, VA: American Geological Institute.

Dutro, J. T., Jr., Dietrich, R. V., & Foose, R. M. (1989). *AGI data sheets for geology in the field, laboratory, and office.* Alexandria, VA: American Geological Institute.

Mathematics

Borowski, E. J., & Borwein, J. M. (1991). *The Harper-Collins dictionary of mathematics.* New York: HarperCollins.

Physics

Daintith, J. (2005). *A dictionary of physics.* New York: Oxford University Press.

Statistics

Everitt, B.S. (2002). *The Cambridge dictionary of statistics.* New York: Cambridge University Press.

Everitt, B. S. (1995). *Cambridge dictionary of statistics in the medical sciences.* New York: Cambridge University Press.

Porkess, R. (1991). *The HarperCollins dictionary of statistics.* New York: HarperCollins.

Online Dictionaries

Date last accessed: October 2009
http://www.medbioworld.com/MedBioWorld/
TopicLinks.aspx?type=Reference%20Tools&&category=
(All)&&concept=Medicine

MedBioWorld˜ is the largest medical and bioscience resource directory on the Internet. Research and reference tools include dictionaries and glossaries, search engines, databases, clinical trials, medical guidelines, and education and training.

http://cancerweb.ncl.ac.uk/omd/

This site contains definitions of more than 46,000 terms not only
from medicine but also from biochemistry and plant biology.

http://thesaurus.reference.com/ Link to a visual Thesaurus that one
can subscribe to.

http://www.bartleby.com/141/index.html

Stunk and White, the famous short but excellent style guide, is
available in full online.

http://www.nlm.nih.gov/medlineplus/mplusdictionary.html

National Library of Medicine

http://www.medicinenet.com/script/main/hp.asp

Webster's new world medical dictionary authored by MedicineNet.

http://allserv.rug.ac.be/~rvdstich/eugloss/welcome.html

This system contains the electronic form of eight glossaries in which
you can find 1,830 technical and popular medical terms in eight
of the nine official European languages: English, Dutch, French,
German, Italian, Spanish, Portuguese, and Danish.

**http://www.glossarist.com/glossaries/science/life-sciences/biology
.asp**

This Comprehensive Directory of Biology serves as a dictionary and
glossary and contains listings of biological terms and terminology.

**http://infotree.library.ohiou.edu/byform:dictionaries/health-and-
life-sciences/biology/**

This good biological dictionary contains some 23,000 definitions,
online glossaries, and acronym dictionaries in the field of biology.

http://www.biology-online.org/dictionary.asp

This online dictionary of biology terms introduces and combines
various basic aspects of the biological and earth sciences.

http://www.userlab.com/Downloads/SE.pdf

This pdf file contains details on using simplified English for an
international audience.

ESL Dictionaries and Other Sources

Konstantinidis, G. (2005). *Elsevier's dictionary of medicine and
biology: In English, Greek, German, Italian, and Latin.* New York:
Elsevier Science.

Longman dictionary of American English (2nd ed.). (1997). White
Plains, NY: Longman.

Long, T. H. (Ed.). (1984). *Longman dictionary of English idioms* (Rev.
ed.). Harlow, England: Longman.

The Oxford dictionary for scientific writers and editors. (1991).
Oxford, England: Clarendon Press.

SUMMARY

WRITING PRINCIPLE 1: Write with the reader in mind.
WRITING PRINCIPLE 2: Use precise words.

WRITING PRINCIPLE 3: Use simple words.
WRITING PRINCIPLE 4: Omit unnecessary words and phrases.
WRITING PRINCIPLE 5: Avoid too many abbreviations.
WRITING PRINCIPLE 6: Use correct terminology and
    nomenclature.
ALSO: Watch out for misused words. Avoid sexism.

PROBLEMS

Problem 2-1 Precise Words

Find the nonspecific terms in the following sentences. Replace the non-specific choices with more precise terms or phrases. Note that it is not necessary to change the sentence structure, just replace the individual words. Guess or invent something if you have to.

1. All OVE mutants showed enhanced iP concentrations.
2. Plants were kept in the cold overnight.
3. Some of the discovered exoplanets have an orbital period of less than 5 days.
4. In general, retinal explants treated with $0.5\,\mu g/ml$ BSA exhibited increased growth rates compared with retinal explants treated with $0.5\,\mu g/ml$ BSA.
5. We field ionized those atoms not ionized by the HCP and detected the electrons produced with the rapidly rising field ionization pulse shown in Fig. 2B.
6. Briefly, cells were incubated with Mytomycin C for 3.5 hours, harvested with trypsin, and frozen in aliquots.
7. To provide proof of concept for our hypothesis, we studied a virus in its host cells.
8. Apart from the discussed main band, weaker emissions were observed.
9. The current was dramatically affected when temperature was increased.
10. (Last sentence in an Introduction) The present paper reports on continuing experiments that were performed to clarify this surprising effect.
11. Only some of the region under study exhibits larger reddening.
12. Heating arises after recapture and subsequent equilibration following from the lowest $\bar{T} = 0.04$, which is obtained by imaging the gas shortly after release from the trap.
13. A second calculated transition state that places the water molecule above the TFA ring is higher in energy than the first transition state.
14. The band showing vibrational splitting of 192/cm in Ne with the most intense peak at 444 nm can be identified with the A-> X transition of the dimer Ag_2.

15. We studied the performance of different functionals with the size of the cluster.
16. The afterglow of the blast wave was markedly brighter than we expected.
17. To determine the molecular events that ensue from this initial epithelial cell contact, we performed an analysis of proteins recruited to lipid rafts generated during a rapid infection with *P. aeruginosa*.
18. It is also possible that, due to the high expression levels of MVP in airway cells, the residual MVP present after siRNA treatment was sufficient for these other signaling and response pathways.

Problem 2-2 Simple Words

Improve the word choice in the following examples by replacing the underlined terms or phrases with simpler word choices. Again, do not change the sentence structure, just change the words.

1. These data <u>substantiate</u> our hypothesis.
2. We <u>utilized</u> UV light to induce *Arabidopsis* for mutations.
3. The differences in our results compared to those of Reuter et al. (1995) <u>can be accounted for by the fact that different conditions were used</u>.
4. <u>In our opinion, it is not an unjustifiable assumption that</u> the vibrational spectrum of CO_2 is temperature dependent.
5. <u>For the purpose of</u> examining cell migration, we dissected mouse brains.
6. Our results <u>are in accordance</u> with Seuter et al. (1988) who measured iP in the culture medium of *Physcomitrella* transformed with the agrobacterial isopentenyltransferase gene.
7. <u>We performed a systematic study of</u> the vibrational spectrum of CO_2 using various isotopomers.
8. <u>An example of this is the fact that</u> branching ratios differ substantially.
9. Protonema filaments had been freshly <u>disintegrated</u> by a blender resulting in filaments that were 10 to 20 cells long.
10. In Swaziland, the number of HIV infected children increased <u>by an order of magnitude</u> in the past decade.
11. It was recently shown that type IIS restriction endonuclease *FokI* <u>realizes cleavage</u> of double-stranded DNA.

Problem 2-3 Commonly Confused/Misused Words

Consider the pairs of confused and misused word choices provided for the following sentences. Using the provided word choices, fill in the correct words. It is okay to use Appendix of this book or a dictionary. Be sure you understand the difference in word choice.

1. **like, as:**
   Plasmids were isolated _____ described by Beates (17).
   Our observations for C1P1-GpP localization were _____ those of Andrews et al. (1989).

Tropospheric ozone (O3) is a naturally occurring greenhouse gas formed _____ a product of photochemical reactions.

The energy is transferred to the lattice before the electrons heat up to temperatures, _____ they do in Cu.

2. **enhance, increase:**

Metal atoms _____ the relative intensity of the band at 476 nm.

Soluble silicon in plants also has an active function in _____ host resistance to plant diseases.

3. **while, whereas:**

Colonies of DH5 alpha cells transformed with the AB construct were able to degrade naphthalene, _____ negative control cells were not.

The first enzyme was added _____ the DNA mixtures were incubating at 37 °C.

Tropical forests growing on highly weathered soils exhibit conservative P-cycling processes, _____ conservative N-cycling properties are more common on younger soils.

Temperatures above 20 °C favor the production of the 619 nm species, _____ at 15 °C, an increase of intensity of the 476 nm band is observed.

4. **varying, various:**

_____ water levels in a pond are often the result of climate conditions.

Each student received _____ concentrations of NaCl solution for the experiment.

_____ animals rely on darkness to hide, to catch prey, to mate, or to interact.

Different varieties of semiconductors layered in solar cells respond to photons of _____ energies to produce electricity.

Electrodes can be of _____ sizes.

5. **effect, affect:**

Nutrition concentration was the most important factor _____ population size.

Although the cows were given steroids, the drugs had little

_____.

Ozone causes cellular damage inside leaves that adversely _____ plant production.

Energy supplies by electrons and ions or chemical reaction (e.g., oxidation) with impurities are just some of the _____ that might be responsible for luminescence via matrix.

6. **include, consist of:**

Her research interests _____ all areas of biochemistry and structural biology.

Components of Hyperion's crust _____ solid H_2O and CO_2.

7. **that, which:**

Fish _____ live in caves show many adaptations to living in darkness.

At one electrode, hydrogen molecules are stripped of their electrons, _____ are then sent through an external circuit to do work.

Adaptive immune responses recognize novel viral antigens _____ are not invariant but nevertheless are foreign to the infected organism.

8. **represents, is:**

25 mg of ketamine _____ an overdose of anesthetic for mice.

The Born–Oppenheimer approximation of uncoupled electronic and nuclear motion _____ a standard tool of the computational chemist.

9. **infers, implies:**

Both curves are of an identical shape, which _____ a constant front profile as well as a constant velocity.

The Intergovernmental Panel on Climate Change has been criticized for _____ that climate-envelope models are more precise than they actually are.

10. **can, may:**

It _____ appear that Table 1 contains an essentially complete summary of patterns that occur in electrochemical systems.

Huge numbers of species _____ be at risk of extinction from climate change.

Problem 2-4 Redundancies and Jargon
Edit the phrases shown; change any redundancies to a shorter and better expression.

absolutely essential	a proportion of
a large number of	along the lines of
an order of magnitude	as a consequence of
despite the fact that	due to the fact that
during the course of	for the purpose of
give rise to	has the capability of
in a position to	in close proximity to
in connection with	in order to
in the event that	in view of the fact that
it has been shown that	it is worth pointing out that
on the basis of	square in shape
the majority of	

Problem 2-5 Redundancies and Jargon
Improve the word choice of the underlined words in the following examples by removing any redundancies, jargon, and unnecessary words and phrases. Do not change the sentence structure.

1. The doubling rate appeared to be <u>quite short</u>.
2. The following strains <u>were obtained by the courtesy of</u> Dr. U. Miller (University of Minnesota).
3. The data of the analysis on cell cycle parameters <u>are shown</u> in Fig. 1. They have revealed that the cell cycle is controlled by factor X.
4. The comparison of cytokinin overproducing mutants <u>compared to</u> the thiA1 auxotroph revealed comparable rates.
5. After 2 hr of <u>incubation</u> of CO_2 on an Ag(110) surface, we ended the incubation procedure.
6. <u>It is known that</u> homologous recombination is the preferred mechanism of DNA repair in yeast.
7. The effect of temperature on conductivity <u>was examined and found not to</u> change dramatically.
8. Often, jewel weed <u>can be found to grow</u> <u>in close proximity to</u> poison ivy. (Two corrections needed.)
9. Transduction efficiencies *in vivo* were much higher than <u>transduction efficiencies</u> *in vitro*.
10. Upon heat activation, filament size increased, and the number of buds decreased. Both <u>the increase in filament length and the decrease in the number of buds</u> were only seen for cytokinin mutants.
11. <u>In order to</u> test the adsorption performance under various near-field conditions of a waste repository, experiments on the retention of radioiodide by different organo-clays were carried out <u>under the influence of</u> elevated temperatures and high-molar saline solutions.
12. Although transition metals <u>have the capability of</u> forming bonds with six shared electron pairs, only quadruply bonded compounds can be isolated as stable species at room temperature.
13. After infecting a host cell, a herpes viral DNA genome enters the nucleus where it is <u>transcribed from DNA to RNA</u> to both messenger RNAs (mRNAs) and noncoding RNAs (ncRNAs).
14. Large size has often been linked to elevated extinction risk in mammals <u>due to the fact that</u> larger species tend to exist at lower average population densities and are disproportionately exploited by humans.
15. Aerobic phototrophic bacteria that <u>utilize</u> bacteriochlorophyll for light harvesting and charge separation were at first thought to be limited to selected, nutrient-rich environments, but similar organisms were later found to be ubiquitously distributed in the upper ocean.

Problem 2-6 Redundancies and Jargon

Identify and remove the jargon and other redundancies in the following sentences.

1. It is also worth pointing out that collagen synthesis returned to normal 3 days post injury.

2. In spite of the fact that our present knowledge on the subject at this point is far from complete, this macromolecular structure can aid in the design of new antibiotics.

3. A substantial proportion of persons in whom severe neuroinvasive WNV disease develops have long-term disability or die as a result of their infection.

4. After 3 hr, the old medium was dumped, and the same amount of fresh medium was added.

5. The data in Table 1 are very consistent with Brokl's (1999) model.

6. This appears to indicate that factor A possibly may have a tendency to interact with factor B.

7. We conducted nearly identical seismic experiments adjacent to the north walls of both the HDR and BTF scarps (Fig. 1) for the purpose of directly correlating the seismic layer 2A/2B boundary with mapped geologic units in young oceanic crusts.

8. In a considerable number of cases, degradation leads to topsoil loss and a reduction in soil fertility.

9. It is well known that when hydrogen nuclei fuse to form helium, they release energy.

10. There are a variety of newer reforestation methodologies becoming available for areas that are currently deforested.

11. Long-range atmospheric transport of pollutants is generally assumed to be the main vector for arctic contamination based on the fact that local pollution sources are rare.

12. Helper T cells perform critical functions in the immune system by way of the production of distinct cytokine profiles.

Problem 2-7 Abbreviations

Identify the writing principle that is violated for the underlined word choices in the following paragraph by describing as best as you can what the general mistake is.

Both cluster and slap calculations were carried out using the local density approximation (LDA) to the exchange correlation functional (ECF). For comparison, all calculations are repeated with gradient corrections as described by the Becker-Perdew functional (BP), PW91 (see above), and Becke-Lee-Yang-Parr (BLYP) functionals. Our choice to include those functionals is partially motivated by the shortcomings of the BP and PW91 functionals. The nonlocal BLYP functional includes a zero-point energy (ZPE) correction. Configuration-interaction methods (CI) or density-functional theory (DFT) describe the exchange and correlation effects. DFT plane-wave calculations and the BLYP functional have not yet been reported for the H2/Si(001) reaction energetics. We found the equilibrium Si-H bond is 1.52 angstrom within LDA, PW91, and BLYP.

Problem 2-8 Mixed Word Choice

Improve the word choice in these examples.

1. A typical scientist spends many long hours, even on the weekend, in his laboratory.

2. We studied the affect of erythromycin on 5 male and 3 female children in three different essays.
3. A graph displaying this data is shown in the overnext slide.
4. We utilized a Sorval centrifuge to obtain a sucrose gradient.
5. We made a picture of the gel we made.
6. We observed a change in cluster size after several minutes.
7. Isolatability of the Nnkla –1 protein was more difficult than expected.
8. Absorbance was measured at varying time points.
9. To make perfect slides of DNA nicks, an electron microscope is absolutely essential.
10. Every postdoc should be taught how to improve his writing.
11. Onchocerciasis or river blindness is a chronic debilitating disease of man caused by infection with the filarial nematode *Onchocerca volvulus.*
12. To reduce the amount of data points, we tossed out every alternative test point.
13. It has been reported that the distortion of the thiophene ring in quarterthiophene is affected by the medium (crystal or solvent) as well as by the intrinsic properties of molecules.
14. Smaller species should, in general, benefit more from the conservation of important threatened areas, while larger species will tend to benefit most from a conservation approach that also singles out individual species for particular attention.
15. Local nutrient enrichment from guano has been documented. On the contrary, the possibility of contamination in areas near seabird nesting sites has been largely overlooked.
16. Mathematical tools which are not yet common in the field of cell biology need to be developed to perform image analysis.
17. The fate of endosomes, that eventually disappear as new synaptic vesicles reform, remains unclear.

Problem 2-9 Mixed Word Choice
Edit one of the following passages. Pay attention to word choice:

> **Use precise and simple words.**
> **Check for misused and confused terms.**
> **Avoid sexism and redundancies.**

a)
Sulfonamides were among the first manmade agents used successfully to treat diseases. On account of their broad antibacterial activity, these drugs were in earlier times used almost exclusively in the treatment of a wide assortment of diseases. It is most fortunate that other drugs have supplanted sulfonamides as antimicrobial agents because all pathogenic bacteria are capable of developing resistance to sulfonamides. Sulfonamides prevent the synthesis of folic acid that is a coenzyme important in amino acid metabolism. Although sulfonamides are for the most part readily tolerated, it has been observed that they do have some side affects.

b)

Butterfly wing patterns are amazingly diverse. One of the patterns quite commonly found on butterfly wings are eyespots. While much is known about how these eyespots are produced on the developing wings (8–12), very little is known in regard to why such patterns exist in butterflies. It has been reported that the function of eyespots for some species may be for mate recognition or attraction (13). In other species, eyespots may function as an anti-predator defense (14). Overall, eyespot patterns reveal high heritability (15).

CHAPTER 3

Word Location

Although word choice is important for the interpretation of a sentence, readers take the greatest percentage of clues for interpretation not from word choice but from the *location* of words within a sentence. That is, readers expect a certain format in each sentence. If this format is not met, readers are forced to divert energy from understanding the content of a text to unraveling its structure, which increases the possibility of misinterpretation or not understanding. Worse, if readers cannot unravel the structure, they will lose interest. Thus, authors need to pay close attention to word location and to the organizational structure of a text.

3.1   READERS' EXPECTATIONS

The location of words within a sentence
is important for its interpretation.

Your task as an author is not only to choose the right words but also the most effective location for your words. You will have to convince most of your readers to interpret your sentences as you intended. There is always a minority, however, who interpret sentences differently from the majority. For this minority of writers, it is even more important to understand the importance of where in a sentence to place what information.

Consider the following example:

In this sentence, the word "mosquitoes" has been placed at the begin-

---

   Example 3-1   a   Mosquitoes often carry parasites.

topic   stress

---

ning of the sentence in the topic position, and the word "parasites" has been placed at the end of the sentence in the stress position. This positioning tells the reader that "mosquitoes" are the topic of the sentence and that "parasites" is to be emphasized. To most readers, the format of this sentence implies that the author has talked about mosquitoes before and is about to introduce a new topic, "parasites."

Another version of the same sentence presents a different emphasis, as can be seen in the following example:

   Example 3-1   b   Parasites are often carried by mosquitoes.

topic   stress

Although the sentence in Example 3-1b uses the same words as the sentence in Example 3-1a, the word locations have been altered. In Example 3-1b, the familiar topic now appears to be "parasites" at the beginning of the sentence, and the emphasized word is "mosquitoes" at the end or stress position of the sentence. Placing "mosquitoes" at the end of the sentence indicates to the reader that the author is stressing this term. The author may want to stress the term to ensure that the reader immediately understands that the stress is on "mosquitoes" and not on fleas or rats; the author may also want to stress the term to ensure that the reader does not miss the introduction of a new topic. Placing "mosquitoes" in the stress position of the sentence guides the reader's attention.

3.2   COMPETITION FOR EMPHASIS

WRITING PRINCIPLE 7:

Establish importance.

To decide on the best placement of words within a sentence, it is crucial that authors decide what is important, what is less important, and what is not important before they start writing or revising. When you write, important information can then be stressed, less important information can be subordinated, and unimportant information can be omitted. Authors need to recognize that the format and structure they use to present information will lead the reader to interpret it as important or less important.

In general, the end position in a sentence is more emphasized than the beginning position, and the main clause is more emphasized than the dependent clause. Thus, a main clause, a clause that is independent and can stand alone as a complete sentence, carries more weight than a dependent clause, which depends on the rest of the sentence for its meaning.

Consider the following four versions of a sentence. In each of these examples, the main clause has been italicized:

Example 3-2   a   Although vitamin B6 seems to reduce the risk of macular degeneration, *it may have some side effects.*

b   *Vitamin B6 reduces the risk of macular degeneration,* but it may have some side effects.

c   *Taking vitamin B6 may have some side effects,* but vitamin B6 also reduces macular degeneration.

d   Although taking vitamin B6 has some side effects, *vitamin B6 reduces macular degeneration.*

When one piece of news is in the main clause and another is at the end of the sentence or in the stress position, interpretation by the reader will vary. Most people will tend to emphasize the main clause more than the dependent clause, however. Even more, most people will tend to emphasize the information at the end of the sentence more than that at the beginning of the sentence. If readers were to vote on the impact of each sentence, the percent of readers that would recommend taking vitamin B6 would be highest in version *d* and lowest in version *a*. A more detailed analysis shows the following:

Sentence	news in main clause	news in end position	perception of vitamin B6 recommendation (%)
a	– negative	– negative	30
b	+ positive	(–) negative (dep. clause)	40
c	– negative	(+) positive (dep. clause)	60
d	+ positive	+ positive	70

Based on these percentages, readers (e.g., physicians) are most likely to recommend taking vitamin B6 after reading sentence *d* and least likely to recommend it after reading sentence *a* as the strongest statement has the + information in both the main clause and the end position of the sentence. Thus, even in more complex sentences, word placement, if considered carefully, can help authors to guide and influence readers.

Although word placement is more important than word choice for interpretation by the reader, if a word is strong or extreme enough, it can dominate the reader's attention. Let us replace "side effects" with an extreme phrase in the strongest positive sentence above and look at the effect:

Example 3-3   Although taking vitamin B6 may result in <u>serious defor-</u><u>mities or even death</u>, vitamin B6 reduces macular degeneration.

In this example, no matter where you put the extreme "serious deformities or even death," it overpowers the structural location of everything else such as that of the stress position.

3.3   PLACEMENT OF WORDS

Complexity

WRITING PRINCIPLE 8:

Place old, familiar, and short information at the beginning of a sentence in the topic position.

WRITING PRINCIPLE 9:

Place new, complex, or long information at the end of a sentence in the stress position.

If information is placed where most readers expect to find it, it is interpreted more easily and more uniformly. Readers expect to see old information that links backward at the beginning of a sentence (or paragraph) and new information at the end of a sentence (or paragraph) where it is emphasized more. Above all, writing "flows" much better if the information is linked through word location. Readers get confused and misinterpret information when the author does not comprehend their structural needs. The general principle that authors should keep in mind is to provide context for their readers before asking these readers to consider anything new as can be seen in the next examples.

Example 3-4

Macular degeneration is affected by *diet. One of the diet compo-nents* that influences the progression of macular degeneration is

vitamin B6. Although *vitamin B6* seems to reduce the risk of macular degeneration, it may have some ***side effects***.

Note how the information at the end position of a sentence in the preceding example is placed at the beginning, or topic position, of the next sentence, leading to "jumping word location." In each of these sentences, the new information in the stress position of one sentence becomes old, familiar information in any subsequent sentence and is therefore placed at the topic position in the sentence that follows. Consequently, if the passage would continue, most people would expect to find information on "side effects" in any subsequent sentence or paragraph because "side effects" has been introduced in the stress position in the last sentence. When authors pay attention to word location as in the preceding example, their writing has good flow and continuity.

Another way to achieve good flow or continuity is to write a whole paragraph from the point of view of the old information as in Example 3-5:

 Example 3-5   ***Depression*** in the elderly is thought to affect more than 6.5 million of the 35 million Americans who are 65 years of age and older. *It* is considered to be a disorder that is commonly underdiagnosed, undertreated, and mismanaged by pharmacotherapy both in community dwelling seniors and in those residing in nursing facilities. ***Depression*** in the elderly has also been closely associated with dependency and disability that presents in both emotional and physical symptoms, thus amplifying the difficulty in diagnosis. ***Major depression***, dysthymic disorder, and subsyndromal depression tend to be higher in persons over 65 who live in a long-term care facility.

Note how in this example, the topic "depression" is consistently placed in the topic position of each sentence, providing a link back for the reader. In each of the sentences in the preceding example, new information is always placed at the end of each sentence. Thus, every sentence provides new information, although the writer does not expand on it. If passages are consistently written from the same point of view as in the preceding example, good flow is also achieved.

Not all paragraphs will follow these principles of word location as exclusively as shown in Examples 3-4 and 3-5. Many paragraphs display a mixture of the word locations shown in these examples (see also Chapter 6 on Paragraph Construction). That is okay. What is not okay is to jump back and forth between one point of view and another for no apparent reason.

If we apply these principles about old and new information to writing and revising, we quickly realize that although some sentences are easy to write or revise, others are not. It is particularly hard to begin sentences well, especially if they are long and complex.

Which of these two sentences do you prefer?

Example 3-6   a   Outbreaks of limb deformities in natural populations of amphibians across the United States and Canada, especially in wetland associated with agricultural fields, were evaluated in this study.

b   We evaluated outbreaks of limb deformities that occurred in natural populations of amphibians across the United States and Canada, especially in wetland associated with agricultural fields.

Most readers dislike example 3-6a because it starts with a long and complex subject. Example 3-6b, on the other hand, begins simply and moves toward complexity. Readers prefer to see short information at the beginning of a sentence and long information at the end of a sentence. Thus, authors need to also consider the length of terms or information when constructing sentences.

Here is another example whose revised version is much preferred by readers:

Example 3-7   The heavily disordered patterns characteristic of interference arising from multiple regions with different phase drops across the junction were eliminated by X (Fig. 2, B and C).

Revised Example 3-7   X eliminated the heavily disordered patterns characteristic of interference arising from multiple regions with different phase drops across the junction in some samples (Fig. 2, B and C).

Subject

WRITING PRINCIPLE 10:

Get to the subject of the main sentence quickly, and make it short and specific. If possible, use central characters and topics as subjects.

In general, readers prefer to get to the subject/topic of the main sentence quickly. They understand a sentence more easily if the subject of it is readily available. When you open sentences with several words before its subject/topic, readers have a hard time understanding what the sentence is about. Thus, writers should avoid long introductory phrases and long subjects.

Example 3-8   Due to the nonlinear and hence complex nature of ocean currents, modeling these currents in the tropical Pacific is difficult.

Revised Example 3-8   **Modeling ocean currents in the tropical Pacific** is difficult due to their nonlinear and hence complex nature.

The subject in Example 3-8 arrives after the first 11 words, whereas the subject in the revised sentence is immediately available, making the revised sentence more easily understandable.

Readers also prefer to see characters as their subjects. In fact, readers get confused if for no good reason authors do not make characters subjects. The central character is the subject of a series of sentences telling a story. Most readers prefer central characters to be live characters rather than abstract terms, but concepts can also serve as central characters if the corresponding verb describes an action. Consider the following example and its revision:

Example 3-9	The reason for rejection on the part of the biochemists was that the focus of the paper was too broad.

Revised Example 3-9	The biochemists rejected the paper because it was too broad.

For Example 3-9, most readers consider the revised sentence to be much clearer than the original one because the central characters (biochemists) are the subject of the verb. In the revised version, the subject is also short and specific and much more concise.

Let us look at another example:

Example 3-10	The cells were incubated at room temperature for two days.

Here, the topic "cells" is the subject. Any possible character such as a biochemist or a laboratory technician is not mentioned because it is not the topic of interest. Instead, "cells" take the place of the live character. This choice is actually preferred in certain sections of a research paper (or grant proposal) such as in the Materials and Methods section.

To the reader, sentences appear the most clear and direct if the subject is also the topic of the sentence (and paragraph). However, the subject of a sentence does not always state its topic as in the following example:

Example 3-11	**No one** could foresee these results
	Subject *topic*

If a subject is deleted entirely, writers create the biggest problem for readers:

Example 3-12	A decision was made in favor of the use of dyes, nitrofurans, and amidines as disinfectants.

The author may know who is doing what, but the readers do not know and usually need more help than authors think. The sentence in Example 3-12 has different interpretations:

	Revised **Example 3-12**	We decided to use dyes, nitrofurans, and amidines as disinfectants. or They/Researchers decided to use dyes, nitrofurans, and amidines as disinfectants.

Verb Placement

WRITING PRINCIPLE 11:

Avoid interruptions between subject and verb and between verb and object.

Information is more easily interpreted if it is not obstructed. Often sentences are obstructed because the verb does not immediately follow the subject. Readers expect grammatical subjects to be followed immediately by the verb. Anything of length that intervenes between subject and verb is read as an interruption and therefore as something of lesser importance. Certain ESL authors should be especially aware of this principle. English sentences are better understood if their subject and verb are not interrupted.

Consider the following opening sentence of an introduction:

	Example 3-13	<u>Dengue virus</u>, a Flavivirus belonging to the family *Flaviviridae*, which includes over 60 known human pathogens such those causing yellow fever, Japanese encephalitis, tick-borne encephalitis, Saint Louis encephalitis, and West Nile encephalitis, <u>is classified</u> into four different serotypes: Types 1, 2, 3, and 4.

This sentence obstructs the reader because the grammatical subject ("Dengue virus") is separated from its verb ("is classified") by 31 words. Without the verb, readers do not know what the subject is doing or what the sentence is all about. As a result, readers focus their attention on the arrival of the verb and resist recognizing anything in the interrupting material as being of primary importance. The longer the interruption lasts, the more likely it becomes that the "interruptive" material actually contains important information; but its structural location will continue to brand it as merely interruptive.

Often an interruption can be moved to the beginning or to the end of a sentence, depending on whether it is connected to old or new information in the sentence. At other times, the author should consider splitting the information into two sentences or even omitting the interrupting information altogether.

In Example 3-13, the relative importance of the intervening material is difficult to evaluate. The material may be important. In that case, the writer should have positioned it to reveal that importance. Here is one way to incorporate it:

Revised Example 3-13	**Dengue virus is** a *Flavivirus* belonging to the family *Flaviviridae*, which includes over 60 known human pathogens such as those causing yellow fever, Japanese encephalitis, tick-borne encephalitis, Saint Louis encephalitis, and West Nile encephalitis. **It is classified** into four different serotypes: Types 1, 2, 3, and 4.

On the other hand, the intervening material might be a mere aside that diverts attention from more important ideas. In that case, the writer should have deleted it, allowing the sentence to drive more directly toward its significant point:

Revised Example 3-13	**Dengue virus is classified** into four different serotypes: Types 1, 2, 3, and 4.

Only the author could tell us which of these revisions more accurately reflects her or his intentions.

Here is another example in which the subject is separated from its verb:

Example 3-14	<u>Previous measurements</u>, in which abrupt changes in the diffraction patterns produced by large fields were generally asymmetric and could only be eliminated by thermal cycling, <u>are in sharp contrast</u> to measurements we made on the cuprates.

A possible revision is shown next.

Revised Example 3-14	Previous measurements showed abrupt changes in the diffraction patterns produced by large fields, were generally asymmetric, and could only be eliminated by thermal cycling. **These experiments are in sharp contrast to measurements we made on the cuprates.**

Readers also like to get past the verb to the object of a sentence quickly. Therefore, authors should avoid any interruptions between verb and object by placing interrupting passages either at the beginning or at the end of the sentence. In some languages other than English, sentences tend to be complex, and information gets repeatedly interrupted. If English is not your native language, resist the temptation to apply the principles of writing in your native language to writing in English. Avoid interruptions between the verb and its object, as shown in Example 3-15 and 3-16.

ESL advice

Example 3-15	We quantitatively compared, <u>using a model-based approximate Bayesian computation (ABC) method relying on computer</u> simulations, the different introduction scenarios for the Western European WCR populations.
Revised Example 3-15	<u>Using a model-based approximate Bayesian computation (ABC) method relying on computer</u> simulations, we quantitatively compared the different introduction scenarios for the Western European WCR populations.

Example 3-16	We conclude, based on very simplified models of solar variability, that solar variability is insignificant.
Revised Example 3-16	We conclude that solar variability is insignificant **based on very simplified models of solar variability.** or **Based on very simplified models of solar variability, we** conclude that solar variability is insignificant.

SUMMARY

The location of words within a sentence is important for its interpretation.

WRITING PRINCIPLE 7: Establish importance.

WRITING PRINCIPLE 8: Place old, familiar, and short information at the beginning of a sentence in the topic position.

WRITING PRINCIPLE 9: Place new, complex, or long information at the end of a sentence in the stress position.

WRITING PRINCIPLE 10: Get to the subject of the main sentence quickly, and make it short and specific. If possible, use central characters and topics as subjects.

WRITING PRINCIPLE 11: Avoid interruptions between subject and verb and between verb and object.

PROBLEMS

Problem 3-1 Sentence Interpretation

When scientists submit papers for publication, they often dread the response of reviewers. Here are four sentences that could have been written in different structural arrangements by reviewers to deliver the same news. Which statement is the one most likely resulting in the paper being accepted, and which is most likely the one resulting in rejection? Explain why.

1. Overall, although this manuscript is of interest for structural biologists, a more detailed analysis of ABC should be provided.
2. Although a more detailed analysis of ABC should be provided, this manuscript is of interest for structural biologists.
3. This manuscript is of interest for structural biologists, but a more detailed analysis of ABC should be provided.
4. A more detailed analysis of ABC should be provided, but overall, this manuscript is of interest for structural biologists.

Problem 3-2 Word Placement and Flow

Rewrite one of the following paragraphs. Place words such that the reader can easily follow the logic flow of the message.

a)

Vegetative cells are cells that are engaged in active growth and reproduction. Endospores can be produced by some bacteria that can cease vegetative growth. Endospores are highly resistant to heat, chemicals, and radiation, unlike vegetative cells. It is possible that the unique structure of the peptidoglycan layer of the spore is in some way associated with resistance.

b)

Mangrove plantations attenuate tsunami-induced waves and protect shorelines against damage. Human activities on the shorelines most damaged by the great 2006 tsunami had reduced the area of mangroves by 26%. Communities can be buffered from future tsunami events by conserving or replanting coastal mangroves and greenbelts. The conservation of dune ecosystems or green belts of other tree species could fulfill the same buffer as mangroves elsewhere.

c)

A quantum dot is a semiconductor nanostructure that confines the motion of conduction band electrons, valence band holes, or excitons in all three spatial directions. It has a discrete quantized energy spectrum. A small number (on the order of 1–100) of conduction band electrons, valence band holes, or excitons are contained in a quantum dot. Colloidal semiconductor nanocrystals are small quantum dots, which can be as small as 2 to 10 nanometers, corresponding to 10 to 50 atoms in diameter.

Problem 3-3 Word Placement and Flow
Write a paragraph using the list of facts provided. To create good flow, place words carefully at the beginning and end positions of sentences.

- fleas transmit plague *bacillus* to humans
- *bacilli* migrate from bite site to lymph nodes
- name "bubonic plague" arises because buboes = enlarged nodes

Problem 3-4 Word Placement and Flow
1. **Construct a paragraph about thermophiles using the list of facts provided. Pay attention to good flow of the message by considering word placement.**
2. **What does the reader expect to read next after having read the last sentence of your paragraph?**

Thermophiles

- microorganisms
- temperature range for growth between 45 °C and 70 °C
- found in hot sulfur springs
- cannot grow at body temperature
- not involved in infectious diseases of humans
- mechanism to resist elevated temperature unclear

Problem 3-5 Word Placement and Flow
Construct a paragraph about extrasolar planets using the list of facts provided. Pay attention to good flow of the message by considering word placement.

Extrasolar planets (exoplanets):

- are common
- possess a large variety of properties
- easy to detect—require little observational time
- some are on eccentric orbits, typical of some comets in the solar system
- some are in multiple planet systems
- orbital periods can range from 1.2 days to ~10 years
- most recently discovered exoplanets have masses only one order of magnitude larger than Earth
- some behemoths have more than 15 times the mass of Jupiter

Problem 3-6 Subject-Verb-Object Placement
Rewrite the following sentences such that the subject is followed immediately by the verb and interruptions between verb and object are avoided. Place the subject early in the sentence if possible.

1. Onchocerciasis, with approximately 18 million infected cases worldwide and 80 million more people at risk of infection, is now recognized as one of the major public health and socioeconomic

problems in many tropical countries (Murdoch et al., 1996; OEPA, 1998).

2. Since 1995, more than 150 extrasolar planets, most of them in orbits quite different from those of the giant planets in our own solar system, have been discovered.

3. Early experiments revealed, as demonstrated by a strong suppression of the transition temperature with impurities (4), the extreme fragility of the superconductivity in the ruthenate superconductor Sr_2RuO_4 (SRO).

4. Aside from these RNA structures found in bacteria, plants, and fungi, a viral RNA, with a sequence very similar to an ATP binding RNA aptamer, has been found to be able to bind ATP.

5. We recorded, for each detected atom released from the trap, the in-plane coordinates x and y and the time of detection t.

6. After a warm spring, female passerines, as a result of phenotypic plasticity, an individual-level response to temperature, often breed earlier than they do after a cold spring.

7. Allosteric ribozymes, which take small molecules as input and whose output makes cascading difficult, as it is a different form than the input, have been shown to perform logical functions (3).

8. Cytoplasmic incompatibility (CI), which occurs when a *Wolbachia*-infected male mates with an uninfected female, resulting in karyogamy failure and early developmental arrest of the mosquito embryo, has attracted scientific attention as a potential vehicle for gene drive.

9. The first two members, COF-1 $[(C_3H_2BO)_6 \cdot (C_9H_{12})_1]$ and COF-5 $(C_9H_4BO_2)$, which have rigid structures, exceptional thermal stabilities (to temperatures up to 600 °C), low densities, and exhibit permanent porosity with specific surface areas surpassing those of well-known zeolites and porous silicates, can be synthesized using a simple "one-pot" procedure under mild reaction conditions that are efficient and high-yielding.

10. A superconducting quantum interference device (SQUID) is used to measure, hence mapping the phase anisotropy of the superconducting order parameter, the phase difference between different real-space tunneling directions.

11. We modified, to investigate asymmetric thermal propagation in a suitable 1D inhomogeneous medium, carbon nanotubes and boron nitride nanotubes so that they assumed a non-uniform axial mass distribution (Fig. 1).

12. We reported previously that, in addition to poly(A) binding protein, a 50 kDa protein, also strongly associated with polysomal mRNA from *Saccharomyces cerevisiae*, is involved in the mechanism [23].

CHAPTER 4

Technical Sentences

4.1   GRAMMAR AND TECHNICAL STYLE

A paper full of grammatical errors discourages readers as well as reviewers and editors. It may also result in misinterpretation of what has been written. Although logically ordered and clearly expressed ideas are more important than perfect grammatical form, editors, reviewers, and readers will all be grateful if you write not only clearly and concisely but also correctly. Know that editors do not expect perfect English from ESL authors. Nor do they expect the ultimate levels of literacy from native English speakers. If you use good technical style and avoid grammatical errors, however, your paper will be clearer, livelier, and will reach a wider audience.

Many authors (especially native English speakers) are surprised to find certain phrases and sentences of their writing marked by editors because of bad style. A trained writer, however, will be able to recognize common style and grammar problems. Excessive use of third person, passive voice, nominalization, noun clusters, redundancies, and jargon are common causes of wordiness and bad style. Unclear use of tense, pronouns, prepositions, and articles can also confuse readers. We discuss all these problems of grammar and technical style in detail in this chapter.

Note that this handbook is not a book of English grammar. The book captures only the mistakes that are most commonly made by scientific authors, particularly mistakes that tend to reduce the clarity of a scientific manuscript. For additional help with grammar and vocabulary, see, for example, Thurman (2002) or Perelman et al. (1997), which are listed in the references section. A glossary of grammatical terms can be found at the end of this book as well.

4.2   PERSON

WRITING PRINCIPLE 12:

Use the first person.

Use the first person ("I" or "we") for describing what you did—but do not overuse it, do not use it if the journal (or your supervisor) has banned it, or if the focus of the sentence should be on the organism or another topic.

It was once fashionable to avoid using "I" or "we" in scientific research papers because these terms were considered to be subjective, whereas the aim in science is to be objective. However, science is not purely objective. Writing from the point of view of "I" or "we" is appropriate in a scientific research paper wherever judgment comes in as the following examples illustrate.

Example 4-1   a   To determine the mechanism for the direct effect of contrast media on heart muscle mechanics, <u>the study</u> on heart muscles isolated from cats was carried out.

b   <u>The authors</u> show here that two separate parameters are important to describe the physical effects of an earthquake: seismic moment and radiated energy.

These sentences taken from two different Introductions would be more accurate and more vigorous if the first person "we" were used for the subject instead of the third person: "the study" in Example 4-1a or "the authors" in 4-1b. The advantage of using the first person is that using "we" generally forces the author to use the active voice, which is lively.

Revised   a   To determine the mechanism for the direct effect of contrast media on heart muscle mechanics, **we carried out** the study on heart muscles isolated from cats.

b   **We** show here that two separate parameters are important to describe the physical effects of an earthquake: seismic moment and radiated energy.

Although in most of the sections of a scientific document, the use of first person is preferred, this use is more controversial in the Materials and Methods section. There, the first person "we" is not usually the topic. Instead, materials, methods, or organisms are usually the topic. In addition, it often may not have been the author(s) who performed a certain experiment but rather a technician or hired helper. Therefore, in the Materials and Methods section, use of third person is usually preferred. In certain fields, such as in ecology, however, many journals require the use of first person and active voice even in the Materials and Methods section.

4.3   VOICE

WRITING PRINCIPLE 13:

Use the active voice.

Use the active voice rather than the passive voice. If the passive voice is used excessively, writing becomes very dull and dense, as in the following examples:

 Example 4-2   a   Cats <u>are hated</u> by dogs.

b   No change in activity <u>was observed</u>.

These sentences are much livelier and more interesting when active voice is used.

 Revised   a   Dogs hate cats.
Example 4-2

b   We **observed** no change in activity.

Do not remove the passive voice completely, however; use the passive voice when readers do not need to know who performed the action. You may also have to use the passive voice when the emphasis should be on a specific topic or when word location needs to be considered.

 Example 4-3   Viral DNA **was isolated** 24 hours after inoculation.

In addition, you can use the passive voice if this allows you to replace a long subject with a short one, gives you a more consistent point of view (i.e., lets you use the same subject in consecutive sentences), or lets you put emphasis on the terms you want to have emphasized.

 Example 4-4   Volcanic pipes are subterranean geological structures formed by the violent, supersonic eruption of deep-origin volcanoes. Volcanic pipes are composed of a deep, narrow cone (described as "carrot-shaped"). <u>Solidified magma usually fills the cone.</u> Volcanic pipes are made up of kimberlite or lamproite rock and are well known as the primary source of diamonds.

 Revised Example 4-4   Volcanic pipes are subterranean geological structures formed by the violent, supersonic eruption of deep-origin volcanoes. Volcanic pipes are composed of a deep, narrow cone (described as "carrot-shaped"). **The cone is usually filled with solidified magma.** Volcanic pipes are made up of kimberlite or lamproite rock and are well known as the primary source of diamonds.

The passive voice is needed in the preceding revision to keep the focus on the cone rather than shifting to new information. In the revision, word location has been considered.

4.4   TENSE

WRITING PRINCIPLE 14:

Use past tense for observations, completed
actions, and specific conclusions.

WRITING PRINCIPLE 15:

Use the present tense for generalizations and
statements of general validity.

ESL advice

The two main tenses that occur in scientific papers are present tense and past tense. Many scientific authors, especially ESL authors, seem to be confused about when to use past tense and present tense. Many are also unsure if past tense and present tense can be mixed in the same sentence or paragraph. Generally, you should use the past tense for observations, completed actions, and specific conclusions. For example, results described in your paper should be described in past tense because you have done these experiments, and your results are not yet accepted "facts." Therefore, the Abstract, Materials and Methods, as well as Results sections should employ past tense as they refer primarily to your own work.

Example 4-5   a   Higher temperatures **resulted** in less bud formation.

b   The three images **were taken** about 90 minutes apart.

You should use the present tense for generalizations and statements of general validity.

That is, results from already published papers should be described in the present tense as published results are generally assumed to be "facts." Thus, most of the Introduction describes previously established knowledge in present tense.

Example 4-6   a   The newly discovered planet **is** at least as big as Pluto.

b   Most regions where this problem **arises belong** to category X.

If you use past tense for describing results of already published work, you are implying to the reader that you do not consider these results to be "facts" but observations.

Can tense be mixed in the same sentence or paragraph? Certainly, as is apparent in the next example:

Example 4-7   Sultan **observed** that certain species of bacteria **respond** to light stimuli.

This example describes an experiment that has been completed. "Observed" is therefore written in past tense. "Respond," however, is present tense because this part of the sentence is still true and is considered established knowledge once published.

The Discussion relates your work to previously established knowledge. This section is the most difficult to write as it includes both past and present tense. Note that generally, remarks about the presentation of data should be in present tense, and descriptions of assumptions and theory should also be described in present tense in your paper.

Example 4-8   The effect of TXY addition **is shown** in Figure 5.

Present perfect tense is used when observations have been repeated or continue from the past to the present.

Example 4-9   X **has been** of interest for the past three decades.

Only experiments that you plan to do in the future should be described in the future tense. Future experiments are usually not described in research articles but are described in grant proposals.

Example 4-10   We **will examine** if parallel universes exit.

4.5   SENTENCE LENGTH

WRITING PRINCIPLE 16:
Write short sentences. Aim for one main idea
in a sentence.

Short sentences are easier to understand than long sentences. A paper full of long sentences is difficult to follow. The average sentence length in many scientific articles is over 30 words; in most newspapers, it is between 15 and 20 words (one of the reasons that newspaper articles are easier to understand.) Many scientific papers could be strengthened by shorter sentences, although not every sentence should be short; using *only* short sentences does not result in strong writing but leads to choppy, hard-to-follow passages. Some sentences need to be long to communicate complex ideas. What scientific authors should be aiming for is an average sentence length of about 20 to 22 words. This means that some sentences will be longer and some shorter, but the average number of words per sentence overall should be around 20 to 22.

Short, simple sentences tend to emphasize the idea contained in them. The longer a sentence gets, the more difficult it is for the reader to identify what is of primary importance. Therefore, single-clause sentences have more weight, and thus more importance, than multi-clause sentences. Writing a short sentence that highlights the main topic is particularly important at the beginning of a section or paragraph. It ensures that you have the attention of the reader from the outset and lets the reader focus on the main idea.

Similarly, readers assign more importance to sentences that stand on their own (independent sentences) than to a clause that depends on the presence of another clause. Thus, independent sentences have more weight than dependent sentences, which in turn have more weight than phrases. Consider the following example:

Example 4-11 a   Rheumatic fever is an autoimmune disease.

b   *It is generally accepted in the field of medicine that* rheumatic fever is an autoimmune disease.

In the preceding example, the words in the sentence in Example 4-11a, "Rheumatic fever is an autoimmune disease," tend to weigh more when they are in their own sentence than when they appear in some longer sentence such as the sentence in Example 4-11b. In addition, in the sentence in Example 4-11b, the same words appear in a dependent clause, which makes the reader perceive them as less important. For both of these reasons, most readers perceive the sentence in Example 4-11a as "weighing more" than the sentence in Example 4-11b.

Many sentences in scientific papers are needlessly complex. As a general guideline, do not present too many ideas in a single sentence. Instead, make sure your sentences do not contain more than one main idea and that they do not wander. The first step to ensure that your sentences do not contain too many ideas is to decide which details in a sentence are important. Only when you have assigned importance will you be able to subordinate less important information and omit unimportant information. Often, you can consider breaking subordinate sentences into separate sentences.

ESL advice

In certain cultures, people write in very complex, indirect ways. If you have this background, be particularly aware that English sentences that are concise and direct are better understood than sentences that are long and contain many different ideas.

It is a good idea to imagine yourself sitting across from an important reader. Write your paper as if you were *telling* this reader about your work. Remember that the purpose of a scientific paper is to inform, not to impress.

Consider the following example:

 Example 4-12   **Excessively long sentence**

Skin swabs were cultured at the time of the removal of the catheter from seven patients with catheter-related bloodstream infection and in five of these cases (71%), the culture yielded bacteria of the same species with a DNA fingerprint pattern similar to that of the isolates from the catheter and the blood, whereas a different organism grew from skin cultures in the other two patients (29%), suggesting that catheter infection may have originated from contamination of the catheter hub.

(79 words/sentence)

In this example, the first idea ends before "and." The second idea ends before "whereas" and the third idea before "suggesting." All of these ideas should be written in separate sentences.

 Revised Example 4-12   Skin swabs were cultured at the time of the removal of the catheter from seven patients with catheter-related bloodstream infection. In five of these cases (71%), the culture yielded bacteria of the same species with a DNA fingerprint pattern similar to that of the isolates from the catheter and the blood. However, a different organism grew from skin cultures in the other two patients (29%). These observations suggest that catheter infection may have originated from contamination of the catheter hub.

(average of 20 words/sentence)

Whereas the original sentence was 79 (!) words long, the revised version has an average sentence length of 20 words. Therefore, the revised version is much easier to understand. The reason is not that the sentences are shorter, but mainly that the ideas are separated into different sentences.

4.6   VERBS AND ACTION

WRITING PRINCIPLE 17:
Use active verbs.

Verbs are perhaps the most important part of an English sentence. With strong and active verbs, your writing enlivens and energizes. If the action of a sentence is expressed by the main verb, the sentence is natural, direct, and easy to understand. If, instead, the action is expressed in a noun, the verb becomes buried and weak, and the sentence is dense and more difficult to understand.

Abstract nouns derived from verbs and adjectives are called nominalizations. Your readers will perceive your writing as dense especially when you use many abstract nouns. "Academese" tends to be full of nominalizations. Many nonnative-speaking scientists also excessively use nouns in their native language, which they then translate and apply in English. For better scientific style, avoid nominalizations—use active verbs instead.

 Active Verb   **Buried Verb/Nominalization**

Active Verb	Buried Verb/Nominalization
assess	assessment
decide	made the decision
depends on	is dependent on
exist	existence
follows	is following
form	formation
inhibit	inhibition
measure	measurement
remove	removal

In the following example, the action is not in the verb but in the noun.

 Example 4-13   Their <u>suggestion</u> for us was a different <u>analysis</u> of the data.

In the revision, the actions are all verbs, resulting in a much clearer and less dense style.

Revised   They **suggested** that we **analyze** the data differently.
Example 4-13

Other examples of verbs and adjectives and their nominalizations include the following:

Verb	Nominalization
analyze	analysis
attempt	attempt
careful	care
centrifuge	centrifugation
compare	comparison
deduce	deduction
determine	determination
different	difference
discover	discovery
discuss	discussion
dissect	dissection
evaluate	evaluation
elute	elution
explain	explanation

ESL advice

fail	failure
hypothesize	hypothesis
increase	increase
isolate	isolation
move	movement
need	need
react	reaction
separate	separation
speculate	speculation

Note that some nominalizations and verbs are identical such as graph (verb) and graph (noun).

Science is more interesting particularly if the actions of animals and cells are described in active verbs:

 Example 4-14   Earthworms react to light.
Muscles contract.
Blood flows.

Avoid writing with weak verbs. Weak verbs seem abstract and impersonal and result in boring writing. Examples of weak verbs include the following:

| occurred | was seen | was noted | was done |
| was observed | caused | produced | make | get |

When you find yourself writing using one of these verbs, stop and check if you can use an active verb instead.

 Example 4-15 a   An <u>increase</u> in temperature <u>occurred</u>.

In this example, the verb ("<u>occurred</u>") is not active but weak. The subject of the sentence ("<u>increase</u>") expresses the action. This noun is also a nominalization of the verb "increase." As a result of the nominalization, the sentence is complicated and indirect. To revise a sentence whose action is buried in a noun, replace the weak verb with the action of the noun.

 Revised   a   Temperature **increased.**
Example 4-15

In the revised sentence, when the verb is active and strong, the sentence is simpler, more direct, and more efficient than when the action is nominalized.

Consider another example:

 Example 4-15 b   This wavelength <u>caused a decrease</u> in the molar absorption coefficient.

This example contains the weak verb "caused." In the sample sentence, the action is buried in the object ("decrease"), and the true object ("molar absorption coefficient") is sidetracked into a prepositional phrase ("in the molar absorption coefficient").

 Revised   b   This wavelength **decreased** the molar absorption
Example 4-15   coefficient.

Whereas the original sentence just sits, the revised sentence moves because the verb is strong and active.

     Sometimes writers express action in the object of a preposition instead of in the verb. (Prepositions are words such as "of," "for," "on," "in," "to," and "with.")

 Example 4-15 c   <u>With an increase in sperm concentration</u>, the fertilization
                rate improved.

In this example, the action in the first part of the sentence is expressed in the nominalization "increase," which is the object of the preposition "with." The sentence is dense and difficult to read for two reasons: (a) "with" is imprecise: "when" would make the sentence clearer; and (b) the action is buried in the prepositional phrase without a verb.

     To make this sentence easier to read, turn the prepositional phrase into a dependent sentence. Using an active verb makes this sentence much livelier and easier to understand.

 Revised   c   *When sperm concentration **was increased**,* the fertiliza-
Example 4-15   tion rate improved.

Although most nominalizations in scientific writing can and should be turned into verbs, there are exceptions. Keep a nominalization if it refers to a previous sentence or if it names the object of the verb.

 Example 4-16 a   **This observation** led us to conclude . . . .

                  b   An example of this theory is provided by **a delay** in the
                      reaction.

Analysis and Revision

To find and revise sentences that may confuse your readers, analyze your sentences:

1. <u>Underline the first 8 to 10 words in the main sentence</u>, ignoring introductory phrases.
2. For the underlined words, identify the central *character* of the sentence or paragraph.

3. Make the *character* the subject.
4. Look for the action.
5. If the actions are nominalizations, change them into verbs and make the relevant characters their subject.
6. Replace weak verbs with strong, active verbs if necessary.
7. Rewrite the sentence using conjunctions such as "because," "if," "when," "although," "that," or "whether." If necessary, turn a prepositional phrase into a dependent clause.
8. Avoid other nominalizations and abstract nouns in the remainder of the sentence as well—change them to verbs.

 Example 4-17   Despite the <u>identification</u> of the AIDS virus by *research-ers, there has been a* <u>failure</u> to develop a vaccine for the immunization of those at risk.

Analysis and Revision
1. The central character of the sentence is *researchers.*
2. Here the nominalization among the first 8 to 10 words is "failure." (Other nominalizations are "identification" and "immunization.")
3. Change the nominalization to a verb: failure > fail (identification > identify, immunization > immunize).
4. Make the character the new subject of the verb, and turn the prepositional phrase into a dependent clause, leading to the following:

 Revised   Although *researchers* **identified** the AIDS virus, *they* **failed**
Example 4-17   to develop a vaccine to immunize those at risk.

If you revise your sentences using the suggestions and principles presented, you will find your sentences not only more concrete, active, and concise but also more coherent and clearer for the reader.

4.7   NOUN CLUSTERS

WRITING PRINCIPLE 18:

Avoid noun clusters.

Noun clusters are nouns that are strung together to form one term. In English, nouns (and adjectives) can be used to modify other nouns. However, when nouns appear one right after the other, it can be difficult to tell how they relate to each other and what the real meaning of the cluster is. When you add more than one modifier in front of a single noun or place additional modifying nouns and/or adjectives in front of an existing noun pair, you may end up with confusing noun clusters. Avoid clusters of nouns, especially if there are more than two or three nouns in the cluster. These noun clusters are awkward and sometimes downright incomprehensible.

Example 4-18 a   <u>sparse matrix crystallization</u>

b   <u>cultured rat tracheal endothelial cells</u>

c   When the strips were exposed to <u>Leishmaniasis diseased patients' sera</u>, we found the bands of 112 and 45 kDa.

d   Peter Carri is a <u>condensed matter</u> and <u>quantum many-body theoretical physicist</u>.

Instead, use prepositions to link the nouns. The prepositions add clarity to the phrase—they show more fully how the nouns relate to one another—and the meaning of your words becomes clearer.

Revised   a   **crystallization** *by* **sparse matrix**
Example 4-18

b   **cultures** *of* **endothelial cells** *from* **the tracheas** *of* **rats**

c   When the strips were exposed to **sera** *of* **patients** *with* **Leishmaniasis disease**, we found the bands of 112 and 45 kDa.

d   Peter Carri is a **theoretical physicist studying condensed matter and quantum many-body physics**.

Note that not all sequences of nouns are noun clusters. Some noun pairs and clusters—such as "water bath," "cell wall," "egg receptor," and "sucrose density gradient"—are recognized as single words and accepted terms. When you untangle noun clusters, treat such terms as single words. Check a dictionary or other recent journal articles to determine which pairs of nouns are considered words if necessary.

If a technical name is a noun cluster (example: pyrophosphate dependent phosphofructo-1-kinase apo form structure), there are three ways to help your reader:

1. Use hyphens to show the relationship between the words (pyrophosphate-dependent phosphofructo-1-kinase apo-form structure).
2. Explain the name (the structure of the apo-form of pyrophosphate-dependent phosphofructo-1-kinase).

3. Explain to the reader that you will use a shorter name (the apo-form structure of pyrophosphate-dependent phosphofructo-1-kinase, short PPi-PFK apo-form).

Hyphens in noun clusters are most often used for

- Two-word terms that are used together (high-pressure chamber, ATP-dependent)
- Adjectives that consist of three or more words (four-to-one ratio)
- Terms that contain a capital letter or a number and a noun (C-terminal end, 10-fold increase)

4.8   PRONOUNS

WRITING PRINCIPLE 19:
Use clear pronouns.

Pronouns are words that take the place of nouns.

Examples	"it," "none," "they," "these," "those," "their," "them," "this," "that," "which," "who," "whose"

It is essential that you use clear pronouns. Unclear pronouns are one of the most common problems in scientific writing. If the pronoun that refers to a noun is unclear, the reader may have trouble understanding the sentence. An author always knows which term she or he is referring to. A reader is not so lucky. Be sure that the pronouns you use refer clearly to a noun in the current or previous sentence. If there are too many possible nouns the pronoun can refer to, repeat the reference noun after the pronoun.

	Example 4-19	Gram⁺ bacteria do not respond to these drugs. Thus, <u>they</u> were of no interest to us.
	Revised Example 4-19	Gram⁺ bacteria do not respond to <u>*these drugs.*</u> Thus, **these drugs** were of no interest to us. or **These drugs** were of no interest to us because Gram⁺ bacteria do not respond to ***them***.

Sometimes the noun that a pronoun refers to has been implied but not stated. To clarify the reference, explicitly state the implied noun after the pronoun as in the next example:

	Example 4-20	If a specimen is frozen in a bath containing dry ice and acetone, the water of the cell can be removed by sublimation to prevent damage to the cell. <u>This</u> is commonly used for preservation of cultures.

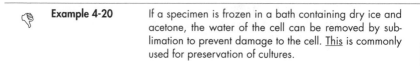

| Revised Example 4-20 | If a specimen is frozen in a bath containing dry ice and acetone, the water of the cell can be removed by sublimation to prevent damage to the cell. **This technique** is commonly used for preservation of cultures. |

Consider another example:

| Example 4-21 | As human impacts on natural environments continue to increase, large species will become extinct faster than small species with similar biological characteristics. <u>This</u> can be illustrated with the use of our model predicting extinction risk from the level of external threat. |

Here, "this" refers to an effect on the decline toward extinction. To avoid repeating so many words, we can use an *implied* term that encompasses the specific terms such as "trend." In the revision, "trend" is preceded by "this" to indicate that the trend was mentioned in the previous sentence.

| Revised Example 4-21 | As human impacts on natural environments continue to increase, large species will become extinct faster than in small species with similar biological characteristics. **This trend** can be illustrated with the use of our model predicting extinction risk from the level of external threat. |

4.9   LISTS AND COMPARISONS

Parallel Ideas

WRITING PRINCIPLE 20:

Use correct parallel form.

Lists and ideas that are joined by "and," "or," or "but" are of equal importance in a sentence and so are ideas that are being compared. These ideas should be treated equally by writing them in parallel form. To write ideas in parallel form, the same grammatical structures are used. These grammatical structures can be single words, prepositional phrases, infinitive phrases, or clauses. If parallel ideas are written in parallel form, the reader does not get distracted by the form but can concentrate on the idea.

The next few examples are sentences that contain parallel ideas joined by "and," "or," or "but," which are written in parallel form.

Example 4-22

	Subject	Verb	Adverb
	The more stable of the two states	**expanded**	**twofold**
but	**the region of the metastable state**	**decreased**	**10-fold.**

In the preceding example, the same parallel form is used for the two ideas that are being compared, namely, the group of words after "but" is in the same grammatical structure as the group of words before "but": in this case, subject, verb, adverb.

Here are a few more examples:

Example 4-23

	Direct Object	Preposition	Object of Preposition
Based on our hypothesis, we expected to see	**a decrease**	**in**	**the infection rate**
and	**an increase**	**in**	**survival of patients.**

Example 4-24

	Subject	Verb
	S. franciscanus sperm	**can fertilize**
S. purpuratus eggs, but	**L. pictus sperm**	**cannot.**

Example 4-25

	Preposition	Object of Preposition
Gene targeting in P. patens occurs both	**in**	**the laboratory**
and	**at**	**its natural habitat.**

Note that in Example 4-25, "in the laboratory" and "at its natural habitat" are in parallel form even though the specific prepositions ("in," "at") are different. For parallel form, it is only important that both terms are prepositions.

Do not confuse the reader by obscuring the logical relationship of parallel ideas.

Example 4-26   Prolonged febrile illness together with subcutaneous nodules in a child could be due to an infection with a Gram+ organism, but it could also be <u>that the child suffers from rheumatic disease</u>.

In this sentence, the groups of words before and after "but" are not parallel, so it is not immediately obvious that the second half of the sentence is giving another possible reason for the illness. Because the second half of the sentence is equal in logic and importance, it should be written in parallel form.

Revised Example 4-26   Prolonged febrile illness together with subcutaneous nodules in a child could be due to an infection with a Gram+ organism, but it could also be **due to rheumatic disease**.

This sentence can be further simplified:

Revised **Example 4-26**	Prolonged febrile illness together with subcutaneous nodules in a child could be due to an infection with a Gram+ organism or due to rheumatic disease.

Coordination

GUIDELINE:
Arrange ideas in a list to read from shorter to longer.

If a sentence lists two or more ideas, these ideas should not only be in parallel form, but they should also be coordinated. Careful writers coordinate ideas that are both grammatically and logically parallel. For good coordination, ideas should be arranged to read from shorter to longer in terms of the number of words contained in the idea. Coordinating ideas in this way makes a sentence more graceful.

Consider the following two sentences:

Example 4-27	a	Loss of coral reefs will affect <u>organisms</u>, such as turtles and sea birds that depend on specific habitats for reproduction, <u>and beaches</u>.
	b	Loss of coral reefs will affect **beaches and organisms** such as turtles and sea birds that depend on specific habitats for reproduction.

Sentence 4-27a seems to end too abruptly with "beaches." Sentence 4-27b has much better flow because here the two parallel ideas have been arranged from shorter to longer.

4.10   FAULTY COMPARISONS

WRITING PRINCIPLE 21:
Avoid faulty comparisons.

Aside from maintaining parallelism in your comparisons, you should avoid grammatical and logical problems when writing comparisons. These problems result in faulty comparisons, one of the most common problems in scientific writing. Faulty comparisons can arise because of ambiguous comparisons and incomplete comparisons. Faulty comparisons may also be due to the overuse of "compared to." Examples for all of these scenarios are shown in the following sections.

Ambiguous Comparisons

The following example is a very typical ambiguous comparison found in scientific papers:

Example 4-28   Our conclusions are consistent with <u>Tamseela et al.</u>

Comparisons such as this are confusing for the reader, as they compare unlike things. To avoid such ambiguous comparisons, make sure that you are comparing like items.

Revised   A   Our conclusions are consistent with **the conclusions of**
Example 4-28   Tamseela et al.

This sentence can be written even simpler by using a pronoun to avoid repetition.

Revised   B   Our conclusions are consistent with **those of** Tamseela et al.
Example 4-28

Incomplete Comparisons

Absolute statements should not be written as comparisons. Information being compared and that with which it is being compared needs to be listed completely and in parallel.

Example 4-29   This study tested 24 subjects compared to Menon's study.

Revised   A   This study tested 24 subjects; Menon's study tested only
Example 4-29   8 subjects.

                  B   In this study, the number of subjects tested (24 subjects)
                       was three times that of Menon's study (8 subjects).

ESL advice

Such incomplete comparisons may confuse readers because their intended meaning is unclear. In certain foreign languages, incomplete comparisons occur often. Avoid these when writing in English.

Here is another example:

Example 4-30   RNA isolation is <u>more difficult</u>.

The question the reader naturally asks when reading this sentence is more difficult than what? To complete the comparison, you need to include the item that RNA isolation is compared with as shown in the revised example:

Revised   RNA isolation is **more difficult than DNA isolation**.
Example 4-30

"Compared to"

Use "than" not "compared to" for comparative terms such as "smaller," "higher," "lower," "fewer," "greater," "more," and so forth.

👎	**Example 4-31**	We found <u>more</u> fertilized eggs in buffer A <u>compared to</u> buffer B.
👍	**Revised Example 4-31**	We found **more** fertilized eggs in buffer A **than in** buffer B.

Note that "in" is repeated in the revised example to keep parallel form.

Avoid using "compared to" with the words "decreased" or "increased" because the meaning is ambiguous.

👎	**Example 4-32**	K_D increased over time <u>compared to</u> K_A.
👍	**Revised Example 4-32**	K_D increased over time, **but K_A did not.**

4.11   COMMON ERRORS

WRITING PRINCIPLE 22:

Avoid errors in spelling, punctuation, and grammar.

In any type of writing, errors should be avoided. Common errors include (a) spelling, (b) punctuation, (c) words that are omitted, (d) a subject and verb that do not make sense together, (e) a subject and verb that do not agree, and (f) unclear modifiers. When one of these errors appears, the reader is slowed down and may even need to reread the sentence to figure out the intended meaning. These errors can be avoided by carefully proofreading and double-checking the manuscript.

Spelling

If you use a computer to prepare your manuscript, make use of a spell checker to avoid some common errors. Such a program will not find all the mistakes, however. For example, the program will not point out words that are wrong but correctly spelled ("from" when you meant "form" or "to" instead of "two").

A spell checker program also will not point out if you spelled the same word in the same way throughout. Compile a word list: Every time you make a decision on spelling, record it, and check your second draft for conformity to the list.

The choice in spelling is sometimes between British and American versions of a word, which is especially confusing for ESL writers. Consult a dictionary to see whether one form is preferred to another or to check which version is British and which is American. Use a dictionary such as *the Dictionary of Contemporary English* that gives both British and American spellings, or search online dictionaries. Stick with one type of spelling or the other, but be consistent in your choice. Although generally written British English is more formal, this formality is not apparent in science writing. It is apparent much more in letter writing and other correspondences. North American English spelling is more common these days than British English spelling in scientific writing. However, it is usually a good idea to see what style your target journal has adopted. Many journals accept and use both British and American spelling if one of these spellings is used consistently throughout an article. Adjust your choice of punctuation rules accordingly as well. Grammar differences between the two language variations are few if any.

Most words that differ markedly between American English and British English are usually not used much in science, if at all. The most common differences in American and British English spelling are listed in Table 4.1 together with scientific words as examples.

Numbers versus Numerals

Various factors determine whether you need to spell out numbers or use numerals. These factors include different style guides, the size of the number, whether it is a scientific number with units, the context of the number, and what it is representing. For example, the APA style manual recommends spelling out single digit numbers (one through nine) and using numerals for 10 and above. However, *The Chicago Manual of Style* recommends spelling out whole numbers one through one hundred. Exceptions to these rules include:

1) Use numerals if scientific values are to be presented, i.e., for values with a unit (e.g., 6 days, 32 hr, 7.8 g) that do not begin a sentence.

2) Spell out any number used to begin a sentence (e.g., "Six hundred and three examinees participated in the experiment."). To avoid beginning a sentence with a number, reword the sentence so that it does not begin with the number (e.g., "Of these, 34 survived to adulthood.").

3) Large numbers, if rounded, are usually spelled out ("Three hundred animals frequented the water hole."); but, very large numbers (millions or more) are normally expressed with a mixture of numerals and spelled-out numbers ("The bird population was estimated to be 1.3 million.").

4) Although in informal writing you can use numerals and the "%" sign, as in "3% of the participants," in formal writing, you should spell out the percentage as, for example, in "5 percent of the participants."

5) For a series of mixed numbers greater and less than ten, use numerals as in "5, 10, and 15 hours."

Table 4.1 Differences between American and British spelling

AMERICAN SPELLING	BRITISH SPELLING	AMERICAN ENGLISH	BRITISH ENGLISH
-am	-amme	program	programme
-ay	-ey	gray	grey
-e-	-ae-, -oe-	edema, anesthesia, leukemia, pediatric	oedema, anaesthesia, leukaemia, paediatric
-er	-re	center, meter, fiber, liter	centre, metre, fibre, litre
-f-	-ph-	sulfur	sulphur
-ing, -able	-eing, -eable	aging, sizable	ageing, sizeable
im-, in-	em-, en-	imbed, insure	embed, ensure
-ize, yze	-ise, -yse	analyze, optimize, emphasize, realize	analyse, optimise, emphasise, realise
-l-	-ll-	signaling, labeling, modeling	signalling, labelling, modelling
-ll	-l	enroll, fulfill	enrol, fulfil
-og	-ogue	catalog	catalogue
-or	-our	tumor, flavor	tumour, flavour
-se	-ce	defense	defence
-um	-ium	aluminum	aluminium
-dg or -g	-dge or gu	aging, argument	ageing, arguement
-ed	-t	dreamed, learned	dreamt, learnt

Capitalization

Often authors are confused as to which words to capitalize. Here are some general guidelines:

1. First word of a sentence

   a. Begin every sentence or sentence equivalent with a capital letter.

   b. When you list items by points and the complete thought can be stated briefly, it is unnecessary to introduce the list with capitals:

 Example 4-33   To start writing a research paper, scientists need

    1. enough data,

    2. valid references,

    3. writing principles.

    c. If the list cannot be stated briefly, introduce each subdivision with a capital letter and end it with a period.

2. Proper nouns

Capitalize all proper nouns within a sentence. Sometimes it may be difficult to distinguish between common and proper nouns. Common nouns do not require capitals within a sentence because they refer to everyday objects in a general sense. Proper nouns describe certain people, groups, or objects or are words derived from these sources. For this reason, the names of months and days are considered proper nouns because they are derived from names of gods and planets. These nouns are capitalized, whereas the seasons of the year are not capitalized (spring, fall). Proper nouns include the following categories:

    a. names of persons and places (countries, regions, counties, cities, and other political and geographical divisions: Canada, Huntington's disease, Celsius)

    b. names of months and days, languages, races, geological and historical periods and events, and documents (November, English, Friday, the Ice Age)

    c. names of organized bodies and the distinguishing names substituted for them (American Chemical Society)

    d. names of institutions, churches, schools, libraries, buildings, hotels, clubs, corporations, ships, and so forth (University of California, National Institutes of Health)

    e. official titles of persons when used without their personal names (the Chief Scientist)

3. Quotations

Use a capital letter for the opening word of a quoted sentence but not for quoted phrases:

    (Their report mentioned only "height, width, and breadth.")

4. Titles

Capitalize every important word in titles and literary titles. Prepositions, articles, and conjunctions are not capitalized unless one of them is the initial word in the title. (*Journal of Biochemistry, Handbook of Scientific Communication*). Note that some styles such as the *Publication Manual of the American Psychological Association* (APA) style capitalize all words in titles of 4 letters or more (e.g., "With").

5. Biological classification

The scientific name of a phylum, class, order, family, or genus is capitalized, but the name of a species or subspecies, or a common name, is not (*Olenellus thompsoni*, but arthropod and trilobite).

6. Parts of a book or report

Capitalize words followed by a number or letter to indicate the parts of a book or report that are used in text references (Appendix A, Chapter 2, Volume 12). Note that the sections of a research article are also capitalized if a specific section is referred to (Materials and Methods, Results, Discussion), but these sections are not capitalized if they are discussed in a general sense (methods section, introduction, discussion.)

Italics

Certain words and phrases in science, generally Latin derivatives, are placed in *italics*. Aside from genus and species names, subheadings, foreign words, emphasized words, and titles of journals, books, and manuals, the following words are typically written in italics. Note that some styles such as APA style do not follow this rule.

in vivo	*in vitro*	*in organello*
de novo	*in vacuo*	

Punctuation

To avoid possible misinterpretation of your writing, pay attention to correct punctuation. In contemporary English, the trend is toward less punctuation and more simplicity.

Here are a few important rules for punctuation:

1. Use simple punctuation.

The best approach to punctuation is usually the simplest. If you are confused over how to punctuate a particular sentence, readers will probably be confused as well. Rewrite the sentence in a form that requires only simple punctuation. Note that unlike some foreign languages, there are very few exclamation marks in English and essentially none in scientific writing.

2. Use a period to end a full sentence.

The trend in scientific writing is to eliminate semicolons and to use only periods (full stops) to end sentences.

☞	**Example 4-34**	*Older style:* TS-25 was a heat inducible <u>mutant; thus</u>, it was given the prefix "TS."
☝	**Revised Example 4-34**	*Newer style:* TS-25 was a heat-inducible **mutant. Thus**, it was given the prefix "TS."

3. Use semicolons to connect two independent sentences.

Place a semicolon between sentences that contain closely related ideas and are not linked by a coordinating conjunction such as "and," "or," and "but." However, remember that the trend in scientific writing is to eliminate semicolons.

☝	**Example 4-35**	Some viruses are **deadly; others** are not.
		or
		Some viruses are **deadly. Others** are not.

(margin) ESL advice

4. Use commas for clarity and emphasis.

Check each sentence for reading errors. Then decide whether a phrase or sentence needs a comma or not. Consider the following example:

Example 4-36	Although samples were incubated at 37 °C for 24 hr we did not observe any change in the growth pattern.

This sentence has more than one possible interpretation depending on where the comma is placed:

Example 4-36	Interpretation A	Although samples were incubated at 37 °C for 24 **hr, we** did not observe any change in the growth pattern.
	Interpretation B	Although samples were incubated at **37 °C, for** 24 hr we did not observe any change in the growth pattern.

It is important to place a comma at the correct location within the sentence to make the meaning of your writing clear. The comma's overall role is to add clarity or emphasis to a sentence.

Commas are needed under the following circumstances:

(a) Whenever a dependent clause or a long adverbial phrase comes before the main statement of the sentence, it needs a comma. Example 4-36 is an example of this rule.

(b) Transitional words and phrases are usually separated by a comma when they are used as an introductory phrase at the start of a sentence.

Example 4-37	Propanol, ethanol, and butanol are all organic alcohols. **However,** aerosol is not an organic alcohol. (or: Aerosol, **however,** is not an organic alcohol.)

However, commas are not placed after or in front of transition words if the flow of the sentence would be interrupted as in the following case:

Example 4-38	The cells absorbed the dye and were therefore blue.

(c) To determine whether to use commas for a clause that is within a sentence, read the sentence without the clause. Punctuation depends on whether the dependent clause within a sentence is essential or not. If the dependent clause is essential to the meaning, do not use commas. If the dependent clause is not essential for the meaning of the sentence, however, set it off by commas.

Example 4-39 a   Some of the **cultures, which** came from Tennessee, were damaged.

b   Of the cultures, 10% were damaged upon arrival. The cultures that were damaged were discarded.

Generally, commas are used with the word "which" but not with "that." See also Appendix on "Commonly Confused and Misused Words" for more information on "which" and "that."

Especially ESL writers are often confused about comma use elsewhere in the sentence. One particular comma placement that gets misused often is between the subject and the verb. Do not separate the subject and verb by a comma. If you disrupt the subject and verb by a dependent clause, use two commas, one before and one after the clause.

5. **Place a comma between the items in a series as well as before the word** *and.*

Commas should separate items in a series. In American scientific writing, a comma is also placed before the *and* in the series unlike in British English and unlike in American literary writing. Often this placement is important to the meaning of the sentence.

Example 4-40 a   *E. coli* can produce septicemia, **meningitis, and** pneumonia.

b   ARE-mRNAs can be rapidly stabilized upon exposure to certain signals including immune stimulation, UV and ionizing irradiation **(5), and** heat shock (10).

6. **Use semicolons and/or numerals to punctuate complex series.**

To separate independent but related clauses without using "and," "or," or "but," use a semicolon, but remember the trend in English is to eliminate this use when possible.

Use semicolons and numerals to separate the items within the sentence when items in a series are unusually long and especially when they contain their own punctuation. Use a colon to introduce the list or series.

Example 4-41   Antibiotic-resistant pathogens often cause infections and are an important source of mortality for patients hospitalized in an ICU. Of concern are especially gram-negative bacilli, such as Klebsiella and Enterobacter, producing extended-spectrum ß-lactamases, multiple drug-resistant *M. tuberculosis,* and fluconazole-resistant Candida spp., especially *Candida krusei, Candida glabrata,* and *Candida albicans* in HIV patients.

Revised Example 4-41   Antibiotic-resistant pathogens often cause infections and are an important source of mortality for patients hospitalized in an ICU. Of concern are especially: **(1)** gram-negative bacilli, such as Klebsiella and Enterobacter, producing extended-spectrum **β-lactamases; (2)** multiple drug-resistant **_M. tuberculosis;_ and (3)** fluconazole-resistant Candida spp., especially Candida krusei, Candida glabrata, and Candida albicans in HIV patients.

7. Avoid weak connectors.

Scientists are extremely fond of coupling pairs of independent sentences. However, the words "and," "but," "for," "or," and "nor" are weak connectors. If weak connectors cannot be avoided, they should be set off by commas. For stronger writing, consider making each statement a separate sentence.

An example of weak writing is given next.

Example 4-42   The amplitude of the potential pattern in the electrolyte decreases with increasing distance from the electrode and expands parallel to the electrode at the same time (Figure 6a, top), <u>and</u> this expansion is also felt at the electrode/electrolyte interface in the form of changed migration current densities, indicating that the migration coupling encompasses a wide range of the electrodes, sometimes even all of the interface.

Revised   The amplitude of the potential pattern in the electro-
Example 4-42   lyte decreases with increasing distance from the electrode and expands parallel to the electrode at the same time (Figure 6a, top). This expansion is also felt at the electrode/electrolyte interface in the form of changed migration current densities. Thus, the migration coupling encompasses a wide range of the electrodes, sometimes even all of the interface.

---

8. Avoid quotation marks.

Whenever you copy another writer's or speaker's exact words, enclose the material in standard (double) quotation marks (although this happens rarely in scientific writing). Use single quotation marks for a quote within a quote. Remember to cite the source of the material, including the page number on which it appears.

When the name of an article or book chapter is cited in the text, enclose it in double quotation marks. Sometimes titles can also be placed in italics.

Example 4-43   An article, "Structure of the 30S Ribosomal Subunit," recently appeared in *Cell.*

In journals, quotation marks are often used around new technical terms, old terms used in an unusual way, or simply to draw attention to a word. (In books, italics or boldface are often used instead.)

9. Avoid hyphenation.

Avoid hyphens if possible. Write "cooperation" and "rearranged," but hyphenate two (or more) word clusters such as "English-speaking," "high-pressure," and terms that contain a capital letter or a number and a noun such as "N-terminal" and "100-fold."

10. Abbreviations.

In scientific English, abbreviations for a single word are spelled with a period as in "Prof.," and "fig." Note that in British English, titles do not have a final period if the last letter of the abbreviation corresponds to the last letter of the word itself (Dr, Mr, Mrs, Ms), whereas American English prefers the period after these abbreviations (Dr., Mr., Mrs., Ms.). Units of measure are usually lowercase without a period (kg, min), even when they are acronyms (ppm [parts per million]). Other acronyms not referring to units of measure are usually capitalized without periods (DNA) but can also be a mixture of lowercase and uppercase letters (iRNA, ssDNA.) Units of measure named after people are capitalized without period (C [Celsius], F [Fahrenheit]).

Subject-Verb Correspondence

The subject and verb of a sentence must agree. A singular subject must have a singular verb, and a plural subject must have a plural verb.

 Example 4-44   The *blood, urine, and stool* of each patient <u>was</u> examined.

The subject of the sentence is "blood, urine, and stool," not "patient." Because the subject is plural, the verb must be plural as well.

 Revised   The *blood, urine, and stool* of each patient **were**
Example 4-44   examined.

Many scientific authors are confused about such words as "data," "spectra," and "media." "Data," "spectra," and "media" are the plural forms of "datum," "spectrum," and "medium" and thus should be treated as plural nouns. (Note that some dictionaries do accept use of both singular and plural verbs for these words.)

 Example 4-45   Our <u>data</u> **suggest** that Klein's hypothesis is correct.

Other singular and plural forms of words commonly used in the biomedical sciences are shown in Table 4.2.

Table 4.2 Singular and plural nouns forms

SINGULAR	PLURAL
alga	algae
analysis	analyses
bacterium	bacteria
criterion	criteria
datum	data
flagellum	flagella
genus	genera
hypothesis	hypotheses
larva	larvae
matrix	matrices
medium	media
nucleus	nuclei
phenomenon	phenomena
serum	sera
spectrum	spectra
stimulus	stimuli

Subjects and verbs should not only agree, they should also make sense together.

 Example 4-46   The *concentration* of the protein was <u>measured</u>.

Unlike temperature or amounts, concentration cannot be measured. It has to be calculated or determined.

 Revised   The *concentration* of the protein was **determined**.
Example 4-46

Unclear Modifiers

Unclear modifiers, such as dangling modifiers or misplaced modifiers, are words or phrases that modify an element of a sentence in an ambiguous manner because they could either be modifying the <u>subject</u> or the object of the clause. Avoid dangling or misplaced modifiers.

In general, you should place modifiers near the word or words they modify, especially when a reader might think that they modify something different in the sentence.

Dangling modifiers most frequently occur at the beginning of sentences (often as introductory clauses or phrases) but can also appear at the end. They often have an -ing word (gerund) or an infinitive phrase near the start of the sentence. To revise a sentence that contains a dangling modifier, name the appropriate or logical doer of the action as the subject of the main clause or change the phrase that dangles into a complete introductory clause by naming the doer of the action *in that clause.*

Example 4-47   Having tested positive for HIV, we disqualified the patients for participation in the study.

This modifier is unclear because it modifies "we." It sounds as if "we" tested positive for HIV.

Revised
Example 4-47   Having tested positive for HIV, **the patients** were disqualified for participation in the study.
or
Patients that tested positive for HIV were disqualified for participation in the study.

Following is another example:

Example 4-48   While incubating at 30 °C, we added another 10 ml of buffer K to the samples.

"While incubating at 30 °C" appears to modify "we." Because "we" were not incubating, this modifier is an unclear modifier.

Revised
Example 4-48   **While the samples were incubating at 30 °C**, we added another 10 ml of buffer K to them.
or
We added another 10 ml of buffer K to the samples **while they were incubating** at 30 °C.

SUMMARY

WRITING PRINCIPLE 12: Use the first person.

WRITING PRINCIPLE 13: Use the active voice.

WRITING PRINCIPLE 14: Use past tense for observations, completed actions, and specific conclusions.

WRITING PRINCIPLE 15: Use the present tense for generalizations and statements of general validity.

WRITING PRINCIPLE 16: Write short sentences. Aim for one main idea in a sentence.

WRITING PRINCIPLE 17: Use active verbs.

WRITING PRINCIPLE 18: Avoid noun clusters.

WRITING PRINCIPLE 19: Use clear pronouns.

WRITING PRINCIPLE 20: Use correct parallel form. If possible, arrange ideas in a list to read from shorter to longer.

WRITING PRINCIPLE 21: Avoid faulty comparisons.

WRITING PRINCIPLE 22: Avoid errors in spelling, punctuation, and grammar.

PROBLEMS

Problem 4-1 First Person
Rewrite these sentences using first person instead of third person.

1. It is one of the aims of the present paper to study the performance of different functionals with the size of clusters.
2. The authors thank Dr. T. J. Guiermo for useful discussions.
3. It is recommended by the authors of the present study that a triple regimen of antibiotics is given to infants with HIV.
4. Based on our results, it is concluded that the measurements on translational cooling in desorption contain information about the dynamics of vibrational energy transfer.
5. No delay in replication was observed.
6. To remove PAH, transgenic *E. coli* were constructed.
7. We also studied the effect of Mg^{2+} and K^+ concentrations on $[^{14}C]$ ATP-pmRNA complex formation. It was found that binding is dependent on magnesium concentration with an optimum at 6 mM and an apparent linear decrease at higher concentrations (Figure 2D).
8. Genetic drive mechanisms are being investigated in the laboratory and field, including the incompatibility mediated by Wolbachia symbionts, which will allow us to apply this technology for disease control.
9. The twisted conformation is predicted to have only one C–F bond in an anomeric position.
10. The authors propose a mechanism of electrocyclic rearrangements initiated by N-silylation.
11. J. D. Beckerle and colleagues studied vibrational energy relaxation on metals; but in this report, the authors establish a new class of non-equilibrium surface reactions.

Problem 4-2 Active and Passive Voice
Rewrite the following sentences in the active voice. Condense them if you can.

1. The following results were obtained:....
2. Isolated tissues were examined for parasites.
3. The collected specimens and videotape recorded during 25 dives were analyzed by the team.
4. Variations in day length are gauged by plants.
5. Recovery of several cytosine deaminases occurred after 2 days.
6. Insight into the dynamics of the excitation process can be obtained from two-pulse correlation measurements.
7. No colonies were seen on blood agar.
8. Phosphorylation is stimulated by the addition of KA-1 to the reaction.

9. Additional arguments will be provided by discussing the electronic structure of the adsorbate system and the absolute temperatures reached during the desorption process.
10. The structure was refined at a resolution of 1.15 Å.
11. Preliminary findings were unveiled by expedition members.
12. The same dosage of antitumor drugs was administered to patients for 60 days.
13. The emerging field of spintronics was launched by the advent of giant magnetoresistance (GMR).
14. Another way of treating the translational cooling is provided by the principle of detailed balance.
15. Computational studies and searches in the CSD (Fig. S1) and the PDB were carried out to investigate the conformational preferences of the aromatic OCH3, OCH2F, OCHF2, and OCF3 groups.

Problem 4-3 Active versus Passive Voice

Decide whether to use active or passive voice in the sentences in italics. Explain your answer.

1. Deep Water Coral, also known as cold water coral, are most often stony corals belonging to the phylum Cnidaria. One of the most common cold water corals is *Lophelia pertusa. Fishing methods such as deep water trawling affect* Lophelia pertusa *negatively.* These methods tend to break corals apart and destroy reefs.
2. The photon is the basic unit of light and all other forms of electromagnetic radiation. It is also the force carrier for the electromagnetic force. The photon has no mass and thus can produce interactions at long distances. *Both wave and particle properties are exhibited by the photon.* For example, a single photon may undergo refraction by a lens or exhibit wave interference but also act as a particle giving a definite result when its location is measured.
3. The bulk of inorganic compounds occur as salts. Important classes of inorganic compounds are the oxides, the carbonates, the sulfates and the halides. *High melting points characterize many inorganic compounds.* Inorganic salts typically are poor conductors in the solid state and their solubility in, e.g., water and ease of crystallization varies.

Problem 4-4 Use of Tense

Decide whether to use past or present tense in the following sentences. If there is more than one answer, explain why.

1. In our study, tree size _____ (**increase**) with reduction of pesticides.
2. The larva of the monarch butterfly _____ (**feed**) on milkweed.
3. In our study, we _____ (**find**) that there are significant flattenings of the PES along the reaction Pth with increasing cluster size, as Fig. 5 _____ (**indicate**).

4. In a study by Fermi, the vibrational levels _____ (**diverge**) into the so-called Fermi dyad.

5. Females of this species _____ (**have**) a unique projection from the dorsal part of the thorax.

6. Recent work by deValt (2001) _____ (**show**) that *nematodes* _____ (**reproduce**) readily in the laboratory.

7. A phagemid construct encoding 2 transgenes in tandem _____ (**can**) be used to transform *E. coli* cells.

8. Table 3 _____ (**show**) that in our study, polychaetes _____ (**be**) most abundant at depths of 10 to 16 m.

9. Baumann (1997) _____ (**find**) that this bacterium _____ (**is**) highly sensitive to pH.

10. We _____ (**observe**) chemoattraction within 30 min of Shh addition.

11. Identification of restriction factors _____ (**give**) us more knowledge of the host cell restriction system.

12. Our results _____ (**suggest**) that the Asn5 glycosylation site _____ (**play**) a critical role in the function of the KCEN 1 protein.

13. We _____ (**investigate**) high-quality single crystals of vanadium oxide.

14. We _____ (**do**) not observe rods in the diffraction patterns but instead well-defined Bragg spots, suggesting that at least 10 interatomic layers in the surface-normal direction _____ (**be**) contributing to the interferences.

15. Coupled climate-carbon cycle models _____ (**suggest**) that Amazon forests _____ (**be**) vulnerable to droughts, but satellite observations _____ (**show**) a green-up in intact evergreen forests of the Amazon in response to a short, intense drought in 2005.

16. The K_a value for ß-cellobiosyl (approaching 10^3 M^{-1}) ____ (**be**) comparable to that for many carbohydrate/protein interactions.

17. Hundreds of species of reef-building corals _____ (**spawn**) synchronously over a few nights each year, and moonlight _____ (**regulate**) this spawning event.

18. Arctic wind anomalies _____ (**be**) part of a global-scale pattern of highly unusual circulation in 2007, the causes of which _____ (**be**) as yet unclear.

19. To map pathways of motion, all observed Bragg diffractions of different planes and zone axes _____ (**be examine**) on the femtosecond to nanosecond time scale.

20. ABSTRACT: The average translational energy _____ (**be**) significantly lower than expected for the calculated surface temperature.

21. SUMMARY: In conclusion, we _____ (**find**) significant deviations from the equilibrium behavior in fs-laser induced desorption of CO/Ru.

Problem 4-5 Sentences Length

Rewrite the following overlong or overloaded sentences by breaking them up into shorter sentences.

Example 1

These results show that ATP and GTP are capable of binding to the same site in the mRNP but higher specificity is shown for ATP than for GTP, while the other nucleotides tested (CTP, UTP, AMP) did not stimulate the initiation of the translation, on the contrary, they produced some inhibition being most pronounced for UTP and CTP, suggesting that these nucleotides compete with ATP and GTP for the same site in the mRNP.

(71 words per sentence)

Example 2

The Dengue virus, a *Flavivirus* belonging to the family *Flaviviridae*, which includes over 60 known human pathogens such as those causing yellow fever, Japanese encephalitis, tick-borne encephalitis, Saint Louis encephalitis, and West Nile encephalitis, is classified into four different serotypes, Types 1, 2, 3, and 4, and infection with any of these four serotypes can result in dengue fever (DF) or dengue hemorrhagic fever (DHF), which is particularly a concern during heterologous secondary dengue virus infections.

(74 words per sentence)

Example 3

Overall, in the TNF signaling pathway, h-IAP1 seems to enable its own expression and once expressed, h-IAP1 exerts its anti-apoptotic effect downstream in the TNF α signaling pathway by stimulating proteolytic cleavage of terminal caspase-3, -7 and caspase-9, thereby inhibiting protein degradation and caspase-mediated apoptosis.

(46 words per sentence)

Example 4

Reports show that the binding of thiamine derivatives to the mRNA that codes for the enzymes that catalyze the thiamine biosynthesis in *E. coli*, facilitate the stabilization of a mRNA conformation that inhibits translation (22b).

(34 words per sentence)

Example 5

Two combinations of half-wave plates and polarizing beam splitters, together with on/off photodetectors in the reflected channel, form the photon-subtraction modules that are placed in the path of the thermal light field, respectively, before and after the parametric crystal that, with the on/off photodetector in the trigger channel, forms the photon-addition module.

(52 words per sentence)

Example 6

An international team of researchers says that some key elements of modern behavior were in place by 164,000 years ago, pushing back the appearance of complex tools and manipulating symbols by 25,000 to 40,000 years as evidenced by complex stone bladelets and ground red pigment—advances usually seen as hallmarks of modern behavior—coupled with the shells of mussels, abalone, and other invertebrates in a cave in South Africa, which showed that these ancient clambakes are the earliest evidence of humans that also include marine resources in their diet.

(89 words per sentence)

Example 7

Attosecond pulses are created when intense laser pulses of femtosecond duration are focused into a gas sample, resulting in a process known as high-harmonic generation to produce light at a range of frequencies that are precisely phased together, creating a train of very short, coherent pulses.

(48 words per sentence)

Example 8

Negative global coupling changes pattern formation only qualitatively if the system oscillates and can result in the formation of standing waves with wave Number 1 just like for N-NDR-systems, but for parameter values for which the migration coupling leads to the formation of Turing-like structures, the interaction of both instabilities can yield a complex wave pattern characterized by two wave numbers, n = 1 and n > 1.

(64 words per sentence)

Example 9

For a remote reference electrode but close counter electrode (or equipotential area), the migrational coupling is local and thus, the fronts move at a constant rate, whereas the fronts are increasingly accelerated, if the distance between working electrode and counter electrode is larger.

(44 words per sentence)

Problem 4-6 Active Verbs
Put the action in the verbs in the following sentences.

1. Measurements were performed on reaction products by a mass spectrometer.
2. An increase in transplant rejection occurred.
3. Removal of mitochondria was achieved by HPLC.
4. Two measurements of amyloid plaques were obtained for each brain.
5. Elevations in WBC count did not occur when aspirin and streptokinase were given.

6. If a correction of the symmetric stretch levels is not made, misleading results occur. (Two corrections needed!)

7. Our results showed protection of the dogs by the vaccine.

8. A PI 3-kinase signaling cascade exerts a controlling regulation of PIP2 levels.

9. Stanozolol caused prolongation of appetite.

10. Increasing cluster size causes a reduction of E_g.

11. Agglomeration of clusters of Ag or Cu in a noble gas matrix causes the emission of visible light.

12. Sequencing of the A-B construct was performed on an ABI377 automated DNA sequencer using Big Dye Sequencing deoxyterminators (ABI v3.0).

13. There was no significant difference in growth rates between the Shh-treated and control samples during the 1 hr prior to protein addition.

14. We performed plasmid isolation to sequence the A-B insert in the pBluescript (STRATAGENE) vector.

15. Rfp2 overexpression caused changes in expression of many genes that regulate cell cycle.

16. To determine whether cell migration is occurring, we dissected mouse brains.

17. Light inactivation of COP1 was achieved prior to its nuclear depletion.

18. Electron liberation occurred through the photoelectric effect.

19. In all instances, the result of the binding of these soluble molecules to their specific receptors is the activation of intracellular signaling cascades.

20. NCs experience no interactions with the very small stray fields present in the ferromagnetic column matrix.

21. Buffalo, elephant, and black rhino abundance all show a rapid decline after 1977.

22. Current displays free flow in a superconductor because the electrons form pairs that travel unhindered through the material.

23. A powerful methodology developed in our laboratory enables structural characterization of increasingly complex systems.

24. To accomplish these objectives, preparation of solvated gas-phase catalytic intermediates is required.

25. We must understand the central puzzle of how O=O bond formation occurs.

26. In the samples containing MMS, a slowing of replication was observed.

27. Water oxidation leads to the production of biochemical fuel and oxygen.

Problem 4-7 Noun Clusters

Untangle the noun clusters in the following sentences by adding the appropriate preposition(s) and other words as needed. Start at the end of the cluster and work your way to the beginning.

1. The results are compared using a high-quality plane-wave basis set.

2. The negative penicillin skin test result group showed no allergic rash.
3. We analyzed a particle flux time series.
4. The spinach culture callus tissue produced no shoots in 1999.
5. With this unconventional technique, we could easily define the aromatic hydrocarbon liquid crystal transition temperatures.
6. In 10 of the 15 patients, we observed a chronic depression syndrome.
7. Previous work suggested that vitamin A-rich fish oil diets protect mice against certain mosquitoes.
8. The quantity of the TCR signal is directly related to the time for which a T-cell remains in contact with the APC-expressing MHC-peptide complex.
9. Our study involved human Epstein–Barr virus (EBV) transformed lymphoblastoid cell lines.
10. In particular, their extraordinary low density (1.00–0.20 g/cm^3) and high surface area (500–4500 m^2/g) make them ideal storage and separation of gases candidates.
11. Inelastic neutron scattering (INS) and diffuse reflectance infrared spectroscopy do not provide adequate information on structural adsorption site details.
12. The different mobility samples spin susceptibility behaves critically near the transition.
13. Ecological community abundance distributions need to be considered in these studies.

Problem 4-8 Pronouns

In the following sentences, the pronoun (underlined) could refer to more than one noun. Revise these sentences to make the meaning clear either by restating the noun, by changing the sentence structure, or by repeating one or more words from the previous sentence.

1. When bacterial cells are cured of viruses, they become avirulent.
2. A few microorganisms such as *Mycobacterium tuberculosis* are resistant to phagocytic digestion. This is one reason why tuberculosis is difficult to cure.
3. Patients often suffer from infections of *Corynebacterium* and from toxins released by them.
4. An action potential triggers calcium channels to open at the synapse. This causes docked synaptic vesicles to fuse to the membrane and release their neurotransmitter content.
5. When the *E. coli* ompA gene was expressed in Sodalis, it displayed a pathogenic phenotype.
6. Another more costly and labor-intensive approach is to perform biochemical analysis of the fats in yeast. This has been done for some yeast mutants (47). This involves thin layer chromatography, coupled to gas chromatography.
7. Results indicate that binding decreased about 4-fold when the temperature increased from 4 to 17 °C. (Fig. 1). This suggests that the

binding among the particles may be governed by interactions such as hydrogen bonds or van der Waals forces that weaken when temperature rises.

8. p53 co-immunoprecipitates with ARF-GFP and HDM2 in the presence of E6 (Fig. 1), <u>which</u> can also be seen for total cell lysates (Fig. 1).

9. The identification of the restriction factor helps us to understand the function of this and similar factors. <u>It</u> gives us more knowledge of the host cell restriction system.

10. The goal of this study was to identify the molecular mechanisms by which rosiglitazone induces primary adipocytes to increase adiponectin. To achieve <u>this</u>, we tested both the mouse 3T3-L1 adipocyte cell line and Ob/Ob mice.

11. Adipose tissue from the same rosiglitazone-treated animals secretes an increased amount of adiponectin. <u>This</u> indicates that there may be additional factors signaling adiponectin secretion.

12. CTLA-4 inhibits the number of activated T-cells, and this effect is maximal at low TCR strength. <u>This</u> is contradictory to earlier observations where CTLA-4 expression increases when TCR strength is increased.

13. The breakthrough was achieved because new methods and concepts could be applied, which had been developed in nonlinear dynamics to describe the spontaneous formation of order in various disciplines. <u>This</u> in turn was only possible because the underlying laws are universal at a certain abstract level.

14. A current-voltage-characteristic with a negative impedance is a necessary prerequisite for this type of oscillator. However, in the stationary current-voltage-plot, <u>this</u> is visible only in a subsystem.

15. Obviously, the disturbance is dampened in the first picture, whereas a spatiotemporal structure forms in the second example. <u>This</u> can best be described as a superposition of the homogeneous oscillation and a sine-shaped standing wave.

16. The concentration profile perpendicular to the electrode does not adjust spontaneously when the concentration changes at the electrode. <u>This</u> is caused by a high reaction rate.

17. Almost all sulfur atoms are also engaged in lone pair bonding to a phenyl edge (25). Most p-MBAs are linked through chains of such interactions extending from one pole of the nanoparticle to the other (Fig. 4D). <u>This</u> exemplifies the "self-assembly" of a thiol monolayer on a gold surface (26).

18. Orienting C–F bonds antiperiplanar to the lone pairs of the now sp3-hybridized O results in an anomeric $nO-\sigma^*CF$ conjugation with concomitant lengthening of the C–F bonds (35). <u>This</u> reduces the conjugation between O and the aromatic π system and eliminates the energetic preference of a planar, in-plane conformation.

19. Our previous models tended to underproduce Na compared with the abundances of HE0107-5240. <u>This</u> has been improved in our current presupernova models.

Problem 4-9 Parallelism
Correct the faulty parallelism in the following sentences.

1. The pathogenesis observed in other cells, such as circulating mono-cytes, may differ from endothelial cells.

2. The stability of these particles appears to be regulated by RNA heli-case activity and protein phosphorylation.

3. Dengue hemorrhagic fever is particularly a concern during heterolo-gous secondary dengue virus infections where cross-reactive antibod-ies are implicated in immune pathology as well as antibody-enhanced infection.

4. MAPs proteins are able to destabilize RNA helices in the presence of ATP, but DNA helices remain stabilized.

5. Diabetes can be affected both by exercise and diet.

6. The fat bodies were washed twice with TBS/Tween, incubated in secondary antibody for 2 hr, and then they were mounted onto glass slides.

7. Aptamer complexes could be used as highly sensitive biosensors, detection probes, efficient inhibitors of cancer, or they can be power-ful tools for accelerating the process of drug discovery.

8. The pellet was rinsed with 200 µl ethanol, air dried, and it was then resuspended in 10 µl dH_2O.

9. These peaks were dependent on kinetic heat flow, not state transition heat flow.

10. Spatial coupling acts on the activator variable and also the inhibitor variable.

11. Other research may indicate that the translating ribosome has the ability to open up downstream mRNA secondary structure and in this way a block in translation initiation is overcome.

12. Whether the crisis for liquids would be avoided in some way at lower temperatures than measurements have been made is con-troversial and is of limited relevance to describing the observed behavior.

13. The thiol monolayer is stabilized not only by gold-sulfur bonding but also interactions between p-MBA molecules.

14. Most sulfur atoms, bonded to a phenyl ring and gold atoms in two dif-ferent shells, are also chiral centers.

15. To keep the time-varying term to a minimum, we study the balance over two long periods, from January to June and July to December.

16. The internal pressure must not only depend on volume but also the rate of filling.

17. We provide a broad overview of the project, including the target population, the project's primary aim, the problem areas the project addresses, and where the services are being delivered.

18. NbCA1 acts as a negative regulator and a suppresser of SGT1, RAR1, and HSP90.

Problem 4-10 Comparisons

Correct the faulty comparisons in the following sentences. Avoid the overuse of "compared to," ambiguous comparisons, and incomplete comparisons.

1. The kinetics of the protein G accumulation were similar to endogenous proteins (19).
2. Thus, the pattern observed for the zot gene probe proved to be much more diverse compared to that of the CT genotype.
3. Optimizations for the MMA-co-MVE copolymerization system were better for the GA simulations in comparison to experimental data.
4. The dendrogram showed that *V. cholerae* O22 is more closely related to *V. cholerae* O139.
5. Mayer's hypothesis on the spread of this disease is similar to our present study.
6. In comparison to eubacteria, archaebacteria are more ancient.
7. We observed a peak for mutant A that was higher than the other mutants.
8. Compared to the other mammals, the male dolphin was larger.
9. The different propagation behavior of electrochemical fronts compared to reaction-diffusion systems is due to a different range of spatial coupling.
10. There are several connections of the Au102 nanoparticle structure with previous works.
11. The Austro-Tai populations were found to be a unity not only in culture but also in genetic structure, compared with a large number of data from other groups in East Asia.
12. VSV vectors induce much stronger CD8 T cell responses compared to other viral vaccine vectors, including poxviruses (25).
13. 'A' stars are less numerous compared to their solar-type equivalents.
14. The cosmic ray anisotropy was more complex.
15. High concentrations in previous studies (800 times more active sites than in this study) cause a signal decrease of no more than 12%, even when the concentration of polarized nuclei is 2,600-fold higher compared with our setup.
16. The New Haven cohort had the same immunological response as Philadelphia and Miami.

Problem 4-11 Common Errors

Rewrite the following sentences such that they make sense. Ensure that the subject and verb make sense together, that the subject and verb agree, that no helping words are omitted, and that modifiers are clear.

1. Altogether, we measured for 5 days.
2. This structure controls the packaging of the DNA into the phage head and the release after recognizing the receptor on the host.
3. Fig. 7 was obtained during the oxidation of hydrogen at the Pt electrode.

4. Our data indicates that more work needs to be done to prove our hypothesis.

5. The liver, gall bladder, and spleen of each patient was examined.

6. We confirmed that five genes were overexpressed and one gene down-regulated in GEHIJs during B-3 infection.

7. Contrast medium was infused at a steady rate into the injection port, and the flow calculated from the observed change in CT number at equilibrium.

8. These complexes form in situ.

9. The Tibet air shower array experiment was at Yangbajing (90.522 E, 30.102 N; 4,300 m above sea level) in Tibet, China.

10. We used a novel programme to analyse the tumor.

11. Upon doping with oxidants, the HBC graphitic nanotubes become electrically conductive.

12. The study of Brownian motion and its generalizations has had a profound impact on physics, mathematics, chemistry, and biology.

13. Having determined that an anisotropy exists, the full data set was rescanned from 1 January 2004 to 31 August 2007.

Problem 4-12 Punctuation
Correct the punctuation in these sentences.

1. Moreover their results can be explained by the fact that the DNA repair defect of the H2A tail deletion mutant is mainly kinetic.

2. Patients with lnd QT syndrome, can experience seizures arrhythmia and sudden death by ventricular fibrillation.

3. As can be seen in Fig. 3 the amount of amplified product increases in parallel fashion to the amount of inclusions observed.

4. The overall relative contributions of anthropogenic versus biogenic sources of VOCs have not been clearly established, and there is geographic variation in the relative contributions of these broad source groups.

5. In conclusion our results suggest, that the pathways, that regulate glucose uptake, have even more components than previously thought.

6. Arabidopsis transformation will be done using Agrobacterium mediated T-DNA transformation thereby allowing in vivo expression of our transgene.

7. The goal was to select the best most efficient deaminases possible.

8. The book, *Handbook of Scientific Writing,* is very helpful.

9. The density matrix of a thermal state the most classical state of light is diagonal in the Fock state basis and accordingly exhibits no phase dependence.

10. The accumulation of oxidizing equivalents was analyzed by gas-phase ion chemistry laser spectroscopy and computational methods.

11. Two key roadblocks had to be overcome i) we needed to store the energy of four photons and ii) we needed to design efficient catalysts.

12. Thus this system of pathways is termed pioneer metabolism.
13. 'A' stars are less numerous than their solar-type equivalents the T-Tauri stars and in general are located farther away from Earth.
14. An insulator has an energy gap separating filled and empty bands of electronic states and thus is electrically inert, because a finite energy is required to dislodge an electron.

CHAPTER 5

Special ESL Grammar Problems

Anyone who wants to be successful in science nowadays has to publish internationally and hold talks at conferences. This fact, more often than not, poses a problem especially for people whose native language is not English. English as a second language (ESL) speakers not only have to formulate their research papers and talks in English, they also need to realize that the English language is a living thing, deeply influenced by Internet communication and constantly changing, also in science.

ESL authors face special grammar problems when writing in English. Depending on how much similarity your language has with English and on how intensely you have studied it, these problems may be few or many. Although this book points out special ESL problems throughout its chapters, Chapter 5 is devoted to the most common problems encountered by ESL authors, discussing these in more depth.

5.1   PREPOSITIONS

GUIDELINE:

Use correct prepositions.

Prepositions are "little words" that link nouns, pronouns, and phrases to other words in a sentence, indicating their temporal, spatial, or logical relationship to the rest of the sentence.

Examples	about	above	after	at	below	by	for
	from	in	into	like	of	off	on
	out	over	than	through	to	up	with

Most verbs can be used with more than one preposition, but you should be sure to choose the preposition that reflects your intended meaning. If you are unsure which preposition to use, consult a dictionary. However, be careful: The meanings of corresponding prepositions in English and other languages do not always coincide. Even native speakers are prone to incorrect use of prepositions.

A few words and expressions need special attention:

compared	*Compared* takes the preposition *to* when it refers to unlike things. It takes *with* when two like things are examined.

 Example 5-1   a   The human brain is sometimes *compared to* a computer.

b   When we *compared* our results *with* those of Pauling's, et al....

comparison	The use of the word comparison is cause for much confusion, as it can take various prepositions: *comparison of A and/with B*, or comparison *between A and B*
different	Do not use *different than* when you should use *different from*. *Different than* is acceptable only in sentences such as the following:

 Example 5-2   a   We obtained *different* activity values for protein RZ-1 *than* Roberts et al....

But

 Example 5-3   Our activity values were *different from* those of Peters et al.

following	Do not use *following* as a preposition. *Following* can be used as an adjective, as in "The following results," but should be avoided when it can be replace with *after*, which is unambiguous.

 Example 5-4   *Following* centrifugation, the supernatant was removed.

 Revised Example 5-4   <u>*After*</u> centrifugation, the supernatant was removed.

The most commonly (mis)used prepositions in scientific writing include the following:

in connection **with**	compared **to/with**
in contrast **to**	correlated **with**
similar **to**	analogous **to**

5.2   ARTICLES

GUIDELINE:
Use correct articles.

The English language has two kinds of articles, definite (*the*) and indefinite (*a, an*). ESL writers often mix them up, leave them out, or put them in when they are not needed.

Every time you use a noun in English, you need to decide which sort of article to use, if any. Use of the article depends on the noun that follows it. A few general guidelines are shown in Table 5.1.

Extensive information about how to determine which article to use, if any, can be found at the following two Web sites (as of October 2009):

http://www.rpi.edu/web/writingcenter/esl.html
http://owl.english.purdue.edu/handouts/esl/eslart.html

Other great resources are Lynch (2008) and Greenbaum (1996).

5.3   VERBS

Plural and Singular Verb Forms

GUIDELINE:
Use correct plural and singular verb forms.

For many authors, especially ESL authors, it is not always obvious which subject words are treated as singular and which as plural. To clarify the use of singular and plural verbs, use Table 5.2.

Table 5.1 Use of articles in English

ARTICLE	USE	EXAMPLE
Definite "the"	to show which *particular* item you mean	**The** culture that was contaminated was discarded.
		The Nobel prize winner in 2006 in medicine…
	in connection with the phrases:	…in **the** presence of…,…in **the** absence of…
	before a noun that names something that has already been mentioned	The technician looked at **the** sample we had measured earlier.
		The methods described previously…
Indefinite "a" "an"	when the noun is one of a group or new to the reader	**A** scientist needs to write well.
		She used **an** aliquot to determine DNA concentration.
	when the noun is generalized	**A** protein can be made up of many different amino acids.
No Article	before a plural noun that has not yet been referred to	She collected samples for DNA isolation.
	before a proper or abstract noun	Berlin was a divided city.
		Degradation occurs after 24 hrs.
		In conclusion, we determined that…
"the," "a," "an"	if you use proper or abstract nouns as common nouns	**The** Berlin that was divided is gone.
		The degradation we observed occurred after 24 hrs.
		The conclusion of the discussion should summarize the results.
		A conclusion about X cannot be drawn.

Also pay attention to what is the true subject of a sentence:

 Example 5-5   **The role** of cytokinins **has** been studied in detail.

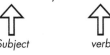

Subject   verb

Here, "role" is the subject, not "cytokinins." "Cytokinins" is the object of the preposition "of" and describes or qualifies "role." Objects of prepositions are never the subject of a sentence. Thus, the verb should be singular as it corresponds to "role."

Table 5.2 Use of singular and plural verb forms

RULE	SUBJECT	EXAMPLE
Certain indefinite pronouns are singular	one, no one, anyone, each, everything, something, someone, everybody	**Each** fragment **was** purified.
Certain indefinite pronouns are plural	both, few, many, several	**Both** fragments **were** purified.
Some indefinite pronouns can be singular or plural	most, some, none, part, any, all	**Some** fragments **were** purified. **Some** of the fraction **was** lost.
Collective nouns are singular	audience, class, group, committee, team, politics, news	This **group** of enzymes **is** known as…
Some collective nouns can be singular or plural	staff, faculty, (fractions such as) one third	**One third** of the patients **were** men. **One third** of the sample **was** used.
Certain abstract nouns are singular despite their plural appearance	news, measles, mumps, physics	**Measles has** essentially been eradicated in the Western world.
Compound subjects joined by "and" are plural	Subject joined by "and"	**Temperature and time are** both important parameters in DNA digestion.
For compound subjects joined by "either, or"; "neither, nor"; "or"; and "not only, but also," the verb must agree with the closest subject	Subject joined by (either, or …) (neither, nor …) (not only, but also …)	Neither the heart nor **the lungs were** inflamed. Not only the boys, but also **the girl was** infectious.

Irregular Verbs

GUIDELINE:

Use correct form of irregular verbs.

Irregular verbs often pose a problem in English writing or speaking, especially for ESL authors. Make sure you use the correct verb forms so mistakes such as the following do not arise:

👎	**Example 5-6**	Our insert was <u>cutted</u> by EcoRI.
👍	**Revised Example 5-6**	Our insert was **cut** by EcoRI.

The most common irregular verbs found in science are listed in Table 5.3. A more complete list of irregular verbs is available from http://www.englishpage.com/irregularverbs/irregularverbs.html (last accessed

Table 5.3 Irregular verbs and their forms

INFINITIVE	PAST TENSE	PAST PARTICIPLE (USED TO FORM PAST PERFECT TENSE AND PASSIVE VOICE)
arise	arose	arisen
become	became	become
begin	began	began
bend	bent	bent
bring	brought	brought
choose	chose	chosen
cut	cut	cut
deal	dealt	dealt
draw	drew	drawn
find	found	found
get	got	got/gotten
give	gave	given
grow	grew	grown
hide	hid	hidden
keep	kept	kept
lead	led	led
let	let	let
lie	lay	lain
lay	laid	laid
lose	lost	lost
prove	proved	proven
put	put	put
read	read	read
rise	rose	risen
run	ran	run
see	saw	seen
seek	sought	sought
send	sent	sent
set	set	set
shake	shook	shaken
show	showed	shown
shrink	shrank	shrunk
spend	spent	spent
spin	spun	spun
spread	spread	spread
take	took	taken
write	wrote	written

October 2009) and http://ww2.college-em.qc.ca/prof/epritchard/pastverb. htm (last accessed October 2009). Other great resources are Lynch (2008) and Greenbaum (1996).

Endings of Verbs

GUIDELINE:
Do not omit endings of verbs.

ESL speakers who do not pronounce endings of words fully often omit endings of verbs in writing as well. Do not omit "-s," "-es," "-ed," or "-d" endings to use the third-person ending of a verb form or to express the past tense or past participle form of a verb.

👎	**Example 5-7**	a	This journal usually *publish* articles in the field of neuro-science only.
		b	Before the substrate was added, we *determine* if the enzyme was temperature sensitive.
👍	**Revised Example 5-7**	a	This journal usually **publishes** articles in the field of neuroscience only.
		b	Before the substrate was added, we **determined** if the enzyme was temperature sensitive.

Gerund and Infinitive

GUIDELINE:
Follow a verb with the correct gerund or infinitive form.

Using gerunds and infinitives is a great way to add action to scientific writing and make science come alive. A gerund is a verb form ending in "-ing" (describing, running). Gerunds are used as nouns in English. An infinitive is the base form of the verb and is preceded by the word "to," such as in "to describe" or "to run." Some verbs may be followed by a gerund, others by an infinitive, and still others may be followed by either. In addition, for certain verbs, a noun or pronoun must be placed between the verb and the infinitive that follows it. Some more detailed, general guidelines are listed in Table 5.4.

Note also that verbs used after helping words that are a form of *to do* use the infinitive form and do not show the singular *-s*. This mistake is commonly made by ESL writers.

👎	**Example 5-8**	Eco RI does not *cuts* RNA.
		Did the sequence *affects* digestion?
👍	**Revised Example 5-8**	Eco RI does not **cut** RNA.
		Did the sequence **affect** digestion?

Table 5.4 Use of gerund and infinitive

FORM	SAMPLE VERBS	EXAMPLE
Verb + gerund	avoid, discuss, imagine, practice, recall, resist, suggest, tolerate	The authors avoided **describing** the problem in detail. Dissecting **requires** steady hands.
Verb + infinitive	agree, decide, want, ask, expect, have, offer, refuse, claim, hope, plan	We decided **to determine** the minimum combustion temperature.
Verb + gerund or infinitive	begin, continue, like, prefer, start	We continued **measuring**… We continued **to measure**…
Verb + noun/pronoun + infinitive	advise, have, instruct, remind, require, tell	Macromolecules require specific conditions **to form** crystals.
Verb + noun/pronoun + infinitive without "to"	let, make ("force"), notice, see, watch	The bear let us **approach**… Notice you **write** better after reading this book.

5.4   ADJECTIVES AND ADVERBS

GUIDELINE:

Distinguish between adjective and adverb.

Many ESL writers have trouble distinguishing between adjectives and adverbs. Adjectives modify nouns and pronouns and are generally placed in front of the noun or pronoun they modify. Adjectives may also be used to complement a subject and are then placed following a linking verb. Linking verbs are verbs that suggest a state of being or feeling rather than an action.

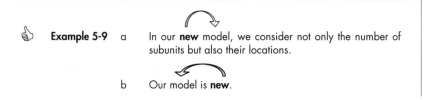

Example 5-9 a   In our **new** model, we consider not only the number of subunits but also their locations.

b   Our model is **new**.

In Example 5-9a, "new" is the adjective that modifies "model." In Example 5-9b, the adjective complements the subject, describing an attribute of the subject. It follows the linking verb.

Adverbs modify verbs, adjectives, or other adverbs. They are usually found at the beginning or at the end of a sentence, before or after a verb, or between a helping verb and a main verb.

👎 **Example 5-10** a   Complete conversion from A to B is **rarely** seen under denaturing conditions.

b   Acetylcholine was released in **precisely** controlled amounts from synaptic vesicles.

Adverbs should not be placed between a verb and its direct object.

👎 **Example 5-11**   Neural cells reinternalize *continually* synaptic vesicles.

⬆   ⬆
Verb   *direct object*

👍 **Revised**   a   Neural cells reinternalize synaptic vesicles **continually**.
Example 5-11

b   Neural cells **continually** reinternalize synaptic vesicles.

5.5   NOUNS AND PRONOUNS

GUIDELINE:
Ensure that every sentence has a subject.

In some languages such as Spanish and Italian, the subject of a sentence can be omitted. However, this is not the case in English unless the sentence is imperative. English sentences, and clauses within sentences, need to have a subject even if it is as simple as "it."

👎 **Example 5-12** a   The editor refused their manuscript because have already published a similar paper.

b   In January, is 25 °C in this region.

c   Is clear that the control experiment was inappropriate.

👍 **Revised**   a   The editor refused their manuscript because **they** have already published a similar paper.
Example 5-12

b   In January, **it** is 25 °C in this region.

c   **It** is clear that the control experiment was inappropriate.

5.6   GRAMMAR REFERENCES

Books

Greenbaum, S. (1996). *The Oxford English Grammar*, (1st ed.). Oxford University Press, USA.
Hacker, D. (2004). *Rules for Writers*. (5th ed.). Bedford/St Martin's.

Lunsford, A. A. (2004). *The Everyday Writer.* (3rd ed.). Bedford/St. Martin's.

Lynch, J. (2008) *The English Language: A User's Guide,* Focus Publishing/R. Pullins Company, Newsburyport, MA.

Schrampfer Azar, B. and Hagen S.A. (2005). *Basic English Grammar,* (3rd ed.) Pearson Longman.

Online References

Following is a list of Web sources for practicing grammar as well as sites for specific grammar problems:

For grammar, punctuation, and capitalization (last accessed October 2009):

http://owl.english.purdue.edu/owl (The Owl at Purdue [online writing lab])

http://grammar.ccc.commnet.edu/grammar (Guide to Grammar & Writing)

http://www.chompchomp.com (Grammar Bytes [grammar exercises & rules])

http://stipo.larc.nasa.gov/sp7084/ (A handbook for technical writers and editors)

http//dianahacker.com/rules (Complete reference for student writers and researchers; includes special section for ESL writers as well as many electronic exercises)

http://www.englisch-hilfen.de/en/

For article use (last accessed October 2009):
http://www.rpi.edu/web/writingcenter/esl.html
http://owl.english.purdue.edu/handouts/esl/eslart.html

For verb forms (last accessed October 2009):
http://www.englishpage.com/irregularverbs/irregularverbs.html
http://www.englishpage.com

Grammar practice (last accessed October 2009):
http://bcs.bedfordstmartins.com/successfulwriting/ (Successful College Writing)

http://bcs.bedfordstmartins.com/smhandbook/ (Bedford/St. Martins Handbook)

http://www.bedfordstmartins.com/lunsford/everyday_writer/ (Everyday Writer)

http://dianahacker.com/bedhandbook (The Bedford Handbook)

http://dianahacker.com/rules (Rules For Writers)

http://dianahacker.com/writersref (A Writer's Reference)

http://www.bedfordresearcher.com/ (The Bedford Researcher)

http://www.esl.net (English as a Second or Other Language [ESOL] Assistance)

SUMMARY

Use correct prepositions.
Use correct articles.
Use correct plural and singular verb forms.
Use correct forms of irregular verbs.
Do not omit the endings of verbs.
Follow a verb with the correct gerund or infinitive form.
Distinguish between adjective and adverb.
Ensure that every sentence has a subject.

PROBLEMS

Problem 5-1 Prepositions
Add the correct preposition to the phrases provided. It is okay to use a dictionary.

1. In connection _____
2. Compared _____
3. In contrast _____
4. Search _____
5. Correlated _____
6. A comparison _____ A _____ B
7. Similar _____
8. To look forward _____
9. Results shown _____ Fig. 3
10. Through the decrease _____
11. Analogous _____
12. Implicit _____
13. Theorize _____
14. Different _____
15. Attempt (n.) _____ attempt (v.) _____
16. _____ respect _____

Problem 5-2 Articles
Add the correct definite or indefinite article where needed.

1. Oscillatory behavior was reduced in _____ presence of a ring electrode.
2. Sea urchin fertilization is _____ model system in developmental biology.
3. Therefore, _____ hypothesis we presented previously is further confirmed.
4. Back to _____ nature.
5. _____ inorganic chemistry proves to be a very interesting subject.
6. Can you name _____ inorganic molecule?
7. Conversion of _____ CK riboside to the base can be regarded as the last step in the pathway.

8. The data for oxidation of hydrogen at the Pt electrode were obtained in _____ absence of chloride ions.
9. _____ theoretical studies cited earlier depict a variety of different patterns in oscillating media.
10. Here we discuss _____ physical-chemical mechanisms leading to pattern formation in electrochemical systems.
11. *Title*: _____ Role of _____ Physical Activity on _____ Severity of Diabetes.
12. *Title*: _____ Orbital Reconstruction and _____ Covalent Bonding at _____ Oxide Interface
13. *Title*: _____ Correlation of _____ Highest-Energy Cosmic Rays with _____ Nearby Extragalactic Objects
14. *Figure 4*: _____ effect of _____ temperature on _____ ctx expression.
15. _____ lichens consist of algae or cyanobacteria and fungi, which coexist in symbiosis.
16. _____ rainfall in _____ Sahara desert occurs rarely.
17. _____ average rainfall in _____ Sahara desert is less than 25 mm per year.
18. Under _____ light microscope, _____ organelles are visible in _____ eukaryotes.
19. At the end of its "life," _____ mRNA molecule is degraded.
20. _____ first probe was chosen upstream of _____ CA1 sequence.
21. _____ NbCA1 promoter was amplified using _____ BD GenomeWalker Universal Kit (BD Bioscience-Clontech, Palo Alto, CA).
22. _____ experimental data show _____ strong interference pattern that qualitatively resembles the pattern induced by a double slit.

Problem 5-3 Verb Forms

Add the correct verb form. Ensure that infinitive and gerunds are used correctly.

1. The analysis was limited to _____ (determine) the change in the population number.
2. We sought to _____ (improve) the yield by adding specific catalysts.
3. Impurities in our samples were barriers to _____ (obtain) chemically pure products using standard isolation techniques.
4. Ongoing volcanic eruptions limited the amount of time that we could dedicate to _____ (collect) samples.
5. The reviewers recommend _____ (add) more information about the bear habitat.
6. As with chemical reactions, the concept of concerted versus consecutive nuclear motions becomes central to _____ (understand) the elementary steps of the mechanism.
7. In the mid-1990s, Bottomly began _____ (report) on immune response to foreign substances.
8. Plants are able _____ (perceive) pathogen attacks and subsequently induce defense responses.
9. The K616 strain failed _____ (grow) on calcium-depleted media.
10. We were surprised _____ (find) so few differences among mutations that determine drug resistance.

Problem 5-4 Verb Forms
Add the correct verb form. Ensure that singular and plural, irregular verb forms, and endings of verbs are used correctly.

1. The conservation of the sequence in similar places in the genomes _____ (suggest) that they play a key role in genome function.
2. Higher-order stimulated Raman scattering (HSRS) in hydrogen or deuterium ____ (have) been _____ (show) to exhibit a broad and coherent HSRS spectrum.
3. High-mobility electron systems with tunable density have _____ (lead) to prominent advances in science and technology over the past decades.
4. In this study, it _____ (be) clear that our amplifying circuit element is simple, fast, modular, and robust.
5. Frequency and fitness _____ (be) greater for annuals in light gaps and biennials in the understory.
6. *S. multiplicata* females did not _____ (choose) hetero-specific mates regardless of water level.
7. Until now, there _____ (be) no experimental confirmation of this hypothesis.
8. Microspheres of like charge were _____ (bind) together with nanoparticles of opposite charge to form clusters of two to nine colloids.
9. Each of the experiments ____ (be) performed at two different photon energies.
10. One third of the mice ____ (be) infected with *P. chabaudi*.
11. Ultrafast electron microscopy (UEM) and ultrafast diffraction _____ (have) been the methods of choice for studies of molecular and phase transitions.
12. Virus A or virus B ____ (be) thought to cause the disease.

Problem 5-5 Adjectives and Adverbs
Add the correct adjective or adverb.

1. X behaved _____ (different, differently) than expected.
2. The simulations demonstrate how _____ (sensitive, sensitively) the formation of the first stars depended on the detailed properties of the still mysterious dark matter.
3. To be highly conserved, genes must play a role so important to survival that evolution keeps them _____ (intact, intactly), weeding out deleterious mutations.
4. Species richness in plants is correlated _____ (biological, biologically) and _____ (geohistorical, geohistorically) and increases ecological opportunities.
5. Diffraction restricts the ability of most electromagnetic devices to image or _____ (selective, selectively) target objects smaller than the wavelength.
6. Mainstream climate science needs to look more _____ (close, closely) at geoengineering.

7. Hurricane Katrina's impact on U.S. Gulf Coast forests was quantified by linking _____ (ecological, ecologically) field studies and _____ (empirical, empirically) based models.

8. It has been reported that the virus linked to the collapse of honeybee colonies may have arrived in the United States via _____ (recent, recently) imported Australian bees.

9. The flux of cosmic rays at Earth decreases very _____ (rapid, rapidly) with energy, from a few particles per square centimeter per second in the low-energy region to less than one particle per square kilometer per century above 10^{20} eV.

10. The analysis offers initial results of an _____ (ambitious, ambitiously) international project that will eventually compare the complete genomes of several dozen TB strains from around the globe.

Problem 5-6 Mixed ESL Errors

Be sure that the following sentences make sense; that correct prepositions, articles, and verb forms are used; that every sentence has a subject; and that adverbs and adjectives are distinguished.

1. Is not clear if the difference in our results was due to the temperature difference in our measurement.

2. We expected observing a large difference in the output.

3. The environment surrounding invasive breast tumors exhibit significant changes.

4. Star formation in the early universe was very differently from that of the present.

5. The graphene bilayer's band structure is predicted giving rise to several unconventional phenomena, such as the Klein paradox and the Vaselago lensing effect.

6. Despite the internationally recognized uniqueness and importance of Peruvian rainforest, the impacts of human activities throughout the region remain poorly understood.

7. At the center is Pt atom, surrounded by a 12-atom Pd icosahedron, and second shell is a 42-atom icosahedron with either 3 or 4 Pt atoms; third and fourth shells are high-symmetry structures with 60 and 50 atoms, respectively.

8. We hypothesize that concept of metamaterial-inspired nanoelectronics ("metactronics") can bring the tools and mathematical machinery of the circuit theory into optics, electronics, metamaterials, and nanoscale devices.

9. The production of an individual platelet begins when a hematopoietic stem cell begins to differentiate. Is complete when a fragment of a bone marrow megakaryocyte is released into the vasculature and begins to circulate.

10. In support for the latter model, platelets formed from proplatelet processes *in vitro* are functional.

11. The fossil's small size suggests that the individual belongs to *H. habilis* or a new species, but the more modern traits, such as long

legs and modern body proportions, have proved the individual to be in *H. erectus* and have showed this species was adapted for long-distance locomotion.

12. Although the other species in the Gorilla genus, Eastern Gorilla, are far less numerous than Western Gorilla, IUCN ranks the former one level lower at "endangered" because it is outside the current area of Ebola outbreaks.

13. Our results indicate that disruption of NbCA1 function accelerate HR cell death.

14. Tanzania's Lake Natron is only know breeding site for East Africa's lesser flamingos.

15. This is first report on cellular calcium signaling during disease resistance response in plants.

Problem 5-7 Mixed ESL Errors
In the following paragraphs, fill in the correct article, verb form, adverb, or adjective.

a) Dengue fever, transmitted most often by the bite of an infected *Aedes aegypti* mosquito, was often seen as an obscure, only _____ (occasional/occasionally) fatal disease of tropical countries, and progress toward a vaccine and drugs to treat it has been slow. Since the disease _____ (become) more virulent and _____ (spread) into new geographic areas, vaccine research has taken on a new urgency. A vaccine can't come a moment too soon. It is estimated that the number of dengue cases tops 50 million _____ (annual/annually), which is similar ____ (to/from/with) the number of malaria cases. The warm, crowded cities of Latin America and Asia provide an ideal habitat for the main vector, *A. aegypti*, which breeds in stagnant water and likes _____ (feed) on humans in quick succession.

b) We have discovered _____(a/an/the/[no article]) imported virus that may be associated _____(to/with) the sudden disappearance of honey bees in the United States, _____(know) as colony collapse disorder (CCD). This syndrome caused _____ (surprising, surprisingly) losses of up to 90% of hives in some apiaries. The suspect is a pathogen called Israel acute paralysis virus (IAPV). Our team _____ (find) the virus in most of the affected colonies they tested but in almost no healthy ones. If ____ (a/an/the/[no article]) virus _____ (prove) to be the cause of CCD, it could have (international, internationally) economic implications. Since 2005, U.S. beekeepers, especially those struggling _____ (keep) up with the insatiable demand for almond pollination in California, have imported several million dollars' worth of bees from Australia. We report that we have found IAPV in imported Australian bees.

c) The basic anatomy of East Asia's dust storms is fairly _____ (well, good) established. The common term "sand storms" is incorrect, however, as sand particles are too heavy _____ (get) lifted high into the atmosphere. Thus, little of _____ (a/an/the/[no article]) dust that afflicts East Asia comes from deserts, where erosion over the millennia has carried away most of the smaller particles. Instead, the dust originates in dry lakebeds and arid lands on desert fringes. In springtime, the crust of the undis- turbed soil is broken _____ (up, off) by plowing and livestock. The dust gets lifted into the air by updrafts created through the temperature difference between a cold atmosphere and a surface warmed by spring sunlight. Subsequently, the airborne dust is carried south and east by winds. It gets sucked into _____(the/ [no article]) upper atmosphere when it reaches low-pressure pockets at the mountain ranges that ring northern China and Mongolia. Easterly winds transport _____(the/[no article]) particulate matter to East Asia, and sometimes across the Pacific Ocean to North America.

d) ___ (A/The/[no article]) typical laser produces coherent radia- tion and emits light in a narrow, low-divergence beam and with a well-defined wavelength. In contrast, a light source such as _____ (a/an/the/[no article]) incandescent lightbulb emits over a wide spectrum of wavelength. Lasers have _____ (become) a multi- billion dollar industry. The most widespread use of ___ (the/no article) lasers is in optical storage devices such as compact disc and DVD players in which the laser (a few millimeters in size) scans the surface of the disc. However, laser use is not limited to _____ (apply) this technology to optical storage devices. Another common application of lasers _____ (be) as bar code readers or laser pointers. In industry, lasers are used for cutting steel and other metals and for inscribing patterns (such as the letters on computer keyboards). Lasers are also _____ (common/commonly) used in various fields in science, especially spectroscopy, _____(typical/typically) because of their well- defined wavelength or short pulse duration. In addition, lasers are _____ (seek) for military and medical applications.

CHAPTER 6

From Sentences to Paragraphs

6.1   PARAGRAPH STRUCTURE

Aside from paying attention to words and sentences, authors need to construct paragraphs carefully. If paragraphs are not clearly constructed, a paper that has perfect word choice, word location, and sentence structure can be difficult to understand. Let us look at a paragraph in which the author has not paid much attention to the needs of the reader:

 Example 6-1   **1**Volcanic ash adsorption poses a great environmental hazard. **2**The deposition of this ash and the subsequent draining of its volatiles is a rapid route by which elements and ions are delivered to the ground (3–5). **3**Due to similar magma types, there appears to be some similarity in the compositions of leachates derived from volcanoes in the same regions. **4**The greatest hazard to the environment is posed by magmas with relatively high halogen content, and many hazardous leachate fluoride concentrations are found in volcanoes with high F/SO_4^{2-} ratios. **5**Finer particle sizes, <2 mm across, appear to experience enhanced adsorption, with the implication that leachate hazards may be high even where ashfall is limited (7, 8). **6**Enhanced growth of sulphuric acid droplets in high humidity conditions can increase gas accumulation, which increases the probability of contact with ash particles (12). **7**The measuring and reporting of leachate results should be standardized.

If you catch yourself reading this paragraph more than once, you are not alone. You may think that you have not paid enough attention and start reading it over. Some readers may even consider themselves not smart enough to understand the topic. The individual sentences are intelligently composed and free of grammatical errors. The sentences are also not overly long or complex. The vocabulary is professional but not beyond the scope of the educated general reader. Nonetheless, most of you arrive at the end of the paragraph without fully understanding what the author is saying. The problem lies not with you but with the author because the author has not composed the paragraph with the reader in mind.

When you take a closer look at this paragraph, you can see that the paragraph has not been organized properly. It is neither coherent nor cohesive or consistent. In fact, important information, such as transitions and logical connections, seem to have been left out. Even more, sentences seem to be put together at random and not in any logical order. For readers to follow the logic of a paragraph, its sentences have to be organized. Let us look at the revised version of Example 6-1:

Revised **Example 6-1**	A	**1**Volcanic <u>ash adsorption</u> poses a great <u>environmental hazard</u>. **2**The <u>deposition of this ash</u> and the subsequent leaching of its volatiles is a rapid route by which elements and ions are delivered to the ground (3-5). **3′***Adsorption can be influenced by magma type, particle size, and humidity conditions (7).* **3***For example,* there appears to be some similarity in the compositions of <u>leachates</u> derived from volcanoes in the same regions due to similar <u>magma types</u>. **4***In fact,* the greatest <u>hazard to the environment</u> is posed by <u>magmas</u> with relatively high halogen content, and many hazardous leachate fluoride concentrations are found in volcanoes with high F/SO_4^{2-} ratios. **5***Aside from magma type,* <u>finer particle sizes</u>, <2 mm across, appear to experience enhanced <u>adsorption</u>, with the implication that <u>leachate hazards</u> may be high even where *ash deposition* is limited (7, 8). **6***Furthermore,* enhanced growth of sulphuric acid droplets in <u>high humidity conditions</u> can increase *adsorption,* which increases the probability of contact with ash particles (12). **7***Ideally,* the measuring and reporting of <u>leachate</u> results should be standardized.

After looking at this revision, you may now recognize the lack of organization and links in the original paragraph. You can see that an important missing link was sentence 3′. It ties sentences 3 through 6 together and links them to the beginning of the paragraph. Adding the transitions "for example," "in fact," "aside from," "furthermore," and "ideally" logically connects the ideas in the paragraph. All these connections were left unarticulated in the original paragraph. Replacing "ashfall" with "ash deposition" and "gas accumulation" with "adsorption" helps to keep the reader focused on the topic because terms are used more consistently throughout the paragraph. We can see that most of our difficulty in understanding the original paragraph was due not to any deficiency in our reading skills but rather to the author's lack of knowledge and understanding of our needs as readers.

Although the revision greatly improved the paragraph, we could revise it even more:

👍 **Revised Example 6-1**   B   **1′**Volcanic <u>ash adsorption</u> poses a great <u>environmental hazard</u> *because adsorbed volatiles can be rapidly deposited and subsequently leached into the ground.* **2′***Adsorption can be influenced by magma type, particle size, and humidity conditions* (7). **3**For example, similar <u>magma types</u> derived from volcanoes in the same regions exhibit similar compositions of <u>leachates</u>. **4**In fact, the greatest <u>hazard to the environment</u> is posed by <u>magmas</u> with a relatively high halogen content and by <u>magmas</u> with high F/SO_4^{2-} ratios in which many hazardous leachate fluoride concentrations are found. **5***Aside from magma type,* <u>finer particle sizes</u>, <2 mm across, appear to experience enhanced <u>adsorption</u>, with the implication that <u>leachate hazards</u> may be high even where *ash deposition* is limited (7, 8). **6***Furthermore,* <u>high humidity</u> results in enhanced growth of sulphuric acid droplets, which increases <u>adsorption</u> by increasing the probability of contact with ash particles (12). **7′***Unfortunately, the use of different leachate analysis techniques currently prevents useful comparison between data.* **8***Ideally,* the measuring and reporting of <u>leachate</u> results should be standardized.

The flow of the paragraph has been further improved in Revised Example 6-1B. Sentence 1 and 2 have been combined into sentence 1′, making their relationship clearer through the use of "because." Another link, sentence 7′, has been placed before the last sentence to more logically connect it to the rest of the paragraph. If these additions and transitions truly reflect what the author had in mind is only known to the author himself or herself, however.

To construct a paragraph clearly, each paragraph needs to be written such that it tells a story. Readers should be able to follow the story of each paragraph regardless of whether they understand the science. A well-constructed paragraph must not only be organized, it must also be coherent. In addition, important ideas should be emphasized, and subtopics should be signaled.

6.2   PARAGRAPH ORGANIZATION

WRITING PRINCIPLE 23:

Organize your paragraphs.

A paragraph is a group of sentences on a single topic. The sentences within a paragraph are not put together randomly, however. In a well-written paragraph, the sentences need to be logically organized and positioned.

Sentence Positions

Every paragraph contains two important power positions: the first sentence and the last sentence. Usually, the first sentence introduces the topic of the paragraph, whereas the last sentence may be used to summarize, draw a conclusion, or emphasize something of importance.

These power positions are not equal. The first position in a paragraph is considered more powerful than the last position because it gives the reader a direction of where the paragraph is going. Within the first and the last sentence of the paragraph, the psychological geography of the sentence structure is particularly important. The beginning of the sentences should describe familiar information, whereas the stress position within these sentences should highlight significant words to be emphasized (see also Chapter 3).

Topic Sentence

WRITING PRINCIPLE 24:

Use a topic sentence to provide an overview
of the paragraph.

Generally, a well written paragraph gives an overview first and then goes into detail. Note that most of the time, it is clearest to have only one message per paragraph. The overview is usually provided by the first sentence, the so-called topic sentence. The topic sentence states the central topic or message of the paragraph and guides the reader into the paragraph. The end or stress position of the topic sentence highlights the topics that the author wants readers to follow in the rest of the paragraph. The paragraph then develops that message by using examples, definitions, justifications, contradictions, or by analyzing and solving a problem.

Although a topic sentence may appear anywhere in the paragraph, it is usually the first sentence, that is, the first power position. The first sentence of a paragraph may also contain a transition from the previous paragraph or section. Some paragraphs may even contain more than one topic sentence. If a topic sentence is placed at the end of a paragraph, it receives extra emphasis. This sentence may introduce the topic of the next paragraph(s). It can also serve as a summary or conclusion. In a well-written paper in which all the topic sentences are in the first power position, a reader can simply scan the topic sentences alone and know what the paper is about without having read it entirely.

The Middle of the Paragraph

GUIDELINE:

Arrange the details in the remaining sentences.

Details within the paragraph are organized depending on the purpose of the information contained in the paragraph. The pattern of the organization may be listing details from most to least important, least to most important, in an announced order, or chronologically. Other paragraphs may be written in a compare-and-contrast pattern or in a problem-and-solution pattern.

The following examples both begin with a topic sentence. The details found in the remaining sentences are organized logically and consistently to explain the message provided by the topic sentence.

 Example 6-2   Volatile organic compounds (VOCs) are emitted from a variety of manmade and natural sources. Manmade sources include motor vehicles, chemical plants, refineries, factories, consumer and commercial products, and other industrial sources. Natural sources responsible for biogenic VOC emissions include oak, citrus, eucalyptus, pine, spruce, maple, hickory, fir, and cottonwood. The overall relative contributions of manmade versus natural sources of VOCs have not been clearly established, but the relative contributions of these source groups vary depending on geography.

The topic of the preceding paragraph is "VOCs." The pattern of organization, that is, the order of the remaining sentences, is not random but proceeds in the order the items are listed: *manmade* first, then *natural*.

Consider another example:

Example 6-3   For the preparation of postmitochondrial fractions, placentas were obtained after delivery. Their membranes were removed, and the organs were washed extensively at 4 °C with buffer A (10 mM Tris-HCl, pH 7.5, 20 mm Mg(acetate)$_2$, 100 mM K(acetate), 0.4 mM EDTA). The organs were then shock frozen in liquid nitrogen as 20 g aliquots and subsequently stored at −80 °C. For postmitochondrial preparation, 20 g of frozen tissue was suspended in 20 ml of buffer A containing 200 mM sucrose through homogenization in a blender. To separate cell debris and nuclei, the homogenate was centrifuged at 1,500 × g for 2 min at 4 °C. To obtain the postmitochondrial fraction, the supernatant was then recentrifuged at 20,000 × g for 20 min.

Example 6-3, which is from a Materials and Methods section of a pre-publication journal article, also has a topic sentence ("For the preparation of postmitochondrial fractions ..."). The remaining sentences of the paragraph are organized in a chronological pattern in which the reader can follow step by step how the fractions were prepared from the original tissue. The subjects of the sentences in the preceding example are all tissue or fraction related. The paragraph begins with a goal ("For the preparation of ...") and ends with a statement that indicates "mission accomplished."

Aside from the general organization of the paragraph, the author also needs to pay attention to word location and to keeping a consistent order of topics and a consistent point of view. These points are addressed in more detail in the next sections.

Topic Order

WRITING PRINCIPLE 25:

Use consistent order.

Although parallel form is most often used at the sentence level (see Chapter 4), in scientific papers, it can also be used in a larger context. Parallelism helps locate information in paragraphs, sections, chapters, and so forth. Good parallel form puts related ideas together in the same grammatical form and style and thus provides consistency throughout the paper.

If you list items in a topic sentence and then describe them in the remaining sentences of a paragraph, you should not only use parallel form but also keep the same order. For example, if the items in the topic sentence are "dogs," "cats," and "birds," the remaining sentences of the paragraph should explain first "dogs," then "cats," and last "birds." This way the reader's expectation is fulfilled. Make sure you include all the items mentioned in the topic sentence. Avoid interrupting the order of your items by filling in with other information. Also, do not add any items *not* mentioned in the topic sentence.

An example of a paragraph in which consistent order is used is Example 6-4.

 Example 6-4

1In response to a foreign macromolecule, five different immunoglobulins can be synthesized: IgG, IgM, IgA, IgE, or IgD. 2IgG is the main immunoglobulin in serum. 3IgM is the first class to appear following exposure to an antigen. 4IgA is the major class in external secretions such as saliva, tears, and mucus. 5Thus, IgA serves as a first line of defense against bacterial and viral antigens. 6IgA is transported across epithelial cells from the blood side to the extracellular side by a specific receptor. 7IgE protects against parasites. 8The role of IgD is not known.

In this example, the topic sentence lists five items. The remaining sentences of the paragraph explain these items in the same order as they are introduced in the topic sentence and use exactly the same key terms (IgG, IgM, IgA, IgE, and IgD). Note that other information has been filled in between sentences 4 and 7. Interruption such as that can make paragraphs difficult to read because reading about the last items is delayed. If such an interruption is longer, consider placing the additional information into a separate paragraph.

Point of View and Person

Use consistent point of view.

ESL advice

Be consistent in your point of view and person. Switching from one style (point of view or person) to another within a document disorients the reader. ESL authors should pay special attention to this principle, as many tend to switch the point of view within paragraphs for no apparent reason.

The point of view is consistent when the same term, or the same category term, is the *subject* of successive sentences that deal with the same topic. The point of view is inconsistent when the topic is the same, but the subjects of the sentences are different or when the person is switched (e.g., from third to second person). An inconsistent point of view makes similarities and differences difficult to see for the reader.

Here is an example.

Example 6-5

1This study suggests that *patients* with a prolonged febrile illness should always be a consideration for tuberculosis, especially if family members have been born in a country where tuberculosis is endemic. **2**Tuberculosis presents not only as fever, **3**but <u>you</u> may also have lymphadenopathy and arthritis. **4***Patients* with disseminated tuberculosis usually have evidence of pulmonary or hepatic disease.

The subjects of sentence 1 and 4 are "patients." However, in sentences 2 and 3, the point of view is not consistent: Neither the first nor the second subject of the sentence is *patients*. The first subject of the sentence is *tuberculosis*. For the second subject, there is a switch in person to *you*. Switches like this are very disruptive to the paragraph and disorienting for the reader.

Revised
Example 6-5

1This study suggests that *patients* with a prolonged febrile illness should always be a consideration for tuberculosis, especially if family members have been born in a country where tuberculosis is endemic. **2Patients** with tuberculosis present not only with fever, **3**but may also have lymphadenopathy and arthritis. **4***Patients* with disseminated tuberculosis usually have evidence of pulmonary or hepatic disease.

6.3   PARAGRAPH COHERENCE

Cohesion and Word Location

WRITING PRINCIPLE 27:

Make your sentences cohesive.

Within a paragraph, the sentences not only need to be logically organized, they also need to be cohesive. Sentences are cohesive if they fit neatly and logically together. When authors arrange sentences to be cohesive, they consider word location. Good word location creates good "flow" of a paragraph.

 Example 6-6 1Important pathogens can be found in the genus

Yersinia. 2*Yersinia* contains several **species**. 3**One species**,

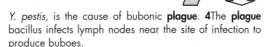

Y. pestis, is the cause of bubonic **plague**. 4The **plague** bacillus infects lymph nodes near the site of infection to produce buboes.

The reader conceives these sentences as cohesive because the information provided at the beginning of sentences 2, 3, and 4 relates to the one at the end position of the sentences directly preceding them (see also Chapter 3 on word location.)

If more and more new information is added before the relationship between the two sentences is clear, the continuity of a paragraph is broken. Consider the following example:

 Example 6-7 a *Yersinia* contains several **species**. The cause of bubonic plague, also known as the "black death," is one **species**, *Y. pestis.*

  b *Yersinia* contains several **species**. **One species**, *Y. pestis,* is the cause of bubonic plague, also known as the "black death."

The flow of Example 6-7a is not as smooth as that of Example 6-7b because new information ("bubonic plague," "black death") is introduced before the information of the previous sentence ("*Yesinia,*" "species") is repeated. Example 6-7b is much more direct, and the continuity between the sentences much smoother due to the word location of the key term "species." In this example, the information of the previous sentence has been placed at the beginning of the new sentence due to jumping word location.

Placing information provided at the end of a sentence at the beginning of the next sentence is not the only way to provide paragraph cohesion.

Cohesion can also be created by providing a consistent point of view as in the next example.

 Example 6-8   **Rhubarb** is a frequently used Chinese herbal medicine. It is used to treat various ailments including constipation, inflammation, and cancer.[1,2] As a drug, **rhubarb** is made up of the roots and rhizomes of three members of the *Polygonaceae* family, *Rheum officinale, R. palmatum,* and *R. tanguticum.* Different **rhubarb** species show substantial differences in purgative effects and chemical compositions. However, **they** are similar in physical appearance and thus difficult to distinguish.

Here information in the topic position of each sentence refers back to the topic position of the first (or topic) sentence. In other words, a consistent point of view is kept—the subject in each sentence is the same term or category term.

Many, if not most, paragraphs contain a mixture of these two types of word placement. For some sentences, information in the topic position may refer back to the end position of the previous sentence. For other sentences, information may be written from the point of view of the old information and refer back to the topic position of the topic sentence or subtopic sentence. An example of such a "mixed" paragraph is shown next.

 Example 6-9   The **prophylactic** administration of drugs can sometimes be detrimental to patients, especially if prophylaxis is **continued** for several days. **Continued** exposure of the host's microbial flora to antibiotics often leads to resistant strains, and this can lead to **superinfection**. **Superinfection** can be avoided, however, by prophylactically using probiotics together with the antibiotics to restore normal flora. **Prophylaxis** is recommended when the host is subjected to a **treatment**, not involving antibiotics, that can lead to serious disease. One such **treatment** is **surgical procedure**. In **surgical procedures, prophylaxis** is usually directed at preventing staphylococcal infection. In **surgical procedures** that are considered "clean," antibiotics are not recommended.

This paragraph flows well. Notice that every sentence in the paragraph has at least one link to a previous or subsequent sentence. Some sentences will have both, a link to the previous sentence and one to a sentence following, and in a few instances there may be three links, such as when a sentence introduces a subtopic.

Constructing paragraphs in which mixed word locations occur is okay—as long as the links do not get too muddled to risk losing the reader as in the next example:

Example 6-10   Animals, particularly domestic animals, are important reservoirs and sources of disease to humans. Salmonella species are normally found in the intestinal tract of animals such as poultry and cattle. When humans ingest contaminated food, the salmonellae can cause disease called Salmonellosis. In terms of animal disease transmission, humans often represent a dead end because the disease cannot be transferred from human to human. Salmonellosis may be acquired from animal but the infected human can also serve as a source of disease to other humans.

In this example, old and new information has been misplaced and is therefore not linked clearly. As a result, there are too many links between sentences. Readers get confused, as they no longer can distinguish between what is the topic and what is the stress in each sentence or in the paragraph as a whole. As a result of such misplaced information, the paragraph is not very cohesive and does not "flow" well. By checking your own writing for what has been placed in the topic and stress positions in your sentences, you can perceive where potential problems exist and then improve your writing to better meet the reader's expectations.

Misplacement of old and new information, as seen in the preceding example, is one of the main problems in professional writing. Information is usually misplaced because most writers want to capture any important new thought before it escapes. As most writers write linearly, new information is wrongly placed at the beginning of the sentence and old information ends up at the end of a sentence. Only during the revision stage are logical links between sentences considered but may not be caught if the author does not revise or not revise sufficiently enough. Authors who misplace information are attending more to their own need for unburdening themselves of their information than to the reader's need for interpreting what is written.

Another extreme in paragraph construction occurs when an author creates no links between sentences. Links are most often ignored when an author assumes that the reader is familiar with the topic or that logical jumps are clear. Missing links may not be obvious to the author but usually confuse and frustrate readers as in the following example:

Example 6-11   **1**A range of giant mammals, birds and reptiles lived on Earth during the Pleistocene Epoch. **2**These creatures included the woolly mammoth, sabre-toothed cat and giant deer in the Northern Hemisphere, and giant marsupials like *Diprotodon* in Australia **3**Palaeontologists are very interested in ice age mammals. **4**The part played by **humans** in the extinction of the megafauna is very unclear. **5**Many researchers believe that the migration of **humans** into various parts of the world contributed to the extinction of many large animals. **6**Environmental changes were also associated with the onset of the last glacial cycle.

In the preceding example, either no links or no clear logical links have been established between sentences 2, 3, and 4, nor between sentences 5 and 6. Here, word location has not been considered, and sentences cannot be linked because critical terms, and thus critical information, are missing. Are giant mammals the same as ice age mammals? Is the Pleistocene epoch meant by megafauna? Does this epoch correspond to the last ice age? The connection between these terms is not clearly established, maybe because the author assumes that readers are familiar with this terminology. Instead, readers are left guessing at the missing connections between sentences.

Coherence and Continuity

Cohesive flow is the first of two steps toward creating continuity for your readers. The other step is to make your paragraphs and passages coherent. A coherent paragraph consists of a series of sentences that lead logically from one to the next, thus creating continuity. The ideas in a sentence need to be linked such that the story flows smoothly from sentence to sentence (and paragraph to paragraph). Readers consider a paragraph to be coherent if they can quickly find the topic of each sentence and if they see how the topics are a related set of ideas.

Along with topic sentences and word location, *key terms* and *transitions* are the main techniques used to create the logical framework, and thus coherence, of a paragraph and of a paper. Repeating and linking key terms ensures that the topic of the work cannot be missed and that relationships between topics are clear. In addition, transitions create continuity by indicating the logical relationship between sentences and/or paragraphs, particularly for sentences that cannot be linked by word location.

Key Terms

WRITING PRINCIPLE 28:

Use key terms to create continuity.
Repeat them exactly and early, and link them.

Key terms are words or short phrases used to identify important ideas in a sentence, a paragraph, and the paper as a whole. Usually, key terms are used to identify your main points in the topic sentence. Key terms should be clearly defined and identical throughout the paragraph (and the document). They can be technical terms, such as "kinase" or "HIV-1," or nontechnical terms such as "mechanism" or "decrease."

Key terms should not be changed but should be kept the same consistently. If you deliberately repeat key terms, your main points are emphasized and you create continuity. For clear continuity, repeat key terms *exactly*. If a key term is not repeated exactly and another term is used instead, it may be difficult to see the relationship between the two terms. Although readers within the field may be familiar with the relationship, readers outside the field may not be.

Example 6-12   To assess original conditions of crystal nucleation and growth in metamorphic rocks, it is necessary to analyze <u>crystal distribution</u> quantitatively. <u>Density</u> could potentially provide insight into the time scale of mineral growth following the thermal peak of metamorphism.

How does "density" relate to the previous sentence? Readers unfamiliar with this particular topic may not know that "crystal distribution" and "density" here mean the same thing. Readers may be confused when two different terms are used. To avoid confusing readers, write with the nonspecialist in mind. Do not change key terms.

Revised
Example 6-12   To assess original conditions of crystal nucleation and growth in metamorphic rocks, it is necessary to analyze <u>crystal distribution</u> quantitatively. **Crystal distribution** could potentially provide insight into the time scale of mineral growth following the thermal peak of metamorphism.

In the revised example, "crystal distribution" is the main key term that holds the paragraph together. Because it is repeated exactly, the relationship between the two sentences is clear, and even readers outside the field of protein chromatography will understand the passage.

Linking Key Terms

When you need to shift from a category term to a more specific term or the other way around, key terms should be linked so continuity is not lost and the paragraph stays coherent. To link key terms, use the category term to define the specific term.

 Example 6-13   Infectious diseases that arise due to travel may be caused by <u>gram-positive organisms</u>. **One such organism**, *Staphylococcus aureus*, can cause cellulitis, purulent arthritis, and suppurative lymphadenitis.

If key terms are not linked, as in the next example, readers stumble and get lost.

 Example 6-14   So far, we have only looked at electrode reactions that are connected to an N-shaped I/φ_{DL} characteristic whereby φ_{DL} is the autocatalytic variable. The S-shaped, current-potential characteristic has a negative differential resistance but shows an opposing behavior with respect to pattern formation.

A key term or category term has to be included in the second sentence to create a link between the sentences. The term "I/ϕ_{DL} characteristic" is the key term in the first sentence. "The S-shaped, current-potential characteristic" described in sentence 2 is actually another type of the "I/ϕ_{DL} characteristic." Although this relationship may be clear to the author, readers may not know how these terms are linked. Linking these key terms by a suppressed "which is…" clause will make the relationship clear to the reader.

 Revised   So far we have only looked at electrode reactions that are
Example 6-14   connected to an N-shaped I/ϕ_{DL} characteristic whereby φ_{DL} is the autocatalytic variable. **Another type of I/ϕ_{DL},** the S-shaped, current-potential characteristic, has a negative differential resistance but shows an opposing behavior with respect to pattern formation.

The term "which is" could be included: "Another type of I/ϕ_{DL}, which is the S-shaped, current-potential characteristic,…" However, because the definition is clear without "which is," it can be omitted. Leaving out the "which is" creates an appositive, a very useful way to define a scientific term while conserving words.

The definition should not be written as a separate sentence because this separation would break the continuity. To maintain continuity, it is important to include the definition as a part of an existing sentence. When key terms are linked and repeated consistently and early, good word location almost always falls right into place.

Transitions

WRITING PRINCIPLE 29:

Use transitions to indicate logical relationships bewteen sentences.

To ensure continuity within a paragraph, a writer must use key terms and also make use of other techniques such as transitions. Transitions ensure that the reader understands what each sentence says and indicate how the sentences and paragraphs are logically related to each other and to the story. Transitions should be placed at the beginning of a sentence for strongest continuity, usually set off by a comma.

If transitions are missing, the logical relationship between sentences can be unclear and may even be nonexistent. The importance of adding transitions is shown in Example 6-15 and its revisions.

Example 6-15   To determine the effects of solid-solution ratio on K12 adsorption at fixed pH 7, K12 adsorption isotherm experiments were conducted. _____ aqueous carbonate concentrations were measured.

The logical relationship between the first and second sentence of this example is not immediately obvious to the reader. Possible transitions that one could fill in between these two sentences include the following:

Revised Example 6-15   To determine the effects of solid-solution ratio on K12 adsorption at fixed pH 7, K12 adsorption isotherm experiments were conducted. **In addition,** aqueous carbonate concentrations were measured.

To determine the effects of solid-solution ratio on K12 adsorption at fixed pH 7, K12 adsorption isotherm experiments were conducted. **For this purpose,** aqueous carbonate concentrations were measured.

To determine the effects of solid-solution ratio on K12 adsorption at fixed pH 7, K12 adsorption isotherm experiments were conducted. **First,** aqueous carbonate concentrations were measured.

To determine the effects of solid-solution ratio on K12 adsorption at fixed pH 7, K12 adsorption isotherm experiments were conducted. **Subsequently,** aqueous carbonate concentrations were measured.

When the transition is missing between these two sentences, most readers may guess that the intended relationship is "in addition," but readers should not be guessing.

Here is another example:

Example 6-16   We determined whether the increased endotoxin susceptibility of *AUF1*$^{-/-}$ mice is due to deregulation of pro-inflammatory cytokine expression. We measured the serum TNFα level in *AUF1*$^{-/-}$ mice after LPS challenge.

In Example 6-16, the logical relationship between the first and second sentence is also not clear because a transition is missing. Once the transition is added in, the relationship between the two sentences becomes obvious.

Revised
Example 6-16   We determined whether the increased endotoxin susceptibility of *AUF1*$^{-/-}$ mice is due to deregulation of pro-inflammatory cytokine expression. **For this purpose,** we measured the serum TNFα level in *AUF1*$^{-/-}$ mice after LPS challenge.

Some transitions that are used in more casual conversation should be avoided in scientific writing:

besides (but not besides X …)	additionally	as a matter of fact
suddenly	admittedly	basically	ergo	at once

Use transitions and conjugations, but only where appropriate. Note that logical relationships can also be clear without transitions. It is not necessary to place a transition or conjugation in every sentence, as logical relationships are often apparent from the word location within sentences.

Use transitions to link ideas, but do not overuse them. In addition, ensure that you are using the correct transitions, especially when you are an ESL author. ESL writers are prone to using transitions whose meaning is quite different from the one they intend. Check a dictionary rather than guessing when using transitions. Do not trust a thesaurus, however. Certain transitions are considered outdated. Rather, consult a scientific editor, current peer-reviewed articles published by native English speakers in well-respected journals, or a book on scientific style to check whether a transition is still in use in contemporary scientific writing. Avoid outdated terms such as the following:

hitherto	aforementioned	henceforth
notwithstanding	firstly (use "first" instead)	secondly	lastly

Common transitions include words, phrases, or even sentences. Whereas transition words are standard terms that indicate logical relationships between sentences, transition phrases are either infinitive or

Table 6.1 Transition words, phrases, and sentences

USE	EXAMPLE		
	TRANSITION WORDS	TRANSITION PHRASE	TRANSITION SENTENCE
Addition	again, also, further, furthermore, in addition, moreover	In addition to X, we… Besides X,…	Further experiments showed that…
Concession	clearly, evidently, obviously, undeniably		Granted that X is…
Comparison	also, likewise, similarly, etc.	As seen in… In the same way,	When A is compared with B… As reported by… When compared to…
Contrast	but, however, nevertheless, nonetheless, still, yet	In contrast to A… On one hand; on the other hand… Despite X… Unlike X,… On the contrary,…	One difference is that… Although X differed…
Example	for example, specifically	To illustrate X…	An example of X is that…, That is,…
Explanation	here, therefore, in short	Because of X… In this experiment…	One reason is that… Because X is…
Purpose	for this purpose,	For the purpose of… To this end,… To determine XYZ, we…	The purpose of X was to…
Result	consequently, generally, hence, therefore, thus	As a result of…	Evidence for XYZ was that… Analysis of ABC showed that…
Sequence/ time	after, finally, first, later, last, meanwhile, next,	After careful analysis of X…	After X was completed,…
	now, second, then, while, subsequently	During centrifugation,…	When we determined X…
Summary	in brief, in conclusion, in fact, in short, in summary	To summarize (our results),…	As a summary of our results shows,…
Strength of transition			

prepositional phrases. A transition sentence uses a subject and verb. The subject in the transition sentence, and the object in a transition phrase, usually repeat a key term. Note that transition sentences are stronger than transition phrases, which in turn are stronger than transition words. The farther away from the main point of a paragraph or section, the stronger the transition should be to link back to the main point (see Table 6.1).

6.4   CONDENSING

WRITING PRINCIPLE 30:

Make your writing concise.

A well-written paragraph needs not only to be organized, coherent, and consistent, but it also needs to be concise. Wordiness is a common problem in scientific writing. Many journals limit the number of words in various parts of an article or in the article as a whole.

If you need to condense a paragraph (or paper), do not despair. A wide range of methods is available to make a paragraph more concise without having to remove important material from a paper. Be aware, however, that clarity is always more important than brevity.

Condensing often needs to be done in combination with other techniques:

1. Emphasizing important information
2. De-emphasizing or omitting less important information
3. Replacing or omitting words and phrases

Establishing Importance

GUIDELINE:

Establish importance.

In many scientific papers, important information is outweighed by unimportant information. Because of this imbalance, it is often difficult for the reader to find the real "meat" of the paper. De-emphasizing or omitting less important information is probably the most important technique in condensing.

As a first step in condensing, you need to decide what is important, what is less important, and what is unimportant information in your manuscript. The next steps are to emphasize the important information and to de-emphasize the less important information. Unimportant information adds nothing but clutter and distracts the reader. It should be omitted.

To emphasize your important information, either place it in a power position or signal it directly by using statements such as "Most important, ..." or "The key finding of this study was...." Less important information can be de-emphasized by subordinating it. This can be done, for example, by placing it in a subordinate clause as shown in the next examples.

👎	**Example 6-17**	Information about the relation between WBC count and hospital case fatality rates is limited. A link between WBC count and increased long-term mortality after acute myocardial infarction may exist.
👍	**Revised Example 6-17**	*Although information about the relation between WBC count and hospital case fatality rates is limited*, a link between WBC count and increased long-term mortality after acute myocardial infarction may exist.

👎	**Example 6-18**	The Hainan aborigines are an ethnic group living at the entrance route to Southeast Asia and have been influenced little by relocation and migrations of other ethnic groups.
👍	**Revised Example 6-18**	The Hainan aborigines, *an ethnic group living at the entrance route to Southeast Asia*, have been influenced little by relocation and migrations of other ethnic groups.

When less important information has been reduced or omitted, and important information has been emphasized, the reader will be able to see the forest for the trees.

Words and Phrases That Can Be Omitted

GUIDELINE:

Omit "overview" words, phrases, and sentences.

Another important technique in condensing is to replace or to omit unnecessary words and phrases. Aside from redundancies and jargon, all of which are discussed in Chapter 2, other words and phrases can be omitted to condense a document. Certain ESL writers whose native language uses many flowery phrases should be particularly aware of what words and phrases can be omitted in a scientific document.

Scrutinize your paper for pointless words and phrases ruthlessly. Dissect every sentence. As a rule, when equivalent alternatives exist, choose the shortest one. In scientific writing, every word should count.

Verbs to Omit
Omit verbs such as

describes	noted	was done	reported
noticed	observed	occurred	seen

Just state the facts.

ESL advice

👎 **Example 6-19**   Jones et al. reported that intracellular calcium is released when adipocytes are stimulated with insulin.

(15 words)

👍 **Revised**   Intracellular calcium is released when adipocytes are
   Example 6-19   stimulated with insulin (Jones et al., 1996.)

(10 words)

Phrases to Omit

Omit phrases and sentences that tell your reader what a sentence/paragraph is about.

👎 **Example 6-20**   Products were verified by gel electrophoresis. The results are presented in Figure 1.

(13 words)

👍 **Revised**   Products were verified by gel electrophoresis (Fig. 1).
   Example 6-20

(6 words)

👎 **Example 6-21**   To assess the purity of the gene product, many techniques were employed.

(12 words)

👍 **Revised**   (Omit)
   Example 6-21

(0 words)

Other expressions to omit: This section describes...

                                    In regard to...

                                    As far as X is concerned...

                                    The experiment was done by...

                                    Figure 6 shows that...

"It...That" Phrases

Aside from jargon and other redundancies, you should replace or omit "It...that" phrases. Most of these phrases are pointless fillers and can be omitted entirely. If the idea in the phrase is essential, replace the phrase with a shorter version.

Examples of "It...that" phrases:

It is interesting to note that...   **omit**

In light of the fact that...   **replace (because)**

It is possible that...   **reword (... may..., perhaps, possibly)**

It has been reported that...   **omit or replace (Taylor reported that...; or [reference])**

Positive Versus Negative Expressions

GUIDELINE:

Aviod writing in the negative.

Changing negative expressions to positive expressions usually results in shorter sentences. Moreover, readers prefer to read positive things, not negative things. Avoid writing in the negative. Write instead in the positive. Above all, avoid double negatives, which can easily confuse readers.

Examples of changing from negative to positive:

negative	positive
do not overlook	note
not different	similar
not infrequently	frequently
not many	few
not the same	different
not unimportant	important

Excessive Detail

GUIDELINE:

Omit excessive detail

Detail that can be inferred or is unimportant should be omitted.

	Example 6-22	a	Using a 1 ml tip, we removed 750 µl of the aliquot into a new Eppendorf tube.
		b	Resuspended cells were transferred to a 0.4 cm cuvette (Bio-Rad) and electroporated using a Bio-Rad Gene Pulser.
	Revised	a	We removed 750 µl of the aliquot.
	Example 6-22	b	Resuspended cells were electroporated using a Bio-Rad Gene Pulser.

Intensifiers and Hedges

GUIDELINE:

Do not overuse intensifiers or hedges

Intensifiers are adjectives, adverbs, or verbs that are used to strengthen nouns or verbs such as the following:

always	basic	central	certainly	clearly
crucial	prove	quite	show	very

Hedges are cautious adjectives, adverbs, or verbs such as the following:

actually	appear	could	essentially	indicate	many	
may	most	often	perhaps	possibly	seem	some
suggest	usually					

ESL advice

Intensifiers and hedges are often overused, especially by ESL writers. Aside from making a sentence wordy, you will sound arrogant and too aggressive if you overuse intensifiers and too cautious or timid if you overuse hedges. Omitting intensifiers and hedges will not only avoid these appearances but also shorten sentences.

	Example 6-23	Our results <u>may</u> <u>indicate</u> that siRNA dublex <u>possibly</u> caused an RNA interference effect.
	Revised **Example 6-23**	Our results **suggest** that siRNA dublex causes an RNA interference effect. **or** siRNA dublex **may** cause an RNA interference effect.

	Example 6-24	Figure 5 <u>clearly</u> shows that the protein was absent in the fraction.
	Revised **Example 6-24**	Figure 5 shows that the protein was absent in the fraction.

Figure Legends

Figure legends in particular can often be condensed. Many journals prefer telegram-style figure legends. Usually, articles can be omitted and prepositional phrases can be shortened or omitted not only in figure titles but also in their explanatory notes.

	Example 6-25	Figure 3. Shown here are tangential sections as seen through a barrel cortex of a trimmed mouse. A-C show the ipsilateral barrel cortex; D-F depict the contralateral barrel cortex. The scale bar is 0.2 mm.
	Revised **Example 6-25**	Figure 3. Tangential sections through a barrel cortex of a trimmed mouse. A-C, Ipsilateral barrel cortex; D-F, Contralateral barrel cortex. Scale bar, 0.2 mm.

SUMMARY

WRITING PRINCIPLE 23: Organize your paragraphs.

WRITING PRINCIPLE 24: Use a topic sentence to provide an overview of the paragraph.
Arrange the details in the remaining sentences.

WRITING PRINCIPLE 25: Use consistent order.

WRITING PRINCIPLE 26: Use consistent point of view (same subject for different sentences within a paragraph).

WRITING PRINCIPLE 27: Make your sentences cohesive.

WRITING PRINCIPLE 28: Use key terms to create continuity. Repeat key terms exactly and early, and link them.

> WRITING PRINCIPLE 29: Use transitions to indicate logical relationships between sentences.
>
> WRITING PRINCIPLE 30: Make your writing concise.
> - Establish importance
> - Omit overview words, phrases, and sentences
> - Avoid writing in the negative
> - Omit excessive detail
> - Do not overuse intensifiers and hedges

PROBLEMS

Problem 6-1 Paragraph Organization

The following paragraph is about the two phases of *P. infestans* infection. Although the first sentence of the paragraph introduces the potato pathogen, the second sentence of the paragraph does not logically link back to the first sentence nor does it clearly introduce the two phases. Rewrite the second sentence to link it to the first sentence and to lead into the description of the two phases. Put parallel ideas into parallel form.

Phytophthora infestans, which precipitated the Irish potato famines in the mid-19th century, remains the most economically important potato pathogen. Up to 36 hr postinoculation, *P. infestans* forms haustoria and requires living plant tissue. After this biotrophic phase of the infection, a necrotrophic phase ensues in which infected host tissue becomes necrotic (1).

Problem 6-2 Paragraph Organization

The following paragraph is about types of glutamate receptors. Although the paragraph has a topic sentence, the remaining sentences of the paragraph do not follow logically from the topic sentence. Add a sentence after the topic sentence to fulfill the expectations of the reader. (What does the reader expect to read about after reading the topic sentence?) Reorganize sentence 2 to make it parallel to sentence 3.

1Ionotropic glutamate receptors fall into two general categories (7, 8). 2When NMDA receptors are activated by N-methyl-D-aspartate (NMDA), ion channels are opened, allowing ions to rush into the cell and thus cause an excitatory postsynaptic potential. 3non-NMDA receptors like PCP bind to kainate and quisqualate only and prevent ion flow and an excitatory postsynaptic potential, even in the presence of NMDA.

Problem 6-3 Paragraph Organization

The following paragraph is about the features characterizing copper oxide superconductors. Although sentences 2 and 3 describe the two features of the copper oxide superconductors, the paragraph does not have a topic sentence. Write a clear topic sentence for this paragraph. The topic sentence should state the message of the paragraph (the features of copper oxide superconductors). In your topic sentence, try to make the topic the subject of the sentence.

Copper Oxide Superconductors

2One feature is, of course, their unprecedented high transition temperatures. 3The other feature is that their normal-state properties are not those of ordinary metals; they are not consistent with the traditional Fermi-liquid quasiparticle picture that is a cornerstone of our understanding of the metallic state.

Problem 6-4 Paragraph Organization

The following paragraph is about the migration of salmon. Although sentences 1 and 2 describe the two possible methods salmon may employ to find their way, the paragraph does not have a topic sentence. Write a clear topic sentence for this paragraph. The topic sentence should state the message of the paragraph (the different methods salmon use to find their way during their homeward journey in the fall). In your topic sentence, make the topic the subject of the sentence. Also, make sentence 2 parallel to sentence 1.

1Salmon may use a magnetic or sun compass to orient themselves. 2As described by Brown et al. (15), olfactory cues learned as smelts may also help salmon to find the river and tributary of their birth. 3Salmon may reenter fresh water in spring, summer, or fall, but spawning occurs in the fall, and the life cycle of the salmon begins anew.

Problem 6-5 Paragraph Organization

1. **Consider the word placement in the following paragraph. Make the point of view consistent (the subject of every sentence should be the same).**
2. **Add a transition at the beginning of sentence 4 to indicate the logical relationship of sentence 3 to sentence 4.**
3. **In addition, ensure that parallel ideas are written in correct parallel form.**

1Interleukin I (IL-I) is a mediator produced by activated macrophages and many other cell types (1). 2IL-I initiates inflammation and thus stimulates various cells. 3In T-cell activation, one of the co-signals is IL-I. 4IL-I plays a role in activating B cells and the production of acute phase proteins, muscle catabolism, bone resorption, and fever is stimulated by IL-I.

Problem 6-6 Paragraph Consistency

Keep the organization of the following paragraph in consistent order. In addition, use a transition to indicate the relationship between sentences 2 and 3.

1Both GAD-positive cell bodies and processes were found in the ventral lateral posterior nucleus and thalamic reticular nucleus. 2Almost all of the neurons in the thalamic reticular nucleus appeared to contain GAD-immunoreactivity. 3Only small round cells in the ventral lateral posterior nucleus were GAD positive.

Problem 6-7 Paragraph Consistency

Ensure consistent order in this paragraph by adding another sentence before sentence 7 (What does the reader expect after reading sentence

2?). In addition, place sentences 7 and 8 in parallel form and order to sentences 4 through 6, and ensure consistent form throughout the paragraph.

1Lipopolysaccharide endotoxin, a component of the bacterial cell wall, stimulates macrophages to produce pro-inflammatory cytokines. 2Two such cytokines have been shown to be critical mediators of septic shock (7). 3The first of these mediators is the tumor necrosis factor α (TNF-α). 4Its excessive production leads to a destructive immune response, ultimately resulting in lethal organ failure and death (1-4, 7). 5TNF-α possesses a signal peptide and is first synthesized as a membrane-associated protein. 6A member of the metalloproteinase family, the TNF-α converting enzyme (TACE), cleaves the membrane-associated form of TNF-α and generates soluble TNF-α (8, 9). 7Unlike TNF-α, interleukin-1β (IL-1β) is synthesized as a precursor molecule, which is then cleaved and activated by the cysteine protease caspase-1. 8IL-1β acts in early immune responses as a chemotactant for lymphocytes and as a signal for eosinophil and basophil degranulation (10).

Problem 6-8 Paragraph Consistency

Ensure that the order of the key terms in the topic sentence is consistent with that of the remaining sentences. Ensure also that the same key terms are used consistently and that correct parallel form is kept when signaling the subtopics.

Apart from seawater acidification, the ocean in a high CO_2 world will experience other changes, including higher surface temperatures, enhanced stratification, and decreased mixed-layer depths. Although the direct effect of rising temperatures on net community carbon-to-nutrient ratios is unclear, at the organism level, most experimental studies indicate an increase in the C:N ratio with increasing temperature. Decreased mixed-layer depth tends to increase phytoplankton carbon-to-nutrient ratios, which would augment the direct CO_2 effect on C:N:P stoichiometry observed in this study. Increased stratification, in contrast, would decrease the supply of nutrients to the surface layer, reducing overall primary and export production. Thus, concurrent with a decrease in the strength of the biological pump caused by a lower nutrient supply, the pump's efficiency is likely to increase at increased CO_2.

Problem 6-9 Paragraph Consistency and Transitions

In the following paragraph, the three decay channels are not consistently introduced. Make the paragraph consistent. Also, add transitions to sentences 4 and 5 that are parallel to the transition used in sentence 3.

1The emission properties of a nanoscale optical emitter can be significantly modified by the proximity of a nanowire that supports surface plasmons. 2In principle, three distinct decay channels exist. 3First, direct optical emission into free-space modes is possible, with a rate modified from that of an isolated quantum dot owing to the proximity of the metallic surface.

4The optical emitter can also be damped nonradiatively owing to Ohmic losses in the conductor. 5Most important, the tight field confinement and reduced velocity of surface plasmons can cause the nanowire to capture the majority of spontaneous radiation into the guided surface plasmon modes, much like a lens with extraordinarily high numerical aperture.

Problem 6-10 Key Terms
The key terms in the following paragraph are not repeated exactly. Revise the passage such that key terms (life span, fat storage, and the nematode *C. elegans*) are repeated exactly. (Note: A nematode is a type of worm.)

A model system to study the underlying mechanism that connects Type II diabetes, life span, and obesity is the nematode *C. elegans*. Worms are a great system to study the connection of diabetes, age, and obesity because they have a well-conserved insulin signaling pathway that affects life span and fat storage, suggesting that there is an underlying molecular mechanism that connects insulin signaling, aging, and obesity. In addition to the known insulin signaling pathway, the entire genome of the nematode has been sequenced, and many molecular and genetic tools are available that will allow a comprehensive identification of the genes that affect insulin-like signaling, aging, and fat regulation in worms.

Problem 6-11 Key Terms
The relationship between the two sentences in the following paragraph is not clear because key terms are not linked. Given that heroin and morphine are examples of opioid drugs, link the underlined key terms in the paragraph.

Opioid drugs are avidly self-administered by both humans and laboratory animals (2, 3). Heroin and morphine mimic the endogenous opioid neurotransmitters, known as endorphins, by binding to one or more of the mu or delta opioid receptors in the brain (4).

Problem 6-12 Key Terms
The following paragraph is about systolic blood pressure in children.

1. **Ensure that this key term is introduced in sentence 1 and used consistently and exactly throughout the paragraph.**
2. **The paragraph describes contrasting results (sentences 2 and 3). Signal the results. Rewrite the paragraph so that the contrast is in perfect parallel form.**
3. **Signal the contrast by using a transition word or phrase at the beginning of sentence 3.**
4. **Fix the faulty comparison in sentence 4.**

1We measured blood pressure in 5000 individuals aged ten to 18 years using a random zero sphygmomanometer. 2The mean systolic blood pressure in girls was 116 mmHg at age 15. 3Fifteen-year-old boys had a pressure of 128 mmHg. **4**Thus, blood pressure in boys was about 10% higher compared to that in girls.

Problem 6-13 Transitions
In the following paragraph, the logical relationship between sentences 3 and 4 to sentence 2 is not clear. To make the relationship between sentences 2, 3, and 4 clear, add a transition word or phrase at the beginning of sentences 3 and 4. Make sentence 4 parallel to sentence 3.

1Substantial arteriosclerotic lesions can be produced by a diet rich in cholesterol (1). **2**Aside from these lesions, other tissues are also effected (2-7) by high cholesterol levels. **3** _____ The liver undergoes severe fatty degeneration (2). **4** _____ Pathological changes occur in the rabbit eye (3,5).

Problem 6-14 Transitions
In the next paragraph, the logical relationships between sentences 1 and 2 and sentences 3 and 4 is not clear. To make the relationships clear, add a transition word at the beginning of sentences 2 and 4.

1Marine coastal ecosystems are among the most productive and diverse communities on Earth (*1*) and are of global importance to climate, nutrient budgets, and primary productivity (*2*). **2** _____, the contributions that coastal ecosystems make to these ecological processes are compromised by human-induced stresses including overfishing, habitat destruction, and pollution (*3–5*). **3**These stressors particularly impact benthic (bottom-living) invertebrate communities because many species are sedentary and cannot avoid disturbance. **4** _____, marine coastal ecosystems are likely to experience a proportionally large change in biodiversity should present trends in human activity continue (*6–8*).

Problem 6-15 Paragraph Construction
Construct a paragraph on cancer cells using the list of facts in the order provided. In your writing, pay attention to writing a good topic sentence and to using good word location. Employ paragraph consistency, key terms, and transitions. Consider also other writing principles such as parallel form and correct pronouns and prepositions.

CANCER CELLS
- Are malignant tumor cells
- Differ from normal cells in three ways:
    - They dedifferentiate—for example: ciliated cells in the bronchi lose their cilia
    - Metastasis is possible—travel to other parts of body. New tumor growth
    - Rapid division—do not stick to each other as firmly as do normal cells

Problem 6-16 Paragraph Construction

Construct a paragraph on reactive oxygen species (ROS) using the list of facts provided. In your writing, pay attention to writing a good topic sentence and to using good word location. Employ paragraph consistency, key terms, and transitions. Consider also other writing principles such as parallel form and correct pronouns and prepositions.

REACTIVE OXYGEN SPECIES (ROS)

- Important roles in cell signaling at low levels
- Include oxygen ions, free radicals, and peroxides
- Highly reactive due to the presence of unpaired valence shell electrons
- Form as a natural by-product of the normal metabolism of oxygen
- During times of environmental stress, ROS levels can increase dramatically
- High levels can result in significant damage to cell structures, cumulating in oxidative stress

Problem 6–17 Paragraph Construction

Construct a paragraph on superconductivity using the list of facts provided. In your writing, pay attention to writing a good topic sentence and to using good word location. Employ paragraph consistency, key terms, and transitions. Consider also other writing principles such as using parallel form and correct pronouns and prepositions.

SUPERCONDUCTIVITY

- Occurs at extremely low temperatures
- Quantum mechanical phenomenon
- Electrical resistance of a superconductor drops abruptly to zero when the material is cooled below its "critical temperature."
- Occurs in certain materials, such as tin and aluminum and various metallic alloys
- An electrical current flowing in a loop of superconducting wire can persist indefinitely with no power source.
- Does not occur in noble metals like gold and silver, nor in most ferromagnetic metals

Problem 6-18 Paragraph Construction

Construct a paragraph on combination drug therapy using the list of facts provided in any order. In your writing, pay attention to writing a good topic sentence and to using good word location. Employ paragraph consistency, key terms, and transitions. Consider also other writing principles such as using parallel form and correct pronouns and prepositions.

COMBINATION DRUG THERAPY

- Applied in treatment of tuberculosis
- Can be expensive

- Can increase risk of toxicity and superinfection
- Of important clinical value
- Used against organisms for which resistance to one drug readily develops

Problem 6-19 Paragraph Construction

In the following paragraph, the author has not paid much attention to good word location. Analyze the paragraph with respect to word location by drawing arrows to determine how key words relate the sentences to each other. Then, revise the paragraph by improving word location.

1Chikungunya is a painful viral disease of the tropics. 2It is rarely fatal but can cause severe fevers, headaches, fatigue, nausea, and muscle and joint pains. 3Medical entomologists have worried about the astonishing ascent of the Asian tiger mosquito. 4Many viral diseases are transmitted by *Aedes albopictus*. 5It was first found in the United States in secondhand tires imported from Asia in Houston, Texas, in 1985. 6Plants shipped in water containers, such as the popular Lucky bamboo, often carry the eggs of the mosquito. 7In Italy, almost 200 people were infected by the mosquito this summer. 8Scientists wonder whether *A. albopictus* has the potential to touch off much larger outbreaks in Europe and the United States.

Problem 6-20 Point of View

1. **Make the point of view in sentence 2 close to the point of view in sentence 1 by linking the key terms in sentences 1 and 2 (use the term "materials" as a link).**
2. **In sentence 2, make the topics the subjects and put the action in verbs.**
3. **Add a transition at the beginning of sentence 2 to indicate the logical relationship of sentence 2 to sentence 1.**

1When the temperature falls at constant pressure, most pure materials pass from gas to liquid to solid. 2 _____ A large number of phases is displayed by rod-like molecules called "liquid crystals" (ref.).

Problem 6-21 Transitions/Coherence

In the following paragraph, sentences 2, 3, and 4 are parallel. The subtopics of sentences 2 and 4 are signaled ("At lower temperatures," "... at even lower temperatures"). However, the subtopic of sentence 3 is not signaled.

1. **Write a signal at the beginning of sentence 3. Your signal should be parallel to the signals in sentences 2 and 4.**
2. **Add a transition word in sentence 5 that connects it to the rest of the paragraph.**

1For typical liquid crystals, the high temperature liquid phase is isotropic, meaning that the positions and the orientations of the molecules are scattered about at random. 2At lower temperatures, the substance undergoes a transition into the so-called "nematic" phase in which the molecules tend to orient in the same direction but in which positions are still scattered. 3 _____The substance passes into the "smectic" phase in which the molecules orient in the same direction, and their positions tend to fall into planes when still lower temperatures are reached. 4Finally, at even lower temperatures, the molecules freeze into a conventional solid. 5 _____Lower temperatures show more and more qualitative order.

Problem 6-22 Condensing
The following paragraph is the central message of an abstract describing a novel function of ATP. Distinguish between important, less important, and unimportant information contained within these sentences. Condense the paragraph by omitting unimportant information and subordinating less important information. In addition, remove the noun cluster in sentence 3.

1Translation of the genetic message into proteins is one of the most energy-consuming processes in living organisms. 2Among other nucleotide-dependent mechanisms, ATP has been demonstrated to be important for initiation of translation. 3Here we provide a description of our detection of an additional and novel function of ATP in the translational mechanism, which uses a wild type *S. cerevisiae*-derived exogenous mRNA-dependent *in vitro* translation system which it is derived from.

(67 words)

Problem 6–23 Condensing
Condense the following paragraph.
   After you have finished condensing, improve sentence structure as needed.
   Make the topic the subject; put the action in the verb.
   Keep sentences short.
   Be sure that pronouns are clear.
   Use parallel form for parallel ideas.

It should be noted here that the characterization of superconductors involves two quite unexpected features. One is, of course, their unprecedented high transition temperature. In addition, it is quite clear that their normal state properties are basically not like ordinary metals, and they are not at all consistent with the traditional Fermi-liquid quasiparticle picture that is a cornerstone of our understanding of the metallic state.

(65 words)

Problem 6-24 Condensing
The following paragraph is an abstract that has far exceeded its permissible length of 100 words. Shorten the abstract to 100 words or less by establishing importance. Omit unimportant information, and subordinate less important information.

Tourette syndrome (TS) is characterized by chronic motor and vocal tics. Habit reversal therapy (HR) is a behavioral treatment for tics which has received recent empirical support. The present study compared the efficacy of HR in reducing tics, improving life-satisfaction and psychosocial functioning in comparison with supportive psychotherapy (SP) in outpatients with TS. [...] Thirty adult outpatients with DSM-IV TS were randomized to 14 individual sessions of HR (n=15); or SP (n=15). HR but not SP reduced tic severity over the course of the treatment. Both groups improved in life-satisfaction and psychosocial functioning during active treatment. Reductions in tic severity (HR) and improvements in life-satisfaction and psychosocial functioning (HR and SP) remained stable at the 6-month follow-up. [...] Our results suggest that HR has specific tic-reducing effects although SP is effective in improving life-satisfaction and psychosocial functioning. Assessments of response inhibition may be of value for predicting treatment response to HR. *(149 words)*

(With permission from Elsevier.)

Problem 6-25 Condensing
Condense the following paragraph. Try to make your revised paragraph less than 35 words.

Our results indicate that between 5 and 25 °C, undoped, high-quality diamond as well as diamond covered with chemically bound hydrogen show no conductivity. Undoped, high-quality diamond also shows no conductivity at higher temperatures. However, diamond covered with chemically bound hydrogen clearly shows a pronounced conductivity at temperatures between 5 and 25 °C.

(51 words)

Problem 6-26 Condensing
The following paragraph is an abstract that has exceeded its permissible length of 150 words. Shorten the abstract to 150 words or less by establishing importance, omitting unimportant and redundant information, and subordinating less important information.

Radionuclide X contamination of ground and surface water is a serious problem in many parts of the world. Radionuclide X interacts with particles, and these particle interactions govern the transport of radionuclide X and ultimately the fate and distribution of radionuclide X in the subsurface. The transport of radionuclide X in subsurface groundwater systems is strongly affected by radionuclide X adsorption to iron oxides. To scale batch adsorption results for use in predictive transport models in the field, the effects of the solid-solution ratio—the ratio of the mass of adsorbent solid to the volume of solution—must be known. Aside from the solid-solution ratio, carbonate interactions with radionuclide X can substantially affect radionuclide X adsorption. To predict radionuclide X transport in groundwater systems, it is important to understand the effects of total carbonate concentration on radionuclide X adsorption in closed systems. In this study, we examined the effects of solid-solution ratio and total carbonate concentration on radionuclide X adsorption onto a heterogeneous natural subsurface soil in batch adsorption experiments. Our experimental results show that radionuclide X adsorption, even normalized to adsorbent mass, was affected by solid-solution ratio at a fixed total carbonate concentration and that increasing total carbonate concentration could decrease radionuclide X adsorption due to the formation of aqueous radionuclide X-carbonate complexes.

(213 words)

PART TWO

Planning and Laying the Foundation

CHAPTER 7

The First Draft

The publication of new results in scientific journals is a central component of the scientific process. Communicating your findings makes them available to others and serves as an indication of your expertise and productivity. Without publications, you are unlikely to get funded or to advance in your field.

7.1 THE WRITING PROCESS

The writing process encompasses several stages, which include: prewriting, drafting, revising, editing, evaluating, and publishing.

Prewriting
In the prewriting stage, you decide on the intended audience for your document and choose the journal to which to send your manuscript. The choice of journal will dictate the format of your paper, its length, and its structure.

Drafting
In the drafting stage, you usually review the literature to learn what has been done and what is known in the field of your topic. You have to organize your thoughts. You may consider creating an outline or sorting bullet points you may have jotted down during your literature search or while collecting data. Arrange material in subsections from general to specific or from abstract to concrete. Then, you write the main body of the text: Introduction, Materials and Methods, Results, and Discussion. The abstract is usually written last.

Revising
The key to strong scientific writing is revising your work. Expect to do multiple revisions. Ask your colleagues for input, and ask them to review your drafts as well. View your writing as a team effort in which different players get to participate.

Editing
After you have drafted and revised your paper, copyedit it for grammar, spelling, and punctuation. Ensure that it is in the correct format.

Evaluating
When you have a final draft, submit it to your journal of interest for anonymous peer review. The editor of the journal will select appropriate reviewers and ask for their opinion of your manuscript in writing. When the editor has received back the evaluations of the reviewers, he or she will inform you of the final decision on your manuscript.

Publishing
Once accepted, your manuscript will be published in the corresponding target journal.

The following sections of this chapter discuss the prewriting stage as well as the first part of the drafting process for scientific documents. Subsequent chapters provide detailed information on specific, diverse sections of scientific documents, revisions, and submissions.

7.2 PREWRITING

No experiment will mean much unless it is communicated. But how do scientists trained in research move from doing experiments to communicating what they mean? Having to write down their thoughts and interpretations can intimidate even the best scientist and writer.

Before writing your first draft, there are several things to consider. First, you need to have accumulated enough data or ideas to write about. Second, you have to know who your readers should be and choose a journal to which to submit. You also need to collect your references and decide on authorship. Last, but not least, you need to get mentally ready.

Collecting data, researching references, organizing your thoughts, and writing takes time and energy. Do not waste any of these—make sure that your research warrants publication in a journal. Only write an article if you have something new, important, useful, and comprehensible to tell.

Audience and Journal Choice

GUIDELINE 1:
Search for the best match of topic, audience, and journal.

Once you have a clearly defined message for your paper, you need to decide on where to publish your work. This decision will depend on several factors such as your audience, the impact factor of the journal, the speed of publication, the prestige of the journal, and its acceptance rate. Take the time to look through potential journals and determine who their audience is and what their mission is. Do not hesitate to ask your colleagues or your supervisor for advice about which journals might be appropriate. You may also want to check where other papers of a similar topic have been published.

You also need to decide on the best format for your findings. Not every study leads to a full article. Other choices include, for example, case reports or letters to the editor.

When you consider your potential readers, know that most publications will not attract nearly as many readers as you might wish. Usually you need to choose a journal in a very specialized area. In short, you need to search for the best match of topic, audience, and journal. Start by looking, for example, at Current Contents, Index Medicus, Journal Citation Reports, Biological Abstracts, Cumulative Index to Nursing and Allied Health Literature (CINAHL), SCOPUS, or the General Science Index to get an idea of specific journals that cover your research area (see also Chapter 8.2 on Selecting References.)

Consider consulting the Journal Citation Reports in your academic library. This will tell you which journals are cited most, that is, the impact factor of a journal. Of course, most scientists would like nothing better than to publish in one of the highest impact journals, but be aware that these journals also have very high rejection rates, some as high as 90%. If you send out a manuscript to one of these journals, you may be waiting for weeks just to receive a rejection. If you feel unsure of where to send your paper, inquire about it by contacting the editor and asking if a manuscript such as yours would be considered for publication. This is the quickest way of getting an answer.

Generally, journals that are published more frequently have a faster turnaround time between acceptance and publication. Some may publish articles online first before publishing them in print, as may some universities after your paper has been accepted.

The ever increasing international readership in science today seeks both an electronic version as well as a prestigious ranking. Beware, however, of journals that publish only electronically, as many of these electronic journals do not put their articles through a rigorous review process and thus are not as prestigious as journals with printed versions. Beware also of new journals; their circulation may be very small.

Instructions to Authors

GUIDELINE 2:

Obtain *Instructions to Authors* and follow them.

When you have decided on a target journal, you should obtain the journal's specific *Instructions to Authors*. Read these instructions. Mark important details such as length of abstract, sections/headings of an article, writing style, format of references, and electronic format. Follow the instructions carefully when you are writing, and compare your format to recently published articles in your target journal. Reading these articles will give you an even better idea of the style and format of papers in this journal.

Instructions to Authors are generally available on the Web site of your target journal as well as in printed volumes of the journal. Web sites that list *Instructions to Authors* for many different journals include the following:

http://mulford.meduohio.edu/instr/ (for biological and medical journals)

http://www.icmje.org (core set of instructions for many biomedical journals)

http://www.inter-biotec.com/biowc/uniform/uniform.html (uniform requirements for manuscripts submitted to biomedical journals)

http:// www.ch.cam.ac.uk/c2k/cj/alpha.html (for chemistry journals)

http://www.lib.auburn.edu/scitech/resguide/chemistry/chemjn.html (for chemistry journals)

http://www.ch.cam.ac.uk/c2k/cj/physical.html (for physics journals)

7.3   AUTHORSHIP

GUIDELINE 2:

Decide on authorship before starting to write.

One of the most common concerns of young scientists is who to include as an author on a publication and in which order. Every person that contributed substantially to the research, to the experimental design, and to writing the paper should be included as a coauthor. People that only assist in data collection or manuscript editing should not be included as coauthors.

Having coauthors can be beneficial but also problematic. Coauthors can help in designing the experiments and in seeing them through as well as in interpreting the data and in writing the paper. Disagreement over authorships, however, can easily result in wrecked friendships. A good piece of advice is to always decide on who is going to be an author as soon as you can—if possible, *before* you start writing the paper (if not before you start the research itself to divide the work).

Deciding on the order of authorship can be difficult. Most people want to be first author (or as close to the first author position as possible) because typically only the name of the first author appears in citations and reference lists. Generally, the person who contributed the most to the research is the first author; the head of the laboratory is typically the last author. The person who did most of the writing is usually also the first or last author, and one of them is usually identified as the corresponding author, the author to whom all correspondence should be directed. Being the corresponding author does not give you much recognition, and if you decide to become the designated corresponding author, you should ensure that you can be reached through your corresponding address.

Aside from the first and last author, other authors are normally listed in the order of their contribution to the study. If more than one person contributed equally to the work, authors may be listed in alphabetical order or may be identified as having contributed equally by a footnote. If more than one paper arises from the study, the reverse order of authors can be used in the second paper. The order of authors, however, may also depend on the policy or preference of the laboratory head.

You may run into people, at worst a senior researcher (investigator or head of department), who ask you to include in the author list a person who has not contributed to the research, possibly even himself or herself. You can try to avoid this, for example, by diplomatically stating that there are stringent requirements put forth by your target journal. However, you may be forced to include that person for political reasons. If that happens, at the very least, ask that person to contribute to writing the article.

ESL advice

In some departments or foreign countries, it may be common practice to always include the head of the department on a paper. In other cultures, the principal investigator may always want to be placed in the first author position, possibly due to financial or political reasons. Know though that you should give the first author place preferably to a student or postdoctoral fellow. These young people usually not only did most of the work but also still have to establish themselves and will have a much better chance of doing so if they receive the recognition due to them. Furthermore, people in your field will know who the principal investigator is, and those not in your field will expect to find that principal investigator's name at the end of the author list.

Potential conflicts can arise when more than one person is writing the manuscript. Problems that arise from multiple authors writing a paper may include not only inconsistencies in language and style but also weak transitions and illogical formats. An example of a paragraph that has been written by more than one author—and in two very distinct styles—is shown in Example 7-1:

 Example 7-1   Structured abstract

Numerous antibiotic classes, including macrolides, the strep-togramins, and the oxazolidinones, bind to the 50S ribo-somal subunit, demonstrating that this is an excellent target for antibiotic drug discovery. Biochemical studies had previ-ously determined that the site of action for all these antibiotic classes was the peptidyl transferase center of the 50S. The crystal structures, however, brought light into the center: they delineated how different antibiotic classes bind to or engage distinct, though often overlapping or adjacent spaces, mak-ing the 50S ribosome a "target of targets." Consider the opportunities! Suddenly, one could determine an appropri-ate position on and a trajectory from an existing antibiotic scaffold to boost affinity or overcome target-based resis-tance. Likewise, one could study the spaces between two adjacent binding sites and either bridge the two or utilize that as a starting point for a new scaffold. At once, ribo-somal drug hunters were freed from the me-too approach of tuning ever-so-slightly the same molecular scaffolds. And so, it was back to the future, targeting the ribosome.

[With permission from Future Medicine Ltd]

It is best to designate one writer (preferably the first author or best writer) for doing all of the writing or revisions. Alternatively, one person could be designated as the coordinator. This person should oversee the logical framework of the paper as well as its style and consistency. All coauthors should at least read and approve the final version before submission to a journal. Coauthors also have to agree to any changes made before publica-tion, and the primary author should confer with the coauthors as to the correct spelling of their names.

7.4   DRAFTING A MANUSCRIPT

General Format

GUIDELINE 3:

Follow the IMRAD format.

Scientific papers have a set format, the Introduction, Methods, Results, and Discussion (IMRAD) format. This format reflects the order of the core sections in academic journal articles, which usually include:

Title page
Abstract

IMRAD format

Papers that do not follow this format or differ significantly from it disorient scientists.

Although this format is the order of a scientific paper when published, it is not necessarily the best order in which to write your manuscript. You can start writing your paper in several different ways. You can write an outline first (simple or detailed), or you can write the easiest section of the manuscript first. For many authors, the easiest section is the Materials and Methods section. For others it may be the Results section or the Introduction. A few may even choose to write the abstract first—considering it a short outline. The bottom line is this: It does not matter where you start or in what order you write the paper. All that matters is what the paper looks like when it is finished. So do what works best for you.

Writing the First Draft

1. PRINCIPLE OF DRAFTING:

Worry about the writing principles in the revisions.

2. PRINCIPLE OF DRAFTING:

Start by writing less.

The prospect of writing a manuscript can seem like an overwhelming task, especially for beginning writers. The key is to take the writing process one step at a time. Start writing your document in separate sections.

Writing the first draft is hard work and difficult. Not only do you need to be mentally prepared, you also need all your materials (notebook, references, figures, tables, coffee) at hand, and you need to have a block of time without interruptions. The main reason why writing the first draft is difficult is that you have only a rough idea of what you want to say. In fact, you will not know exactly what it is you want to say until you are writing. If you do not know exactly what to say when you begin, do not be discouraged. The exact words and even the exact sentences are not important at this stage. What is important is that you start—somewhere.

Center your writing on the overall question/purpose of the paper and on its answer. Depending on the question and answer, you need to decide what to include in the paper and how to organize it. Be aware, however, that your question and answer might change as you are writing. When you write, you will discover the best and most precise way of formulating your question and answer. In the revisions, you can then adjust your writing accordingly.

Do not expect to finish your first draft in one setting. You may need a week to write the first draft—or more. To make writing less overwhelming, start by writing less. For example, write a very brief Introduction of maybe just one paragraph. You will fill in a lot more during your revisions as ideas come to you or as your colleagues comment. Similarly, in the Discussion, start off by writing 2 paragraphs, not 12.

The most important thing for writing first drafts is momentum. Do not worry about any writing principles or any other rules of writing. You can always add or take out any words, sentences, or even paragraphs later. Do not worry about whether your pronouns are clear, whether you use good parallel form, or whether your paragraphs are in the right order. Deal with the writing principles during the revisions. Just keep moving.

Put your ideas on paper or into the computer so that you have something to work with. Remember, your first draft is not written in concrete. It should contain all you can think of and want to say in one place. Later revisions will allow you to examine, rethink, and rearrange.

Collecting References

GUIDELINE 4:
Collect, organize, and study your references.

To write a scientific document, you will need references. Due to the constant and large flow of information, often compounded by multidisciplinary collaborations, it is almost impossible to keep up with all the publications relevant in your field. Although you may be on a notification list for all the important new material, you still have to devote much time to reading and fully processing published research results.

Even if you are staying abreast with current literature, you may find yourself doing a literature search before you are writing an article. Search online databases such as MEDLINE or Pub Med for relevant references. Also ask your colleagues and your advisor for references and reprints, and start reading review articles. They can provide you with great ideas for your introduction and discussion as well as point you to other important references in the field.

An important aspect of obtaining high-quality references in your research field is that these can provide great examples of how successful papers are written. Study them closely. Also note the wording and terminology, especially standard phrases that you could use in your own writing, particularly if English is not your native language.

From the start, organize your references and reprints and compile a reference database in EndNote or in ReferenceManager. These computer programs can save you much time and frustration later on in the writing process, especially when it comes to reformatting references. For more information on references, see also Chapter 8.

ESL advice

7.5 OUTLINING AND COMPOSING A MANUSCRIPT

Study Question and Outlines

GUIDELINE 5:

Pay attention to order and organization.

A first step as you begin the process of writing is to order and organize the information you wish to present. Some people work well from an outline, others do not. Yet other people start to write first and discover and arrange the important points in the process of doing so. Whatever process works for you, you will come to realize that scientific writing requires special attention to order and organization.

Research papers and proposals are divided into sections, and you need to know what information will go into each. Even if you do not normally work from an outline, you may at least want to create a list of the major points to be included in each section before you begin to write.

Each section of a research paper, review paper, or proposal has a set internal structure. These structures are described in detail in subsequent chapters in this book (see Chapters 10–17 for research papers, Chapter 18 for review articles, and Chapters 19–26 for grant proposals). To get started, use an outline of these internal structures. For example, the Introduction section of a research paper contains the known or background information of your topic, the unknown or problem area that your paper aims to explore, the overall purpose of your paper, and a general description of your experimental approach. The known/background information can be further subdivided into general and specific data, leading to the following general topic outline for an Introduction.

Example 7-2   *Topic outline for Introduction*

A. Known/Background
   1. General
   2. Specific reported data
B. Unknown/Problem
C. Research purpose/Question
D. Experimental approach

One of the most important ideas to write down relatively early is the purpose or question of your research. I recommend writing it on a Post-it note and sticking the note onto the side of your computer screen. In this way, you can always recall your research question immediately to remind you what the central focus of your paper should be.

Example 7-3   *Question of study*

Question:	We wanted to determine how long hummingbirds feed their chicks after they leave the nest.

If you decide to go the route of a more detailed outline, fill in the topics of your outline with ideas as they come to you and with material you have collected and researched. An outline can be written in full sentences, using reference citations and a list of references on a separate paper for ease of keeping track later, or it may be written in bullet point form. Be aware, however, that arrangements of data, ideas, and outlines may change; and if you find that something is missing or incomplete, you may have to go back to the laboratory bench.

The following example is an outline for an Introduction.

 Example 7-4   Full sentence outline

Question:

> To determine how changing CO_2 concentrations affect flowering plants ...

A. *Background:*
  1. General: Influence of climate change
     (a) Gas composition of air in past (Liu and Froschauer 1, p. 80; Dohler et al., 2005)
     (b) Gas composition of air has changed due to increasing temperature of atmosphere (Peters et al., 1990)
        i. O_2 concentration has decreased due to climate change (Joh et al., 1999)
        ii. CO_2 concentration has increased due to climate change (Kim et al., 2000)
  2. Specific reported data: Changing O_2 and CO_2 concentration affects plant populations (Honeck et al., 2003)
     (a) Maple fertility decreases with decreasing O_2 concentration (Weinberger & Shewitz, 2000a)
     (b) Maize growth rate and size increases with increasing CO_2 concentrations (Herbert et al., 2002)
B. *Unknown/Problem:*
  Not known how increasing CO_2 concentrations affect the growth of flowering plants
C. *Research Purpose/Question:*
  To determine how changing CO_2 concentrations affect flowering plants
D. *Experimental Approach:*
  We observed and evaluated the growth rate and size of six different flowering plant species under various CO_2 concentrations.

Note that the author of this outline has kept track of reference sources. This approach will come in handy later in the drafting process.

From Data and Ideas to Composition

GUIDELINE 6:
Organize your data and ideas before writing.

Some people will tell you not to write your manuscript until you have collected all of your data. Others will tell you to write as you go—everyday for some time. In all scenarios, the underlying prerequisite to start writing is that you must have accumulated enough data and ideas to start a manuscript.

It helps to have a well-organized laboratory notebook where you record your experiments and references daily. In this notebook, you should record the sources of your materials, the protocols used for your experiments, the raw and derived data, and any references.

When you have enough data, lay them out in front of you. Preparing good figures and tables is essential to a research paper. Therefore, include all the necessary information to make your figures and tables fully informative.

At the same time, write down any ideas as they come to you or as you come across them when looking at your notebook or searching the literature. An efficient way to keep track of your ideas and sources is writing them down as bullet points. Then, arrange and rearrange your bullet points to determine their best possible presentation. If you have an outline, you may also try to fill them into this outline.

An example of this approach is the following collection of information on soil fertility.

Example 7-5   Bullet points

- Fertile soil—rich in nutrients
- Nutrients are necessary for basic plant nutrition including nitrogen, phosphorus, and potassium
- Ideal soil pH is 6.0 to 6.8
- Good soil structure creates well drained soil
- Good soil contains a range of microorganisms that support plant growth
- Good soil contains large amounts of topsoil
- Depletion = if soil fertility components are removed and not replaced
- Soil depletion leads to poor crop yields
- Depletion can be due to intense cultivation and inadequate soil management
- Fertile soil contains sufficient minerals (trace elements) for plant nutrition including boron, chlorine, cobalt, copper, iron, manganese, magnesium, molybdenum, sulfur, and zinc
- Soil depletion is most widespread in tropical zones/ where nutrient content of soils is low
- Depletion often a combination of growing population densities, large-scale industrial logging, slash-and-burn agriculture, and ranching
- Fertile soil contains soil organic matter
- Nutrient rich organic topsoil takes hundreds to thousands of years to build up
- Organic matter improves soil structure and soil moisture retention
- Overtillage damages soil structure
- Overuse of synthetic fertilizers and herbicides leaves buildups that inhibit microorganisms
- Salinization of soil can deplete it

A collection of bullet points, such as shown in the preceding example, can then be grouped in various ways, for example, by topic and subtopics, by pro and con, or chronologically. You as the author will need to be the one to decide which arrangement will best get the main message of the paper across.

In the preceding example, the information can be roughly divided into two main topics, fertile soils and soil depletion, as shown here:

Revised Example 7-5A

Fertile soil
- Fertile soil—rich in nutrients
- Good soil contains a range of microorganisms that support plant growth
- Good soil contains large amounts of topsoil
- Good soil structure creates well drained soil
- Fertile soil contains sufficient minerals (trace elements) for plant nutrition including boron, chlorine, cobalt, copper, iron, manganese, magnesium, molybdenum, sulfur, and zinc
- Fertile soil contains soil organic matter
- Nutrients are necessary for basic plant nutrition including nitrogen, phosphorus, and potassium
- Ideal soil pH is 6.0 to 6.8
- Nutrient rich organic topsoil takes hundreds to thousands of years to build up
- Soil organic matter improves soil structure and soil moisture retention

Soil depletion
- Depletion = if soil fertility components are removed and not replaced
- Soil depletion leads to poor crop yields
- Depletion can be due to intense cultivation and inadequate soil management
- Soil depletion is most widespread in tropical zones/where nutrient content of soils is low
- Depletion often a combination of growing population densities, large-scale industrial logging, slash-and-burn agriculture, and ranching
- Overtillage damages soil structure
- Overuse of synthetic fertilizers and herbicides leaves buildups that inhibit microorganisms
- Salinization of soil can deplete it

Now that we have somewhat more of an order of the information into topics, you may see other ways to subdivide these topics further, as shown in Revised Example 7-5B.

Revised Example 7-5B

Fertile soil
- General information on fertile soil
  - Nutrient rich organic topsoil takes hundreds to thousands of years to build up
  - Good soil structure creates well drained soil
  - Ideal soil pH is 6.0 to 6.8

- Fertile soil content
  - Many nutrients necessary for basic plant nutrition including nitrogen, phosphorus, and potassium
  - Range of microorganisms that support plant growth
  - Large amounts of topsoil
  - Sufficient minerals (trace elements) for plant nutrition including boron, chlorine, cobalt, copper, iron, manganese, magnesium, molybdenum, sulfur, and zinc.
  - Soil organic matter, which improves soil structure and soil moisture retention

Soil depletion
- General information on soil depletion
  - Depletion occurs if soil fertility components are removed and not replaced
  - Most widespread in tropical zones/where nutrient content of soils is low
  - Soil depletion leads to poor crop yields
- Causes of soil depletion
  - Often a combination of growing population densities, large-scale industrial logging, slash-and-burn agriculture, and ranching
  - Can be due to intense cultivation and inadequate soil management
  - Overtillage damages soil structure
  - Overuse of synthetic fertilizers and herbicides leaves build-ups that inhibit microorganisms
  - Salinization of soil

Once your bullets are grouped into their corresponding topics and subsections and have been arranged by importance or chronologically, you can then write up these ideas in paragraph form (see also Chapters 3 and 6 on how to construct good paragraphs and how to consider word and sentence location). At any stage, do not be afraid to add onto or remove information from your lists if needed.

Our example on soil fertility could be composed into the following first draft:

Revised Example 7-5C

Nutrient rich organic topsoil takes hundreds to thousands of years to build up. Its pH range lies ideally between 6.0 and 6.8. Fertile soil contains many nutrients necessary for basic plant nutrition including nitrogen, phosphorus, and potassium. It also contains a range of microorganisms that support plant growth and large amounts of topsoil. In addition, fertile soil contains sufficient minerals (trace elements) for plant nutrition including boron, chlorine, cobalt, copper, iron, manganese, magnesium, molybdenum, sulfur, and zinc. Soil organic matter improves soil structure and soil moisture retention.

Soil depletion occurs if soil fertility components are removed and not replaced. It is most widespread in tropical zones and/or where nutrient content of soils is low. Soil depletion usually leads to poor crop yields.

> Depletion of topsoil is often a combination of growing popula-
> tion densities, large-scale industrial logging, slash-and-burn agri-
> culture, and ranching. Depletion of topsoil can be due to intense
> cultivation and inadequate soil management. Other factors con-
> tributing to soil depletion include overtillage, overuse of synthetic
> fertilizers and herbicides, and salinization of soil.

In this example, the point, "Good soil structure creates well drained soil," has been taken out, as it did not fit into the paragraph well. This omission leaves us with a very short first paragraph for which another sentence or two may be needed to balance it better. Alternatively, the first paragraph could also be combined with the second paragraph, as shown.

After you have written your paragraphs based on the data and information you collected, arrange the resulting paragraphs according to your tentative outline. Sometimes, you may find yourself rearranging your outline. That is okay. Work until you come up with a logical sequence for your paragraphs that ensures a smooth flow of ideas from one paragraph to the next. Subsequent revisions (described in Chapter 16) will serve to fine tune your draft.

7.6   WRITER'S BLOCK?

TO OVERCOME WRITER'S BLOCK:
- Know that you are not alone
- Organize your materials
- Buy yourself an ugly, cheap notebook
- Make your own rules
- Use provided sample sentences as starting points
- Learn to imitate
- Write as if you are explaining your work to a friend
- Write one paragraph at a time
- Set yourself some deadlines
- Involve coauthors
- Percolate

If you have trouble putting your thoughts into words or finishing your manuscript, you may suffer from writer's block. Most people experience writer's block as a temporary condition in which they may not be able to think of a word or have trouble putting their thoughts into a sentence. For some people, this condition can be a serious impairment. Writer's block is probably responsible for thousands of unwritten or unpublished manuscripts. Know that you are not alone.

Although native English speakers also experience writer's block, if you are an ESL author, you most likely are even less confident about writing in a foreign language. Lack of confidence can create a major roadblock—but the problem is not without solution. Here are several suggestions that can help you get over writer's block:

1. Organize your materials—your notebook, your figures and tables, your references.
2. Buy yourself an ugly, cheap notebook. Pretty notebooks often make people feel as if they have to write polished things in them. If you use an ugly notebook, you are better able to be yourself, to doodle, to write down even things that do not make sense. In other words, ugly notebooks let you write down anything that comes to your mind. They also let you try out different phrases, theories, and conclusions.
3. Make your own rules. Do not follow any preset rules on where or how to start writing. Start wherever you want, the abstract, the Conclusion, the Reference List. Then, write what you want, not what you think you should.
4. Use the sample sentence fragments provided in this book as starting points for the different parts of your paper—just add or substitute your data and ideas.
5. Learn to imitate. Look at how other authors in the field have worded their Introduction, Materials and Methods, or Results, but do not copy them word by word. Instead, use their ideas (use citations if you have to, paraphrase, fill in your data).
6. Write as if you are explaining your work to a friend or parent. Be informal and use simple words. Do not write as if you are trying to impress someone in your field.
7. Write one paragraph and one section at a time. Often people feel overwhelmed thinking of all the writing in a paper at once.
8. Set yourself some deadlines or have someone else set them for you. Many people do not get tasks done unless they feel some pressure, be that self-imposed or from a superior. Setting weekly or daily objectives can help to finish tasks. Make writing a priority, and be aware that you probably need to double any estimated time for completing a writing task.
9. Involve your coauthors. Ask them for comments on drafts. Also ask them to help in writing different parts of the manuscript—realize, though, that only one author should put the final version together to ensure consistency in style and format.
10. Percolate. Think about your ideas. Daydreaming is often your brain being hard at work. Ideas and theories often need time to be formed. Spend time letting your thoughts wander (when you are driving to work, when you cannot sleep at night, during a boring seminar).

FOR ESL AUTHORS

GUIDELINE 7:
Do not apply the same principles of style as are used in your own language.

If you are an ESL author, that is, if English is not your first language, choose one of the following methods to write your manuscript:

1. Write the paper in English to the best of your ability (the best method).
2. Write the first draft in your own language and translate it into English yourself.
3. Employ a translator who is familiar with the terminology of your field.

Whichever method you choose, do not apply the same principles of style when you write in English as are used in your language. If you need to, borrow technical phrases such as the ones provided in this book—but never copy whole sentences or paragraphs from articles by English or American scientists in well-edited journals. Instead, learn to imitate the style and wordings of others. (See also Chapter 8 on plagiarism and paraphrasing.)

Ask for help from a scientific colleague, friend, or correspondent whose native language is English. Have a native speaker—preferably a scientist—check any translations. Editors and reviewers will gladly correct minor mistakes, but the English must first be good enough for them to understand what you are trying to say.

7.8   OUTSIDE HELP

GUIDELINE 8:
Hire a scientific editor.
Collect sample phrases for references.
Consult books on scientific/technical writing.

Although editors of journals may correct minor mistakes, they usually will not write or edit an article for you. You have to take care of this problem yourself. There are three solutions to writing and editing your manuscript—all of which should be done, ideally in combination, before you submit your paper to a journal:

1. Hire a scientific editor.
   a. You may ask a scientific editor to write your article outright (know that this may be quite expensive, as this person will have to do much research before being able to write).

b. After completing your manuscript as much as possible, ask a scientific editor to revise it for grammar, spelling, punctuation, style, format, composition, and impact. This person should not be just any editor but ideally an editor *and* a scientist within your field or a related one. These people may be hard to come by but are usually well worth your money. Native-speaking English colleagues make good editors only to a point. They will be invaluable for correcting grammar, spelling, and punctuation; but unless they have done extensive editing and are experts in the field of scientific writing, they often fail when it comes to style, format, or impact. Here, a trained scientific editor can provide expert help.

ESL advice

Start working with an editor early in the writing process, and view the revision process as a joint effort. Agree on the level of editing and on charges before you begin to work with the editor. Ask your peers if they could recommend an editor, or inquire with the American Medical Writers association or the Council of Biology Editors for a list of scientific editors. Some journals also have a list of scientific editors you can approach, or you may find one on the Web by searching under the appropriate keywords, possibly in conjunction with names in the acknowledgement sections of published articles.

Be sure to distinguish between *science editors/writers* and *scientific editors/writers*. Science editors/writers are trained mostly in writing or journalism with an emphasis on science, but they generally do not hold a PhD degree in any scientific field. Scientific editor/writers usually hold a PhD degree in a scientific field and have specialized in editing and writing scientific documents. Ideally, you should look for a scientific editor/writer to help you with a manuscript and especially with any proposal.

2. Collect sample phrases for references. The best source of such sample phrases are articles that have been written in highly known and respected scientific journals by native speakers within the past 5 years. Again, I am not talking about copying entire passages to be placed into your manuscript, but individual sample phrases and expressions that can be applied to writing your research article. For example, performing a polymerase chain reaction (PCR) experiment can be described in only so many ways (see also Chapter 8). When you are unsure how to word such an experiment in English, use your collected sample phrases to see how native speakers in known journals describe such an experiment. Lists of sample phrases are great, as they can be used repeatedly when composing research papers.

ESL advice

3. Consult books on scientific/technical writing for problems of grammar, technical style, good format and composition, and possible sample sentences. This book, for example, presents ample practical sample phrases and wordings for scientific authors and addresses scientific authors within the international scientific community. Another useful book for sample wordings of research papers is *Langenscheidt Scientific English für Mediziner und Naturwissenschaftler* or *Scientific*

ESL advice

English—*L'inglese scientifico per relazioni e conferenze in medicina, biologia e scienze naturali* for German or Italian native speakers. Additional books are described in the References section.

SUMMARY

PREWRITING AND DRAFTING

1. Search journals for the best match of topic, audience, and journal.
2. Obtain *Instructions to Authors* and follow them.
3. Decide on authorship before starting to write.
4. Follow the IMRAD format.
5. Worry about the writing principles in the revisions.
6. Start by writing less.
7. Collect, organize, and study your references.
8. Pay attention to order and organization.
9. Organize your data or ideas before writing.

TO OVERCOME WRITER'S BLOCK

- Know that you are not alone.
- Organize your materials.
- Buy yourself an ugly, cheap notebook.
- Make your own rules.
- Use provided sample sentences as starting points.
- Learn to imitate.
- Write as if you are explaining your work to a friend.
- Write one paragraph at a time.
- Set yourself some deadlines.
- Involve coauthors.
- Percolate.

FOR ESL AUTHORS

Do not apply the same principles of style as are used in your own language.

OUTSIDE HELP

- Hire a scientific editor.
- Collect sample phrases for references.
- Consult books on scientific/technical writing.

PROBLEMS

Problem 7-1 Outline
Use the following bullet points to create a meaningful outline for a few paragraphs on polar ice reduction.

- Sea level is projected to rise between 13 and 94 cm over the next 100 yr

- There is continued climate warming
- Controversial geologic evidence suggests that current polar ice sheets have been eliminated or greatly reduced during previous Pleistocene interglacials
- Modern polar ice sheets have become unstable within the natural range of interglacial climates
- Sea level may have been more than 20 m higher than today during a presumably very warm interglacial about 400 ka during marine isotope stage 11
- Conflicting evidence for warmer conditions and higher sea level during marine isotope stage 11
- Microfossil and isotopic data from marine sediments of the Cariaco Basin support the interpretation that global sea level was 10 to 20 m higher than today during marine isotope stage 11
- The West Antarctic ice sheet and the Greenland ice sheet were absent or greatly reduced during marine isotope stage 11
- Warm marine isotope stage 11 interglacial climate with sea level as high as or above modern sea level lasted for 25 to 30 ky
- Variations in Earth's orbit around the sun are considered to be a primary external force driving glacial–interglacial cycles
- Anthropogenic climate warming will accelerate the natural process toward reduction in polar ice sheets
- Increased rates of sea level rise related to polar ice sheet decay is a potential natural hazard

Problem 7-2 Outline
Use the following bullet points to create a meaningful outline for a few paragraphs on green chemistry: Fischer–Tropsch refining.

- Crude oil is not an infinite resource
- At present, transportation fuels are primarily produced from crude oil
- Liquid fuels can be produced from non-crude-oil carbon sources by direct means, such as direct coal liquefaction, or indirect means, such as biomass gasification followed by hydrocarbon synthesis
- Current transportation sector of the economy is carbon based
- Fischer–Tropsch (FT) synthesis is a key technology for gas-to-liquid (GTL), coal-to-liquid (CTL), and biomass-to-liquid (BTL) conversion
- Assumption: Fischer–Tropsch syncrude can be refined similarly as crude oil
- Syncrude has to be further refined to produce transportation fuels
- No refining technologies have been specifically developed for refining of Fischer–Tropsch syncrude
- Little attention has been paid to the refining of Fischer–Tropsch syncrude

- When syncrude is treated as if it is crude oil, refining becomes inefficient, and it violates green chemistry principles
- Responsible syncrude refinery design is necessary from a commercial and an environmental point of view
- There will be increasing scope for transportation fuel production from other carbon sources
- Product from Fischer–Tropsch synthesis is a synthetic crude oil, syncrude
- There are significant differences between Fischer–Tropsch syncrude and crude oil
- Syncrude refining can be more efficient than crude oil refining (2)

Problem 7-3 Constructing a Paragraph
Compose a short passage using the following bullet points.

- The active ingredient in most chemical-based mosquito repellents is DEET (N,N-diethyl-meta-toluamide)
- Recent research suggests that DEET products, used sparingly for brief periods, are relatively safe (2)
- DEET is absorbed readily into the skin and should be used with caution
- Common side effects to DEET-based products: rash, swelling, itching, and eye-irritation, often due to overapplication (4)
- DEET alternatives: eucalyptus oil containing cineol, 1:5 parts diluted garlic juice, soybean oil, neem oil, marigolds, Avon Skin-so-soft bath oil mixed 1:1 with rubbing alcohol, juice and extract of Thai lemon grass (7,8)
- DEET was developed by the U.S. military in the 1940s
- Some research points to toxic encephalopathy associated with use of DEET insect repellents (5)
- True citronella has not been proven to be a very effective mosquito repellent (3)

See Chapters 3 and 6 for additional problems on paragraph constructions.

CHAPTER 8

References and Plagiarism

8.1 ABOUT REFERENCES

Most of the time, your scientific findings will build on previous studies. Although direct quotations are rarely used in scientific writing, paraphrased versions of source material are very common. In all cases, whenever you use the ideas and findings of others, the source needs to be cited in the text and listed in a Reference List at the end of a research article. References not only give appropriate credit to the contributions of others but also direct readers who want further information to other literature of interest. In addition, references provide editors with a list of potential reviewers and show how familiar you are in your area of specialty.

8.2 SELECTING REFERENCES

REFERENCE GUIDELINE 1:
Select the most relevant references.

REFERENCE GUIDELINE 2:
Verify your references against the original document.

Most authors identify more than enough references for their manuscripts. Potentially relevant references can easily be obtained by searching online databases such as MEDLINE®, SCOPUS®, BIOSIS, and the Web of Science using, for example, keywords, author, source, author's affiliation, cited author, cited work, cited year.

Top science databases include:

MEDLINE®	MEDLINE, produced by the National Library of Medicine, covers more than 4,000 journal titles and is international in scope. Broad coverage includes basic biomedical research and clinical sciences.
SCOPUS™	SCOPUS provides broad international coverage of the sciences and social sciences, indexing 14,000 journals.
CINAHL	CINAHL covers English-language journals in nursing and allied health fields, indexing 1,200 journals.
PsycINFO®	The PsycINFO database covers psychology and related disciplines. Coverage is worldwide, indexing over 1,300 journals and dissertations.
BIOSIS	BIOSIS, the online version of Biological Abstracts and Biological Abstracts-Reports, Reviews, Meetings contains literature references from all of the life sciences. This is the premier database for coverage of botany research.
Web of Science	The ISI Citation Databases collectively index more than 8,000 peer-reviewed journals. It provides web access to Science Citation Index Expanded, which covers 6,300 international science and engineering journals.
Current Contents®	Current Contents Science Edition covers all the Science editions of the Current Contents Search® database in one package.

The more difficult part is to separate immediately relevant references from all the rest. Unlike review articles, which contain many references because they have to cover extensive information, research papers should contain only immediately relevant references and should use a minimal number of references so as not to overwhelm the reader (20–40 on average).

The most relevant references consist of the most significant and the most available references. Such references are generally journal articles followed by books and PhD theses. Note that PhD theses are typically not widely available. Similarly, abstracts for meetings, conference proceedings, personal communications, and unpublished data may be cited in the text in parenthesis, but these references are usually not listed in the Reference List at the end of a paper because they are not considered relevant or available. Thus, they should not be used to draw any strong conclusions, only to support findings.

To keep the number of references low, cite original articles and select the most important, the most elegant, or the most recent paper on a subject rather than listing any and all papers published on the topic. To validate specific findings, use a primary source, which is the original, peer-reviewed publication of a scientist's new data, results, and theories. For general overview of a topic, you may also use secondary or tertiary sources. A secondary source cites, builds on, discusses, or generalizes primary sources, such as review articles. A tertiary source generalizes and analyzes primary sources while attempting to provide a broad overview of a topic such as in textbooks (see also Chapter 18 Section 18.2).

When you write a research paper, review articles, or grant proposals, consider citing review articles where possible. Know, however, that review articles sometimes contain faulty references and misrepresent or misinterpret data. Always verify the original article if you are citing a specific piece of information from a review article or tertiary source.

Verifying References

The citations within the text, the Reference List, and the information you cite must be accurate. References tend to have a surprisingly high rate of error. Therefore, you need to verify them against the original document. Make sure you have read all references you cite to prevent false representation of the reference or the information within. In addition, ensure that every reference in the text is included in the Reference List and every reference in the Reference List is cited in the text. Ensure also that citations and references follow the required format as described in the *Instructions to Authors*.

If you quote directly from a published paper, use quotation marks and check that every word and punctuation mark is exactly as it was in the original. If you paraphrase the ideas, make sure that your statements do not misrepresent the authors' ideas (see also Section 8.5).

8.3   MANAGING REFERENCES

REFERENCE GUIDELINE 3:
Manage your references well.

Computer programs that help you manage your references are very useful. These include, but are not limited to, EndNote or ReferenceManager. I strongly recommend using one of these programs to help organize, keep track of, and format your references. You can save yourself much time and much frustration if you manage your references from the start. There are few aspects of preparing a manuscript that are more irritating than painstakingly typing, changing, or correcting the Reference List. Start using a reference managing program right when you download your references from the library. If you are unfamiliar with such a program, inquire at your library. Most libraries offer short classes on reference programs.

8.4   TEXT CITATIONS

Form and Order

REFERENCE GUIDELINE 4:
Follow the journal's style for details in the reference citation.

REFERENCE GUIDELINE 5:

Cite references in the correct form and order.

References are listed in two formats in a paper: as text citations and in the Reference List. Text citations list references within the text in short version, such as by name and year. The Reference List at the end of a paper displays the full citation of your reference.

In the text as well as in the Reference List, references can be cited in different ways depending on the journal. Check the *Instructions to Authors* of your target journal for information on how to cite references. Follow all the details in these instructions for both text citations and the Reference List.

Some common formats for text citations are "(author, year)," "(number)," and "[number]." Two examples of text citations are shown in Example 8-1a and 8-1b.

👍	**Example 8-1a**	Vit-E, a fat-soluble vitamin, requires micellar formation for absorption and is transported in the intact animal via lymphatics **(Hollander et al. 1976)**.
👍	**Example 8-1b**	Vit-E, a fat-soluble vitamin, requires micellar formation for absorption and is transported in the intact animal via lymphatics **(8)**.

If you cite multiple references for a point in your text, list the references in the text and in the Reference List in chronological order. If your text citations are by name and year, cite multiple references chronologically in the text and by alphabetical order in the Reference List. If your text citations are by number, list multiple references in the text and in the Reference List in numerical order.

Generally, for a paper by one author, cite that author's name:

👍	**Example 8-2a**	... described by Popi (18, 20).

For a paper by two authors, cite both authors' names:

👍	**Example 8-2b**	Daniles and Ebert (9) reported XYZ.

For a paper by three or more authors, cite the first author's name followed by "et al."

👍	**Example 8-2c**	... has previously been reported (Brown et al., 1999a; Brown et al., 1999b; Liu et al., 2003).

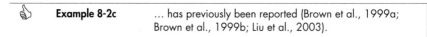

(Note that citation rules may differ depending on the style of your target journal. For example, if your target journal uses APA style, "et al." is only used for citing articles of six or more authors in text citations. Thus, be sure to check the *Instructions to Authors* as well as other articles published in your target journal carefully.)

If you cite abstracts for meetings, conference proceedings, personal communications, or unpublished data in the text, you should give a publication status in parenthesis. Following are a few sample wordings for indicating publication status:

(manuscript submitted)	(manuscript in preparation)
(unpublished data)	(data not shown)
(personal communication)	(manuscript in press)

Placement of Citations within the Text

REFERENCE GUIDELINE 6:

Know where to place references in a sentence.

References can be incorporated into the text in various ways, but incorporation of references should be meaningful and specific. In general, references can be placed either after the idea you are referring to or after the names of the authors depending on what you want to emphasize: the science or the scientist.

Example 8-3	a	Starfish fertilization is species specific **(17)**.
	b	Peterson **(17)** reported that starfish fertilization is species specific.

Do not place references in the middle of an idea or after general information of a study, such as after "in a recent study" or "has been reported."

Example 8-4	In a current study **(16)**, proteins expressed by the Epstein–Barr virus have been classified into three general groups according to their kinetics and synthesis requirements.
Revised Example 8-4	In a current study, proteins expressed by the Epstein–Barr virus have been classified into three general groups according to their kinetics and synthesis requirements **(16)**.

References do not necessarily have to be placed at the end of a sentence, however. To distinguish between your work and the work of others, for example, put the citation after the corresponding finding, not after your work.

Example 8-5   The Mo+PyF101 M-MuLV used in this study was obtained from the supernatant of the 25-3 cell line confluently infected with the virus **(3)**.

Revised
Example 8-5   The Mo+PyF101 M-MuLV **(3)** used in this study was obtained from the supernatant of the 25-3 cell line confluently infected with the virus.

Also note that references for different points in one sentence have to be cited after the appropriate point rather than grouping all the references together at the end of the sentence.

Example 8-6   Two methods of tailoring the adhesion properties of thermoset stamps exist: One approach is plasma-deposited fluorinated film and coating with a self-assembled monolayer (SAM) of fluoroalkyltrichlorosilane; the other approach is using C4F8 as a feed gas to deposit Teflon-like antisticking layers and form the SAM layer from the gas phase <u>(3, 4)</u>.

Revised
Example 8-6   Two methods of tailoring the adhesion properties of thermoset stamps exist: One approach is plasma-deposited fluorinated film and coating with a self-assembled monolayer (SAM) of fluoroalkyltrichlorosilane **(3)**; the other approach is using C4F8 as a feed gas to deposit Teflon-like antisticking layers and form the SAM layer from the gas phase **(4)**.

Purpose for Using References

As a scientific author, you will use sources for varying purposes: to support, refute, compare with, or highlight your own findings or ideas.

Table 8.1 provides you with sample wording when using references for various purposes.

Tone and Style

At all times, be professional and courteous when citing other's work. Above all, do not insult the author(s) of previous studies.

Example 8-7   a   Unbelievably, Brown failed to consider …

b   The study by Aday is without merit.

c   Clearly, Chumsky's theory is wrong.

Insults will get you nowhere, and you run the risk of having your work rejected if you write nonprofessionally and annoy your editor or reviewers. Remain objective and neutral as in the next examples:

Revised Example 8-7	a	Different theories on this topic exist (27–29).	
	b	Our study differs from that of Aday (35) in that …	
	c	X has been the topic of much controversy (4,5,8).	

Table 8.1 Use of Citation

	TO SUPPORT YOUR FINDINGS (OR THOSE OF OTHERS)	TO REFUTE FINDINGS OF OTHERS	TO COMPARE YOUR FINDINGS WITH OTHERS	TO HIGHLIGHT YOUR FINDINGS
Sample wording	Our results are consistent with those reported previously (5, 8, 10–14).	Different findings have been reported previously (33).	Our findings are comparable to those of Vignanery's (9).	Unlike other previous findings (23–25), our work presents …
	Similar results have also been observed by Alton et al. (11).	Our findings show that the previously proposed theory A (17) is not supported by …	When compared to recent findings described by (8), our study shows …	Our study expands on work of Hui's (33), which reported Y, and highlights the importance of …
	Mayer and Bims also determined … (Here we confirm Lohn's theory of …)	Not all studies agree on this finding. For example, Peters et al. determined … (5).	Although Dauh et al. reported recently XXX (6), our study also takes into consideration …	Our study sets itself apart from other studies (33–35) in that …

8.5   PLAGIARISM

REFERENCE GUIDELINE 7:

Ensure that you are not plagiarizing.

Failing to indicate the source of information in scholarly scientific work is called plagiarism and is a form of academic misconduct. You are obligated, as an ethical obligation to other writers and as a defense for yourself, to acknowledge all borrowings you take from other sources, even if you do not copy the exact words used in the original.

To avoid plagiarism, you need to know what constitutes it. Plagiarism includes:

- Quoting material without acknowledging the source. (This is the most obvious kind of plagiarism.)
- Borrowing someone else's ideas, concepts, results, and conclusions and passing them off as your own without acknowledging them—even if these ideas have been substantially reworded.

- Summarizing and paraphrasing another's work without acknowledging the source.

The rules apply to both textual and visual information.

If you are using the World Wide Web as a source of information, you must also cite that source. In many cases, it might be a good idea to obtain permission from the Web site's owner before using graphics or text.

Know that you do not have to document facts that are considered common knowledge. Common knowledge is information that can be found in numerous places and is likely to be known by a lot of people such as the information found in Example 8-8.

Example 8-8   A vast number of endemic species exists on the Galapagos Islands.

However, information that is not generally known and ideas that interpret facts have to be referenced as in Example 8-9.

Example 8-9   Based on a recent study, the blue iguanas of the Grand Cayman Islands are an endangered species (9).

The finding that "blue iguanas of the Grand Cayman Islands are an endangered species" is not a fact but an *interpretation*. Consequently, you need to cite your source.

Following are some other examples of common knowledge that do not require a citation:

Example 8-10 a   Because nitrogen (N) is an essential component of proteins, enzymes, and other biologically important compounds, the natural cycling of N in freshwater ecosystems needs to be studied in detail.

b   Historical eruptions are almost always preceded and accompanied by "volcanic unrest," providing early warning of a possible impending eruption.

Statements that contain information and interpretations that result through the work of others need to be cited, as shown in the next examples.

Example 8-11   a   While low nitrogen (N) availability can limit primary productivity, excess N has been linked to eutrophication and health concerns (19).

b   The eruption of Mount Pinatubo in 1991 was preceded by a relatively short progression of precursory activity before its full-blown eruption (22).

Know that imitation and borrowing by themselves are not plagiarism. Drawing on other people's ideas is perfectly reasonable and in fact unavoidable when you write academic documents—but you must acknowledge the source.

8.6   PARAPHRASING

REFERENCE GUIDELINE 8:
Keep track of ideas and references.

REFERENCE GUIDELINE 9:
Know how to paraphrase.

Paraphrasing means taking another person's ideas and putting those ideas in your own words. This is the skill you will probably use most when incorporating sources into your writing. Although you use your own words to paraphrase, you must still cite the source of the information in the text at the end of the sentence or idea.

It is important that you distinguish between paraphrasing and plagiarizing. Changing a word or two in someone else's sentence or changing the sentence structure while using the original words is not paraphrasing but plagiarizing. The best way to avoid plagiarism is to do the following when collecting and using information in scientific writing:

- Keep track of references and write down the information you intend to use whenever you come across a passage that you think may be useful for your document.
- Keep a detailed list of sources in a reference managing program: for example, in Endnote or Reference Manager.
- If you copy something word by word, put it in quotation marks, but know that writing in the sciences uses direct quotations only rarely. When you want to use details from the original but not necessarily all of them and not necessarily in the same order as the original, you need to paraphrase.

- Write down the most important ideas in your own words using bullet points.
- Take notes with the book closed. This way you are forced to put the ideas into your own words.
- Double-check that the reference and information is correct by going back to the original when you compose your document.

Following is an example of a paragraph that instead of being paraphrased has been plagiarized:

Example 8-12   **Plagiarized paragraph**

Original:
Healthy older adults often experience mild decline in some areas of cognition. The most prominent cognitive deficits of normal aging include forgetfulness, vulnerability to distraction and other types of interference, as well as impairments in multi-tasking and mental flexibility. These cognitive functions are the domain of the prefrontal cortex, the most highly evolved part of the human brain. Prefrontal cortical cognitive abilities begin to weaken even in middle age, and are especially impaired when we are stressed. Understanding how the prefrontal cortex changes with age is a top priority for rescuing the memory and attention functions we need to survive in our fast-paced, complex culture.

Plagiarized sample:
In healthy older adults often some areas of cognition decline. The most noticeable cognitive declines of normal aging include forgetfulness, vulnerability to distraction, and problems in multi-tasking. These cognitive tasks are localized in the prefrontal cortex, which is the most highly evolved portion of the brain. Already in middle age prefrontal cortical cognitive functions start to decrease. Such functions are also particularly affected during any type of stress. Studying memory and attention is important to understand how the prefrontal cortex changes with age. It is particularly important to understand these changes in our current fast-paced life style.

The preceding passage is considered plagiarism for two reasons: (a) the writer has only changed around a few words and phrases or changed the order of the original's sentences, and (b) the writer has failed to cite a source for any of the ideas or facts.

An acceptable way of paraphrasing the preceding sample paragraph would be the following:

Revised
Example 8-12

Paraphrased sample:
A recent study shows that the process of aging is accompanied by a decline in cognitive abilities, deficits in working memory, and compromised integrity of neural circuitry in the brain (3). Impairments in these executive operations compromise the regulation of thought, emotion, and behavior, and ultimately jeopardize independence and quality of life. Elucidating mechanisms underlying the decline of neural circuitry is critical for the treatment of debilitating cognitive deficits, such as those found in the elderly.

This is acceptable paraphrasing because the writer accurately relays the information in the original using his or her own words. The writer also lets the reader know the source of the information.

Here is another example:

Example 8-13

Plagiarized paragraph

Original:
It has currently been recognized that both the type and characteristics of the rust layers formed on the steel surfaces are very important because they can determine their protective properties. According to a recent theoretical model developed by Hoerlé et al. (2), the long-term corrosion behavior of iron exposed to wet–dry cycles is largely controlled by the characteristics of the rust layers. Additionally, the differences between the corrosion behavior observed for different types of steels have been related to the rust layer characteristics. Okada et al. (8) have carefully investigated, by using detailed variable temperature Mössbauer spectrometry, the protective rust formed on both weathering and mild steels after 35 years of exposure to a Japanese semirural type atmosphere. They reported that the rust on both steels is composed of goethite (major component), lepidocrocite (minor component) and traces of magnetite.

(With permission from Elsevier)

Plagiarized sample:
Both the type and characteristics of the rust layers formed on the steel surfaces are very important because they determine their protective properties. Recently, Hoerlé et al. developed a theoretical model (2) that the long-term corrosion behavior of iron exposed to wet–dry cycles is largely controlled by the characteristics of the rust layers. The differences between the corrosion behavior observed for different types of steels have also been related to the rust layer characteristics. Using detailed variable temperature Mössbauer spectrometry, the protective rust formed on both weathering and mild steels after 35 years of exposure to a Japanese semirural type atmosphere have been determined by Okada et al. (8) who reported that the rust on both steels is composed of goethite (major component), lepidocrocite (minor component) and traces of magnetite.

In this example, although the author cited the sources, the passage is plagiarized because the writer has only changed around a few words and
phrases. An acceptable, paraphrased version of the same passage would
be the following:

Revised
Example 8-13

Paraphrased sample:
Rust constituents can determine the performance of
steels and influence their life expectancy. The characteristics of the rust layers control the long term corrosion behavior of iron exposed to wet–dry cycles (2).
For example, after 35 years of exposure in Japan, rust
formed on steels under various conditions is composed
of mainly goethite, some lepidocrocite, and traces of
magnetite (8).

Unlike elsewhere in a scientific research paper, many portions of the
Materials and Methods section will sound extremely similar to each other,
mainly because there are only so many ways one can describe procedures
whose technique and setup is essentially identical with the exception of
the variables. Using very similar phrases in such passages, and substituting your variables, would not be considered plagiarism. Therefore, do not
desperately try to invent new wordings to describe the same procedure.

Here are some examples of passages that would not be considered
plagiarized:

Example 8-14

Method description in paper A:
Real-time fluorescence quantitative PCR was performed
in an Applied Biosystems Prism 7000 instrument in
the reactions containing an Applied Biosystems SYBR
green master mix reagent and oligonucleotide pairs
to the endogenous control gene 'A' and cDNA of 'B'.
The reagents were denatured at 95 °C for 10 min,
followed by 40 cycles of 15 s at 95 °C and 60 s at
60 °C. The primer sequences (5'–3') were 'A' forward
5'-GACACCTATGCCGAACCGTGAA-3'; 'A' reverse,
5'-CTGAGTATCAGTCGGCCTTGAA-3'; 'B' forward
5'-GTTCGACGACATCAACATCA-3'; 'B' reverse
5'-TGATGACGTCCTTCTCCATG-3'.

Method description in paper B:
PCR amplification of 'X' sequences was done using the
GC RICH PCR System (Roche, Mannheim, Germany).
All non-'X' sequences were amplified using Taq DNA
polymerase (Promega, Madison, WI). Primers were
designed using published sequences for 'X-1' (GenBank:
Xxxxxx) and 'x-13' (GenBank: Xxxxxx) (Table 1). PCR
thermal cycling conditions were: 2 min at 50 °C, 10
min at 95 °C, followed by 40 cycles of 15 s at 95 °C
and 1 min at 60 °C. PCR reactions were run with molecular weight standards on 0.8% agarose gels containing
ethidium bromide and visualized by UV light. The primers used were: 5'-GGCTCACCAGCATCATATACG-3'
and 5'-GGCTACAATGACGACGTCA-3'.

Method description in paper C:
Real-time PCR were performed using the TaKaRa SYBR PCR kit and ABI Prism 7000 sequence detection system according to the manufacturer's specifications. The primers for amplification were *abc* (5'-CGCTCCTCTG-CATCTAATCAG-3' and 5'-GACACTTAGCACGCACT-CA-3') and *def* (5'-GCATCTTCAAGTAAGGACTATC-3' and 5'-GACTTTCACAGTACCAGATT-3'). Total reaction volume was 50 µl including 25 µl SYBR Premix Ex Taq with SYBR Green I, 300 nM forward and reverse primers and 2 µl cDNA. The thermal cycler program was 1 cycle at 95 °C for 10 s, followed by 40 cycles at 95 °C for 5 s and 60 °C for 30 s. The PCR products were detected by electrophoresis through a 2% agarose gel stained with ethidium bromide.

For these passages and for similar ones that occur mainly in the Materials and Methods section of a research paper, it may not be a bad idea to collect sample phrases from other articles for your reference. Know though that I am not advising you to copy entire passages to be placed into your manuscript but individual sample phrases and expressions that can be applied to writing your research article.

8.7   REFERENCES WITHIN A SCIENTIFIC PAPER

REFERENCE GUIDELINE 10:

Know where to place references in a scientific paper.

Abstract	Avoid placing any citations in the Abstract. Start citing sources in your Introduction.
Introduction	Cite the most relevant references only. Although the amount of background information needed depends on the audience, do not review the literature. Limit yourself to the most recent, the most important or original, and the most elegant references. Consider citing review articles when possible.
Materials and Methods	Cite original references for Materials or Methods used in your study. Include references of methods published in widely accessible journals instead of repeating details of those methods. (For example: "Growth was measured and analyzed according to Billings (1988).")
Results	Usually statements that need to be referenced, such as comparisons with previous reports, are not written in the Results section. These statements are made in the

Discussion

Discussion section. If a short comparison does not fit smoothly into the Discussion, however, it may be included in the Results section, and then it needs to be referenced. In the Discussion, although your findings are the main topic, your results have to be discussed in a broad context. That means that you have to include references to compare and contrast your findings, studies that provide explanations, or those that give your findings some importance.

8.8   THE REFERENCE LIST

Listing References

REFERENCE GUIDELINE 11:

Follow the journal's style for details in the reference list.

Your Reference List at the end of your paper should contain a list of only the literature cited in the text. Usually, abstracts for meetings, conference proceedings, personal communications, and unpublished data are not available to the public and are therefore not listed in the Reference List, even if they were used as citations in the text. If the work has been submitted and accepted for publication, but has not yet been printed, list it in the Reference List with author name and title followed by the journal name (and volume if known), and the term "(in press)." If a paper has been submitted to a journal but has not been accepted, it may or may not get listed in the Reference List. If you list such work, use the author name, year of the version, and title followed by the term "Manuscript submitted for publication."

Most journals maintain individual styles for their references. Be sure to check and follow the reference style of your target journal listed in the *Instructions to Authors* very carefully. Reference styles vary on such details as to whether titles of articles are included, whether last page numbers are included, where authors' initials are placed (before or after the last name), where the year of publication is placed (after the authors' names, after the journal title, at the end of the reference), and how items are punctuated. Computer programs that put references in various styles are available (EndNote or Reference Manager) and very useful. Note that in the Reference List names are inverted: Last name, then first name or initial, then middle initial. This custom differs from that used in some foreign countries.

Typically, if you used the "(author, year)" system in the text, references are listed in alphabetical order and are not numbered in the Reference List.

ESL advice

Example 8-15a   Bailey, S.E., Olin, T.J., Bricka, R.M. and Adrian, D.D. (1999). A review of potentially low-cost sorbents for heavy metals. Water Res. **33**:2469–2479.

Das, N.C. and Bandyopadhyay, M. (1992). Removal of copper(II) using vermiculite. Water Environ. Res. **64**:852–857.
Hani, H. (1990). The analysis of inorganic pollutants in soil with special regard to their bioavailability. J. Environ. Anal. Chem. **39**:197–208.
Lackovic, K., Angove, M.J., Wells, J.D. and Johnson, B.B. (2004). Modelling the adsorption of Cd(II) onto goethite in the presence of citric acid. J. Colloid Interface Sci. **269**:37–45.

If you used the (number) system in the text, references in the list are numbered in the order in which each reference is first cited in the text.

Example 8-15b

1. Lackovic, K., Angove, M.J. Wells, J.D. and Johnson, B.B. (2004). Modelling the adsorption of Cd(II) onto goethite in the presence of citric acid. J. Colloid Interface Sci. **269**:37–45.
2. Bailey, S.E., Olin, T.J., Bricka, R.M., and Adrian, D.D. (1999). A review of potentially low-cost sorbents for heavy metals. Water Res. **33**:2469–2479.
3. Hani, H. (1990). The analysis of inorganic pollutants in soil with special regard to their bioavailability. J. Environ. Anal. Chem. **39**:197–208.
4. Das, N.C. and Bandyopadhyay, M. (1992). Removal of copper(II) using vermiculite. Water Environ. Res. **64**:852–857.

References to books or book chapters also require great attention to detail. Here too, there are various formats depending on your target journal.

Example 8-16

Ege, Seyhan N. (1984). Organic Chemistry. D.C. Heath and Company, Lexington, Massachusetts/Toronto. pp. 203–229.

General Conventions

Generally, when you list references in the Reference List at the end of a paper, journal names are abbreviated in the listing. (Note that this depends on the style preference of your target journal; if APA style is used for example, then journals listed in the Reference List are not abbreviated.) Journal abbreviations have been standardized for most journals according to the American National Standards Institute. The most common journal title word abbreviations are listed in Table 8.2.

Other abbreviations can usually be obtained from listings in databases or from other reference lists.

8.9 COMMON REFERENCE STYLES

If no citation instructions are given, consider using one of the more commonly accepted formats, such as the Modern Language Association (MLA), Council of Scientific Editors (CSE; formerly CBE), or American Psychological Association (APA) style.

Table 8.2 Journal abbreviations

JOURNAL NAME	ABBREVIATION
Abstracts	Abstr.
Academy	Acad.
American	Am.
Annals	Ann.
Applied	Appl.
Archives	Arch.
Botany, Botanical	Bot.
Chemical, Chemistry	Chem.
Clinical	Clin.
Current	Curr.
European	Eur.
International	Int.
Internal	Intern.
Journal	"J." or "J"
Medical, Medicine	Med.
National	Natl.
Natural, Nature	Nat.
Research	Res.
Science, Scientific	Sci.
Society	Soc.
Technical	Tech.
United States	U.S.
"-ology" words ("Biological")	"-l." ("biol.")
One word titles ("Cell")	Not abbreviated

👍   **Example 8-17**   **MLA Style**

In Text   … as reported previously (1).

Bibliography   **Books**
Okuda, Michael, and Denise Okuda. Star Trek Chronology: The History of the Future. New York: Pocket, 1993.

Journal Article
Jefferson, Thomas A., Stacey, Pam J., and Baird, Robin W. "A review of Killer Whale interactions with other marine mammals: predation to co-existence." Mammal Review 21.4 (2008):151–80.

	Example 8-18	CSE Style	
		In-Text	...observed by Hinter (2008) (McCormac and Kennedy 2004) (Meise et al. 2003) (Hinter 2008)
		Bibliography	**Books** McCormac JS, Kennedy G. 2004. Birds of Ohio. Auburn (WA): Lone Pine. p. 77–78.
			Journal Article Meise CJ, Johnson DL, Stehlik LL, Manderson J, Shaheen P. 2003. Growth rates of juvenile Winter Flounder under varying environmental conditions. Trans Am Fish Soc 132(2):225–345.

	Example 8-19	APA Style	
		In Text	... as reported by Juls (1999). Research by Wegener and Petty (1994) supports ... (Juls, 1999) (Wegener & Petty, 1994) (Kernis et al., 1993)
		Bibliography	**Books** Calfee, R. C., & Valencia, R. R. (1991). *APA guide to preparing manuscripts for journal publication.* Washington, DC: American Psychological Association. **Journal Article** Scruton, R. (1996). The eclipse of listening. *The New Criterion, 15*(30), 5–13.

Online Sources for Common Reference Styles

Certain online sources are a great resource for various scientific style guides (last accessed October 2009).

http://www.liu.edu/cwis/cwp/library/workshop/citation.htm
http://library.osu.edu/sites/guides/csegd.php#articlesone
http://owl.english.purdue.edu/owl/resource/560/02/
http://www.bedfordstmartins.com/online/citex.html
http://bailiwick.lib.uiowa.edu/journalism/cite.html

8.10   CITING THE INTERNET

REFERENCE GUIDELINE 12:

Know how to cite and list references from the Internet.

Before you cite sources form the Internet, you should consult the *Instructions to Authors* and recent issues of your target journal. Use of Web citations is not always accepted, but this is a developing area, so check your target journal's policy early. Generally, you as the author can decide which style to choose. However, make sure that the style does not conflict with that asked for by your target journal.

To cite and list a reference from the Internet or the World Wide Web, use the following form:

Example 8-20

Author's name (last name first) Title. Available from: URL: http://Internet address or World Wide Web address.

There are also other good reference styles for citing Internet sources such as the MLA, CSE (formerly CBE), or Chicago styles.

Example 8-21

MLA Style

Online document
Author's name (last name first). Document title. Date of Internet publication. Date of access <URL>.

Book
Bryant, Peter J. "The Age of Mammals." *Biodiversity and Conservation.* 28 Aug. 1999. 20 Oct. 2009 <http://darwin.bio.uci.edu/~sustain/bio65/lec02/b65lec02.htm>.

Article in an electronic journal (ejournal)
Joyce, Michael. "On the Birthday of the Stranger (in Memory of John Hawkes)." *Evergreen Review* 5 Mar. 1999. 20 Oct. 2009 <http://www.evergreenreview.com/102/evexcite/joyce/nojoyce.html>.

Example 8-22

CSE Style

To document a file available for viewing and downloading via the **World Wide Web**, provide the following information:
  ○ Author's name (if known)
  ○ Date of publication or last revision
  ○ Title of document
  ○ Title of complete work (if relevant)
  ○ URL, in angle brackets
  ○ Date of access

Book
Bryant P. 1999 Aug 28. Biodiversity and conservation. <http://darwin.bio.uci.edu/~sustain/bio65/index.html>. Accessed Oct. 20, 2009.

Article in an electronic journal (ejournal)
Browning T. 1997. Embedded visuals: student design in Web spaces. Kairos: A Journal for Teachers of Writing in Webbed Environments 3(1). <http://english.ttu.edu/kairos/2.1/features/browning/bridge.html>. Accessed 2009 Oct. 20.

Example 8-23   **Chicago Style**

To document a file available for viewing and downloading via the **World Wide Web**, provide the following information:

- Author's name
- Title of document, in quotation marks
- Title of complete work (if relevant), in italics or underlined
- Date of publication or last revision
- URL, in angle brackets
- Date of access, in parentheses

Book   Peter J. Bryant, "The Age of Mammals," in *Biodiversity and Conservation* August 1999. <http://darwin.bio.uci.edu/~sustain/bio65/index.html> (October 20, 2009).

Article in an electronic journal (ejournal)
Tonya Browning, "Embedded Visuals: Student Design in Web Spaces," *Kairos: A Journal for Teachers of Writing in Webbed Environments* 3, no. 1 (1997), <http://english.ttu.edu/kairos/2.1/features/browning/index.html> (October 20, 2009).

8.11   FOOTNOTES AND ENDNOTES

Journals usually do not like to see footnotes or endnotes because they are often expensive to reproduce. Exceptions are footnotes for figures or tables. If explanatory notes in the text still prove necessary to your document, insert a number formatted in superscript following a punctuation mark or the name of the author.

Example 8-24   Solange Brault[2] and Hal Caswell[2]

List all footnotes on the bottom of the same page of your document on which your superscript appeared.

Example 8-25   [2]Order of authorship determined by toss of a coin.

8.12   ACKNOWLEDGMENTS

ACKNOWLEDGMENTS GUIDELINE:

List all the people whose help was important
but not enough to warrant authorship.

General

Acknowledge any organizations or individuals who provided grants, materials, financial or technical assistance. If possible, specify the type of support you received. Also acknowledge people who contributed ideas, information, critical writing or editing, and advice to your work.

Thank everyone who went out of their way to help you, but do not acknowledge those who did no more than their routine work. People named in the acknowledgments should give their permission to be named and should approve of the wording of your acknowledgment. This can save friendships, avoid embarrassing situations, and ensure future collaborations.

Typically, start by listing intellectual contributions, then move on to technical support, provision of materials, helpful discussions, and revisions and preparations of the manuscript. Last, list any funds, grants, fellowships, or financial contributions.

 Example 8-26 a   We thank Dr. J. Holzheimer for his technical advice on crystallization assays. We are also grateful to Drs. Thomas Hugh and Fred Grant for their critical review of the manuscript. This Study was supported by grant # XXX from the National Institutes of Health.

            b   The authors thank Brad Burdick who went out of his way in collecting the HPLC data. The laboratory of Dr. F. Verus provided invaluable data and technical support. Dr. Peter Gress, Belinda Gross, and Paul Sepega contributed immensely to the preparation of the manuscript. This work was made possible in part by a grant from the ABC Foundation. The authors have no conflicting financial interests.

Competing Interest Statement

Some acknowledgment sections contain a statement about conflict of interests as a last sentence in the paragraph. Conflict of interest exists when an author (or the author's institution), reviewer, or editor has financial or personal relationships that inappropriately influence (bias) his or her actions. Such relationships are also known as dual commitments, competing interests, or competing loyalties. These relationships vary from those with negligible potential to those with great potential to influence judgment; but not all relationships represent true conflict of interest. Conflicts can occur for other reasons, such as scientific judgment, financial relationships, personal relationships, academic competition, and intellectual passion. All participants in the peer review and publication process must disclose all relationships that could be viewed as presenting a potential conflict of interest. Disclosure of these relationships is particularly important in connection with editorials and review articles, because it can be more difficult to detect bias in these types of publications than in reports of original research. Conflict of interest is usually mentioned in the Acknowledgment section or in a separate section just before or after the Acknowledgments.

Possible ways to phrase a conflict of interest statement include:

 Example 8-27   a) The authors declare no competing financial interests.

               b) No potential conflict of interest relevant to this article was reported.

     c) Dr. A. reports serving as a consultant to C. Clinic,
       and serving as a consultant for a patent-infringement
       case involving U.S. patent xx/yyy,zzz, in the treat-
       ment of …
     d) Dr. B reports receiving lecture fees from X and grant
       support from Y. No other potential conflict of interest
       relevant to this article was reported.
     e) Y University owns a patent, xx/yyy,zzz, that uses
       the approach outlined in this article and which has
       been licensed to Z.
     f) The authors have nothing to disclose.

SUMMARY

REFERENCE GUIDELINES:

1. Select the most relevant references.
2. Verify your references against the original document.
3. Manage your references well.
4. Follow the journal's style for details in the reference citation.
5. Cite references in the correct form and order.
6. Know where to place references in a sentence.
7. Ensure that you are not plagiarizing.
8. Keep track of ideas and references.
9. Know how to paraphrase.
10. Know where to place references in a scientific paper.
11. Follow the journals style for details in the Reference List.
12. Know how to cite and list references from the internet.

ACKNOWLEDGMENTS GUIDELINE:
   List all the people whose help was important but not enough
to warrant authorship.

PROBLEMS

Problem 8-1

**When scientists submit papers for publication, references have to be
listed as required by the *Instructions to Authors*. The following reference section belongs to a paper submitted to Molecular and General
Genetics (MGG). Their *Instructions to Authors* states the following:**

Journal articles:

- The list of references should only include work cited in the text
  and that has been published or accepted for publication ("in
  press").
- In the text, references should be cited by author and year (e.g.,
  Hammer 1990a; Hammer 1990b; Hammer 1994; Hammer and
  Sjöqvist 1995; Hammer et al. 1993). Note that no comma separates author and year, and that only the last name of the first

author is indicated, all other authors are abbreviated by *et al.* If there are two authors, both authors are listed by last name.

- References should be listed in alphabetical order in the Reference List.
- Journals should be abbreviated in accordance with Bibliographic Guide for Editors and Authors (BGEA). Note: a correct reference contains one single period (following the article title) and commas only to separate authors.

Examples of the correct styles are shown below:

Article:

Gallie DR, Young TE (2004) The ethylene biosynthetic and perception machinery is differentially expressed during endosperm and embryo development in maize. Mol Gen Genomics 271:267–281

Book:

Pechan P, de Vries G (2004) Genes on the Menu. Springer, Berlin Heidelberg New York

Internet:

Doe J (1999) Title of subordinate document. In: The dictionary of substances and their effects. Royal Society of Chemistry. Available via DIALOG. http://www.rsc.org/dose/title of subordinate document. Cited 15 Jan 1999

Check the following reference list to ensure that the references are listed as required. Circle the ones that need to be adjusted to fit the required style.

Bollag RJ, Waldmann AS, Liskay RM (1989) Homologous recombination in mammalian cells. Annu Rev Genet 23:199–225

Capecchi MR (1989a) Altering the genome by homologous recombination. Science 244:1288–1292

Capecchi MR (1989b) The new mouse genetics: altering the genome by gene targeting. Trends Genet 5:70–76

GrimmC, Kohli J (1988) Observations on integrative transformation in Schizosaccharomyces pombe. *Mol Gen Genet* 215: 87–93

Halfter U, Morris PC, Willmitzer L (1992). Gene targeting in Arabidopsis thaliana. Mol. Gen. Genet. 231:186–193

Kempin SA, Liljegren SJ, Block LM, Rounsley SD, Yanofsky MF (1997) Targeted disruption in Arabidopsis. Nature 389:802–803

Lee KY, Lund P, Lowe K, Dunsmuir P (1990) Homologous recombination in plant cells after Agrobacterium-mediated transformation. Plant Cell 2:415–425

Offringa R, de Groot JA, Haagsman HJ, Does MP, van den Elzen PJM, Hooykaas PJJ (1990) Extrachromosomal recombination and gene targeting in plant cells after Agrobacterium-mediated transformation. EMBO J 9:3077–3084

Offringa R, Franke-van Dijk MEI, de Groot MJA, van den Elzen PJM, Hooykaas PJJ (1993) Non-reciprocal homologous recombination between Agrobacterium transferred DNA and a plant chromosomal locus. Proc Natl Acad Sci USA 90:7346–7350

Paszkowski J., Baur M., Bogucki A., Potrykus I. (1988). Gene targeting in plants. EMBO J 7:4021–4026

Puchta H, Hohn B (1996) From centimorgans to base pairs: homologous recombination in plants. Trends Plant Sci 1:340–348

Puchta H, Dujon B, Hohn B (1996) Two different but related mechanisms are used in plants for the repair of genomic double strand breaks by homologous recombination. Proc Natl Acad Sci USA 93:5055–5060.

Riele H te, Maandag ER, Berns A. Highly efficient gene targeting in embryonic stem cells through homologous recombination with isogenic DNA constructs. Proc Natl Acad Sci USA (1992) 89:5128–5132

Rossant J, Joyner A (1989) Toward a molecular-genetic analysis of mammalian development. Trends Genet 5:277–283

Roth D., Wilson J. (1988). Illegitimate recombination in mammalian cells. In: Kucherlapati R, Smith GR (eds) Genetic recombination. American Society for Microbiology, Washington DC, pp 621–653

Schaefer DG, Bisztray G, ZryÈd JP (1994) Genetic transformation of the moss Physcomitrella patens. Biotech Agric For 29:349–364

Struhl K (1983). The new yeast genetics. *Nature* 305:391–397

Rothstein R (1991) Targeting, disruption, replacement and allele rescue: integrative DNA transformation in yeast. Methods Enzymol. 194:281–301

Timberlake WE, Marshall MA Genetic engineering in filamentous fungi. Science 244:1313–1317 (1989)

Problem 8-2

In the following paragraphs, which are part of an Introduction, check that references are cited correctly in the text for the same paper as in Problem 8-1 given the same instructions for publication in MGG.

The integration of foreign DNA into eukaryotic genomes occurs mainly at random loci by illegitimate recombination, even when the introduced DNA has homology to endogenous sequences (Roth et al. 1988; Bollag et al. 1989; Capecchi 1989a; Puchta et al 1994; Puchta and Hohn 1996). Exceptions to this generalization include *Saccharomyces cerevisiae* (Rothstein R., 1991), *Schizosaccharomyces pombe* (Grimm and Kohli, 1988) and some filamentous fungi (Timberlake and Marshall 1989). Direct integration by homologous recombination (gene targeting) is rare compared with non-homologous integration, but would provide a powerful technique for the molecular genetic study of higher organisms and the determination of gene function. Although gene targeting is now a well established tool for the specific inactivation or modification of genes in yeast and mouse embryonic stem cells (Struhl 1983; Capecchi 1989b; Rossant and Joyner 1989; te Riele et al. 1992), the development of similar techniques for plant systems (Puchta and Hohn 1996) is still at a very early stage. Some groups have reported genomic integration by homologous recombination in plant cells using DNA introduced by direct gene transfer or *Agrobacterium*-mediated DNA transfer (Paszkowski et al. 1988; Offringa 1990, 1993; Halfter et al. 1992). The highest rates of gene disruption so far published for flowering plants are between 0.01 and 0.1% (Lee et al. 1990; Miao and Lam 1995; Kempin et al. 1997), and these are too low to allow routine gene disruption. However, it has recently been shown (Schaefer and ZryÈd 1997) that integrative transformants resulting from homologous recombination can be obtained following polyethylene glycol (PEG)-mediated direct gene transfer into protoplasts of the moss *Physcomitrella patens*, at an efficiency comparable to that found for *S. cerevisiae*.

(With permission from Springer Science and Business Media)

Problem 8-3

When scientists write papers for publication, they often forget to reference information. In the following passage, indicate where a citation should be made.

The carbon ($\delta^{13}C$), nitrogen ($\delta^{15}N$), and sulfur ($\delta^{34}S$) isotope ratios of humans, other animals, and microbes are strongly correlated with the isotope ratios of their dietary inputs (1–5). There are limited differences ($\leq 1\text{‰}$) between heterotrophic organisms and their diet in either the $\delta^{13}C$ or $\delta^{34}S$ values. Hydrogen (δ^2H) and oxygen ($\delta^{18}O$) isotope ratios of organic matter, however, are more useful, because δ^2H and $\delta^{18}O$ values of precipitation and tap waters vary along geographic gradients. Although differences in the δ^2H and $\delta^{18}O$ values of scalp hair have been noted in humans, less is known about diet–organism patterns of δ^2H and $\delta^{18}O$ values. Four potential sources can be important: dietary organic molecules, dietary waters, drinking waters, and atmospheric diatomic oxygen. Hobson *et al.* provided evidence that δ^2H values of drinking water were incorporated into different proteinaceous tissues of quail. Other research showed that the δ^2H values of bird feathers and butterfly wings (both are largely keratin) and water in the region in which the tissue was produced are highly correlated (14, 15). Kreuzer-Martin *et al.* showed that ~70% of the oxygen and ~30% of the hydrogen atoms in microbial spores (~50% proteinaceous) were derived from the water in the growth medium, whereas the remainder was derived from the organic compounds supplied as substrate.

(With permission from National Academy of Sciences, U.S.A.)

Problem 8-4

In the following paragraph, which is part of an Introduction, check that the text citations have been placed appropriately.

Ostracodes are small bivalved Crustacea that form an important component of deep-sea meiobenthic communities along with nematodes and copepods (10). Crustaceans (10, 11) are dense and diverse in the deep sea and are one of the most representative groups of whole deep-sea benthic community. Pedersen et al. as well as Jackson et al. reported (12–14) that ostracode species have a variety of habitat and ecology preferences (e.g., infaunal, epifaunal, scavenging, and detrital feeders), representing a wide range of deep-sea soft sediment niches. Furthermore, Ostracoda is the only commonly fossilized metazoan group in deep-sea sediments. Thus, fossil ostracodes are considered to be generally representative of the broader benthic community. The distribution and abundance of deep-sea ostracode taxa in the North Atlantic Ocean are influenced by several factors (14, 15), among them, temperature, oxygen, sediment flux, and food supply. Several paleoecological studies suggest (1, 16) that these factors influence deep-sea ecosystems over orbital and millennial timescales.

(With permission from National Academy of Sciences, U.S.A.)

Problem 8-5
Paraphrase the following passage.

A drought is defined in the glossary of meteorology as "a period of abnormally dry weather sufficiently prolonged so that the lack of water causes a serious hydrologic imbalance in the affected area." A drought is the consequence of a natural reduction in the amount of precipitation received over an extended period of time, usually a season or more in length, although the onset and end of a drought are generally difficult to determine. Drought is a normal, recurring feature of climate, which occurs in high, as well as low rainfall areas. Although it is a frequent and often catastrophic feature in semi-arid climates, it is less frequent and disruptive in humid regions and an even less meaningful concept for deserts. The severity of a drought can be significantly aggravated by other climatic factors, such as by high temperatures and winds or by low relative humidity. Resulting imbalances may include crop damage, water supply shortage, and increased fire hazards. These effects often accumulate slowly over a considerable period of time and may linger for years after drought termination. There remains much disagreement within the scientific and policy community about exact drought characteristics, and therefore not much progress has been made in drought management in many parts of the world.

Problem 8-6
Paraphrase the following passage.

A few simple mutations of the deadly H5N1 avian influenza virus might give it the ability to spread readily among humans. Once in the human population, the virus could spark a global pandemic that could kill tens of millions. The virus started spreading in earnest among birds in late 2003. It spread to various countries and appears to be here to stay. In 2007, the virus surfaced in poultry flocks in eight new countries, including in Bangladesh, Poland, and Ghana. Outbreaks returned in 23 countries stretching from Indonesia to the United Kingdom. In some countries, such as in Indonesia and Nigeria, outbreaks are now more or less continuous. Indonesia is the hardest-hit country, reporting 42 cases and 32 deaths in 2007. As long as the virus is circulating in birds, sporadic human cases will continue to occur, and most of them will be fatal. Although poultry trading is the primary means of spreading the virus, the role wild birds play in long-distance spread is still unclear. Therefore, scientists should take advantage of new lightweight transmitters that enable satellite tracking of migratory species.

Problem 8-7

Given the following information from another article, incorporate this information in the proposed introduction on earthquake predictions shown following the original.

Original from another article:

Most great earthquakes occur at submarine subduction zones where one tectonic plate slides at a gentle angle (10°–30°) beneath another. In this paper I look at the probabilities of the largest earthquakes and, in the context of recurrence times and our short history of observation, argue that we cannot rule out an Mw ≥ 9 earthquake at any subduction zone. Simulations using recurrence times of the maximum size earthquakes (called M9) at subduction zones suggest that 1–3 M9 earthquakes should occur within any 100 yr span. The five M9 events that have occurred since 1952 probably represent temporal clustering and not a long-term average.

(With permission from the Geological Society of America)

Proposed new Introduction on earthquake predictions

Humans have long sought to automate emergency response procedures for earthquakes in an effort to protect against loss of life and reduce property damage. Impending earthquakes have been argued to be associated with such varied phenomena as seismicity patterns, electromagnetic fields, weather conditions, gas content of soil or ground water, water levels, animal behavior, and lunar phases. Thus far, earthquake prediction is controversial because data are sparse and there is little evidence or verified physical theory to link observable phenomena to subsequent seismicity. However, probabilities of occurrence have been calculated, particularly for earthquakes of magnitude larger than 5. For example, ... [insert paraphrased citation here].

Problem 8-8

The following passage comes from a research paper publication. Read through the original to get an understanding of its central points and then determine whether the student paraphrases are plagiarized or paraphrased versions.

Original

Killer Whales are well-known as predators of other marine mammals, including the large Sperm and baleen whales. Members of all marine mammal families, except the river dolphins and manatees, have been recorded as prey of Killer Whales; attacks have been observed on 20 species of cetaceans, 14 species of pinnipeds, the Sea Otter, and the Dugong. Ecological interactions have not been systematically studied and further work may indicate that the Killer Whale is a more important predator for some populations than previously believed. Not all behavioural interactions between Killer Whales and other marine mammal species result in predation, however. Some involve 'harassment' by the Killer Whales, feeding by both species in the same area, porpoises playing around Killer Whales, both species apparently 'ignoring' each other, and even apparently unprovoked attacks on Killer Whales by sea lions. These non-predatory interactions are relatively common. We conclude that interactions between Killer Whales and marine mammals are complex, involving many different factors that we are just beginning to understand.

(Jefferson, T. A., Stacey, P. J., and Baird, R. W. (2008).
Mammal Review 21(4), 151–180)

Student Version A

Killer Whales prey on other marine mammals such as the large Sperm and baleen whales, and on members of all marine mammal families, except the river dolphins and manatees. Killer Whales have also attacked cetaceans, pinnipeds, the Sea Otter, and the Dugong. Some interactions between Killer Whales and other marine mammal species do not result in predation, for example, 'harassment' by the Killer Whales, feeding by both species in the same area, porpoises playing around Killer Whales, and even attacks on Killer Whales by sea lions. These non-predatory interactions are relatively common. Thus, interactions between Killer Whales and marine mammals are complex, and involve many different factors that we are just beginning to understand, [and] further work is necessary to understand the ecological interactions of Killer Whales.

Student Version B

Ecological interaction of Killer Whales with other marine mammals can be predatory or non-predatory. Predatory interactions involve Killer Whale attacks on all marine mammals. Non-predatory interactions involve Killer Whales harassing marine mammals, feeding in the same area, playing around each other, and even attacks on Killer Whales by the other species (Jefferson et al. 2008). Although these interactions have been observed, further work is needed to fully understand Killer Whale interactions with other marine mammals.

Problem 8-9

1. **From a source related to your research topic, choose a 7 to 12 sentence-long passage. Avoid introductory or concluding paragraphs, as these are often very general.**
2. **Paraphrase the original and include the citation in the text as well as in a Reference List using CBE format.**

CHAPTER 9

Figures and Tables

Figures and tables are meant to demonstrate evidence visually and therefore should be designed for strong visual impact. As readers often look at figures and tables to see whether the rest of a paper is worth reading, each figure and table should be capable of standing on its own without reference to the text.

9.1   GENERAL GUIDELINES

ILLUSTRATION GUIDELINES:

1. Figures and tables should be able to stand on their own.
2. Decide whether to present data in a table, a figure, or in the text.
3. Use the fewest figures and tables needed to tell a story.
4. Design figures and tables to have strong visual impact.

Decide first whether to present your findings in a figure, in a table, or in the text. Editors and reviewers will stringently judge if a figure or table is useful. Each figure or table should therefore be important enough to be included in a document. Your reader can also pull the story together more easily from 5 or 6 figures and tables than from 15 or 16. Be aware that figures and tables take longer to create and may cost more to produce than text.

Figures and tables must clearly and accurately show what the text states. For example, if the text describes a new organism, the important features of this organism should be immediately visible in the figure. Similarly, if the text says that when X was done, Y increased, then in the figure, Y should look as if it increased. If the increase is not obvious, the figure is unconvincing.

At the same time, figures and their legends as well as tables and their titles must be independent of the text and of each other, that is, they must be able to stand on their own. Figures and tables need to be numbered in the order they appear in the text. In addition, the name of the variable, the unit of measurement, and the values should be the same in the text and in the illustration. If possible, provide statistical information in figures and tables.

For a very detailed description of "dos" and "don'ts" for figures and tables, see also Rubens (1992, pp. 327–423.)

9.2   IMPORTANCE OF FORMATTING AND PLACEMENT OF INFORMATION

GUIDELINE 1:

Prepare figures and tables with the reader in mind—place information where the reader expects to find it.

In the same way that text should be formatted for the reader, figures and tables should be formatted to meet the reader's expectations. Most readers can understand the intended meaning of what is presented only if the illustration has been formatted for this interpretation.

Scientists can present data in different formats. See Example 9.1, including Tables 9.1, 9.2, and 9.3, for various possibilities for a set of data.

Example 9-1   **a**   0 °C, 0.011% hermaphrodites
6 °C, 0.011% herm.
25 °C, 51/1000   18 °C, 0.028%
water T = 30 °C , 0.124%   T = 10 °C , 2/100
35 °C, 0.152%

b   Table 9-1: Sample Data Display 1

% HERMAPHRODITES N = 120	WATER TEMPERATURE (°C)
0.011	0
0.011	6
0.020	10
0.028	18
0.051	25
0.124	30
0.152	35

c   Table 9-2: Sample Data Display 2

Water temperature (°C)	0	6	10	18	25	30	35
% Hermaphrodites	0.011	0.011	0.020	0.028	0.051	0.124	0.152

d   Table 9-3: Sample Data Display 3

WATER TEMPERATURE (°C)	% HERMAPHRODITES
0	0.011
6	0.011
10	0.020
18	0.028
25	0.051
30	0.124
35	0.152

Although the exact same information appears in all formats, most readers prefer Example 9-1d because it is the easiest to interpret. The reason for the easier interpretation is twofold: (a) the data is written as a table, and, even more important, (b) this table is structured such that the familiar context (temperature) appears on the left, whereas the interesting results appear on the right in a less obvious pattern.

Usually, information in a table is more readily available for readers than information in the text (unless there are very few data). However, some tables are easier to interpret than others. For example, readers find tables much harder to follow if they are presented as shown in Example 9-1b (Table 9.1), or if the table is horizontally arranged as in Example 9-1c (Table 9.2). Readers interpret information more easily if it is placed where they expect to find it. Because we read from left to right, we prefer the familiar context on the left and the new information on the right as in Example 9-1d (Table 9.3).

9.3   FIGURE OR TABLE?

FIGURES GUIDELINE 1:
Use figures to show trends and relationships and to emphasize data.

When you have chosen to use an illustration rather than text, you may then need to decide in what format to present them, figure or table. It helps if you spread your data out on the table and arrange them in all possible combinations. Look for patterns. There are better and worse presentations of data in a paper. Go with a simple pattern if possible.

Generally, choose figures when trends or relationships are more important than exact values or when hidden relationships or trends need to be revealed. Choose tables to report precise numerical information, to compare component groups, or when data are not enough to produce a satisfactory graph.

Consider the next two illustrations (Example 9.2; Table 9.4). Which of the following presentations would you prefer given the same data?

Example 9-2   a   Table 9-4: Sample Data Presentation A

TIME (DAYS)	HORMONE A	HORMONE D54
0	200.5	455.8
5	187.1	356.7
10	166.5	321.9
15	201.1	400.6
20	289.8	500.7
25	204.1	489.9

TIME (DAYS)	HORMONE A	HORMONE D54
30	189.9	389.4
35	288.9	513.4
40	205.1	499.3
45	182.9	298.5
50	278.8	533.2
55	223.4	498.5
60	199.6	250.6

Example 9-2   b

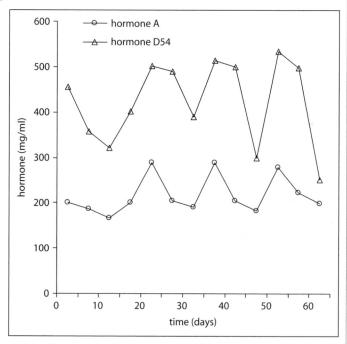

Figure 9.1: Sample data presentation B. The same data as in Example 9-2a are presented as a line graph instead of a table.

Most readers would prefer Example 9-2b (Figure 9.1) because the trend and relationship of the data is more obvious, and exact numbers seem not as important.

Consider another example in Figure 9.2.

Example 9-3

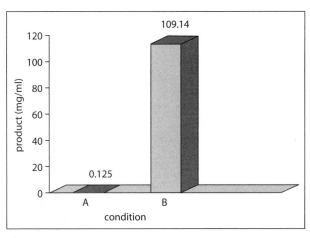

Figure 9.2: Bar graph

Example 9-3 would best be depicted in the text because there are only two data points, and exact values seem important.

9.4   GENERAL INFORMATION ON FIGURES

General Advice

When you prepare a figure, be sure that it carries your point better than the text or a table would. Never use figures simply because they are available. Consider what kind of figure you need. You can choose to present your data in a **photograph**, a **drawing**, a **diagram**, or a **graph**. Select the size and format to fit the journal, poster, or slide.

Overall, figures must be simple and clear enough for readers to get the message immediately. Good figures clarify relationships of complicated data sets and highlight trends or patterns that may not be immediately evident in the text.

The requirements for figures that will appear in print are different from those for oral presentations or posters (see Table 9.5 and also Chapter 27 and 28). Although it is best to design for each medium separately, if you plan carefully, you may be able to use the same artwork for a journal article, an oral presentation, and a poster presentation.

Table 9.5 Differences between figures in text, slides, and posters

	TITLE	LEGEND
Figure in text	—	✓
Figure on slide	✓	—
Figure on poster	✓	✓
Table in text, on slide or poster	✓	—

Producing Figures

FIGURES GUIDELINE 2:

Prepare professional figures.

Figures submitted to professional journals must meet high standards. Inadequate figures spoil many articles, and few journals redraw or reletter figures submitted by authors. If you do not have your figures designed professionally, it is a wise idea to get help from your colleagues or to consult a recommended book or maybe take a course on how to prepare professional figures.

You can prepare your own figures by using various computer software: Excel or Lotus are great programs to convert your data and to export it to graphing programs such as Charisma®, CorelDRAW®, Harvard Graphics®, KyPlot®, SigmaPlot®, and DeltaGraphPro® or to slide-making programs such as PowerPoint®, Adobe®Persuasion®, or Astound®. Other great draw, paint, and graphics programs include Adobe®Illustrator®, Adobe®Photoshop®, ChemDraw®, Visio®, and SmartSketch®. You may even be able to try out or use some of these programs as free beta versions that can be downloaded from the Web.

If you use a computer graphics program for your figures, the output must be of good enough quality to be reproduced well by conventional printing techniques. Lines must be clearly drawn, with good contrast against a plain white background. Axes, curves, and lettering must be solid and smooth. Symbols must be clear, large, and distinctive. Your figures need to be labeled well, and all components must be identified. Do not clutter your illustrations. Include scale bars on maps, micrographs, and anatomical drawings.

If an artist or photographer who makes figures for you is employed by your organization, note that copyright of the figures belongs to the organization. Thus, you should be able to use the figures without obtaining further copyright permission from the maker. If you employ a freelance artist or photographer, however, the copyright belongs to him or her unless a written agreement is made transferring copyright and other rights to you. An agreement of this kind makes you the owner of a "work made for hire."

It is a good idea to start preparing figures at an early stage, before you even think of drafting the text. First, check whether the journal limits the number or size of figures you may submit. Then, draft your figures and decide which ones to use. Design any others you think you will need, and draft legends for all of them. When you create the final versions of the figures at the revision stage, note that many journals require you to submit figures electronically as tagged image file format (TIFF) files.

Remember that the data you present may be interpreted in more than one way. You can highlight, exaggerate, or even misrepresent a given set of data depending on the way you choose to display it. Thus, constructing a graph is more than just plotting points. It requires that you understand the rationale behind the quantitative methods. It also requires that you are as objective and honest as possible.

Misleading Readers

FIGURES GUIDELINE 3:

Do not mislead readers.

You should make readers aware of certain trends. However, as a scientist, you have to be responsible enough not to mislead your readers. It may be very tempting to distort information and trends by, for example, deleting data points or by massaging line fits. Resist these temptations. Do not mislead readers.

If you can, provide statistical information in your figures and figure legends. This will lend validity to your data. If a point represents the mean of a number of observations, indicate the magnitude of the variability by a vertical line centered at each point. State whether standard error or standard deviation is used, and specify the number of observations, sample size, p-values and other statistical information.

Generally, begin at zero for the scales used for the axes of a graph whenever possible. Choose these scales carefully, and mark them clearly. When two or more graphs or other figures are to be compared, draw them to the same scale. If possible, place them side-by-side. Furthermore, remember the limitations of your data. The extrapolation of a line or a curve beyond the points shown on a graph may mislead both the writer and the reader.

9.5   TYPES OF FIGURES

Photographs

Photographs record results that rely on visual inspection, such as X rays or electron microscopic presentations. The major advantage of a photograph is realism. If you decide that it is essential to show the actual appearance of a subject or object, choose photographs of the best possible quality. Photographs intended for journal publication should be sharply focused, with good contrast. See also the CBE Scientific Illustration Committee (1988, pp. 219-233) for a more detailed description of photographs.

Never assume that the reader will recognize anything in a photograph. Label everything that is relevant. Some kinds of records do not come out well in photographs. It may be more effective to draw a diagram of a piece of equipment or describe it in the text rather than to include a photograph.

If you are thinking of including color photographs, find out first whether your target journal accepts them (see the *Instructions to Authors* or consult the editor). Some journals do not print color photographs; others may not accept them unless the author agrees to pay some or all of the costs of reproduction, which are usually high.

Always include scale bars on micrographs. This is preferable to giving magnifications in legends because the scales will remain correct if the figure is reduced during the production process or is later printed at some other size. Alternatively, if you want to state the original magnification, give it in the legend and add the photographic reduction when it is known (see Figure 9.3). Example 9-5 (Figure 9.4) is another example of a photograph together with a line graph.

Example 9-4

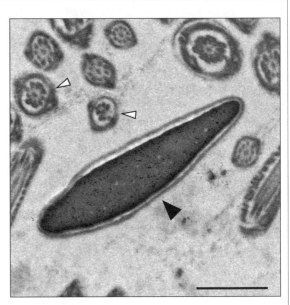

Figure 9.3: Electron micrograph of mouse testis embedded in epoxy resin. Sperm was immuno-labeled with 10 nm gold post embedding. Black arrow, longitudinal section of sperm head; white arrows, cross sections of sperm tails. Bar, 1 μm.
(With permission from Rudolf Lurz)

Example 9-5

Class Description   NDVI-cycle (phenology)

Medium dense deciduous forest

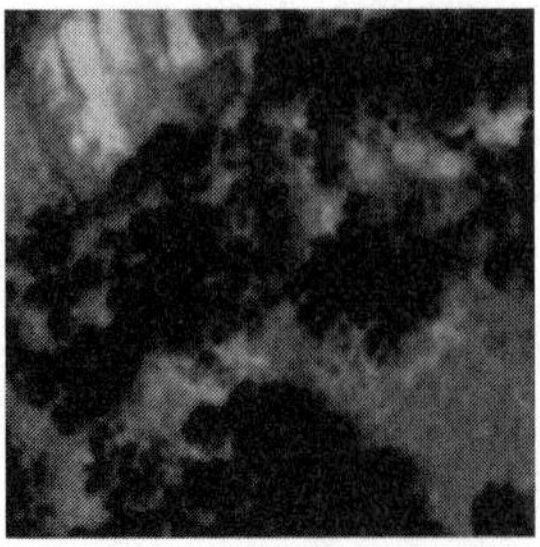

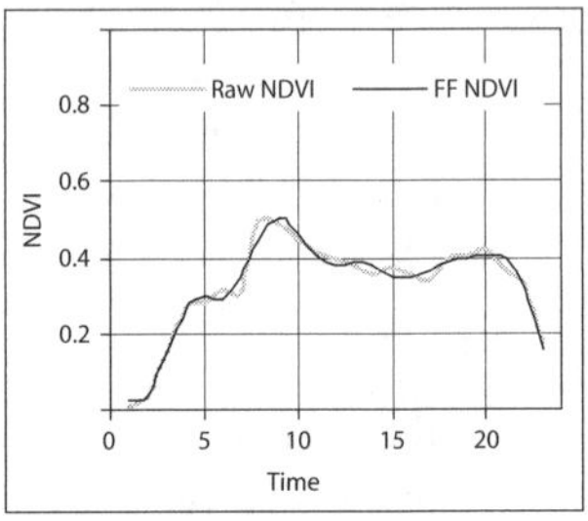

Figure 9.4: Phenological cycles from 250 m Modis NDVI time-series (version 4). Width of snapshot is about 150 m. Source for snapshot is Google Earth.
(With permission from Roland Geerken)

Drawings and Diagrams

Drawings and diagrams illustrate basic principles or otherwise explain text material. They include flow charts, diagrams, block diagrams, maps, and line art. Because these figures are much more artistic than graphs and usually more difficult to produce, you should seriously consider getting professional help with these types of figures.

The advantage of drawings and diagrams is that you present unusual perspectives while controlling the amount of detail. This type of artwork must show the specific details of key features while omitting others to avoid distracting the reader. An example of such a diagram is shown in Example 9-6 (Figure 9.5). A common mistake in scientific writing is to present a figure that is much more complex than the accompanying text. When this is done, illustrations confuse rather than inform readers.

Example 9-6

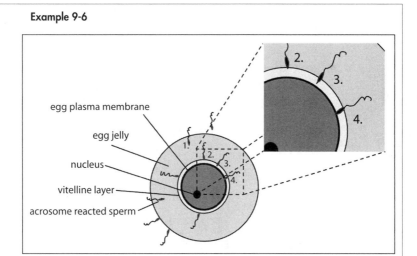

Figure 9.5: Diagram of *S. purpuratus* egg and sperm fertilization. 1. The egg triggers the sperm acrosome reaction. 2. The sperm attaches to the egg vitellin layer at the acrosomal process. 3. The acrosomal process extends as it penetrates the vitellin layer. 4. The plasma membranes of the egg and sperm fuse.

A specific type of diagram is the block diagram. A block diagram depicts objects or materials in action, as for example the block diagram shown in Example 9-7 (Figure 9.6). It is one of the most often used visuals in engineering and science and shows how things react with each other, be they electrical, mechanical, chemical, biological, or some combination of these.

Example 9-7   Typical block diagram

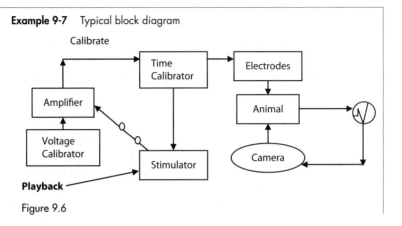

Figure 9.6

Guidelines for Developing Technical Block Diagrams

- Use no more than 8 to 10 blocks.
- Use short functional names for each block, and use exactly the same key terms as in the text.
- Do not clutter your diagram. Show only major actions and interactions to reduce confusion.
- Use arrows to indicate direction.
- When describing the diagram in the text, follow the direction of flow on the diagram and describe the entire diagram. The same applies when you present such a figure in a talk or poster.

Graphs

GUIDELINE 2:

Use line g\raphs for dynamic comparisons.
Use scatter plots to find a correlation for
a collection of data.
Use bar graphs when the findings can be subdivided
and compared.
Use pie charts to compare parts of a whole.

Present your data in graphs if the data show pronounced trends, relationships, or patterns. Graphs can be presented as bar graphs, line graphs, pie charts, box plots, or scatter plots. Often, these may be two- and three-dimensional.

Line Graphs

Line graphs are the most popular type of graph in scientific papers. Line graphs are used for dynamic comparisons, often over time.

Do not try to cram too much into one figure—but do not waste space either. Three or four curves should be the maximum in a line graph, especially if the lines cross each other two or three times. When curves must cross, show which lines run where by making them of different thickness or different patterns. Draw curves as straight lines between data points or as best fit, smoothed curves.

Scatter Plot
Scatter plots are useful for finding correlations for a collection of data. Scatter plots can be produced in two or in three dimensions. A line of best fit can be drawn through this collection to find any correlation between the variables, and a corresponding equation for the correlation between the variables can be determined by established best-fit procedures.

Bar Graph
Another common graph in science is the bar graph. Bar graphs tend to be more effective than line graphs for general audiences. If you use a bar graph, use a vertical rather than a horizontal bar graph because most readers are accustomed to the former. Use bar graphs in preference to line graphs when there is no evidence of a continuum between the experimental points or when the findings can be subdivided and compared in different ways. Make the bars the same width, and the space between bars one-half the bar width.

Pie Chart
Yet a different kind of graph is the pie chart. Whereas bar graphs allow you to compare different whole quantities, pie charts allow you to compare parts of a single whole. They are most effective when the segments are arranged from large to small, with the largest segment starting at the top (see Figure 9.7).

Box Plot
Box plots can be useful to display differences between data sets, especially in descriptive statistics. Box plots can be drawn either horizontally or vertically. This type of plot graphically depicts groups of numerical data through their five-number summaries: the smallest observation, lower quartile (Q1), median (Q2), upper quartile (Q3), and largest observation. The spacing between the different parts of the box help indicate the degree of dispersion or spread (see Figure 9.8).

Example 9-8   a   Sample pie chart

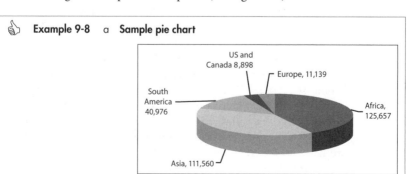

Figure 9.7: Annual death due to disease A, by region.

👍 **Example 9-8   b   Sample box plot**

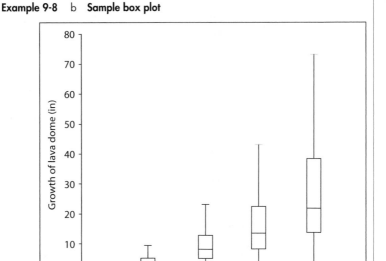

Figure 9.8: Box plot. The plot displays the median, interquartile ranges (box), and extreme values (whiskers) from collected data.

9.6   FORMATTING GRAPHS

FIGURES GUIDELINE 4:

Place the independent variable at the x-axis and the dependent variable at the y-axis.

Readers expect to see information in figures at certain places. For line graphs and bar graphs, readers expect to find the independent variable at the *x*-axis and the dependent variable at the *y*-axis. If information is placed as expected, readers are much more directed and do not have to spend extra time trying to understand illustrations.

The important information (the data) should be immediately recognizable. Information will be immediately obvious if you emphasize it by using different line weights. For example, in line graphs, curves should be the darkest lines; letters in axis labels should be less dark; and axes, tick marks, error bars, keys, and curve labels should be least dark. Make plotted points stand out well. If they fall on an axis line, break the axis on each side of the point. Plan graphs so that they need as little lettering as possible. Draft short but informative descriptions for the axes. Use the same symbols when the same entities occur in several figures, and use the same coordinates for different figures if values in them are to be compared. If you measured two variables in different ways, do not compare them on the same axes. Draw the curves for the two variables separately, using one common axis where appropriate.

An example of a well constructed line graph is shown in Example 9-9 (Figure 9.9).

 Example 9-9

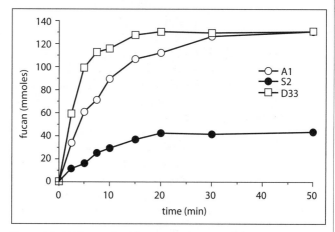

Figure 9.9: Well organized line graph. Curves and data points are easily distinguishable. Data stand out well, and axis are designed and labeled clearly.

As an additional example, a well constructed scatter plot is also shown in the next example (Figure 9.10).

 Example 9-10

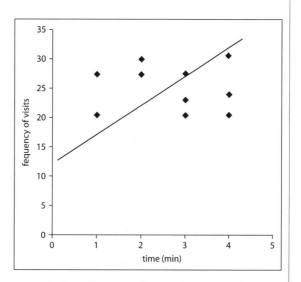

Figure 9.10: Well-presented scatter plot. Data stand out well, and axes are designed and labeled well.

Data points in scatter plots often overlap. That is okay. A scatter plot allows you to find a clear correlation for the points, such as the linear relationship in Example 9-10.

Readability

FIGURES GUIDELINE 5:
Make each figure easy to read.

For good first impressions, make each figure easy to read. The lettering should be large enough to be legible *after* the graph is reduced. Most journals prefer the font type Arial or Arial narrow no smaller than size 8 after reduction. You can check legibility by reducing the figure to publication size on a photocopier.

Symbols should be large enough to be seen easily—make them two or three times the width of curves in the graph. The shapes should also be easy to distinguish. The easiest data point symbols to distinguish are filled and open circles. If you need three or four symbols, use filled and open triangles, squares, and so forth in addition to the circles. Keep hexagons and squares away from the circles, however, because it is difficult to distinguish these symbols after reduction. If data points overlap, draw them overlapping. If the points coincide, one solution is to use just one symbol for these points. Do not use X, +, 0, or *—these symbols are not distinctive enough.

Do not clutter your graphs. If there is no room for curve labels or a key on the face of the graph, define the curves in the figure legend. In addition, avoid using grid lines within the graphs. They usually result in clutter. Grid lines are useful when plotting points but only rarely afterward. Furthermore, do not draw boxes around line graphs unless the journal specifically requires this. Figure 9.11 in the next example is an example of a cluttered graph.

Example 9.11

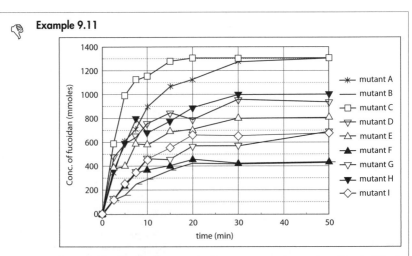

Figure 9.11: Cluttered line graph. Curves are superimposed, making it difficult to distinguish individual curves and data points. Grid lines add to the clutter.

Lines and Curves

GUIDELINE 3:
Differentiate points, lines, and curves well.

Differentiate curves by using different symbols for points joined by the same type of line or by using the same symbol joined by different types of lines. When only two curves need to be differentiated, use different types of lines.

Where appropriate, draw a vertical line to show the standard deviation or the standard error of the mean for each data point. These lines are usually drawn in pairs in line graphs, one above and one below a data point. For bar graphs, it is really only necessary to draw the top line of each pair. Make the lines of the error bars thinner than other lines in the main body of the data, and do not let them overlap. In the legend, tell readers what the vertical line represents, and state how many observations each mean is based on.

Rather than using a key, label curves and other components directly where possible (but check your target journal's practice). Keep labels well away from lines and curves, and position them horizontally.

Axis Labels and Scales

GUIDELINE 3:
Label axes and scales well.

Label the axes of graphs as briefly and simply as possible. Write labels for the "y" axes horizontally or parallel to the axis, reading upward. Write the label for the "x" axis parallel to it, and center the "x" axis label below the horizontal line.

For the axis label, name the variable and give the unit of measurement in parentheses (use SI abbreviations). Use lowercase lettering for labeling axes. Reserve capital letters for the first letter of the first word, for any other words usually written with an initial capital, and for abbreviations. Explain all abbreviations in the legends.

Choose scales for axes carefully. For example, for large-scale numbers, use $\times 10^y$. If an axis does not start at zero (or 1 for log scales), mark a break in the scale. Do not extend axis lines beyond the last marked scale point, and do not end them with an arrow pointing away from zero. Mark scale calibrations (tick marks) clearly. Put them outside the axis, and mark and number only as many as are necessary for clarity (see Figure 9.12).

Example 9-12

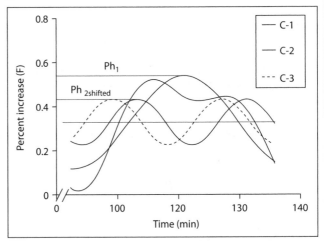

Figure 9.12: Clearly designed graph. Curves are easy to distinguish, the graph is uncluttered, and axes are easy to read.

9.7 EXAMPLES OF GRAPHS

 Example 9-13

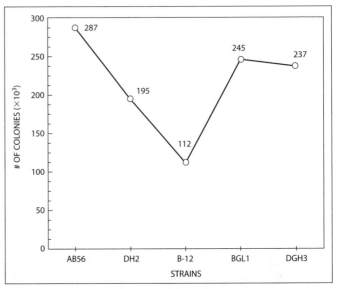

Figure 9.13: Graph with some common faults: (1) excessive numbering and tick marks on the y axis; (2) calibrations (tick marks) face inward where they are either unnecessary or may conflict with the data; (3) heavy black frame is distracting; (4) labels on the x- and y-axis are capitalized, and thus difficult to read; (5) background color is unnecessary; (6) a line graph is plotted, although exact values seem to be compared and of importance.

Revised Example 9-13

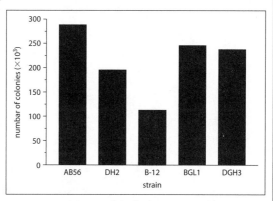

Figure 9.14: A more appropriate presentation of the findings in Example 9-13: (1) The number of tick marks has been reduced; (2) tick marks face outward; (3) the box around the graph has been removed; (4) the *x*- and *y*-axis labels are easier to read written in lowercase letters; (5) the background color has been removed; last but not least, (6) data are compared in a bar graph rather than a line graph.

Example 9-14

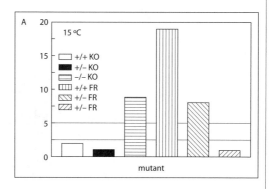

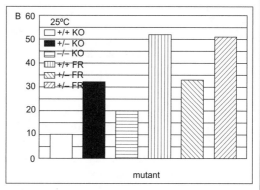

Figure 9.15A, B: Graph with several common faults: (1) scales in A and B are different, distorting the information; (2) horizontal grid lines are distracting and make the graph appear cluttered; (3) key within the figure adds to the clutter; (4) box around the graph is unnecessary; and (5) the temperature labels are emphasized too much.

👍 **Revised Example 9-14**

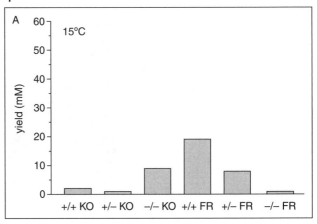

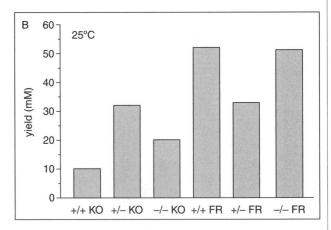

Figure 9.16A, B: A better presentation to that of the data in
Figure 9.15A, B, Example 9-14. In the revised figure, the two graphs
are drawn to the same scale, there is economy of line and lettering,
and labels are correctly positioned on both the x- and y-axis and
arranged to read upward. Grid lines have been removed, and
information in the key has been incorporated in the x-axis labeling. In
addition, temperature indications are much less pronounced.

An alternative to Example 9-14a is to graph the data in two different
graphs with the same scale for the y-axis but depicting the y-axis interrupted for the graph with the data displaying the upper end of the range.
Another way to depict widely ranging data on the same graph is by using
an insert to enlarge data with much smaller values, such as shown in
Example 9-14b (Figure 9.17).

👍 **Revised Example 9-14b**

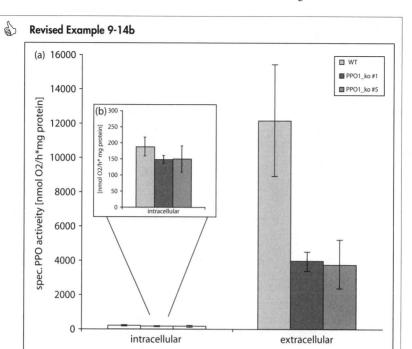

Figure 9.17 Graph with widely ranging data. Insert enlarges bars for data of much smaller values.   *(With permission from Hanna Richter)*

👎 **Example 9-15**

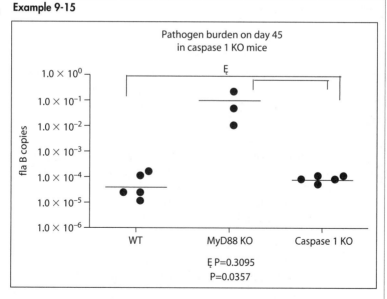

Figure 9.18A: Graph with faults. Data plotted as individual points results in clutter. Horizontal average lines are distracting as are the brackets labeled E on top. Title is unnecessary for journal publication. EP and P values below the graph are confusing and may be better placed into the figure legend.

Revised Example 9-15

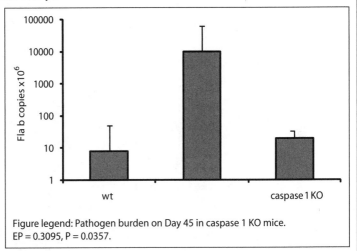

Figure legend: Pathogen burden on Day 45 in caspase 1 KO mice.
EP = 0.3095, P = 0.0357.

Figure 9.18B: A more appropriate presentation of findings in Example 9-15.
The data are now represented in a bar graph. Error bars indicate the spread.
The title, as well as values for EP and P, has been placed in the figure legend.

Note that for the preceding graph, if there is considerable spread in the
original measurements, a box plot or even the original points, similar to
the original Example 9-15 (Figure 9.18A), may be more appropriate.

9.8   FIGURE LEGENDS

FIGURES GUIDELINE 6:

The figure legend should be a description of the figure content.

A figure legend is a descriptive statement that is printed below or next to
a figure in a published article. A legend is needed so that the figure can
stand on its own without reference to the text. The legend should be an
objective description of the contents of the figure. Include enough information that the reader can understand the figure easily. The figure with
its legend should be a complete unit of communication.

A figure legend typically has different parts:

- A brief title
- Description of figure contents
- Definitions of symbols, line, or bar patterns (written in telegraph
  style)
- Abbreviations not defined earlier
- Possibly statistical information
- Possibly experimental details

Some journals do not follow this format. Other journals request only a title. Yet others want complete experimental details in the legend and not in the Methods section of the paper. If the journal gives explicit instructions, follow them.

The Figure Title

The title is the first item in the figure legend. For a scientific paper, it does not appear on the figure itself (unlike for a slide, poster, or certain state or federal reports.) The title should be brief and should identify the specific topic or the point of the figure. It should be written as an incomplete sentence and use the same key terms as are used on the graph and in the text of the paper. It should not contain abbreviations.

Description of Figure Content

Draft a set of legends in the style usually used in your target journal. Keep the draft legends and the figures in their preliminary form beside you while you write the first draft of the paper. When you have the final version of the figure, rewrite the legend if necessary.

The descriptive material in a legend may include letters or symbols to explain special abbreviations and symbols shown in the figure. Abbreviations and symbols should be consistent between the figure and its legend, consistent between legends, and consistent with the text as well. Use parallel wording for similar illustrations. Do not write the phrase "see text for explanation."

Example 9-16 shows two different figure legends. The legend in Example 9-16a describes the figure content and that in Example 9-16b describes the experimental details and provides some statistical information in form of a p-value.

 Example 9-16 a) Figure X: Western blot of protein Z deletion analogs. Lane 1 and 10, molecular weight standards. Lane 2, protein Z. Lane 3–6, ammonium sulfate pellets of the processed deletion analogs that exhibited agglutination. Lane 3, N18; lane 4, N74; lane 5, C166; lane 5, C121. Arrows indicate position of protein Z and its deletion analogs.

b) Fig. Z: *In vitro* X activity of wild type and X knockout plants. Intra- and extracellular protein extracts were prepared from cultures grown under standard conditions for 7 days. X activity was determined polarographically using 4-methyl catechol as a substrate. P= 0.0386.

9.9   GENERAL INFORMATION ON TABLES

General Advice

TABLES GUIDELINE 1:

Prepare tables rather than graphs when it is
important to give precise numbers.

Like figures, tables exhibit data visually, and the information presented should be independent of the text. Tables present data more exactly than a graph but usually do not show trends within your data. Tables also present facts more concisely than text does while allowing a side-by-side comparison of them. Design separate tables for separate topics. Do not use tables just to show off how much data you have collected. Limit the number of tables, and arrange the data to allow for easy interpretation. Do not repeat data in tables if you plan to put the same data in the text or in a figure, but state the most important data in the text as well.

Check the *Instructions to Authors* to see whether the journal limits the number or size of tables. You will probably make rough versions of tables as your work proceeds. Before drafting the text, decide which tables you need for the paper, and redesign them with publication in mind.

For more detailed information on tables see Browner (1999, pp. 71–88) and Rubens (1992, pp. 336–364).

Producing Tables

Tables can easily be prepared electronically. You can use spreadsheets such as Excel to convert your data into a table. Alternatively, you can use Word or WordPerfect to create a table directly in a word file.

9.10   FORMATTING TABLES

General Format

TABLES GUIDELINE 2:

Keep the structure of your table as simple
as possible.

TABLES GUIDELINE 3:

Place familiar context on the left and new, important information on the right.

Keep the structure of your tables as simple as possible. Use tables published in your target journal as models for drafting your tables.

A typical scientific table consists of a title, column headings, row (or side) headings, the body (the rows and columns containing the data), and, usually, explanatory notes. Many tables in scientific papers follow the pattern shown in Example 9-17a (Table 9.6) and 9-17b (Table 9.7), with row entries, column headings, and the body of the table.

Because readers read from left to right, they interpret tables more easily if you place familiar information on the left and new, important information on the right.

Example 9-17a

Table 9-6 General format of a table

	COLUMN HEADING	
Side Heading	Column Subheading (units)[a]	Column Subheading (units)
Row 1	8.16* (0.84)	n/a
Row 2	7.28 (0.59)	9.45 (0.29)
Row 3	9.68 (0.12)	8.59 (1.55)

Standard errors of the mean are given in parentheses.
[a]Footnote a. *Explanation/footnote.

 Example 9-17 b

Table 9-7 Another general format of a table

	COLUMN HEADING	
Side Heading	Subheading (units)[a]	Subheading (units)
Row 1	n.d.	34
Row 2	n.d.	163
Row 3	25	96

n.d. = not determined.

Organize your tables logically to make mental comparisons easy for readers (e.g., pretreatment measurements should precede posttreatment ones.) In addition, arrange the column headings and the data in a logical order, and make sure that your listings and numbers are aligned. It is also easier to follow similar components when they are arranged vertically, not horizontally. Note that some journals allow or require brief descriptions of the methods to follow the title of a table.

Table Titles

TABLE GUIDELINE 4:

Design table titles to identify the specific topic.

Draft a concise title for each table. State the point of the table or say which items it compares, and perhaps indicate the experimental design.

 Example 9-18 a   Table B Characteristics of intestinal flora

b   Table C Soil analysis

The titles in the preceding example are not sufficient. Titles should be specific and informative without being wordy.

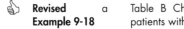

Revised	a	Table B Characteristics of intestinal flora found for 21
Example 9-18		patients with HIV-l
	b	Table C Soil analysis of six farm fields near New Haven, CT

The titles in the revised example are much better titles because they are much more specific. Aim for consistency in style and length of titles.

Column and Row Headings

TABLES GUIDELINE 5:

Label dependent variables in column headings and independent variables in row headings.

Columns and rows in a table typically have a heading. Column headings should label dependent variables and row headings, independent variables.

Each type of information should have its own vertical column, and each column should have its own heading. Column headings consist of headings that identify the items listed in the columns below them, subheadings if necessary, and units of measurement as needed. Do not combine two types of information in one column. For example, under a column headed "Drug," only the names of the drugs should appear, not both the drugs and the doses. Do not omit the heading that states the name of the first column on the left nor the one that states the name of the dependent variable. To keep column headings brief, and thus save space in the table, use short terms or abbreviations in the column headings and subheadings, and explain the abbreviations in footnotes. Note that more abbreviations are used in tables than in the text. If the abbreviation is not defined in a footnote, however, readers who do not know the meaning (and there are always some readers who do not) frustratingly have to search through the text to find the definition.

Capitalize column headings the same as the table title, but use a smaller font. Center one line column headings, and write all headings horizontally. If headings are too long, turn them 90 degrees counter-clockwise.

Row headings typically present independent variables. Keep these headings short as well, and left justify or right justify them. Capitalize the first letter of the first word, and include units where needed.

Table Size

A table must contain enough data to justify its existence, but not too much to overwhelm the reader. If your proposed table has only one or two rows of data, preferably present your findings in one or two sentences in the text instead of constructing a table. Similarly, if your table lists descriptions in words rather than numbers, consider whether you really need a table—a few sentences in the text may be better. Table 9.8 in the next example could easily be converted to text.

 Example 9-19

Table 9-8 Antibiotic targeting of various organisms

ANTIBIOTICS TARGET	ORGANISMS	CELLULAR
I	*Bacillus*	ribosomes
I	*Saccharomyces*	mitochondria
II	*Bacillus*	ribosomes
III	*Streptococcus*	cell walls
III	*Saccharomyces*	mitochondria

 Revised Example 9-19   *Bacillus* ribosomes were targeted by antibiotics I and II, *Streptococcus* cell walls were targeted by antibiotic III, and *Saccharomyces* mitochondria were targeted by antibiotics I and III.

Sometimes an excessively large or small table is necessary or desirable. For example, a large table may be needed to give background data for a large number of subjects or to present individual data sets for all subjects, animals, or specimens. Similarly, sometimes a small table may be desirable to present data, such as for the most important point in the paper, even if the values would take up less space in the text. The reason is that a table has more visual impact.

Avoid oversized tables. Try to design tables that fit easily into one or two journal columns. Do not try to cram a table onto the page by using narrow spacing or tiny print. The copy editor needs room to insert typographical instructions, and too small print is unsuitable for publication. Make the lettering suitable for reduction to 50% to 75% of the original size. If a table seems too big, ask yourself first whether all the columns and rows are needed. For example, columns that contain only one value can be reported in the text or in a footnote. Columns of less important data can be omitted. If you cannot eliminate any columns, try splitting a very wide table into two smaller tables. Alternatively, if your collection of data is very large or likely to interest only a few readers, send your data to an archive as supplementary material for storage separately from the published paper (see *Instructions to Authors* of your target journal). Make sure, however, that the tables you submit for publication contain enough values for referees to assess what you have done and for other scientists to check their results against yours.

9.11   OTHER KINDS OF SUPPLEMENTARY INFORMATION: FORMULAS, EQUATIONS, PROOFS, AND ALGORITHMS

Aside from figure and tables, many scientific fields make use of formulas, equations, proofs, or algorithms. Professional presentation of formulas and equations is an art. Although in this book I do not delve into much detail on these, I describe these additional types of supplementary information in general terms. For a more thorough treatment of formulas and equations, see, for example, Ebel, Bliefert, and Russey (2004).

Formulas and Equations

FORMULAS AND EQUATIONS GUIDELINES:

1. Treat equations and formulas as part of the text.
2. Never assume that the reader is familiar with the problem you are solving.
3. State any assumptions you have made.
4. Explain why and how you arrived at a solution.

Often scientific documents contain formulas or equations. Chemical formulas may have to be created using specialized programs such as ChemSketch or ISIS/Draw. Usually, formulas are set apart in a separate line and indented or centered.

Example 9-20

Mathematical formulas or equations are often created as graphic objects that can be pasted electronically into a word-processing file. The best known programs for mathematical formulas and equations in a word processing context are Tex and Latex. Other programs that enjoy much popularity with mathematicians are FrameMaker, MathType, and MathML. Many word-processing programs contain equation editors. In newer versions of Microsoft Word, the equation editor is available under the Insert menu (select Object and then Equation).

When writing equations, use mathematical notation correctly, and learn how to use symbols properly. Brief equations or definitions may be included in the text if space allows it, although you should not start a sentence with a formula.

Example 9-21 Eligible individuals are vaccinated at a rate of $\kappa_a \cdot (a)\phi(a)$, and a vaccine reduces the infection incidence by $1-\eta = 10\%$ for severe infections and by $1-\xi = 35\%$ for mild infections.

Lengthy equations or mathematical expressions, and those with larger than normal height, should be set apart in a separate line, indented or centered, and isolated from the surrounding text. If you will refer back to them later in your document, number them. Equation numbers are placed at the right margin. An example of such an equation is shown next.

 Example 9-22 The average growth spurt per animal over the age interval i to i + 1 years can be estimated by (34)

$$\lambda_i = -\ln\left(\frac{1-p_{i+1}}{1-p_i}\right).$$
(Eq 3.2)

Even if an equation is displayed, the equation is treated as if it were grammatically part of the text. Punctuation rules apply accordingly. For a sequence of equations, align the = symbol in each line, or, if either side is long, align the = symbol with the first operator in the first line.

Note that you should never assume that the reader is familiar with the problem you are solving. Instead, give an overview of all important details in the problem. State any assumptions you have made, and explain why and how you arrived at a solution. If you are representing quantities and functions with arbitrary letters, define these letters in your formulas.

Equations are often used in proofs. If these proofs are long or if space is limited for the journal article, you may be asked to place these and other elaborate calculations into an appendix. An example of such a proof is shown next.

 Example 9-23 Proof (in an Appendix)

We use mathematical induction to prove that a skewed distribution as described by Equation 1 in the text is an equilibrium consequence of a variable number of tandem repeats (VNTR) model assuming a minimum size of $\alpha > 0$, a maximum size of Ω, and insertions and deletions that are equally probable and that proceed at a rate proportional to the number of tandem repeats.

Let C_j be the count of genotypes of repeat number j. Because insertion and deletion mutations from each size class are assumed to occur at a rate μ times the number of repeats, the average counts in each size class change each $1/\mu$ generations as $C_j' = C_j + (j-1)C_{j-1} - 2jC_j + (j+1)C_{j+1}$. At equilibrium, $\bar{C}_j' = \bar{C}_j$, so that

$$(j-1)\bar{C}_{j-1} - 2j\bar{C}_j + (j+1)\bar{C}_{j+1} = 0.$$
(A1)

Because $\alpha > 0$, C_1 is a boundary, and transitions from C_1 to C_0 are prohibited. Therefore, from Equation A1 for the case of $j = 1 - \bar{C}_1 + 2\bar{C}_2 = 0$, . Thus,

$$\bar{C}_2 = \tfrac{1}{2}\bar{C}_1 .$$
(A2)

Substituting Equation A2 into Equation A1 for the case of $j = 2$, $\bar{C}_1 - 4\left(\frac{1}{2}\bar{C}_1\right) + 3\bar{C}_3 = 0$. Thus,

$$\bar{C}_3 = \tfrac{1}{3}\bar{C}_1. \tag{A3}$$

Similarly, substituting Equation A3 into Equation A2 for the case $j = 3$,

$$\bar{C}_4 = \tfrac{1}{4}\bar{C}_1. \tag{A4}$$

A pattern is apparent. To prove that for all positive integers j, $\bar{C}_j = \tfrac{1}{j}\bar{C}_1$, we assume that for some positive integer k,

$$\bar{C}_{k-1} = \tfrac{1}{k-1}\bar{C}_1, \tag{A5}$$

$$\bar{C}_k = \tfrac{1}{k}\bar{C}_1, \tag{A6}$$

and show that when Equations A5 and A6 are true, it is also true that $\bar{C}_{k+1} = \tfrac{1}{k+1}\bar{C}_1$.

From equations A1, A5, and A6, $(k-1)\left(\tfrac{1}{k-1}\bar{C}_1\right) - 2k\left(\tfrac{1}{k}\bar{C}_1\right) + (k+1)\bar{C}_{k+1} = 0$. Thus,

$$\bar{C}_{k+1} = \tfrac{1}{k+1}\bar{C}_1, \quad Q.E.D. \tag{A7}$$

The result, $\bar{C}_j = \tfrac{1}{j}\bar{C}_1$, is demonstrated for the case $j = 1$ by identity and for integral values of j between 1 and 4 by Equations A2 to A4. It is demonstrated for all integers $j > 4$ by the satisfaction of Equations A5 and A6 for the case $j = 4$ and the proof of Equation A7.

If there is a maximum number of repeats for a VNTR of Ω, beyond which the VNTR may not grow, the equilibrium frequency of size class i, $\bar{P}_i$, is then

$$\bar{P}_i = \frac{\frac{1}{i}}{\displaystyle\sum_{j=\alpha}^{\Omega}\frac{1}{j}} = \frac{1}{i\displaystyle\sum_{j=\alpha}^{\Omega}\frac{1}{j}} \tag{A8}$$

(With permission from Macmillan Publishers Ltd.)

Algorithms

Mathematics, computer science, informatics, computational biology, and other related fields often use algorithms for calculation and data processing. An algorithm is a sequence of well-defined, finite instructions for completing a task.

Algorithms can be presented in the text, as a flow chart, or as a list of instructions. When you want to display what the implementation looks like, it is clearest for readers if, like figures and tables, algorithms are labeled and numbered and set off from the main text, for example, by being boxed. However, if you are discussing an algorithm more abstractly, then a written description may be better as not all descriptions lend themselves to a pseudocode representation (structured English for describing algorithms).

Pseudocode describes the entire logic of the algorithm and allows the designer to focus on the logic without being distracted by details of language syntax. Pseudocode is not a rigorous notation. In fact, each individual designer may have their own personal style of pseudocode, but it is helpful to follow a similar style.

The logic within the pseudocode must be decomposed to the level of a single loop or decision, but not all lines within an algorithm are treated equally. All variables and functions are italicized, and all commands are bold faced. In addition, certain commands that control flow, such as functions, classes, and methods, have associated indentation levels as do **try/catch** blocks and the basic constructs that use the commands **for, if, while,** and **do** (counting loops, selections, repetitions). Constructs can be embedded within each other. If such nested constructs are used, they should be clearly indented from their surrounding constructs, and individual loops and branches should be kept at the same indentation distances as is shown in the following example.

👍 **Example 9-24 General algorithm**

Algorithm 2 mergeSubtrees(StateList *leftList*, StateList *rightList*, node *root*)

Require: *leftList* and *rightList*: the lists of partial states, *root*: a tree node.

Ensure: Set of valid, non-zero probability states combining elements in *leftList* and *rightList*.

1: *mergedList* ←*emptyList*
2: **for** all partial states *l* in *leftList* **do**
3: **for** all partial states r in *rightList* **do**
4: **if** *compatible*(*l* , r) == true **then**
5: *m = merge*(*l* , r)
6: **if** *root* == *initialroot* **then**
7:    *mergedList*.add(m)
8: **else**
9:    **for** op ∈ {C; D; I; C*;D*; I*} **do**
10:     **if** *isPossibleUpstream*(m,op) **then**
11:       *mergedList*.add(*addAncestorBranch*(m,op))
12:     **end if**
13:    **end for**
14: **end if**
15:   **end if**
16: **end for**
17: **end for**
18: **return** *mergedList*

(With permission from Mary Ann Liebert, Inc. Publishers)

References for Formulas, Equations, and Algorithms

References

Alley, M. (1996). *The craft of scientific writing* (3rd ed.). New York: Springer.

Connolly, P., & Vilardi, T. (1989). *Writing to learn mathematics and science.* New York: Teachers College Press.

Steenrod, N., Halmos, P.R., Schiffer, M. M., & Dieudonne, J. A. (1973). *How to write mathematics.* Providence, RI: American Mathematical Society.

Higham, N. J. (1998). *Handbook of writing for the mathematical sciences.* Philadelphia: SIAM.

Knuth, D. E., Larrabee, T., & Roberts, P. M. (1989). *Mathematical writing.* Washington, D.C.: Mathematical Association of America.

Kovac J., & Sherwood, D. W. (2001). *Writing across the chemistry curriculum: An instructor's handbook.* Upper Saddle River, NJ: Prentice Hall.

Krantz, S. G. (1997). *A primer of mathematical writing.* Providence, RI: American Mathematical Society.

Lester , J. D. (1976). *Writing research papers: A complete guide.* Glenview, IL: Scott, Foresman.

Miller, A., & Solomon, P. H. (2000). *Writing reaction mechanisms in organic chemistry.* New York: Academic.

Rubens, P. (1992). *Science and technical writing.* New York: Holt.

Online Sources

A few online sources provide further insight into formulas, equations, and algorithms (last accessed October 2009).

http://books.google.com/books?id=MfKmWlzF73UC&pg=PA234 &lpg=PA234&dq=Miller,+A+and+Solomon+P.H.+-+ Science+%E2%80%93&source=bl&ots=E4c3aayM4M&sig= NM0peu4N2mphVcnjHZZp-SFnwUs&hl=en&sa=X&oi=book_ result&resnum=1&ct=result

http://ems.calumet.purdue.edu/mcss/kevinlee/mathwriting/ writingman.pdf

http://www.math.niu.edu/~behr/Teaching/writing.html

http://www.mit.edu/afs/athena.mit.edu/course/other/mathp2/www/ piil.html

http://www.mit.edu/people/dimitrib/Ten_Rules.pdf

http://web.mit.edu/jrickert/www/mathadvice.html

http://www.geneseo.edu/~mclean/Dept/JournalArticle.pdf

http://versita.com/UserFiles/File/Authors/CEJP/CEJP_ PaperWritingGuide.pdf

SUMMARY

GENERAL ILLUSTRATION GUIDELINES:
1. Figures and tables should be able to stand on their own.
2. Decide whether to present data in a table, a figure, or in the text.

3. Use the fewest figures and tables needed to tell a story.
4. Design figures and tables to have strong visual impact.

Also: Prepare figures and tables with the reader in mind—place information where the reader expects to find it.

GUIDELINES FOR FIGURES

1. Use figures to show trends and relationships and to emphasize data.
2. Prepare professional figures.
3. Do not mislead readers.
4. Place the independent variable at the x-axis and the dependent variable at the y-axis.
5. Make each figure easy to read.
   • Differentiate points, lines, and curves well.
   • Label axes and scales well.
6. The figure legend should be a description of the figure content.

GUIDELINES FOR TABLES

1. Prepare tables rather than graphs when it is important to give precise numbers.
2. Keep the structure of your table as simple as possible.
3. Place familiar context on the left and new, important information on the right.
4. Design table titles to identify the specific topic.
5. Label dependent variables in column headings and independent variables in row headings.

GUIDELINES FOR FORMULAS AND EQUATIONS

1. Treat equations and formulas as part of the text.
2. Never assume that the reader is familiar with the problem you are solving.
3. State any assumptions you have made.
4. Explain why and how you arrived at a solution.

PROBLEMS

Problem 9-1 Figure or Table?
What format (table, graph, text) would you use for the following examples? Justify your choices.

1. You have gathered a series of data concerning ice location, thickness, and consistency in Greenland. How should you present this information?
2. You have examined mortality rates for male and female dogs exposed to the West Nile virus in all the counties of Connecticut. Should you use a table, graph, or figure?

3. You have assessed the effectiveness of a new Lyme disease vaccine after inoculating white rats with various doses of the vaccine and have measured changes in their blood pressure throughout a 2-week period. You also have a really nice photograph of one of your control rats. What should you publish and in what form?

4. You have written a paper about a new species of a bacterial pathogen implicated in a case of bacterial pneumonia. You have a chest roentgenogram showing typical findings of pneumonia and an electron micrograph of newly discovered structural details of the bacterium's flagellum. Should you include either or both in your paper?

5. You want to explain a new type of apparatus, a rotating cylindrical annulus apparatus that was used in your experiments. Should you use a schematic or describe it in the text or both?

6. You have constructed a new drug delivery system involving the use of nanoparticles to deliver anticancer agents to specific brain tumors. Should you explain the system using a schematic or describe it in the text?

Problem 9-2 Figure or Table

Given the raw data provided (in triplicate readings), create a table or figure, whichever represents the data best. Time in hours, units of measurement in ml.

Time 0, KA = { 6.7 (1. reading); 8.1 (2. reading); 7.0 (3.reading) }   KB = { 3.9; 3.9; 3.8 }   TFR = { 0; 0; 0.1 }

Time 2, KA = { 7.2; 7.4; 6.9 }   KB = { 5.0; 5.1; 4.7 }   TFR = { 1.2; 2.1; 1.4 }

Time 3, KA = { 5.8; 6.7; 6.1 }   KB = { 5.4; 5.5; 5.3 }   TFR = { 1.4; 1.0; 1.0 }

Time 4, KA = { 5.8; 6.0; 5.7 }   KB = { 5.9; 5.9; 5.8 }   TFR = { 2.0; 1.0; 1.1 }

Problem 9-3
Evaluate the following graph. How could this graph be improved? Make a list.

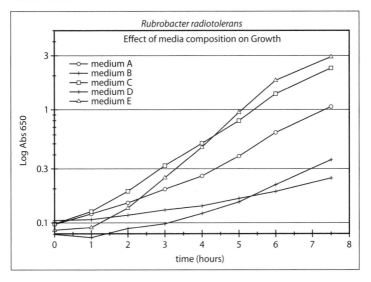

Problem 9-4 Figures
Evaluate the following graphs. Explain why this figure is confusing.

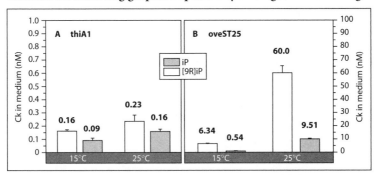

Fig. 1: Thermal induction of cytokinin overproduction in the mutant oveST25. Comparison of cytokinin concentration measured by HPLC-ELISA in the culture medium of thiAl (auxotrophic wt) and its temperature sensitive mutant oveST25. iP (isopentenyladenine) and [9R]iP were measured in the medium of liquid cultures cultivated continuously at 15 °C or 25 °C. The age of the culture was 3 weeks; 2.1-3.5 mg chloronema tissue were used per ml medium.

(With permission from the American Society of Plant Biologists)

Problem 9-5 Table or Figure from the Literature
Find a badly represented table or figure in the scientific literature and discuss why it is bad and how it could be improved.

PART THREE

Manuscripts: Research Papers and Review Articles

A. RESEARCH PAPERS

CHAPTER 10

The Introduction

10.1 OVERALL

INTRODUCTION GUIDELINE 1:
Interest your audience and provide context.

The purpose of the Introduction is twofold: to interest your audience to read the paper and to provide sufficient context or background information for readers to understand your study independently of other previous publications on the topic. Often, the Introduction also gives an overview of what to expect in the paper. Note that the Introduction may repeat some parts of the Abstract, which is ok.

10.2 CONTENT AND ORGANIZATION

INTRODUCTION GUIDELINE 2:
Follow a "funnel" structure.

Include:
Background
Unknown/Problem
Question/Purpose of Study
Experimental Approach
Optional: Results/Conclusion
Significance

INTRODUCTION GUIDELINE 3:
Keep the Introduction short.

Readers have relatively fixed expectations about where in a document they will encounter particular items. Based on the location of these items, they will interpret the text. If writers can become aware of these locations, they can better guide the reader through the document, highlighting and emphasizing various pieces of information depending on the degree of importance. Generally, readers expect the parts of the Introduction to be arranged in a standard structure: a **"funnel,"** starting broadly with background information and then narrowing to what is the question of the paper (see also Zeiger, 2000).

The Introduction should be as short as possible but contain all the information needed to lead into the work. Ideally, an introduction of a typical journal article should be one to two double-spaced pages (about 250–600 words). Check the *Instructions to Authors* of your target journal to ensure that you are within the set word limits for the Introduction.

Most research papers in basic science are investigative. That is, they are based on specific research questions you try to answer or on a particular hypothesis you try to test. Introductions for these papers should contain the following elements:

1. Background	broad and specific background information and previous research in the area
2. Unknown/Problem	problems of previous work and unknown factors in the area
3. Question/Purpose of Study	addition made by your research
4. Experimental Approach	approach taken toward this addition

The Introduction should funnel from broad general background, to knowledge on a specific aspect of the topic, to something unknown or problematic, and then to the research question of the paper and its experimental approach. Although not an absolute necessity, I recommend including your main results and conclusions as well as to state the overall significance of the paper to round up this section. If you include main results and conclusions, place them at the end of the Introduction. Including results and conclusion in the Introduction will let your readers know what to expect and will let them more easily follow the paper. If your paper deals with a controversial topic in your field, however, you may consider withholding main results and conclusions in your Introduction to encourage as many readers as possible to continue reading your paper and argumentation.

10.3   ELEMENTS OF THE INTRODUCTION

Background

INTRODUCTION GUIDELINE 4:

Provide pertinent background information,
but do not review the literature.

Start the Introduction by providing some background information. The amount of background information needed depends on how much the intended audience can be expected to know about the topic. You should start very broad to provide some general context of your work. Then write about the specific aspect of the topic that is of interest, mention the existing research in the area, and discuss current beliefs. A good partial introduction, in which the background starts very broad but then narrows down quickly to the research topic, is shown next (see also Example 10-7 and Revised Example 10-10):

 Example 10-1 A partial Introduction showing good funneling of background

Broad background

In mammals, the auditory hair cells of the inner ear are the sensory receptors of the auditory system. Two functionally and anatomically distinct types of mammalian auditory hair cells exist: inner and outer hair cells. Outer hair cells do not send neural signals to the brain, but they mechanically amplify low-level sound that enters the inner ear (1). The amplification is powered by an electrically driven motility of their cell bodies (2).

Specific background

Unknown/ Problem

The molecular basis of this mechanism is thought to be the motor protein prestin, which is embedded in the lateral membrane of the outer hair cells. Mammalian prestin is an 80 kDa, 744 amino acid membrane protein whose function appears to depend on chloride channel signaling (3,4). Although prestin has been researched intensively, its molecular function has not been fully established.

Note that you should not review the topic when you are writing a research paper. A summary pertinent to the research you are presenting in the paper should suffice. In the following example, the Introduction of Example 10-1 is shown again; but in this version, the author has reviewed the topic. Too much irrelevant information (underlined) has been included. Consequently, the introduction does not clearly funnel down to the topic of interest (the molecular function of prestin), and readers get confused because they do not know what aspect of the background information to focus on.

Example 10-1 B   Introduction that reads like a review

In mammals, the auditory hair cells of the inner ear are the sensory receptors of the auditory system and the vestibular system. <u>The auditory hair cells are located within the organ of Corti on a thin basilar membrane in the cochlea of the inner ear. Their name derives from a structure known as the hair bundle or stereocilia found on the apical surface of the cell, which extends into the scala media within the cochlea. Damage to the hair cells results in sensorineural hearing loss.</u>

Two functionally and anatomically distinct types of mammalian auditory hair cells exist: inner and outer hair cells. In inner hair cells, the stereocilia are deflected mechanically, thus opening gated ion channels and allowing positively charged potassium and calcium to enter the cell. The influx of these ions depolarizes the cell, resulting in a receptor potential that subsequently triggers the release of neurotransmitters at the basal end of the cell. The neurotransmitters in turn trigger action potentials in the nerve, converting the mechanical sound signal into an electrical nerve signal.

Outer hair cells, which have evolved only in mammals, do not send neural signals to the brain, but they mechanically amplify low-level sound that enters the inner ear (1). The amplification is powered by an electrically driven motility of their cell bodies (2). <u>This so-called somatic electromotility consists of oscillations of the cell's length, which occur at the frequency of the incoming sound and in a stable phase relation. Outer hair cells have not improved hearing sensitivity of mammals, but have extended the hearing range from about 11 kHz (maximum in some birds) to about 200 kHz (maximum in some marine mammals). They have also improved frequency selectivity (frequency discrimination), enabling sophisticated human speech.</u>

Background starts broad, but then goes into irrelevant details that are not important for the research topic

The molecular basis of the electrically driven motility of outer hair cells is thought to be the motor protein prestin, which is embedded in the lateral membrane of the outer hair cells. Mammalian prestin is an 80 kDa, 744 amino acid membrane protein whose function depends on chloride channel signaling (3,4). <u>Prestin is compromised by the common marine pesticide tributyltin (TBT) as has been shown by high concentrations of prestin in Orcas and toothed whales.</u> Although prestin has been researched intensively, its molecular function has not been fully established.

The specific background also contains some irrelevant information

The Unknown/Problem

INTRODUCTION GUIDELINE 5:
State the unknown or problem.

After discussing general background and specific aspects of existing research, describe what the problems with the existing research are or what is unknown. The unknown is clearest if you signal it by stating it directly, for example, "X is unknown" or "Y is unclear." You can also use other phrases to state the unknown outright: "has not been established," "or "has not been determined." Alternatively, you can imply rather than state the unknown by using a suggestion or a possibility ("Previous findings suggest that …"; see also Table 10-1 in the section on "Signals for the Reader.")

Use an objective tone when criticizing previous work. Avoid antagonistic phrases:

👎 *Not appropriate*	👍 *Better*
… does not seem to understand …	The results of study X have been questioned.
… failed to …	One study found A, another study found B.
… made the mistake of …	Findings on X are controversial.
… used improper methods …	Although A showed X, our results do not agree …

Also, do not blame individual authors or teams. You may end up creating your own enemies who one day may be reviewers of one of your papers or grant proposals.

The Question/Purpose

INTRODUCTION GUIDELINE 6:
State the central point (question/purpose) precisely.

The most important element in a research paper is the research question or purpose of the work. The question/purpose is the "central point" of your Introduction and of the paper as a whole. It therefore needs to be worded very carefully. If the central point is stated precisely, the reader immediately has an idea of what to expect in the paper. Furthermore, the reader can read the paper in a directed way rather than blindly, and the experiments make more sense.

Because the question/purpose provides an overview of the entire paper, and every paragraph and sentence in your paper relates to it, I recommend for you to write your research question/purpose onto a Post-it note before you start composing your manuscript. Place this note on the side or top of your computer screen where it cannot be overlooked. It will remind you to keep your writing focused on the question/purpose of the paper.

The research question/purpose of a research paper should name the variables studied as well as the main features of the study. Note that the question/purpose is usually not written in the form of a question but as an infinitive phrase or as a sentence, using a present tense verb, as in the next examples:

👍	**Example 10-2**	**Phrasing of question/purpose**
	a	To determine if the triggered cellular processes **affect** the rRNA structure and folding dynamic **in vivo,** …
	b	Here we asked how rheumatic fever **influences** heart rate.

> c   In this study, we show that a sequential scheme of phosphor-
> ylation and dephosphorylation **can** generate circadian oscil-
> lations.
>
> d   Here we examine the effects of total carbonate concentration
> on U(VI) adsorption.

The research question should follow logically from the previous statements of what is known or believed and what is still unknown or problematic. Thus, the topic of the research question should be the same as the
one found in what is known. Equally important, the research question
should be the question the reader would expect after reading about what
is unknown or problematic.

Experimental Approach

INTRODUCTION GUIDELINE 7:

State the experimental approach briefly.

In the Introduction of your research paper, you should also briefly indicate your experimental approach. In general, the experimental approach is
short—usually one sentence, at most, two or three sentences. The experimental approach should be signaled so readers can identify it immediately.
Examples of how to signal the experimental approach are shown in Example
10-3 (see also Table 10-1 in the section on "Signals for the Reader").

Example 10-3   **Signals for the experimental approach**

a   **We analyzed X by** agarose gel electrophoresis.

b   We simulated Tropical Instability Waves **using** a constant coefficient Laplacian friction scheme.

c   The structures of the compounds **were characterized by** UV, IR,
1H NMR, 19F NMR spectra, and HRMS.

Results and Conclusion

After the experimental approach, you may briefly state your main results
and conclusion. Although their inclusion is not a must, know that readers like to read about the main results and conclusion of your work in
the Introduction. Most readers dislike having to read the whole paper,
waiting and searching for the answer to the research question. In some
journals, results and conclusions are delayed until the Discussion section
of the paper. If you have the chance to include results and conclusions in
your Introduction, however, do so. Readers will be thankful for it. See also
Table 10–1 in the section on "Signals for the Reader" for ways to signal
results in the Introduction.

Significance and Implication

Consider stating why your findings are important. Do so by stating what
the significance or implication of the study is, as shown in the following
example.

Example 10-4	**Stating the significance or implication**
a	X is an important addition to...
b	... which aids in the elucidation of...

If you state the significance or implication at the very end of your Introduction, it not only rounds up this section nicely, but also provides the overall perspective of your work for the reader. See also Table 10–1 in the section on "Signals for the Reader" for additional ideas on how to signal the significance or implication of your article.

10.4   SPECIAL CASE: INTRODUCTIONS FOR DESCRIPTIVE PAPERS

> For descriptive papers, include
> Background
> Discovery statement
> Experimental approach—if appropriate
> Description
> Implication
> State an implication at the end of the Introduction
> of a descriptive paper.

Some research papers are not written to answer specific questions or prove a hypothesis but rather to describe a new finding such as a new organism, an unknown disease, or a novel apparatus. This category also includes methods papers. Introductions for these papers follow a slightly different "funnel" structure because elements that readers expect to find in the Introduction of a descriptive paper differ from those of an investigative paper. Elements that should be contained in the Introduction of a descriptive paper include:

1. Background	background information and previous research in the area
2. Discovery statement	new discovery made by your research
3. Experimental approach (optional)	approach taken toward analyzing this discovery
4. Description	description of the new element
5. Implication	importance of the findings

In descriptive papers, the Introduction should funnel from the general background to the specific aspect (if it is known), to the discovery statement, and then to the description and implication. For some descriptive papers, the reason for the discovery or previous problems before discovery can also be included in the Introduction. Most descriptive papers do not include an unknown or problem statement found in other research papers.

The most important element in a descriptive paper is the discovery statement. This statement is usually written as a sentence, which should follow logically from the background information provided. The verb tense of the discovery statement can be either present tense or past tense, depending on how the statement is worded.

 Example 10-5   **Phrasing the discovery statement**

a   Here we **describe** a novel bacterial protease, Ecpro-3.

b   We **discovered** a new toxigenic serogroup, O210, of Vibrio cholerae. O210 can produce cholera toxin (CT) and cause severe symptoms.

c   We **examined** the structure of the bacterial ribosome in complex with tetracycline.

d   Here, we **present** a classification algorithm that classifies vegetation from NDVI time-series according to the shape of the temporal cycle.

After the discovery statement, provide an overview of the novelty being reported (new method, new apparatus, new organism). This description may also contain information on how the discovery was made or tested. It may even point out what the advantages and disadvantages of the discovery are. The description of the discovery should be written in present tense ("X is heat resistant and does not appear affected by high pH."). Note that for descriptive papers, an overview of an experimental approach may not be needed. However, stating the implication is particularly important at the end of the Introduction.

10.5   IMPORTANT WRITING PRINCIPLES FOR THE INTRODUCTION

Tense

In the Introduction, you will find a mix of verb tenses, often in the same sentence. The general rule about verb tense applies (Chapter 4, Section 4.4). When reporting completed actions, such as when you are referring to previously reported studies, use past tense. For statements of general validity and those whose information is still true, that is, if you consider a finding to be a general rule, use present tense. When stating the question/purpose of your research, use present tense as well; but when describing the experimental approach, use past tense. An example of a mixed tense sentence is shown below.

 Example 10-6   **Use of tense**

Herbert et al (9) **found** that peanut butter **can be contaminated** with Salmonella.

This example shows that the completed action ("found") is described in past tense, whereas the information that still holds true ("can be contaminated") is described in present tense.

Note that previously established knowledge in the Introduction should be described in present tense. If you use past tense for describing results of already published work, you are implying to the reader that you do not consider these results to be "facts" but observations.

Using Strong Verbs and Short Sentences

The initial part of the Introduction often is written in more lay terms than the rest of your manuscript. As this part of the Introduction is also more often read by a larger lay audience, it is particularly important to use short sentences and strong verbs instead of nominalizations.

Coherence and Cohesion

INTRODUCTION GUIDELINE 8:
Ensure good cohesion and coherence.

In long introductions, the story line can be difficult to follow, because you are telling the overall story (background, unknown, research question and/or discovery statement, experimental approach) in addition to separate smaller stories within paragraphs.

To ensure good flow or continuity, that is, to ensure that the overall story is clear, you should use all the techniques of cohesion and coherence presented in this book (see Chapter 6):

- Sentence location
- Topic sentences
- Word location
- Key terms
- Transitions

10.6   SIGNALS FOR THE READER

INTRODUCTION GUIDELINE 9:
Signal all the elements of the Introduction.

Generally, all the parts of the Introduction should be signaled so the reader does not have to guess about the information provided. The signals vary depending on how the known, unknown, question, and experimental approach are phrased. Numerous variations on these signals are possible. Some examples are listed in Table 10-1 (additional examples are also highlighted in the bubbles for examples shown previously.)

These examples provide great starting points when you write your first draft and may be particularly useful for those authors who have writer's block or whose native language is not English.

ESL advice

Table 10-1-Signals of the Introduction

BACKGROUND	UNKNOWN	QUESTION, PURPOSE, OR DISCOVERY	EXPERIMENTAL APPROACH	RESULTS	IMPLICATION
X is …	… is unknown	We hypothesized that…	To test this hypothesis, we …	We found…	…consistent with
X affects…	… has not been determined	To determine…	we. …	…was found	…indicating that
X is a component of Y	The question remains whether ….	To study …, To examine…, To assess… To analyze… In this study we examined …	We analyzed… For this purpose, we …	We determined …	… make it possible to …
X is observed when Y happens …	… is unclear	Here we describe …	… by/using …	Our findings were…	…may be used to…
X is considered to be …		Here we report… This report describes…	For this study we…	We observed that…	… is important for…
X causes Y	… does not exist …is not known	We examined whether X is … We assessed if… We determined if … We analyzed Y…	To evaluate…, we… To answer this question, we…	Based on our observations …	Our analysis implies/ suggests … Our findings indicate that …

10.7   COMMON PROBLEMS OF INTRODUCTIONS

The most common problems of Introductions include

- Missing elements (unknown/problem, question, experimental approach) (Section 10.3)
- Obscured elements (Section 10.5)
- Excessive length (Section 10.2)
- Context/background is too narrow (Section 10.3)
- Overview sentences

Missing or Obscured Elements of the Introduction

If one or more of the elements of the Introduction is missing or obscured, readers get confused and frustrated because they come away with the feeling that they have not fully understood what the paper is about. The most common elements that are missing are the unknown and the experimental approach. Sometimes even the question/purpose is not stated or not stated clearly. Obscured elements commonly arise when authors have not clearly signaled the unknown/problem, question/purpose, or experimental approach.

Consider, for example, the following introduction in which the unknown and research question is missing:

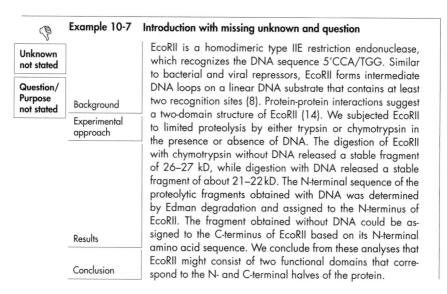

Example 10-7   Introduction with missing unknown and question

Unknown not stated

Question/ Purpose not stated

Background

Experimental approach

Results

Conclusion

EcoRII is a homodimeric type IIE restriction endonuclease, which recognizes the DNA sequence 5'CCA/TGG. Similar to bacterial and viral repressors, EcoRII forms intermediate DNA loops on a linear DNA substrate that contains at least two recognition sites (8). Protein-protein interactions suggest a two-domain structure of EcoRII (14). We subjected EcoRII to limited proteolysis by either trypsin or chymotrypsin in the presence or absence of DNA. The digestion of EcoRII with chymotrypsin without DNA released a stable fragment of 26–27 kD, while digestion with DNA released a stable fragment of about 21–22 kD. The N-terminal sequence of the proteolytic fragments obtained with DNA was determined by Edman degradation and assigned to the N-terminus of EcoRII. The fragment obtained without DNA could be assigned to the C-terminus of EcoRII based on its N-terminal amino acid sequence. We conclude from these analyses that EcoRII might consist of two functional domains that correspond to the N- and C-terminal halves of the protein.

When readers read this Introduction, they are left in the dark as to why the proteolysis was done. The concluding sentence at the very end of the Introduction hints at the reason ("functional domains"), but readers will not know for sure. If the unknown and question would have been stated, readers could have read the Introduction in a directed way.

Revised Example 10-7

Background

Unknown

Question/Purpose

EcoRII is a homodimeric type IIE restriction endonuclease, which recognizes the DNA sequence 5'CCA/TGG. Similar to bacterial and viral repressors, EcoRII forms intermediate DNA loops on a linear DNA substrate that contains at least two recognition sites (8). Although protein-protein interactions suggested a two-domain structure of EcoRII (14), the functional domains of EcoRII remain unclear. To study the functional domains of the endonuclease, we subjected

Experimental approach	EcoRII to limited proteolysis by either trypsin or chymotrypsin in the presence or absence of DNA. This limited proteoplysis can provide information about tightly folded regions of the protein. The digestion of EcoRII with chymotrypsin without DNA released a stable fragment of 26–27 kD, while digestion with DNA released a stable fragment of about 21–22 kD. The N-terminal sequence of the proteolytic fragments obtained with DNA was determined by Edman degradation and assigned to the N-terminus of EcoRII. The fragment obtained without DNA
Results	could be assigned to the C-terminus of EcoRII based on its N-terminal amino acid sequence. We conclude from these data that EcoRII might consist of two functional domains that
Conclusion	correspond to the N- and C-terminal halves of the protein.

The following is another example of an incomplete Introduction in which the unknown has not been stated:

Example 10-8   Introduction with missing unknown

Modeling ocean currents is difficult due to their nonlinear and hence complex nature. Contributing to the nonlinearity are eddies, such as Tropical Instability Waves (TIWs) (1). TIWs appear as monthly oscillations of the currents, sea level, and sea surface temperature of the eastern equatorial Pacific. They are understood as unstable waves feeding on the kinetic and potential energy of the mean currents. These waves not only contribute to the climate variability over the Atlantic but also are influenced by it. Reproducing these eddies in a model is essential for obtaining the correct flow structure and momentum balances (2, 3), both of which influence other ocean features such as energy balances and ocean temperatures, thus affecting global climate.

Not all proposed TIW models resolve these waves effectively. Possible reasons include coarse resolutions of models or the use of a simple friction scheme. Friction is often necessary to slow the strong zonal equatorial currents produced in models. However, using a simple friction scheme to damp these currents has the side effect of suppressing some of the eddy activity (4).

Unknown not stated   Background

Question/Purpose
Experimental approach

To determine the effect of eddies on depth-integrated transports and friction to the system, we employed a high resolution ocean circulation model with a biharmonic Smagorinsky friction scheme (5) that allows TIWs to be well resolved. We found that under such a scheme the depth-integrated currents can be almost completely described by windstress and nonlinearity alone, without the need for a large frictional damping. We also explored how the inclusion of TIWs to our model provides its own friction to the system via the nonlinear advection term. Our results indicate that care must

Results and significance   be taken in resolving eddies near the equator, if the correct dynamics are to be resolved.

(With permission from Jaclyn Brown, PhD)

The revised example is much easier to follow as the unknown is pointed out clearly.

Revised **Example 10-8**	... Not all proposed TIW models resolve these waves effectively. Possible reasons include coarse resolutions of models or the use of a simple friction scheme. Friction is often necessary to slow the strong zonal equatorial currents produced in models. However, using a simple friction scheme to damp these currents has the side effect of suppressing some of the eddy activity (4). <u>It is unclear how eddies act as a friction to the system and how they affect depth-integrated transports.</u>
Unknown	
Question/ Purpose	To determine the effect of eddies on depth-integrated transports and friction to the system, we employed a high resolution ocean circulation model with a biharmonic Smagorinsky friction scheme (5) that allows TIWs to be well resolved....

Excessive Length

This problem usually arises when authors review the topic rather than funnel down stringently from background information to the research question. Readers expect the author to guide them through the pertinent information on the topic. If the author reviews the topic, the reader does not know what topic to focus on or which topic is pertinent for the paper.

In the next example, the author has reviewed the topic. The amount of detail not pertinent to the question of the paper distracts and confuses readers.

Example 10-9	**Introduction that reviews the topic**
	Age-related macular degeneration (AMD) is an eye disease that destroys central vision by damaging the macula. It is the most common cause of vision loss in the United States in the elderly, and its prevalence increases with age [1,2,3]. AMD is caused by hardening of the arteries that nourish the retina. It is likely to be a mechanistically heterogeneous group of disorders. Clinically, AMD can be divided into two forms, atrophic (dry) or exudative (wet) [4,5]. The atrophic form is the most common form, representing approximately 90% of all AMD cases. However, atrophic AMD accounts for only about 12% of the severe vision loss associated with AMD.
	Wet AMD represents approximately 10% of all cases of AMD but is responsible for about 88% of the severe vision loss associated with the disease [5,6]. Both genetic and environmental factors, such as cigarette smoking, nutrition, obesity, and lipid level are likely to play a role in the development and progression of AMD [7–10].
Background too broad—**unclear** **what the** **specific aspect** **of interest is**	The pathogenesis of AMD is related to adverse vascular changes. In fact, AMD and cardiovascular disease may share common antecedents [4]. Specifically, it has been suggested that elevated plasma concentrations of total homocysteine not only increase the risk of vascular disease [5,6] but also that of exudative neovascular AMD [7].

Background too broad—**unclear what the specific aspect of interest is**

Homocysteine is a sulphydryl-containing amino acid derived from the demethylation of methionine. The total plasma homocysteine level can be influenced by genetic defects, renal impairment, and various drugs and diseases [9]. Dietary factors have also been shown to play an important role in the control of homocysteine levels, as homocysteine metabolism is dependent on reactions involving vitamins B-6, B-12, and folate for transsulfuration and remethylation [10]. High levels of plasma homocysteine are toxic to the vascular endothelium by releasing free radicals, thus affecting the vessel walls and increasing the risk for thrombosis [11, 12].

Unknown: "...is unclear."

Question/Purpose: "To examine ..."

Experimental approach: "... we studied ..."

Although a recent study suggested a possible involvement of increased plasma homocysteine levels in AMD [6], the association of homocysteine and AMD is controversial [9], and any relationship between homocysteine levels and AMD is unclear [4]. To examine if there is any the correlation between homocysteine levels and AMD, we studied the relationship between levels of plasma homocysteine and AMD in an independent study of 520 men and women 50 years or older.

In the revision, excess detail has been removed. Readers can now focus on the specific aspect of the previous research that the paper covers.

Revised Example 10-9

Background— now shorter and more focused

Age-related macular degeneration (AMD) is an eye disease that destroys central vision by damaging the macula. It is the most common cause of vision loss in the United States in the elderly, and its prevalence increases with age [1,2,3]. The pathogenesis of AMD is related to adverse vascular changes. In fact, AMD and cardiovascular disease may share common antecedents [4]. Specifically, it has been suggested that elevated plasma concentrations of total homocysteine not only increase the risk of vascular disease [5,6] but also that of exudative neovascular AMD [7]. However, vascular disease has been associated with AMD in some but not in all epidemiologic studies [8].

Specific background— now focused

Specific background

Homocysteine is a sulphydryl-containing amino acid derived from the demethylation of methionine. The total plasma homocysteine level can be influenced by genetic defects, renal impairment, and various drugs and diseases [9]. Dietary factors have also been shown to play an important role in the control of homocysteine levels [10]. High levels of plasma homocysteine are toxic to the vascular endothelium by releasing free radicals, thus affecting the vessel walls and increasing the risk for thrombosis [11, 12].

Unknown: "...is unclear."

Question/ Purpose: "To examine ..."

Experimental approach: "... we studied ..."

Although a recent study suggested a possible involvement of increased plasma homocysteine levels in AMD [6], the association of homocysteine and AMD is controversial [9], and any relationship between homocysteine levels and AMD is unclear [4]. To examine if there is any the correlation between homocysteine levels and AMD, we studied the relationship between levels of plasma homocysteine and AMD in an independent study of 520 men and women 50 years of age or older.

Background Is Too Narrow

If the background information provided in your Introduction is too narrow, most readers will not be enticed to keep reading, because they will feel lost from the start. You will also lose your readers if general background information is absent, too abstract, or too technical. The result is that readers will be missing context, that is, a general overview of the topic is lacking.

Consider the following two openings of an Introduction. Which one would you find more interesting as a reader?

Example 10-10 A   Opening of Introduction

FR2 is a member of the DExD/H-box family of proteins (1). DExD/H-box family proteins possess NTPase and often helicase activity (1). FR2 exhibits NTPase and helicase activity from its C-terminal helicase domain (FR2hel) (2,3). FR2 also binds to HCV NS4A to form the complex FR23–4A. FR2–4A exhibits serine protease activity from its N-terminal protease domain (4,5) and is localized to the surface of the endoplasmic reticulum via NS4A (6).

Example 10-10 B   Opening of Introduction

Hepatitis C, which is caused by the Hepatitis C virus (HCV), infects an estimated 170 million people worldwide and 4 million in the United States. An essential replicative component of HCV is FR2 (1,2). FR2 is a member of the DExD/H-box family of proteins (3). Like other members of this family, FR2 exhibits NTPase (3). In addition, FR2 also displays helicase activity from its C-terminal helicase domain (FR2hel) (4,5), an activity that is often seen in other DExD/H-box family members (3). Furthermore, FR2 binds to HCV NS4A to form the complex FR2–4A, which exhibits serine protease activity from its N-terminal protease domain (6,7) and which is localized to the surface of the endoplasmic reticulum via NS4A (8).

Most readers prefer Example 10-10B because its opening is much more generally understandable than that of Example 10-10A. Composing particularly the first sentences and paragraphs of the Introduction in more lay terms and relating the topic to common problems or topics of public interest will entice more readers than an opening that jumps right into technical specifications and information or one that assumes prior knowledge of a specific scientific field.

Remember also that short sentences carry more weight than long sentences. Writing a short opening sentence in your Introduction will not only be easier to understand for most readers, it will also be more interesting and grab more of their attention.

Overview Sentences

On occasion, authors blend introductions of research and review papers. In such mixed introductions most sentences are informative, but the last sentence is usually an overview of the paper, stating that something will be described or discussed in the text. Such statements are not useful in an introduction for a research article and should be avoided.

10.8   SAMPLE INTRODUCTIONS

This following short, one-paragraph Introduction consists of only three sentences, yet it contains all the elements necessary for the reader to follow: background, unknown, research question, and experimental approach. Because this Introduction funnels down through all the required elements, the reader can logically follow the Introduction of the paper.

Example 10-11   Complete, short sample Introduction

Background	CT-3 has shown marked antiallodynic and analgesic effects in animals (1). However, it has not been determined whether
Unknown	CT-3 also possesses the ability to treat neuropathic pain in
Question/ Purpose	humans. To examine the analgesic efficacy and safety of CT-3 for chronic neuropathic pain in humans, we conducted
Experimental approach	a randomized, double blind, placebo-controlled crossover study on 21 patients for 5 weeks using two daily doses of 10 mg CT-3 or placebo.

Longer Introductions should also contain all these required elements. For longer Introductions, it is especially important to follow the funnel structure because additional smaller stories are incorporated within the paragraphs. If the funnel structure is not followed, the reader may get confused and frustrated. Two examples of longer Introductions that include all the essential elements are shown next:

Example 10-12   Complete Introduction

Broad background	Remote sensing through satellites is often used for investigating land-surface phenologies. These phenologies, measured as Normalized Difference Vegetation Index (NDVI) data, can provide a baseline from which to monitor changes in vegetation associated with events such as fire, drought, climate fluctuations, and climate change. A known source of uncertainty limiting the use of remotely sensed time-series is the degree of noise found in the signals. Typical causes for noise include atmospheric influences, sensor performance, and efficiency of post acquisition cloud- and compositing-algorithms (1). To minimize this noise, various authors have demonstrated the efficiency of Fourier filtering (2–5). There are, however, limits to this method beyond which the representation of the gen-
Problem of research topic	eral shape of a periodic cycle is too inaccurate. Such limits exist particularly when cycles show erratic outliers as compared to easy to correct periodic noise features.
Discovery/ new method	Here, we present an algorithm that classifies vegetation from NDVI time-series according to the shape of the temporal cycle and is minimally affected by atmospheric and sensor effects. The algorithm is derived from examples from the Middle East and Central Asia using 250 m Modis NDVI data and based

| Discovery/
new method | on the Fourier components magnitude and phase that can be applied to periodic cycles. High shape fidelity can be ensured through several user controls. The algorithm is invariant to cycle modifications that may be caused by climate, soil, or topography, but are unrelated to the vegetation type. The output is a highly consistent clustering of NDVI-cycles, which can be linked to distinct vegetation types or land use practices. The algorithm allows vegetation changes to be monitored with the possibility to distinguish between pure coverage fluctuations and actual phenological changes. |
| General
description | |

(With permission from the ISPRS Journal of Photogrammetry and Remote Sensing)

Example 10-13 Complete Introduction

Broad background	The biological activities of proteins depend not only on their amino acid sequences, but also on their discrete conformations. Even slight perturbations of the conformations of a protein may render it inactive (1).... Several reports have been published regarding the isolation and purification of recombinant proteins from inclusion bodies, but the methods employed for renaturation may not be generally applicable (5–14). A more general method for determining renaturation conditions would be of significant value because in most cases very little is known about the renaturation of the protein of interest.
Specific background	Many naturally occurring proteins have been the subject of renaturation studies (13,15). The rate of renaturation and yield of active enzyme depend not only on the mode of denaturation and the enzyme concentration during renaturation, but also on the solvent composition.... Although much is known about the role of denaturants, reducing conditions, and protein concentration on refolding, the effect of solvent composition has not received much systematic study.
Problem of research topic "… not much systematic study."	
Discovery/ new method	Here we report a new sparse matrix method employing 50 different solvent systems for establishing initial solvent conditions that facilitate protein renaturation.... This screening matrix is based on a set of solutions originally selected for their demonstrated utility for protein crystallization (22). Our results suggest that this screening method may be widely applicable in identifying conditions that support renaturation and that the same conditions which promote protein crystallizations may also promote protein renaturation.
General description	
Implication: "Our results suggest … may be applicable …"	

(With permission from Elsevier)

10.9   REVISING THE INTRODUCTION

When you have finished writing the Introduction (or if you are asked to edit an Introduction for a colleague), you can use the following checklist to "dissect" the Introduction systematically. In your revisions and in editing papers, work your way backward from paragraph location and structure to word choice and spelling.

☐   1. Does the topic present something new and interesting?
☐   2. Are all the components there? To ensure that all necessary components are present, on the margins of the Introduction clearly mark

	Investigative Paper	Descriptive Paper
	Background	Background
	Unknown	(reason/problem)
	Question/Purpose	Discovery statement
	Experimental approach	(experimental approach)
	(results/conclusion)	Description
	(significance)	Implication

☐   3. Is the research question stated precisely? (Is it in present tense?)
☐   4. Do all the components logically follow each other? (Is the unknown what one would expect to hear after reading about what is known? Is the research question really the question one would expect to read after reading the unknown? Does the answer really answer the research question?)
☐   5. Is all background information directly relevant to your research question? Did you list only the most pertinent literature and not review the topic? Is the Introduction less than two double-spaced pages?
☐   6. Have all elements been signaled clearly?
☐   7. Did you keep the Introduction short?
☐   8. Are references placed correctly and where needed? (Chapter 8)
☐   9. Is the introduction cohesive and coherent? (see Chapter 6, Section 6.3)
☐ 10. Has sentence location been considered? (see Chapter 6, Section 6.1)
☐ 11. Are topic sentences used well? (see Chapter 6, Section 6.1)
☐ 12. Revise for style and composition using the writing principles of this book:
   ☐   a. Are paragraphs consistent? (Chapter 6, Section 6.2)
   ☐   b. Are paragraphs cohesive? (Chapter 6, Section 6.3)
   ☐   c. Are key terms consistent? (Chapter 6, Section 6.3)
   ☐   d. Are key terms linked? (Chapter 6, Section 6.3)
   ☐   e. Are transitions used, and do they make sense? (Chapter 6, Section 6.3)
   ☐   f. Is the action in the verbs? Are nominalizations avoided? (Chapter 4, Section 4.6)
   ☐   g. Did you vary sentence length and use one idea per sentence? (Chapter 4, Section 4.5)
   ☐   h. Are lists parallel? (Chapter 4, Section 4.9)
   ☐   i. Are comparisons written correctly? (Chapter 4, Sections 4.9 and 4.10)

☐   j. Have noun clusters been resolved? (Chapter 4, Section 4.7)

☐   k. Has word location been considered? (Verb following subject immediately? Old, short information at the beginning of the sentence? New, long information at the end of the sentence?) (Chapter 3, Section 3.1)

☐   l. Have grammar and technical style been considered? (person, voice, tense, pronouns, prepositions, articles?; Chapter 4, Sections 4.1–4.4)

☐   m. Are words and phrases precise? (Chapter 2, Sections 2.2 and 2.3)

☐   n. Are nontechnical words and phrases simple? (Chapter 2, Section 2.2)

☐   o. Have unnecessary terms (redundancies, jargon) been reduced? (Chapter 2, Section 2.4)

☐   p. Have spelling and punctuation been checked? (Chapter 4, Section 4.11)

SUMMARY

INTRODUCTION GUIDELINES:

1. Interest your audience and provide context.
2. Follow the "funnel" structure:
   a. For investigative papers, include:
      Background
      Unknown
      Question/Purpose
      Experimental approach
      Optional: Results/Conclusion
      Significance
   b. For descriptive papers, include:
      Background
      Discovery statement
      Experimental approach—if appropriate
      Description
      Implication
   State an implication at the end of the Introduction of a descriptive paper.
3. Keep the Introduction short.
4. Provide pertinent background information, but do not review the literature.
5. State the unknown or problem.
6. State the central point (question/purpose or discovery statement) precisely.
7. State the experimental approach briefly.
8. Ensure good cohesion and coherence.
9. Signal all the elements of the introduction.

PROBLEMS

Problem 10-1

Why does this introduction seem incomplete? Identify the known, unknown, question/purpose, and experimental approach. Are these elements clearly identifiable? Why or why not?

Astaxanthin is a carotenoid that is found in microalgae, yeast, salmon, trout, krill, shrimp, crayfish, crustaceans, and the feathers of some birds (1,2). Astaxanthin is a natural nutritional component, but it is also used as a food supplement intended for human, animal, and aquaculture consumption (4).

Like many carotenoids, astaxanthin is a colorful, fat/oil-soluble pigment, providing a redish and pink coloration (2). Whereas in certain bird species, all adult members display carotenoid containing feathers rich in color, in many gulls and terns an unusual light pink coloring (or flush) to the normally white plumage can be found in highly variable proportions within and across populations (5). It has been suggested that some gulls turn pink because they acquire unusually high amounts of astaxanthin in their diets at the time of feather growth, especially in areas with farm-raised salmon (5). However, the exact relationship between astaxanthin and plumage is not fully understood. Here we examine this relationship in more detail and discuss its implication.

Problem 10-2

Why does this introduction seem incomplete? Identify the known, unknown, question/purpose, and experimental approach. Are these elements clearly identifiable? Why or why not?

Methane clathrate is a solid form of water that contains a large amount of methane within its crystal structure (a clathrate hydrate). Significant deposits of methane clathrate have been found under sediments on the ocean floors.[1] Methane hydrates are believed to form by migration of gas from depth along geological faults, followed by precipitation, or crystallization, on contact of the rising gas stream with cold sea water.

A "Bottom Simulating Reflector" (BSR) was used to detect the presence of methane clathrates along the ocean floor of the Blake Bahama Outer Ridge. Through seismic reflection at the sediment to clathrate stability zone interface caused by the unequal densities of normal sediments and those laced with clathrates, we were able to identify several deposits of methane clathrates at depths of 500 to 1000 m.

Problem 10-3

Revise the following Introduction. Ensure that the necessary elements (known, unknown, question/purpose, and experimental approach) are present and clearly signaled.

INTRODUCTION

Ehrlichiae are obligatory intracellular bacteria that infect leukocytes and platelets of a wide variety of mammals and are transmitted by ticks (1).

Dogs can be infected by *E. canis*, *E. chaffeensis*, *E. ewingii*, and *Anaplasma phagocytophilum* (*Ehrlichia equi*) (2). Infection with any of these species can cause a severe disease with indistinguishable hematological and clinical anomalies (3).

E. canis was the first species described in dogs (3–5). It is worldwide distributed, particularly in tropical and subtropical regions, and is the causative agent of classical canine monocytic ehrlichiosis, which presents three well characterized clinical phases. Dogs treated during the acute phase of the disease normally recover rapidly. However, since this stage of the infection can evolve with moderate or imperceptible clinical signs, infected dogs can develop the subclinical phase and some of them reach the chronic phase (6).

E. chaffeensis is the etiological agent of human monocytic ehrlichiosis. This potentially mortal disease was reported for the first time the year 1987 in USA. *E. canis* was first thought to be the causative agent due to a crossreaction of sera from patients with an antigen preparation from this species (7). In 1991, the bacterium was isolated and characterized at the molecular level. It was then established that it was a distinct species of Ehrlichia (8). The recognized natural reservoir for this bacterium is the white-tailed deer (9).

Ehrlichiae with tropism for monocytes, lymphocytes, neutrophils and platelets from dogs diagnosed by BCS, have been reported since 1982 (10). The presence of these rickettsiae in monocytes and platelets has been also confirmed using transmission electron microscopy (11,12). In this study we describe primary cultures of monocytes from the blood of a dog with canine monocytic ehrlichiosis. We also identified *E. canis* and *E. chaffeensis* using nested PCR with DNA samples extracted from the primary cultures and from dogs with natural and experimental infections.

(Veterinary Clinical Pathology 37(3). pp. 258–265, 2008)

Problem 10-4

For the following Introduction, ensure that the necessary elements (known, unknown, question/purpose, and experimental approach) are present and clearly signaled.

Septic shock and sepsis syndrome is one of the leading causes of death in hospitalized patients and accounts for 9% of the overall deaths in the United States annually (1–4). While commonly initiated by a bacterial infection, the pathophysiological changes in sepsis are often not due to the infectious organism itself but rather to the uncontrolled production of pro-inflammatory cytokines produced mainly by macrophages. This over-production can result in an overwhelming systemic inflammatory response that leads to multiple organ failure. Lipopolysaccharide (LPS) endotoxin, a component of the bacterial cell wall, stimulates macrophages to produce pro-inflammatory cytokines such as tumor necrosis factor α (TNF-α) and interleukin-1β (IL-1β), both of which have been shown to be critical mediators of septic shock (7). It is the excessive production of these pro-inflammatory cytokines that causes systemic capillary leakage, tissue destruction, and ultimately lethal organ failure and death (1–4, 7).

Thus, the expression of these pro-inflammatory cytokines needs to be tightly regulated during an inflammatory response.

Regulation of mRNA stability is critical for controlling gene expression as the abundance of an mRNA transcript is also regulated by the rate of mRNA degradation (5, 6, 8). The mRNAs encoding most inflammatory cytokines are short-lived, with instability conferred by an AU-rich element (ARE) in their 3′ noncoding regions (5, 6). ARE promotes rapid degradation of mRNAs (6) and in some cases translation arrest (9). Furthermore, ARE-mRNAs can be rapidly stabilized upon exposure to certain signals including immune stimulation, UV and ionizing irradiation (5), and heat shock (10). The stability of ARE-mRNAs is controlled by trans-acting AU-rich element binding proteins (6, 16). Hu ARE-binding proteins, such as HuR, stabilize ARE-mRNAs and inhibit mRNA translation (17–19).

In contrast to ARE, AUF1 (or hnRNP-D) (20, 21), TTP (22), BRF1 (23), and KSRP (24) have been shown to mediate rapid decay of various cytokine mRNAs. AUF1, an ARE-mRNA destabilizing factor, consists of four isoforms (37, 40, 42, and 45 kD) generated by alternative splicing (20, 21). Increased expression of AUF1 has been correlated with rapid ARE-mRNA degradation in various types of cells (25–27), with p37 AUF1 isoform exhibiting the highest destabilizing activity (25, 26). However, implication of AUF1 is based largely on correlations with *in vitro* binding to AREs (28, 29) and ectopic overexpression studies in cell lines (25, 26). The role of AUF1 in the regulation of cytokine expression *in vivo* has not been examined in a pathophysiological context.

To examine the role of AUF1 in promoting inflammatory mRNA degradation *in vivo*, we generated AUF1 null mutant mice and studied their response to LPS-induced endotoxemia as well as to the expression of pro-inflammatory cytokines. We observed that *AUF1*⁻ᐟ⁻ mice were acutely susceptible to endotoxin, showed manifestations typical of endotoxic shock, and had a significantly lower survival rate when challenged by LPS. These phenotypes were associated with over-expression of the pro-inflammatory cytokines TNFα and IL-1β as a result of abnormal stabilization of their mRNAs. Our results provide the first *in vivo* evidence implicating AUF1 in regulating inflammatory response. This regulation occurs through targeted degradation of selective cytokine mRNAs, deregulation of which would contribute to the development of endotoxic shock.

(With permission from Cold Spring Harbor Laboratory Press and Robert J. Schneider)

Problem 10-5

1. This Introduction is too long. Condense this Introduction.

2. Identify the individual elements of the condensed Introduction (known, unknown, question/purpose, and experimental approach).

INTRODUCTION

The establishment of the germline generation after generation is crucial for the propagation of a species. Typically, the germline is set apart from the somatic tissues early in development. To maintain the unique ability of the germline to give rise to all tissues of the next generation, the germline must be protected from somatic differentiation signals, which would restrict the fate of the cells. In many species, germ cells contain specialized cytoplasm, known as germ plasm, which contains proteins and RNAs required for germline development. In some cases, this germ plasm includes distinct ribonucleoprotein particles, known as polar granules in *Drosophila* and as P granules in *C. elegans*. At least some protein components of germ granules are conserved, particularly the homologs of *vasa*, an RNA helicase that is found in germ cells in many organisms from *C. elegans* to mammals.

In *C. elegans*, the germline is set apart from the somatic tissue after only four rounds of asymmetric cell division, which give rise to blastomeres P1, P2, P3, and P4 (1). The germline blastomeres inherit P granules in each round of division. P granules are initially dispersed in the cytoplasm in oocytes and in the early embryo. Beginning in the P2 blastomere, P granules associate with the nuclear membrane, an association that is maintained in the adult germ cells. A protein with similarity to receptor tyrosine kinases, MES-1, controls the asymmetric partitioning of P granules during the divisions of P2 and P3 (2). The blastomere P4 marks the point of germline restriction, as this cell divide gives rise to a pair of primordial germ cells that proliferate to produce all future germ cells in the animal. The germline blastomeres are transcriptionally silent during the early divisions until after the birth of P4 (3). Therefore, genes that function in the germline establishment in these blastomeres are likely to be maternally deposited proteins or maternally deposited mRNAs under post-transcriptional control.

Of the proteins present in P granules that have been previously identified, most contain predicted RNA binding domains, suggesting that regulation of RNAs may be a function of P granules. Some proteins localize exclusively to P granules in both embryonic and adult stages, specifically PGL-1 and PGL-3 (4,5) and the four homologs of *vasa* (-1–4) (6). In contrast, other proteins found in P granules localize to additional cellular locations. For example, PIE-1 is in both P granules and the nucleus, where it represses transcription (7). While many P granule components are required for normal germline development, P granules alone are not sufficient to confer a germline fate. Despite the identification of greater than twenty proteins found in P granules, a molecular function for this complex has not been elucidated. Additionally, genetic relationships between many P granule components have not yet been reported.

In this study, we report the identification and characterization of a novel protein in *C. elegans*. This gene is necessary for the development of the germline of the progeny. We have named this gene *egcd-1*. We show that loss of *ecdg-1* results in a severe decrease in germ cell proliferation and abnormalities in the germline blastomeres. Our genetic analysis indicates that *egcd-1* functions synergistically with another gene important in P granules.

(With permission from Stefanie W. Leacock, PhD)

Problem 10-6 Your Own Introduction

Write a short Introduction based on your work. Be sure to follow the funnel shape. Include known, unknown, question, and experimental approach. Also consider including results, conclusions, and significance of the paper. Be sure to signal the parts of your Introduction.

CHAPTER 11

Materials and Methods

11.1 OVERALL

The purpose of the Materials and Methods section is to describe the experimental approach used to arrive at your conclusions. Although most readers are not interested in the experimental details of your work and will therefore not read this section, some readers will want to repeat part or all of your procedures and will read the Materials and Methods section in great detail. Above all, reviewers will read this section very carefully to ensure that it contains sufficient detail to evaluate or repeat your work. You should write this section with great care because if your experimental approach appears faulty, incomplete, or unprofessional, your paper may get rejected.

11.2 CONTENT

MATERIALS AND METHODS GUIDELINE 1:
Provide enough details and references to enable a trained scientist to evaluate or repeat your work.

MATERIALS AND METHODS GUIDELINE 2:
Include materials and methods, but not results.

The Materials and Methods section should cover

- Materials (drugs, culture media, buffers, gases, or apparatus used)
- Subjects (patients, experimental materials, animals, microorganisms, plants)

For medical studies include

    (a) Total number of subjects

    (b) Number of subjects receiving treatment

    (c) How subjects were selected

    (d) Details such as sex and age if relevant

- Design (includes independent and dependent variables, experimental and control groups)
- Procedure (what, how, and why you did something)

Define the materials and methods as precisely as you can. Do not forget to include your control experiments. Check and follow the detailed specifications found in the *Instructions to Authors* of the journal to which you plan to send your manuscript.

Note that the Materials and Methods section is unavoidably linked to the Results section. In the Materials and Methods section, you need to describe how you obtained the results you report. Vice versa, in the Results section, you need to provide results for everything you describe in Materials and Methods. Do not make the error of mixing in some of the results in this section except for necessary intermediate results that provide the information needed for the next logical experimental step of your study.

References

If your methods have not been reported previously, you must provide all of the necessary detail. If, however, methods have been previously described in a standard journal, provide only that literature reference.

Example 11-1   **Referring to previously described methods**

    a)   Plasmids were isolated according to Braun (19).

    b)   For the numerical calculations involved in this study, we used a state-of-the-art chemistry-climate model (CCM) developed at the National Center for Atmospheric Research (22).

If you modified a previously published method, provide the literature reference and give a detailed description of your modifications.

Example 11-2   **Referring to described methods that were modified**

Plasmids were isolated according to Braun (19) with minor modifications. Instead of dissolving DNA pellets in sterile water, pellets were dissolved in buffer A.

Be sure to quote original references, that is, references that actually provide the method you want to describe. Do not just list a reference that refers the reader to another paper.

Details and Technical Specifications

In the Materials and Methods section, you need to provide sufficient details and exact technical specifications such as temperature, pH, total volume, time, and quantities to ensure that scientists can repeat your work. Include trade names, manufacturer, model numbers, and lot numbers, if essential. Identify organisms with full taxonomic names.

 Example 11-3 a **Providing sufficient detail**

> To lyze the cells, we used 250 µl of SDS-solubilization buffer (10 mM Tris-HCl pH 7.5, 150 mM NaCl, 1 mM EDTA pH 8.0, 1% SDS w/v, 1:100 100 mM Phenylmethylsulfonyl fluoride (PMSF), and 100× Lysophosphatidic acid (LPA)). After lysis, cells were resuspended once the supernatant had been removed.

In Example 11-3a, the author provides good detail of the solubilization buffer but fails to do so for the resuspension buffer.

 Revised   a   To lyze the cells, we used 250 µl of SDS-solubilization buffer
Example 11-3   (10 mM Tris-HCl pH 7.5, 150 mM NaCl, 1 mM EDTA pH 8.0, 1% SDS w/v, 1:100 100 mM Phenylmethylsulfonyl fluoride (PMSF), and 100× Lysophosphatidic acid (LPA)). After lysis, cells were resuspended in **100 µl sterile saline** after removal of the supernatant.

Here are a few more examples that do not provide sufficient detail:

 Example 11-3   **Providing sufficient detail**

b   To classify native species, orchids were collected.
c   To identify genes with a high probability of having differential expression in adenomas and follicular carcinomas, we used statistical methods.
d   All samples were centrifuged.

 Revised   b   To classify native species, orchids were collected **in the**
Example 11-3   **Everglades during the month of January**.
c   To identify genes with a high probability of having differential expression in adenomas and follicular carcinomas, we used **parametric (t test) and nonparametric (Mann–Whitney U test) methods**.
d   All samples were centrifuged at **5000xg for 30 min at 25 °C**.

Whereas some Materials and Methods sections contain too little detail, others include unnecessary extra detail, such as the one shown in the next example:

Example 11-4   **Unnecessary information in Materials and Methods**

CD4⁺ CD44ʰⁱᵍʰ T cell purification

<u>We were interested in examining the gene expression of the anergic portion of T cells in Jak3 KO mice that are CD4⁺ CD44ʰⁱᵍʰ.</u> To obtain the CD4⁺ CD44ʰⁱᵍʰ T cells used for RNA isolation, spleens were removed from Jak3 KO and Jak3 Het mice at 8 to 10 weeks of age. Splenocytes were isolated by homogenizing the tissue with frosted glass slides (Fisher Scientific, Pittsburgh, PA).

In this example, the first sentence of the paragraph is unnecessary. This sentence should have been given in the introduction of the paper. The purpose for the actual portion of the experiment can be found in the second sentence of the paragraph "To obtain …". Thus, the first sentence can be omitted.

Revised   *CD4⁺ CD44ʰⁱᵍʰ T cell purification*
Example 11-4   To obtain the CD4⁺ CD44ʰⁱᵍʰ T cells used for RNA isolation, spleens were removed from Jak3 KO and Jak3 Het mice at 8 to 10 weeks of age. Splenocytes were isolated by homogenizing the tissue with frosted glass slides (Fisher Scientific, Pittsburgh, PA).

Consider another example:

Example 11-5   **Unnecessary information in Materials and Methods**

Cells were scraped out of the wells and resuspended <u>in a 1.5 ml Eppendorf tube</u>.

In the preceding example, we find a description for the tube size used, "in a 1.5 ml Eppendorf tube." Although for some experiments it may be important to mention the manufacturer for certain equipment if it is essential for the success of the experiment, a description of the tube size is usually considered unnecessary detail and should be avoided.

Revised   Cells were scraped out of the wells and resuspended **in**
Example 11-5   **100 µl sterile saline.**

Use of Parentheses

To provide enough details while maintaining good flow in your writing, parentheses are commonly used in the Material and Methods section. Often these technical specifications include manufacturer's names, lot numbers, names of machinery, and additional explanations and specifications.

Example 11-6   **Use of parentheses**

a   20 mg/ml trypsin (TPCK, bovine pancrease) dissolved in Z-buffer (10 mM Tris HCl pH 8.0, 120 mM NaCl, 50% (v/v) glycerol) was thermally denatured at 65 °C for 3 min.

b   To assess trends over time, we calculated relative abundances of species, averaging monthly capture data over 6-month periods (January–June and July–December).

Appendices

MATERIALS AND METHODS GUIDELINE 3:
Place detailed description of procedures or other lengthy details in an appendix.

Rather than putting detailed descriptions of procedures or other lengthy details in the body of the paper, you should place them in an appendix. If so, this must be part of your plan for the paper, not an afterthought submitted with the proofs, because an appendix has to be reviewed with the rest of the paper. Alternatively, you may be able to send lengthy material as supplementary material to an archive recommended by the journal.

Aside from detailed descriptions of procedures, material included in appendices rather than in the main body of a research paper (or proposal) may include detailed calculations, algorithms, proofs, tables, plots, and images or large data sets for meta-analyses and comparisons. Many journals now maintain electronic archives of supplementary material, including original data. This arrangement allows authors to be both thorough overall and concise in the main body of the article.

11.3   ORGANIZATION

MATERIALS AND METHODS GUIDELINE 4:
Arrange experimental details as protocols grouped in chronological order or by subsections.

MATERIALS AND METHODS GUIDELINE 5:
Signal and link the different topics.

MATERIALS AND METHODS GUIDELINE 6:
Explain the purpose for any procedure whose function is not clear.

The Materials and Methods section is usually a long section and typically covers various topics. These topics need to be organized and methods need to be described in logical order, including the sequence of the procedures for each method.

You can organize your Materials and Methods section by separating each group of actions into one or more paragraphs. Paragraphs on the same type of information can then be grouped into subsections. The sequence of events within these subsections is usually written in chronological order or from most to least important. Each subsection has its own subheading, which functions as a signal, naming the particular material, variable, or specific procedure. Although use of subsections is optional, it usually simplifies and clarifies the presentation for the reader. Subsections can include one or more paragraphs.

Example 11-7   **Materials and Methods subsection**

Cultures. Samples taken from the tips and subcutaneous sections of aseptically removed IVs were cultured as previously described (12). To identify the sources of organisms that colonize IVs, swab cultures of surrounding skin were obtained at the time of IV insertion as well as at the time of IV removal. In addition, one or more peripheral blood samples were obtained. Isolated organisms were identified by standard microbiologic methods.

Check your target journal to find out what subheadings are commonly used in your field and construct your subsections accordingly. You may even consider using the same subheadings in Materials and Methods and in the Results section. Examples of common subheadings are

Analysis of X	Antibodies
Cell Cultures	Chemicals and Reagents
Cloning	Data Analysis
Materials	Outcome Measures
Plasmids	Protein Expression and Purification
Sequence Analysis	Statistical Analysis
Study Design	Study Population
Synthesis of Y	Treatment Protocol

Other ways to signal different topics within the Materials and Method section can be by topic sentences or transitions. Topic sentences can be used to signal the topic of a paragraph, especially within a subsection. Transitions are often placed at the beginning of the first sentence of a paragraph to link the paragraph to the previous one before introducing the topic of the remaining sentences in the paragraph. No signals are used if the topic becomes apparent from the subject and verb.

In all cases, it is important to ensure that the reader will understand why each procedure was performed and how each procedure is linked to the central question of the paper. Therefore, you should state the purpose or give a reason for any procedure whose function or relation to the question of the paper is not clear. Also, provide any background information that might be necessary to understand the experiments you performed. Statements of purpose or background are usually placed at the beginning of a paragraph and typically serve as topic sentences and transitions.

Example 11-8   **Topic sentences/Statement of purpose**

a   **To purify prolyl 4-hyderoxylase from human placenta,** full-term human placentae were collected 30 min after delivery.

b   **Next, a trait-by-trait correlation matrix was developed to assist in identifying clusters of related traits.**

11.4 IMPORTANT WRITING PRINCIPLES FOR MATERIALS AND METHODS

Voice

MATERIALS AND METHODS GUIDELINE 7:

In the Materials and Methods section, passive voice is often preferred.

MATERIALS AND METHODS GUIDELINE 8:

Do not switch from one point of view to another for no apparent reason.

The Materials and Methods section is the one section in a research paper where often passive voice is preferred over active voice. The reason is two-fold: It lets you emphasize materials or methods as the topic of your sentences, and readers do not need to know who performed the action.

Example 11-9	Use of voice
	The principal investigator collected the different fungal species from various tepuis in Venezuela.

Revised Example 11-9	Different fungal species were collected from various tepuis in Venezuela.

It is easiest to write your entire Materials and Methods section from one point of view. The disadvantage is that if most sentences are written in passive voice, writing becomes dull. If you are more experienced in writing and are taking into consideration word location and cohesion, you may choose to write in both active and passive voice in the Materials and Methods section to make your writing more smooth, interesting, and clear.

What you should avoid at all costs, however, is changing back and forth from one point of view to another within one paragraph for no apparent reason. Such switches will unnecessarily confuse and distract readers.

Example 11-10	Use of voice
	The assays <u>were performed</u> for 10 min at room temperature. <u>We</u> then <u>added</u> 10 ml of 95% ethanol.

ESL advice

In this example, the author switches from passive to active voice for no apparent reason. These types of switches are seen particularly often for ESL writers. It is easier for the reader to follow a passage if the passive voice is used consistently as in the revised example.

Revised Example 11-10	The assays **were performed** for 10 min at room temperature. Then 10 ml of 95% ethanol **were added**.

Tense

In the Materials and Methods section, the general rule about verb tense (Chapter 4.4) applies. When reporting completed actions, use past tense (see previous Example 11-10 and Revised Example 11-12). However, use present tense for statements of general validity and for those whose information is still true or if you are referring to figures and tables.

Example 11-11   **Use of tense**

a   Because mud volcanoes **emit** incombustible gases such as helium in close proximity to lava volcanoes, we collected gaseous samples from Lusi.

b   Criteria used in selecting subjects **are listed** in Table 2.

Note that in many descriptive papers, especially in computational biology, the Methods section is written in present tense.

Word Choice

MATERIALS AND METHODS GUIDELINE 9:

Choose your words carefully.

In Materials and Methods, exact and specific items are being described. Therefore, avoid jargon and other redundancies.

Example 11-12   **Choice of words**

After 3 hours, the old medium was <u>dumped</u> and the same amount of fresh medium was added.

In Example 11-12, the use of the word dumped is jargon and not appropriate.

Revised   After 3 hours, the old medium **was replaced by an equiva-**
Example 11-12   **lent amount of fresh medium.**

Some jargon terms have been used so often that particularly nonnative speakers view them as "normal" English usage. Examples include "bugs" for bacteria, "overnext" instead of "the one after next," and "western blotting" instead of "western blot analysis." ESL writers should be particularly careful and have a native-speaking scientist or scientific editor review their writing.

Precise use of English is also a must in specific word choice. For example, distinguish between "determine," "measure," "calculate," "quantitate," and "quantify."

determine   to find by investigation, calculation, experimentation, survey, or study

measure   to find the size, length, amount, degree, etc.

calculate	to work out or find out something by using numbers, to compute
quantitate	to measure something precisely
quantify	to measure the quantity of something

Example 11-13   **Precise use of English**

We **measured** the absorbance of XYZ at OD_{560}.
The percent error was **calculated**.
To **determine** absorbance of XYZ at OD_{560} and percent error, samples were weighed, dissolved in 1 ml buffer A, and incubated at 25 °C for 2 hr.

11.5   ETHICAL CONDUCT

MATERIALS AND METHODS GUIDELINE 10:

Follow guidelines on ethical conduct.

ESL advice

Research investigators should be aware of the ethical, legal, and regulatory requirements for research on human subjects and for animal experimentation in their own countries as well as applicable international countries with which they may collaborate. Check requirements before reporting findings. Most journals require that submitted manuscripts reporting the findings of human and animal research conform to respective policies and mandates. Journal editors may ask authors to produce written approval of their research by an ethics committee (see Table 11.1).

Table 11-1 Guidelines on Ethical Conduct

GUIDELINES/ MANDATE	SOURCE	PURPOSE	WEB SITE
Uniform Requirements for Manuscripts Submitted to Biomedical Journals: Writing and Editing for Biomedical Publication (Vancouver rules)	International Committee of Medical Journal Editors (ICMJE)	Statement of ethical principles in the conduct and reporting of research and recommendations relating to specific elements of editing and writing	http://www.icmje.org/
Declaration of Helsinki	The World Medical Association	Statement of ethical principles to provide guidance to physicians and other participants in medical research involving human subjects	http://www.wma.net/e/policy/b3.htm
Guide for the Care and Use of Laboratory Animals	Institute for Laboratory Animal Research of the National Research Council	Book describing the ethical care and use of laboratory animals	http://books.nap.edu/openbook.php?isbn=0309053773

Human Subjects

Depending on the journal, authors may be requested to indicate within the published article (usually within the Materials and Methods section) whether the procedures followed were in accordance with the ethical standards of the responsible committee on human experimentation (institutional and national) and/or with any other mandate, such as the Declaration of Helsinki, for reporting experiments on human subjects. Furthermore, disclosure of patient data in scientific articles usually requires informed consent of the patients concerned. Patients normally are anonymous such that people other than the patients themselves are unlikely to recognize them. If this is not possible, explicit written consent of the patients is required.

Example 11-14   **Phrasing of written consent**

(a) We obtained informed consent from all participating individuals; the study was approved by institutional review boards at the University of X, M. General Hospital, and the University of Y.

(b) All patients were studied under National Institute of Neurological Disorders and Stroke internal review board-approved protocols (NIH 79-N-0089 and NIH 00-N-0140) after providing informed consent.

(c) Written informed consent was obtained on enrollment, and the study was approved by the institutional review boards of the five centers and conducted according to the procedures of the Declaration of Helsinki.

Animal Studies

Authors are also often asked to indicate whether the institutional and national guide for the care and use of laboratory animals was followed when reporting experiments on animals. The place within a scientific article that should contain information of this kind is the Materials and Methods section.

Example 11-15   **Phrasing of compliance with guide for animal care**

(a) Animal procedures were carried out under an institutionally approved protocol in accordance with ethical principles and standards of the Federation of European Animal Science Associations and were approved by the Ethical Committee at Y University.

(b) Procedures involving animals and their care were carried out under the guidelines established by the Medical Research Council in "Responsibility in the Use of Animals for Medical Research" (July 1993) and Home Office Project License No. 30/2198. All experiments conformed to international guidelines on the ethical use of animals, and all efforts were made to minimize the number of animals used and their suffering.

(c) All procedures were approved by Z University Animal Care and Use Committee and followed the guidelines of the National Institutes of Health Guide for the Care and Use of Laboratory Animals.

11.6   COMMON PROBLEMS OF THE MATERIAL AND METHODS SECTION

The most common problems involving the Materials and Methods section of a research paper are

- Insufficient details (see Section 11.2 and Example 11.17)
- Omission of the purpose for an experiment (see Section 11.3)
- Unjustifiably switching between passive to active voice (see Section 11.4)
- Unjustifiably switching between past tense and present tense (see Section 11.4)

11.7   SAMPLE MATERIAL AND METHODS SECTIONS

The following is a good example of a subsection of a Materials and Methods section. It contains sufficient technical detail to allow a scientist to repeat the work. All experimental details have been clearly explained, and technical specifications have been given.

Example 11-16   **Materials and Methods subsection**
Expression and purification of Salmonella Gol1

Exact cell strain is given	Recombinant plasmids containing the gene for wild type or mutant Gol1 were transformed into *E. coli* K-12 TO3 as described previously (6). *E. coli* K-12 TO3 lacks the *gol1* gene and contains the pRARE plasmid that had been iso-
Necessary background is stated	lated from Rosetta cells (Novagen). Rosetta host strains are BL21derivatives designed to enhance the expression of eukaryotic proteins that contain codons rarely used in *E. coli*.
Technical details (concentration, time, temperature) are stated	*E. coli* cultures were grown in the presence of ampicillin and chloramphenicol to mid-logarithmic phase and were then induced by adding IPTG to a final concentration of 0.4 mM. After induction, cell growth proceeded for 12 hr at 30 °C. The protein purification procedure was performed
Reference given for full details	as described previously for the preparation of *Salmonella* Gol4 (7).

Unlike the preceding example, the next example does not contain sufficient detail. It is a very carelessly written Materials and Methods section.

Example 11-17   **Subsection without sufficient detail**
***In situ* hybridization**

Details for PCR amplification missing	*In situ* hybridization was performed as described (24,25). Probes were prepared from cDNA cloned in the pCRII vector (Invitrogen). The cDNA fragment used for the *hira* probe corresponded to nucleotides 4-2585 where numbers indicate nucleotides from ATG to the predicted spliced coding region. The cDNA was amplified by PCR from about 2 ng plasmid using gene-specific primer pairs. To label either a sense or anti-sense

Unclear which probes	single-stranded probe, 800 ng of the PCR product was used with a single gene-specific primer for repeated primer
Unclear how they were applied	extension with digoxigenin-11-dUTP after PCR amplification. Probes were diluted 1:2 in hybridization buffer, then applied to gonads and hybridized at 48 °C for 24 to 30 hr.
Step missing— What happens to probes and hybridization buffer after hybridization?	The gonads were then incubated with alkaline-phosphatase-conjugated anti-Dig (Fab2 fragment; Roche, Indianapolis, IN) at 4 °C overnight. Gonads were stained with BCIP/NBT tablets (Sigma or Roche), mounted, and viewed using a Zeiss Axioplan 2 imaging epifluorescence microscope.

11.8   REVISING THE MATERIALS AND METHODS SECTION

When you have finished writing the Materials and Methods section (or if you are asked to edit a Materials and Methods section for a colleague), you can use the following checklist:

☐   1. Do the listed materials and methods describe all procedures done to obtain the results presented?

☐   2. Are sufficient details and/or references provided?

☐   3. Are protocols logically grouped and organized?

☐   4. Are topics signaled and linked?

☐   5. Did you pay attention to voice (mainly passive)?

☐   6. Did you pay attention to correct use of past and present tense?

☐   7. Did you choose your words carefully?

☐   8. Did you ensure that major results are not stated in the Materials and Methods section?

☐   9. Did you pay attention to ethical conduct?

☐   10. Is the purpose stated for any procedure whose function is not clear?

☐   11. Has sentence location been considered?

☐   12. Is the point of view consistent?

☐   13. Revise for style and composition using the writing principles of this book:

   ☐   a. Are paragraphs consistent? (Chapter 6, Section 6.2)

   ☐   b. Are paragraphs cohesive? (Chapter 6, Section 6.3)

   ☐   c. Are key terms consistent? (Chapter 6, Section 6.3)

   ☐   d. Are key terms linked? (Chapter 6, Section 6.3)

   ☐   e. Are transitions used and do they make sense? (Chapter 6, Section 6.3)

   ☐   f. Is the action in the verbs? Are nominalizations avoided? (Chapter 4, Section 4.6)

   ☐   g. Did you vary sentence length and use one idea per sentence? (Chapter 4, Section 4.5)

   ☐   h. Are lists parallel? (Chapter 4, Section 4.9)

   ☐   i. Are comparisons written correctly? (Chapter 4, Sections 4.9 and 4.10)

   ☐   j. Have noun clusters been resolved? (Chapter 4, Section 4.7)

☐ k. Has word location been considered? (Verb following subject immediately? Old, short information at the beginning of the sentence? New, long information at the end of the sentence?; Chapter 3, Section 3.1)

☐ l. Have grammar and technical style been considered (person, voice, tense, pronouns, prepositions, articles)? (Chapter 4, Sections 4.1–4.4)

☐ m. Are words and phrases precise? (Chapter 2, Sections 2.2 and 2.3)

☐ n. Are nontechnical words and phrases simple? (Chapter 2, Section 2.2)

☐ o. Have unnecessary terms (redundancies, jargon) been reduced? (Chapter 2, Section 2.4)

☐ p. Have spelling and punctuation been checked? (Chapter 4, Section 4.11)

SUMMARY

MATERIALS AND METHODS GUIDELINES:

1. Provide enough details and references to enable a trained scientist to evaluate or repeat your work, but do not include unnecessary detail.
2. Include materials and methods but not results.
3. Place detailed description of procedures or other lengthy details in an appendix.
4. Arrange experimental details as protocols grouped in order or by subsections.
5. Signal and link the different topics.
6. Explain the purpose for any procedure whose function is not clear.
7. In the Materials and Methods section, passive voice is often preferred.
8. Do not switch from one point of view to another for no apparent reason.
9. Use precise words.
10. Follow guidelines on ethical conduct.

PROBLEMS

Problem 11-1

For each pair of sentences provided, select the better version and explain why you chose it.

1. a. As described by Barnes et al. (23), images of coated specimens were obtained at a working distance of 8–9 mm with an accelerating voltage of 15 kV and probe current of 200 pA.

   b. Images of coated specimens were obtained as described by Barnes et al. (23).

2. a. Seedlings were grown in continuous light before being collected.

   b. Seedlings were grown in continuous light at 15 °C for 21 days before being collected.

3. a. Here we provide a description of how subjects were chosen for our study. We selected only healthy males between 60 and 80 years.

   b. For our study, we selected only healthy males between 60 and 80 years.

4. a. Study subjects were presented with a list of potentially traumatic events, and we asked them to use three response categories (*yes, no, unsure*) to indicate if they had ever experienced them.

   b. Study subjects were presented with a list of potentially traumatic events, and were asked to use three response categories (*yes, no, unsure*) to indicate if they had ever experienced them.

5. a. The Stress Index Short Form (SI/SF), which is a 36-item questionnaire, was used to assess stress after natural disasters (26).

   b. The Stress Index Short Form (SI/SF), which was a 36-item questionnaire, was used to assess stress after natural disasters (26).

Problem 11-2

The following sentences have all been taken from prepublication Materials and Methods sections. Each one violates a basic scientific writing principle. Revise the sentences.

1. The analyses were performed on an Agilent series 1100 HPLC instrument (Agilent, Waldbronn, Germany) equipped with a quaternary pump, a diode-array detector (DAD), an autosampler, and a column compartment.

2. Immunoblotting was performed following standard procedure.

3. Adaptitude to changes in light was determined by ...

4. PCR products and long oligonucleotides (in 50% DMSO, 20 mM) were spotted with SmartArray™ Microarrayer (CapitalBio Corp., Beijing, China).

5. After centrifugation, 10 × buffer was added, and the samples were incubated for 2 min on ice.

Problem 11-3

Evaluate the following Materials and Methods subsection. Is it apparent for what purpose this test was performed? Why or why not? Suggest ways to rewrite the passage.

Wisconsin Card Sorting Test (WCST)

Participants sorted cards according to color, shape, or number stimuli depicted on the card. Sorting the cards by color was initially verbally reinforced. After a participant responded correctly for 10 consecutive cards in that color category, the participant continued sorting by form and numbers without verbal stimuli or reinforcement. Subsequently, the ratio of correct responses to errors was computed, and the numbers of trials, errors, perseverative responses, and perseverative errors were analyzed. In the WCST (Berg, 1948), test subjects have to correctly identify, implement, and remember sorting rules.

Problem 11-4

Evaluate the following Materials and Methods subsection. Identify places where not enough technical specifications have been given to repeat the activity assay.

Determination of Trypsin Activity
Trypsin (TPCK, bovine pancrease) (20ug/µl) dissolved in 0.01 M HCl, was denatured by adding 8 M urea, 33 mM Tris, pH 8.0, in a 1:4 (v/v) ratio and then boiling the mixture for 20 min. After denaturation, 5µl denatured trypsin was added to 5µl of each crystallization buffer, and the mixture was incubated at room temperature. After the first incubation, an equal volume of the crystallization buffer (10µl) was added and the mixture was incubated again. Addition of equal volume of buffer together with incubation was repeated twice more (20 and 40µl, respectively). Substrate (0.05g (1%) azocasein in 5ml 10 mM Tris, pH 8) (100µl) was added, and the mixtures incubated. The mixtures were then precipitated with trichloroacetic acid, and centrifuged. The supernatants were read in the spectrophotometer to determine activity.

(With permission from Elsevier)

Problem 11-5

1. **On the left margin of this Methods section, write the topic of each paragraph.**
2. **Which paragraph(s) describe(s) the experiment done to answer the question asked? Write "Experiment" next to this/these paragraph(s).**
3. **Identify—if possible—the topic sentences (circle them).**
4. **Identify—if possible—an example of each technique of continuity:**
   - **repeated key term (box it)**
   - **transition word and transition phrase or clause (underline it)**
   - **consistent point of view**
   - **parallel form**
5. **What organization does <u>one</u> of the paragraphs follow (i.e., most to least important, pro–con, etc. ...)? Are the paragraphs organized well?**

Research Question/Purpose: To construct and test a safe, live, attenuated, oral vaccine candidate, IEM108, immune to CTXΦ infection

Construction of the candidate IEM108. The 1.15-kb XbaI fragment containing the upstream regulatory and coding regions of ctxB recovered from pBR (a pUC19-derived plasmid carrying ctxB and rstR, constructed in our laboratory before) and cloned into the XbaI site of pXXB106, containing E. coli-derived thyA (30), resulting in two new constructs, pUTBL1-5 and pUTB1-6. ctxB and thyA have the same transcriptional direction in pUTBL1-5 and opposite directions in pUTBL1-6 (Fig. 1).

The rstR gene and its upstream sequence were amplified from El Tor strain Bin-43 with primers PrstR1 (CCGAATTCACTCACCTTGTATTCG) and PrstR2 (CGGAATTCTCGACATCAAATGGCATG). The amplified fragment was then cloned into the EcoRI site of pUTBL1-5, yielding new construct pUTBL2. Subsequently, an 0.8-kb PvuI fragment of the bla gene in pUTBL2 was deleted to generate pUTBL3. pUTBL3 was then electroporated into IEM101-T to construct IEM108.

Serum vibriocidal antibody assay. Serum vibriocidal antibody titers were measured in a microassay using 96-well plates. The immunized rabbit sera were inactivated at 56°C for 30 min and diluted 1:5 with PBS before use. The prediluted rabbit sera were added into the first well and then serially diluted threefold in PBS. PBS was added to the last well as a negative control. The plates were incubated for 30 min at 37°C with 25 µl of a solution containing 10^2 CFU of *V. cholerae* Bin-43/ml of culture and 20% guinea pig serum as a complement source in PBS. One hundred fifty microliters of 0.01% 2,3,5-trihenyltetrazolium chloride in LB broth was added to each well, and the plates were further incubated for 4 to 6 h at 37°C until the negative-control wells showed a color change. The reciprocal vibriocidal titer is defined as the highest dilution of serum that completely inhibits growth of Bin-43, i.e., no color change.

Rabbit immunization. Eight adult New Zealand White rabbits (2 to 2.5 kg) were divided into naive, IEM101, and IEM108 groups. The naive group consisted of two rabbits that were not immunized. Each immunization group had three rabbits. After fasting for 24 h, the rabbits in both immunization groups were anesthetized with ether. After the abdominal skin was sterilized with an iodine tincture and alcohol, the abdominal cavity was opened by vertical incision (under sterile conditions). General exploration was performed to find the ileocecal region. This region was ligated to the inner wall of the abdomen. Then 10^9 CFU of vaccine strain IEM101 or IEM108 were injected into the proximal ileum. Finally, the abdominal cavity was closed. The ligature that tied the ileocecal region to the abdominal wall was removed 2 h later, and the rabbits were given water and feed for 28 days. One rabbit of the IEM101 group died after the operation, probably because of heavy anesthesia. Serum samples were collected from the immunized rabbits prior to the immunization and on days 6, 10, 14, 21, and 28 after the vaccination. The serum titers for the anti-CT antibody and vibriocidal antibody were measured as described above.

Rabbit ileal loop assay and protection model. To evaluate the protection efficacy *in vivo*, the immunized rabbits were challenged with pure CT and four virulent *V. cholerae* strains (395, 119, Wujiang-2, and Bin-43) of different serotypes and biotypes (Table 1) 28 days after the single-dose immunization. Rabbits were anesthetized and their abdomens were opened as described above. Their intestines were tied into 4- to 5-cm-long loops, and then 10^5 to 10^8 CFU of challenge strains or 1, 2, 3, or 4 μg of pure CT was injected into each loop. Normal saline was used as negative control. At 16 to 18 h postchallenge, the rabbits were sacrificed and the accumulated fluid from each loop was collected and measured. The ratio of the volume of accumulated fluid (milliliters) to the length of the loop (centimeters) was calculated for each loop in the challenged rabbits.

(With permission from American Society for Microbiology)

Problem 11-6

1. **Which paragraph(s) relate directly to the overall purpose of the study.**
2. **Identify—if possible—the topic sentences (circle them).**
3. **Identify—if possible—an example of each technique of continuity:**
   - **repeated key term (box it)**
   - **transition word and transition phrase or clause (underline it)**
   - **parallel form**
4. **Are the paragraphs of this section organized well?**

Purpose: To evaluate the Child FIRST program, part of the Bridgeport Safe Start Initiative (BSSI), which was established to reduce the incidence and the impact of exposure to violence among children six years old and younger.

Method

Service utilization In collaboration with Child FIRST staff, the evaluation team developed a form that staff used to document the dates, types, and duration (recorded in fifteen-minute increments) of all services provided to children and their families. Examples of services included in-home assessment, classroom assessment and consultation, in-home care coordination, and staff consultation and supervision. Data documented on this form were also used to determine the length of time in the program from entry to discharge.

Family violence and traumatic events The Traumatic Events Screening Inventory–Parent Report Revised, or TESI–PRR (Ghosh-Ippen, Ford, Racusin, Acker, Bosquet, Rogers, et al., 2002), is a twenty-four-item semi-structured interview that determines a history of exposure to traumatic events for children six years old and younger. Parents are presented with a list of traumatic events and asked to use three response categories (yes, no, and unsure) to indicate if the child has ever experienced them.

Twelve of the twenty-four TESI-PRR items were used to screen all children entering the program for a history of family violence and thus determine eligibility for inclusion in the evaluation study: separation from a family member; suicide by someone close to the child; physical assault or physical injury or bruising of child by family member; threat of serious physical harm to child; kidnapping by a family member; witnessing by the child of physical fighting or the use of a gun, knife, or other dangerous weapon by a family member; witnessing of verbal threats to seriously harm by a family member; witnessing of arrest of family member; experiencing inappropriate sexual activity; witnessing inappropriate sexual activity; being yelled at repeatedly in a scary way or told that he or she is no good (verbal abuse); and experiencing neglect. The nonfamily violence items were exposure to serious accident; serious natural disaster; severe illness or injury of someone close; death of someone close; serious medical procedure or life-threatening illness; mugging; animal attack; community violence; direct exposure to war, conflict, or terrorism; exposure to war, conflict, or terrorism via television or radio; or other stressful events. Since the TESI is an inventory-type measurement, it is a reflexive measure in which indicators of internal consistency and other psychometric properties are not suitable.

(With permission from Lyceum Books, Inc.,)

Problem 11-7 Your Own Materials and Methods

Write a Materials and Methods section based on your work. Above all, arrange your sentences and topics logically. Use topic sentences and transitions. Signal subtopics. Use enough technical detail.

CHAPTER 12

Results

12.1 OVERALL

The Results section is the major scientific contribution of your study. Whereas the Introduction provides background information and states the purpose/question of the paper, and the Discussion confers what your data mean when tied in to current knowledge and theories, the Results section represents the core or skeleton that may be of interest much longer than any conclusions drawn from your observations.

12.2 CONTENT

General Content

RESULTS GUIDELINE 1:
Report your main findings as well as other important findings.

RESULTS GUIDELINE 2:
Point the reader to the data shown in figures and tables.

RESULTS GUIDELINE 3:
Include control results.

The Results section presents the results of your experiments and points the reader to the data shown in the figures and tables. You should report only results that are pertinent to the information provided in the Introduction and to the experiments described in Materials and Methods. Exclude preliminary results and results that are not relevant. Do not forget to include control results, however, and if needed, explain the purpose of an experiment shortly. Also incorporate results whether or not they support your hypothesis, and explain any contradicting results if necessary.

Other important findings may consist of additional supportive evidence or alternate measurements as well as additional results that may not be part of the main story if they are meaningful for the paper. Know that not every result that you obtained from your experiments and not every observation you made has to be reported in the Results section. Concentrate on the most relevant findings; but when deciding what to include or not, know the difference between leaving out irrelevant results and suppressing contradictory ones. Do not omit the latter.

If you find that you need to collect more data as you write, do so. It is more important to do a thorough job than to submit an incomplete manuscript quickly.

Interpreting Data

RESULTS GUIDELINE 4:
Interpret your data for the reader.

In the Results section, do not just present data, but summarize and interpret their meaning for the reader by presenting them as results. Only data that have been interpreted will be meaningful for your readers.

To present your results to the reader clearly, you need to distinguish between data and results. Data are values derived from scientific experiments (concentrations, absorbance, mean, percent increase). Results interpret data (e.g., "Absorbance **increased** when samples were incubated at 25 °C instead of 15 °C"). Although most data should be presented in figures and tables, your main findings should be stated in the text as well along with your interpretations of all data. When you state your interpretations/results, ensure that you make reference to your data in figures or tables by referencing the figure or table number in parenthesis.

☞	**Example 12-1**	**Presenting data without interpretation**

Heart rate was <u>100 beats per minute</u> after digitalis was added (Fig. 3).

Unless your readers are physicians, they may not be able to put "100 beats per minute" into any relation, especially if no comparative value is given. You need to let them know whether this is higher or lower than normal.

 Revised   a   Heart rate *increased* **to 100 beats per minute** after
Example 12-1   digitalis was added (Fig. 3).

In the revised example, the data have been interpreted and are presented as a result, making the revised example much more meaningful for the reader.

To put the results into relation for nonspecialists in the field, you need to give comparative values as well.

 Revised   b   Heart rate **increased** *from 60 to* **100 beats per minute**
Example 12-1   after digitalis was added (Fig. 3).

When the magnitude of change is given by a comparative value ("... from 60 to 100 ..."), the data have been interpreted such that it is understandable for most scientists.

Here is another example of providing data without interpretation or explanation:

 Example 12-2   **Presenting data without interpretation**

The sequences for the proteins K 309 and K 415 were compared (Fig.4).

This example fails to interpret the data provided. The author neither explains nor analyzes the data for the reader but simply refers the reader to a figure. As a consequence, the reader does not know if the data are similar or different. Instead, the reader is expected to interpret the data himself or herself. The author should make the point clear so readers do not have to find their own interpretations.

 Revised   When the sequences for the proteins K 309 and K 415 were
Example 12-2   compared, **their C-terminal sections were found to
be 90% homologous (Fig. 4).**

The following is yet another example in which an author presents data but fails to interpret them:

 Example 12-3   **Presenting data without interpretation**

Among the 785 HIV positive participants in the study group, we found 622 men and 163 women.

To interpret the data for the reader, the author needs to first present the interpretation and then the data that supports it.

Revised
Example 12-3

We found that 3.8 times as many men (79.2%) than women (20.8%) tested positive for HIV in our study group.

Presenting Data and Statistical Information

RESULTS GUIDELINE 5:

Place statistical information with data. Do not use
it instead of results.

Understanding statistical information in research articles can be problematic. Often readers blame themselves for not comprehending what has been written. However, the true reason for their comprehension problems lies in the misrepresentation of statistical information.

To avoid confusing readers, make reference to the event you are referring to:

Example 12-4   **Presenting data without reference**

There is a 20% chance of a big earthquake in California.

Sentences such as the preceding one in which no reference class is given result in much misunderstanding among your readers. Readers interpret this sentence in various ways: 20% *of the area* of California has a big chance for an earthquake, or 20% *of the earthquakes* in California are big, or 20% *of the time* the chances for an earthquake are big.

Confusion can be reduced by specifying a reference class, such as time and area, before giving a single event probability.

Even more confusing are sentences that talk about more than one statistical result at a time such as in the following example:

Example 12-5   **Confusing description of statistical data**

The probability of contracting XDR TB is 80% for HIV patients. For people with XDR TB, the probablility that it will be detected through rapid skin tests is 50%. In 10% of the cases, rapid skin tests do not detect XDR TB.

This type of example leads readers to wide misinterpretations, and such misinterpretations, particularly in the medical field, can have severe consequences for patients.

If we were to restate the example using numbers rather than probabilities, the example becomes much easier to understand and more graspable.

Revised	Out of 100 HIV patients, 80 contract XDR TB. Of these, 40
Example 12-5	cases will be detected through rapid skin tests. For 8 out of
	the 80 XDR TB cases (or 1 in 10), rapid skin tests do not
	detect XDR TB.

Readers usually profit from representing statistical information using numbers or frequencies rather than probabilities or sensitivities.

Many students and novice writers come up with tedious lists of statistical test results rather than a description and interpretation of experimental observations for their Results section. When you report statistical information, include descriptive statistics such as mean, standard deviation, confidence intervals, p values, and sample size as well as bivariate analysis such as chi-square or t test or multivariate analysis such as regression analysis. Ensure that you interpret descriptive statistics for your readers. Do not just list them in your Results section. Statistical analysis should serve as reinforcement for your data and should not replace their interpretation. Therefore, preferably place statistical information in your figure legends or tables or in parenthesis following the description of data.

Example 12-6	**Preferred placement of statistical information**
	Vaccination rates among the elderly was higher than among younger participants when the risk of flu was high (61.6% vs. 46.8%; OR = 2.67, 95% CI = 1.94–3.67).

A good resource in providing background material on basic statistics is, for example, J. L. Fleiss, *Statistical Methods for Rates and Proportions,* by John Wiley & Sons, 1981.

12.3   ORGANIZATION

Overall Organization

RESULTS GUIDELINE 6:

Place results that answer the question of the paper at the beginning of the results section.

RESULTS GUIDELINE 7:

Organize the Results section chronologically or from most to least important.

RESULTS GUIDELINE 8:
Emphasize and signal your results. Subordinate secondary information.

Start the Results section by presenting your main findings in the first paragraph. Your main findings are the findings used in providing the overall answer/conclusion of the paper. You may also start the first paragraph with a brief overview of your general observations and then move on to the main findings (see Example 13-20). In the latter case, do not devote more than a few sentences to any overview, and ensure that your main findings still appear in the first paragraph, as it is a power position.

In subsequent paragraphs, present your specific observations. The overall structure of this remainder of the Results section is normally either chronological or from most to least important. Use topic sentences to provide an overview of each experiment. Start each subsection or paragraph by explaining the purpose of the experiment, by giving a short background, or by stating the results of an experiment (see *Organization Within Results Segments* for a more detailed explanation). Consider also including a paragraph or two describing specific details of an observation. Readers will understand papers better if specific details are highlighted or given as examples.

Throughout the Results section, emphasize the data and their meaning. Subordinate control results and methods.

The following is an example of a paragraph in a Results section. The topic sentence of this paragraph does not present the main findings but rather points the reader to a figure, thus emphasizing the figure rather than the results. Note that in general, mentioning of a table or figure is best if it is done in parenthesis rather than in the text, as tables and figures are considered supporting evidence and not results.

☞	**Example 12-7**	**Results paragraph emphasizing figure rather than results**

Topic sentence does not emphasize results	***ESI-MS/MS of free anthraquinones.*** <u>The mass spectra of five free anthraquinones identified in rhubarb are shown in</u>
Generalized results	<u>Figure 3</u>. ESI-MS of anthraquinones appeared to provide more structural information than APCI-MS did as reported previously [28]. In the MS/MS spectrum of chrysophanol, a product ion at m/z 225 was observed, resulting from the direct loss of CO from $[M - H]^-$. The m/z 225 ion was very stable and did not yield any further fragmentation. We believe that the CO elimination may originate from
Details of results	C-10, since the carbonyl group at C-9 has intramolecular hydrogen bonding with the α-hydroxyl groups at C-1 and C-8 and is, thus, difficult to be cleaved.

(With permission from Elsevier)

To emphasize results, the author should have subordinated the reference to the figure.

Revised Example 12-7	*ESI-MS/MS of free anthraquinones.* **The ESI-MS frag**
Topic sentence emphasizes results	**mentation behavior of five free anthraquinones** **identified in rhubarb was different from that in** **APCI-MS reported previously (28) (Figure 3).** In the MS/MS spectrum of chrysophanol, a product ion at m/z 225 was observed, resulting from the direct loss of CO from [M – H]⁻. The m/z 225 ion was very stable and did not yield any further fragmentation. We believe that
Details of results	the CO elimination may originate from C-10, since the carbonyl group at C-9 has intramolecular hydrogen bonding with the α-hydroxyl groups at C-1 and C-8 and is, thus, difficult to be cleaved.

Organization Within Results Segments

RESULTS GUIDELINE 9:
Organize your Results into different segments.
In each segment, state:

Purpose or background of experiment
Experimental approach
Results
Interpretation of results (optional for descriptive papers)

To organize your Results section, think of it in different segments. Each segment pertains to one set of experiments. Many, if not most, segments will be only one paragraph long; others may be longer. You may even consider dividing your Results section into different subsections, making use of the separate segments.

Information in these Results segments or paragraphs needs to be organized. This also includes the first paragraph. Each segment that describes results of a specific individual experiment should contain four essential components:

1. Purpose or background of experiment if needed
2. Experimental approach
3. Results
4. Interpretation of results

Start your segments or paragraphs by providing a topic sentence. This topic sentence usually indicates the purpose of the experiment performed. It may also provide context in form of background information. The purpose is followed by a short statement of your experimental approach (about half a sentence). The purpose may be written in the form of a transitional phrase or clause, for example. Follow the experimental approach immediately with your results for the experiment. Place important or general results first and less important details later in the segment/paragraph. Last, give an interpretation of your results to make them meaningful for the reader. Note that you should signal all of these elements (see Section 12.5).

It is important that you do not simply list your data—instead, *interpret* your data for the reader. Your interpretations should be limited to 1 to 2 sentences in the Results section. Avoid any lengthy interpretations, speculations, or conclusions. Save such detailed discussions for the discussion section.

The following is a well-formed paragraph/segment of a Results section:

👍 **Example 12-8**   **Well-formed Results segment**

Background	**1**Considerable evidence suggests that ATP is needed in the binding of mRNA to the 40S ribosomal subunits (13).
Purpose	**2**To understand the interaction between ATP and mRNP particles better, **3**we incubated the mRNP particles with ^{14}C ATP at optimal concentrations for *in vitro* yeast translation.
Experimental approach	
Results	**4**Results indicate that ^{14}C ATP bound to mRNP particles, but the binding decreased about 4-fold when the temperature increased from 4 to 17°C (Fig. 1). **5**These results suggest that the binding between ATP and mRNP particles may be governed by interactions such as hydrogen bonds or van der Waals that weaken when temperature rises.
Interpretation Results	

In Example 12–8, the first sentence gives a short background, sentence 2 states the purpose/question, and sentence 3 explains the experimental approach. Sentence 4 gives the results, whereas sentence 5 interprets them. Note that data are presented in a figure (Fig. 1), and the description of the figure is subordinated. Thus, the actual results are emphasized.

Special Case: Organization of Descriptions

For descriptive papers, list your results and descriptions by describing what you discovered. These descriptions often do not need an interpretation. Instead, provide conclusions and implications in the Discussion section.

👍 **Example 12-9**   **Results segment in a descriptive paper**

Purpose	To characterize the protease, various known protease inhibitors were tested for their effect on the proteolytic cleavage of the fusion protein. None of the inhibitors,
Experimental approach	

antipain, aprotinin, chymostatin, EDTA, leupeptin, p-chlo-
romercurobenzoate, pepstatin, phenylmethyl-sulfonylchlo-
ride, or soybean trypsin inhibitor, significantly slowed the
rate of cleavage (data not shown). However, the bivalent
cations Zn++, Cu++, and Co++ were found to inhibit the
protease to near completion when added to spheroplasts
before lysis (Fig. 2-6). Benzamidine addition resulted in
only partial inhibition. Addition of any of the inhibitors or
cations after cell lysis did not inhibit the protease activity.

Results

(With permission from Elsevier)

12.4   IMPORTANT WRITING PRINCIPLES FOR THE RESULTS

Word Choice

RESULTS GUIDELINE 10:

Pay attention to word choice.

Words in the Results section should be chosen carefully. Choose the most
precise and descriptive wording that reflects what you want to say, but
keep wording simple. Consider the following example:

Example 12-10	Choice of words
	Mg^{2+} binds to the complex and increases complex forma-tion, reaching an optimum at 4 to 10 mM (Fig. 2D). <u>K^+, on the other hand, has the *opposite effect* on complex forma-tion, *reaching its optimum at 0 mM K^+* (Fig. 2E).</u>

What does the author want to say in the second sentence? What is the
"opposite effect," and how can an optimum be *"reached"* at 0 mM? The
opposite of *increase* is *decrease* or *inhibition,* and an optimum can never
be *reached* at 0 mM, no matter what reagent is added. What the author
intends to state here is that K^+ inhibits the formation. Thus, the second
part of sentence 2 should be omitted entirely because the word *inhibition*
already implies the effect. The revision simplifies and clarifies what the
author is trying to say.

Revised Example 12-10	Mg^{2+} binds to the complex and increases complex formation, reaching an optimum at 4 to 10 mM (Fig. 2D). **K^+, on the other hand,** *inhibits* **complex formation (Fig. 2E).**

ESL advice

In addition to using simple, precise words and avoiding jargon and
repetitive words, you should pay particular attention to the following

specific words and phrases in the Results section. These words are often used carelessly by authors, especially ESL authors, but should be distinguished because of their implied meaning:

Did not

Choose your words carefully. Use neutral descriptions such as "did not" rather than "could not" or "failed to" when reporting results.

Example 12-11   **Choice of words**

We **did not** detect any insulin production. (neutral—no expectation implied.)

Clearly/it is clear/obvious

Omit "clearly" and similarly subjective phrases in the Results. "Clearly" makes authors seem arrogant and they appear as if they are trying to influence the reader.

Example 12-12   **Choice of words**

Figure 6 <u>clearly</u> shows that the growth rate of K103 was reduced when Ca²⁺ was added.

Revised   Figure 6 shows that the growth rate of K103 was reduced
Example 12-12   when Ca²⁺ was added.

Significant

"Significant" in science refers to "statistically significant." If you write, for example, "Flow rate decreased significantly," the reader expects statistical details to follow this phrase. If you report results of statistical significance, specify the significance level. Consult a standard statistics textbook for detailed advice.

If you do not plan to provide statistical details, use "markedly" or "substantially" instead of "significantly." It is best, however, to reserve these words for the Discussion. Also remember that you should quantify these qualitative words by using precise values or referring to data ("Flow rate decreased substantially (23%).").

Tense

RESULTS GUIDELINE 11:
Use past tense for your results, but present tense for descriptive papers.

Results are usually reported in past tense because they are events and observations that occurred in the past.

Example 12-13   **Use of tense**

a Imidazole **inhibited** the increase in arterial pressure.

b Once nectar **was depleted** from the *D. wrightii* flower, all of the moths then **switched** to feeding from the *A. palmeri* flowers.

Exceptions are results of descriptive studies. These results are reported in present tense because the description is still true.

Example 12-14   **Use of tense**

a The *fgk* gene **has** several different introns.

Other exceptions are statements of general validity, that is, if something is still true now or is considered a general rule, it should be written in present tense.

Example 12-14   **Use of tense**

b Our results suggest that learning the association between nectar reward and flower type is primarily olfactory mediated.

12.5   SIGNALS FOR THE READER

To emphasize different portions of the Results section, consider using signals, such as those shown in Table 12.1.

These examples provide great starting points when you write your first draft, and may be particularly useful for those authors who have writer's block or whose native language is not English.

The most important result may be specially signaled to highlight it so the reader cannot miss it:

ESL advice

Table 12-1 Signals for the Results

PURPOSE/QUESTION	EXPERIMENTAL APPROACH	RESULTS	INTERPRETATION OF RESULTS
To determine ...	... we did ...	We found ...	, indicating that ...
To establish if ...	X was subjected to ...	We observed ...	, consistent with ...
Z was tested ...	... by/using ...	We detected ...	, which indicates that ...
For the purpose of XYZ ...	ABC was performed ...	Our results	This observation
	Experiment X showed ...	indicate that ...	indicates that ...
		that ...	A is specific for ...

Example 12-15	**Highlighting important results**
	Most interestingly, almost half of the newlywed couples (45.5%) start out sharing everyday household tasks equally or with husbands doing even a greater share than their wives.

12.6 COMMON PROBLEMS OF THE RESULTS SECTION

The most common problems of Results sections include the following:

- Missing components (purpose of experiment, experimental approach, results, or their interpretation; Section 12.3)
- Inclusion of irrelevant or peripheral information (Section 12.3)
- Excessive experimental details (Section 12.3)
- Inclusion of comparisons, speculations, and conclusions beyond the interpretation of results (Section 12.3)

Missing Components

Of the four components that Results segments should contain (purpose of experiment, experimental approach, results, and their interpretation), beginning authors most often forget to include the purpose and the interpretation of the results. If these or other components are missing in any of the Results segments, findings will not be clear to readers.

In the following example, the interpretation of the results has not been included, leaving the reader wondering what these findings mean:

Example 12-16	**Results paragraph without interpretation of results**
	Successful colonization in the small intestines is indicated by prolonged shedding of vibrios in coproculture.
Experimental Approach —transitional clause	Therefore, to test for colonization, we tested for shedding of vibrios in coproculture. We found that there was
	a significant difference in shedding between P-5 and the
Results—overall	positive control. The coproculture of the rabbits vaccinated
	with P-5 had a shedding time of 8 days, whereas the shed-
Results—details	ding time in the rabbits vaccinated with the positive control
Interpretation missing	was 5.5 days.

When the interpretation is included, as in the following revised version, the reader will gain a much better understanding of what these findings imply.

Revised Example 12-16	
	Successful colonization in the small intestines is indicated by prolonged shedding of vibrios in coproculture.
Experimental Approach— transitional clause	Therefore, to test for colonization, we tested for shedding of vibrios in coproculture. We found that there was

Results—overall	a significant difference in shedding between P-5 and the positive control. The coproculture of the rabbits vaccinated with P-5 had a shedding time of 8 days, whereas the shedding time in the rabbits vaccinated with the positive control
Results—details	was 5.5 days, indicating that the vaccine candidate strain
Interpretation	P-5 efficiently colonizes rabbit intestines.

Irrelevant or Peripheral Information

RESULTS GUIDELINE 12:
Omit peripheral information and irrelevant general statements.

Do not confuse the reader, or yourself, by including irrelevant or peripheral information.

Example 12-17   **Irrelevant and peripheral information**

It took 2 hr to process 22,000 molecules and 32 hr to screen the entire ChemBridge database.

Readers are not interested to read how long your work took. Statements like these should be omitted.

Also omit irrelevant general statements of goals or overview sentences such as the one shown next.

Example 12-18   **Irrelevant overview sentences**

To present our results, we first list all components of the macromolecule together with their optima and then describe the outcome of their individual omission.

Overview sentences only add clutter. You do not need to explain how you will proceed in a research paper if your writing is coherent and well organized.

Aside from overview sentences, you should also avoid repeating all the data shown in figures and tables. Instead, describe your results and point the reader to a figure or table by citing this figure or table in parenthesis after the description of the results.

Example 12-19   **Referring to figures and tables**

A total of 34 stilbenes were identified in this study, <u>and they are listed in Table 3.</u>

Revised Example 12-19   A total of 34 stilbenes were identified in this study **(Table 3).**

Note that your main results should also be described in the text even if they are shown in a figure or table.

Excessive Experimental Details

RESULTS GUIDELINE 13:
Avoid experimental details.

Do not describe experimental approaches in detail again, and do not introduce new experimental setups that were not mentioned in the Materials and Methods section.

 Example 12-20   **Results segment with unnecessary experimental details**

Because *Nocardia* have a unique cell wall and membrane structure, drug permeability should not be overlooked. Growth inhibition can be tested employing the surrogate marker *Nocardia asteroides*, which has high genomic sequence similarity to that of *N. brasiliensis*. To test the inhibitor candidates from the *in vitro* assay, <u>the bacteria were grown in modified media (7H9 medium from BD Co. with 0.2% of glucose and 0.05% of Tween80) (12) for two days. To facilitate dispersion of *Nocardia*, a surfactant Tween80 was added to the liquid culture. Different amounts (1, 5, 10, 15, and 50 µg/ml, respectively) of the five candidate molecules were added to the media (7H10 from BD Co. with 0.2% glycerol) containing agar, mixed thoroughly and allowed to solidify. Media without any inhibitor candidate served as a negative control and media with 50 µg/ml of kanamycin served as a positive control. Plates were incubated at 37°C for 4 days.</u> Two out of the five candidate molecules clearly demonstrated inhibition of the bacterial growth (Figure 7). NB22 demonstrated 100% growth inhibition at 15 µg/ml. NB20 weakly inhibited the growth at 15 µg/ml and at 50 µg/ml. Bacterial growth on the negative control plate was not affected, and no growth was observed for the kanamycin treated plate (Figure 6, top row).

Experimental details should not be described in the Results

In the preceding example, all experimental details should have been described in the Materials and Methods section. These details should be omitted in the Results section because they make it unnecessarily lengthy and distract from the actual results.

**Revised
Example 12-20**

Experimental details
have been omitted

Because *Nocardia* have a unique cell wall and membrane structure, drug permeability should not be overlooked. Growth inhibition **for the final inhibitor candidates from the** *in vitro* **assay were tested** employing the surrogate marker *Nocardia asteroides*, which has high genomic sequence similarity to that of *N. brasiliensis*. Two out of the five candidate molecules clearly demonstrated inhibition of the bacterial growth (Figure 7). NB22 demonstrated 100% growth inhibition at 15 µg/ml. NB20 weakly inhibited the growth at 15 µg/ml and at 50 µg/ml. Bacterial growth on the negative control plate was not affected. No colony was observed for the kanamycin treated plate (Figure 6, top row).

In the revised example, experimental detail has been omitted. In addition, the second and third sentences have been combined. This paragraph presents the results much clearer and is much more concise than the original version.

Comparisons, Speculations, and Conclusions

RESULTS GUIDELINE 14:
Avoid general conclusions, speculations, or comparisons with other studies.

Do not compare your data to that of other studies, speculate on possible mechanisms, or draw general conclusions. Leave these comparisons, speculations, and conclusions for the discussion—but remember to interpret your results briefly at the end of each Results segment, thus setting the stage for the discussion.

Example 12-21

Results segment with partial discussion

Background

Purpose

Biofilm growth. Biofilm formation starts with bacterial attachment to the surface, then progresses to auto-aggregation, micro-colonies formation, maturation, and eventually cell detachment. To determine the dynamics of the biofilm growth and the effect on membrane performance in reverse osmosis, biofilm was grown on membranes for 6, 12, 24, or 48 hours post inoculation. Subsequently, specific biovolumes of the viable cells and dead cells were determined (Table 1). Micro-colonies of bacteria were observed 6 hours after inoculation. After 12 hours some dead cells were observed. 24 hours after inoculation, more dead cells were found, and crevices and holes appeared in the biofilm, probably due to detachment of cells and small aggregates from the biofilm. These rapid changes in biofilm structure do not follow the previously reported biofilm formation stages (12, 43). Possible reasons for this rapid change in biofilm structure could be a depletion of nutrients after 24 hours. In addition, previous studies used rich growth media (12, 43), whereas in our study minimal media was applied.

Interpretation

Omit in the results—place into the Discussion

(With permission from Moshe Herzberg)

 Revised Example 12-21

Background

Experimental Approach

Results

Interpretation

Biofilm growth. Biofilm formation starts with bacterial attachment to the surface, then progresses to auto-aggregation, micro-colonies formation, maturation, and eventually cell detachment. To determine the dynamics of the biofilm growth and the effect on membrane performance in reverse osmosis, biofilm was grown on membranes for 6, 12, 24, or 48 hours post inoculation. Subsequently, specific biovolumes of the viable cells and dead cells were determined (Table 1). Micro-colonies of bacteria were observed 6 hours after inoculation. After 12 hours, some dead cells were observed. 24 hours after inoculation, more dead cells were found, and crevices and holes appeared in the biofilm, probably due to detachment of cells and small aggregates from the biofilm. This rapid change in biofilm structure differs from that observed previously (12, 43).

12.7   SAMPLE RESULTS SECTIONS

 Example 12-22

Overall background/ purpose and experimental approach

Main overall results

Investigative paper: First paragraph indicating overall results followed by more detailed description of individual experimental results

During fertilization experiments performed to determine species specificity of two sea urchin species, *S. purpuratus* and *S. fanciscanus*, we discovered varying degrees of fertilizability when gametes of different individuals of a species were crossed (2, unpublished observation). Furthermore, as many as one third of the sea urchins examined appeared to be infertile. Substantial variations in fertilization were observed for approximately 30% of the individual urchins (Fig. 2).

Background/Purpose	The first set of *S. purpuratus* sea urchins tested for their fer-
	tilization efficiency contained 10 different males (m1–m10)
	and 5 different females (f1–f5). Gametes of each individ-
Experimental	ual male were crossed with those of each female in all
Approach	possible heterosexual combinations and the percent fer-
	tilization was determined. Although each individual cross
	reached at least 80% fertilization (Fig. 3), the amount of
	sperm needed to reach maximum fertilization rates var-
	ied for the different combination of gametes. For example,
	about six times as many sperm from individual m10 were
	required to result in 50% fertilization success of female f4
	than females f1 and f2, and about twice as many sperm
Results	were required for females f3 and f5 than for f1 and f2
	(Fig. 6), suggesting that eggs from individual females are
Interpretation	responsible for the variable success in fertilization.

	Although crosses between gametes displayed signifi-
	cantly reduced fertilization efficiency in one cross, they
	did yield normal levels with gametes of other individuals
	of the opposite sex. Thus, although gametes of female f4
	seemed to require more sperm of m4 and m6 to be fertil-
	ized than other females, the males m1, m2, and m9 were
Additional results	able to fertilize eggs of f4 at normal efficiency, suggesting
	that sperm of different individuals also effect the success
Interpretation	of fertilization.

Note, how in the preceding example, the first paragraph gives an over-view of the entire Results section by listing the main results of the study. The subsequent paragraphs describe individual experimental results in more detail and contain the required elements (background/purpose, experimental approach, results, and interpretation). In the last paragraph, the background/purpose and experimental approach are not repeated, as it is a continuation of the setup described in the preceding paragraphs.

Example 12-23	**Results of a descriptive paper**
	Results
	...On the basis of our difference electron density maps
	and on the available biochemical and functional data, we
	propose the structural basis for the modes of action of the
Overall results provide	antibiotics chloramphenicol, clindamycin, erythromycin,
overview	clarithromycin and roxithromycin.
	Chloramphenicol
	...Chloramphenicol has several reactive groups that
Topic sentence	can form hydrogen bonds with various nucleotides of the

peptidyl transferase cavity: two oxygens of the para-nitro (p-NO2) group, the 1OH group, the 3OH group and the 4' carboxyl group.

One of the oxygens of the p-NO2 group of chloramphenicol appears to form hydrogen bonds with N4 of C2431Dr (C2452Ec) (see Methods for definition), which has been shown to be involved in chloramphenicol resistance14. The other oxygen of the p-NO2 group interacts with O2' of U2483Dr (U2504Ec) (Fig. 1a–c) . . .

More detailed results/ description

Macrolides

Topic sentence

In contrast to chloramphenicol and the lincosamides, macrolides of the erythromycin class do not block peptidyl transferase activity26. Although they bind to the peptidyl transferase ring, the erythromycin group of the macrolides, which includes clarithromycin and roxithromycin, is thought to block the tunnel that channels the nascent peptides away from the peptidyl transferase center 2, 27, 28. Our results confirm this assumption. We could unambiguously determine that the macrolides erythromycin, clarithromycin and roxithromycin all bind to the same site in the 50S subunit of *D. radiodurans*, at the entrance of the tunnel (Fig. 5). Their binding contacts clearly differ from those of chloramphenicol, but overlap those of clindamycin to a large extent (Figs 3a and 4).

More detailed results/ description

(With permission from Nature Publishing Group)

Note how the topic sentences of the preceding example emphasize the results within each paragraph. Also note how the topic sentences weave a nice continuous story throughout the section. Here, the purpose/background and experimental approach are not repeated in every paragraph because they have been provided in the introductory first paragraph. All subsequent paragraphs start immediately with the results, that is, a description of what has been found, as they belong to the same segment.

12.8   REVISING THE RESULTS SECTION

When you have finished writing the Results (or if you are asked to edit a Results section for a colleague), you can use the following checklist to systematically "dissect" the section:

- ☐ 1. Did you report all main findings as well as other important findings?
- ☐ 2. Are your most important results and their interpretation provided in the beginning of the Results section?

☐   3. Are the data for your most important results also mentioned in the text?

☐   4. Is the organization from most to least important within the paragraphs?

☐   5. Does each Results segment or paragraph contain all components (purpose of experiment, experimental approach, results, and their interpretation)?

    ☐   a. Is the purpose of each experiment apparent?

    ☐   b. Is the experimental approach provided?

    ☐   c. Are results interpreted?

☐   6. Are all components (purpose of experiment, experimental approach, results, and their interpretation) signaled?

☐   7. Are results emphasized?

☐   8. Did you place statistical information with data?

☐   9. Is the reader pointed to figures and tables?

☐   10. Are control results included?

☐   11. Are irrelevant statements and peripheral information avoided?

☐   12. Have general conclusions, speculations, or comparisons with other studies been excluded?

☐   13. Are the references placed correctly and where needed?

☐   14. Revise for style and composition based on the writing principles of the book:

    ☐   a. Are paragraphs consistent? (Chapter 6, Section 6.2)

    ☐   b. Are paragraphs cohesive? (Chapter 6, Section 6.3)

    ☐   c. Are key terms consistent? (Chapter 6, Section 6.3)

    ☐   d. Are key terms linked? (Chapter 6, Section 6.3)

    ☐   e. Are transitions used and do they make sense? (Chapter, Section 6.3)

    ☐   f. Is the action in the verbs? Are nominalizations avoided? (Chapter 4, Section 4.6)

    ☐   g. Did you vary sentence length and use one idea per sentence? (Chapter 4, Section 4.5)

    ☐   h. Are lists parallel? (Chapter 4, Section 4.9)

    ☐   i. Are comparisons written correctly? (Chapter. 4, Sections 4.9 and 4.10)

    ☐   j. Have noun clusters been resolved? (Chapter 4, Section 4.7)

    ☐   k. Has word location been considered? (Verb following subject immediately? Old, short information at the beginning of the sentence? New, long information at the end of the sentence?) (Chapter 3, Section 3.1)

    ☐   l. Have grammar and technical style been considered (person, voice, tense, pronouns, prepositions, articles)? (Chapter 4, Sections 4.1–4.4)

    ☐   m. Is past tense used for results and present tense for descriptive papers?

    ☐   n. Are words and phrases precise? (Chapter 2, Sections 2.2 and 2.3)

☐ o. Are nontechnical words and phrases simple? (Chapter 2, Section 2.2)

☐ p. Have unnecessary terms (redundancies, jargon) been reduced? (Chapter 2, Section 2.4)

☐ q. Have spelling and punctuation been checked? (Chapter 4, Section 4.11)

SUMMARY

RESULTS GUIDELINES:

1. Report your main findings as well as other important findings.
2. Point the reader to the data shown in figures and tables.
3. Include control results.
4. Interpret your results for the reader.
5. Place statistical information with data. Do not use it instead of results.
6. Place results that answer the question of the paper at the beginning of the results section.
7. Organize the Results section chronologically or from most to least important.
8. Emphasize and signal your results. Subordinate secondary information.
9. Organize your Results into different segments. In each segment, state
   Purpose or background of experiment
   Experimental approach
   Results
   Interpretation of results (optional for descriptive papers)
10. Pay attention to word choice.
11. Use past tense for your results, but present tense for descriptive papers.
12. Omit peripheral information and irrelevant general statements.
13. Avoid experimental details.
14. Avoid general conclusions, speculations, or comparisons with other studies.

PROBLEMS

Problem 12-1

Using basic writing principles, improve the following sentences, which were taken from prepublication Results sections.

1. Disruption of AUF1 did not cause a significant alteration in the expression level of ARE-binding proteins examined here.
2. Intravascular coagulation was common in kidney, lung, and liver of $AUF1^{-/-}$ mice after endotoxin challenge but not in wild-type mice (Figure 2E).

3. Comparison of the accumulated levels of cytokine secretion after LPS stimulation showed that production of TNFα and IL-1β were strongly increased in AUF1-deficient macrophages.

4. In total, 19.7% (542/2758) of genes located on chromosome 1 and 41.34% (456/1103) of genes located on chromosome 2 were absent from at least one strain, which seems to indicate higher conservation in chromosome 1.

5. The number of variant genes in toxigenic strains is much less (missing 0 to 35 genes for each), exhibiting the highest conservation as described in previous research.

6. The T-test also showed a significant decrease in Cat-315 expression only in the contralateral barrel cortex.

7. Three of the molecules, CB4, CB6, and CB10, exhibited more than 50% inhibition of the enzyme activity.

Problem 12-2
Assess the following partial Results section.

1. **Identify**
   - **the purpose or background of the experiment**
   - **the experimental approach**
   - **the results**
   - **the interpretation of the results**
2. **Are all the parts of a paragraph for the Results section provided? Please explain.**

We found that the H384A mutant reduced the k_{cat} value more than 3-fold. The apparent K_m values were increased 7-fold for Fru 6-P and 3.5-fold for PPi. The increase of the K_m values and the reduction of the k_{cat} value of the H384A mutant suggest that the imidazole group of His384 is important for the binding stability as well as for catalytic efficiency of Fru 6-P and PPi substrates.

Problem 12-3
Assess and revise the following partial Results section. Ensure that all the parts of a paragraph for the Results section are provided and any unnecessary parts are omitted.

To evaluate inhibitory effects of the selected molecules, 10 mM stock solutions of each molecule were prepared in DMSO. A reaction mixture (200 μl) was prepared with the same formula optimized for the enzyme activity assay (0.1 M Tris-HCl pH 8, 0.1 M KCl, 25 mM NaCl, 0.25 mM ATP, and two units of inorganic yeast pyrophosphatase) with 10 μM of the sample molecule. The reaction mixture was incubated for 20 minutes at ambient temperature. Enzymatic reaction was triggered by addition of the substrate B (0.2 mM) and the absorbance of the product was monitored at 290 nm for 10 minutes.

Six out of 15 sample molecules showed appreciable inhibition at 10 μM (Figure 5). Three of the molecules, A3, A6, and A7, exhibited more than 50% inhibition of the enzyme activity and were further diluted to find the minimal inhibitory concentration (MIC). Molecules A3 and A6 exhibited 30% and 45% inhibition of the enzyme activity, respectively, even at 1 μM. DMSO was found not to interfere with the enzyme.

Problem 12-4

Assess the following partial Results section.

1. **Identify**
   - **the purpose or background of the experiment**
   - **the experimental approach**
   - **the results**
   - **the interpretation of the results**
2. **Are all the parts of a paragraph for the Results section provided? Please explain.**

Sensory deprivation during the first 30 days of development leads to decreased Cat-315 expression in layer IV of the barrel cortex.

Because aggrecan-reactive nets are strongly expressed in the postnatal barrel cortex (Fig. 2A–H), we asked whether altering sensory input from the whiskers would alter the expression of PNs and aggrecan. To evaluate this, whiskers were trimmed from the right whisker pad of mice every other day from birth through P30. Tangential sections through layer IV of the barrel cortex were analyzed using stereological methods. Nissl staining of the barrel cortex illustrated that the gross development of the barrels was not altered by this manipulation and did not differ from controls (Fig. 2A,E,I,M). However, our studies revealed a significant decrease in the number of cells with Cat-315-positive PNs in the sensory-deprived barrel cortex of trimmed animals (Fig. 2N–P) compared with the nondeprived barrel cortex of the same animals (Fig. 2J–L) and compared with both barrel cortices of control animals (Fig. 2B–D, F–H; Table 1). There was a statistically significant reduction in Cat-315 staining in trimmed animals compared with control animals using a two-way repeated-measures ANOVA ($p = 0.0282$). The decrease in Cat-315 in the deprived hemisphere compared with the nondeprived hemisphere in trimmed animals was significant using a paired t test ($p = 0.0034$) (Table 1). There was also a significant decrease in the deprived hemisphere of trimmed animals compared with the left hemisphere of control animals ($p = 0.0015$) (Table 1). However, the nondeprived barrel cortex did not differ significantly from the barrel cortex of control animals ($p = 0.9454$) (Table 1).

(With permission from The Journal of Neuroscience)

Problem 12-5

Assess the following partial Results section. Ensure that all essential parts of a Results section paragraph are there. Check that parts are signaled clearly. Improve transitions if needed. Condense where possible.

... Previous studies have shown that, in biological networks, hubs tend to be essential [7,9], and betweenness of a node is correlated with its degree [20]. We found that degree and betweenness are indeed highly correlated quantities in the networks we analyzed (Pearson correlation coefficient of 0.49, $p < 10^{-15}$ for the interaction network; Pearson correlation coefficient of 0.67, $p < 10^{-15}$ for the regulatory network; p-values measure the significance of the Pearson correlation coefficient scores according to t distributions; i.e., many bottlenecks also tend to be hubs). Therefore, we further investigate which one of these two quantities is a better predictor of protein essentiality in both regulatory and interaction networks.

To disentangle the effects of betweenness and degree, we divided all proteins in a certain network into four categories: (1) nonhub–nonbottlenecks; (2) hub–nonbottlenecks; (3) nonhub–bottlenecks; and (4) hub–bottlenecks (see Figure 1). Even though the two quantities are highly correlated, the number of hub–nonbottlenecks and nonhub–bottlenecks is enough for reliable statistics (see Table S1). This is in agreement with the previous observation by Huang and his colleagues, who found that proteins with high betweenness but low degree (i.e., nonhub–bottlenecks) are abundant in the yeast protein interaction network [18].

(PLoS Comput. Biology 3(4), 2007)

Problem 12-6
Assess the following partial Results section:

1. **Identify**
   - **the purpose or background of the experiment**
   - **the experimental approach**
   - **the results**
   - **the interpretation of the results**
2. **Are all the parts of a paragraph for the Results section provided? Please explain.**

To explore the forces behind the strong tendency of husbands to decrease their share of housework, and the low tendency to increase their share of housework in the course of marriage, we looked at the findings from the multivariate event-history models. We first explored the role of economic resources in changing couple's division of housework in the course of marriage (Table 1). We investigate the impact of the spouse's relative economic resources on the likelihood of dividing housework either more equally or less equally in the course of marriage. We find that husbands who work a similar number of hours (Husband=Wife), or lower number of hours (Husband<Wife) than their wives, are less likely to decrease their share in household labor, compared to husbands who work longer hours than their wives (Husband>Wife). Equal earning levels between the spouses also seem to reduce the likelihood for husbands to decrease their share of housework in the course of marriage (model 3b). For couples with an 'atypical' female provider earnings ratio

(Husband < Wife) the effect is not significant, however. It appears that a winning margin in economic resources does more for the husband than for the wife when housework is redistributed.

We also looked at the effects of family formation on the gender division of household tasks to assess how shifts in economic resources play out in this context (Table 2). We found a pronounced and significant effect for both directions of change. During the first year after childbirth, fathers seem to be about twice as likely to decrease their contribution to housework. In the same period fathers' likelihood to increase their share in housework is reduced by almost 50 percent, compared to childless men. This push towards a more traditional division of housework seems to come to a halt when the youngest child reaches age two. We found no indication, however, that parents readjust back to a more egalitarian division of housework when kids grow older and mothers return to their previous jobs. The time-varying economic indicators (models 2ab and 3ab) do not seem to explain these processes at all. Here, we need to be cautious not to interpret lack of significance in the economic indicators as lack of relevance, especially since the share of parents in the 'non-traditional' resource categories is low. The number of mothers out-earning or working longer hours than their husbands is small in this sample and the share of continuously working mothers is low.

(With permission from Daniela Grunow)

Problem 12-7 Your Own Results

Write your own Results section based on the experimental results of your work. Provide the overall question of the paper. Use all basic writing principles studied, and follow the summary on how to write a Results section.

CHAPTER 13

Discussion

13.1 OVERALL

The Discussion is usually the hardest section to define and to write. Many papers are rejected by editors because of a bad Discussion section. Even though the data may be valid and interesting, the interpretation or presentation of it in the Discussion may obscure it. Therefore, good style and clear, logical presentation is especially important here.

13.2 CONTENT

DISCUSSION GUIDELINE 1:
State and interpret your key findings. Provide the answer to the research question.

DISCUSSION GUIDELINE 2:
Summarize and generalize.

DISCUSSION GUIDELINE 3:
Keep in mind who your potential readers will be.

The main function of the Discussion is to interpret your key findings and to draw conclusions based on these findings—in other words, answer the question(s) asked in the Introduction. The discussion should also explain how you arrived at your conclusion, compare and contrast your findings with existing knowledge on the topic, and state theoretical implications or practical applications. It should give the paper significance by summarizing and generalizing results while clearly indicating how your study has advanced knowledge.

In the Discussion, explain what is new in your work and say why your results are important. You should also include explanations for any results that do not support the answers and discuss other results and hypotheses that are relevant to yours. In addition, you may discuss any possible errors or limitations in your methods, give explanations of unexpected findings, and indicate what the next steps might be. Do not refer to every detail of your work again; repeating the Results section in the Discussion is a common mistake of inexperienced writers. Another common mistake is to add another Introduction. Instead, in your Discussion, summarize and generalize.

Adjust your Discussion according to who your potential readers will be and make it no longer than necessary. If you are writing for a very specific group of people, stay within their area of interest. If you are writing for a broad audience, you probably need to discuss much broader implications and provide more generalizations and background.

In general, know that in the related fields of biology and medicine, basic scientists and clinicians read each other's papers. So if you write your paper primarily for a scientific audience, do not ignore the clinical implications of your results; and if you are addressing a clinical audience, try to discuss the scientific significance as well. In this way, your work will have much greater impact.

13.3   ORGANIZATION

DISCUSSION GUIDELINE 4:
Organize the Discussion in a *pyramid structure*:

First paragraph:	Interpretation/Answer based on key findings
	Supporting evidence
Middle paragraphs:	Comparisons/Contrasts to previous studies
	Limitations of your study
	Unexpected findings
	Hypotheses or models
Last paragraph:	Summary
	Significance/Implication

Opposite of the Introduction, which follows a funnel shape, the Discussion follows a pyramid shape. In other words, it moves from specific to general.

The pyramid structure of the Discussion can be divided into

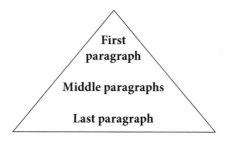

In the first paragraph, tell your readers what your key findings were and what they mean. In subsequent paragraphs, explain how your findings fit into what is known in the field. In the last paragraph, summarize and generalize why the contribution of your study is important overall, in your field, outside your field, and/or for society.

A more detailed explanation of this organizational structure is provided in the next subsections of this chapter.

13.4 FIRST PARAGRAPH

Begin the Discussion with an interpretation of the key finding(s), which present the answer to the question posed in the Introduction. Then support your answer by stating the relevant results, providing explanations, and/or other data. Do not assume that the reader has memorized the results or will search for them. You have to put the story together for the reader.

Because the interpretation of your key findings is the most important statement in the paper, it should appear in the most prominent position: the first paragraph of the Discussion. The interpretation of your key findings should match the question/purpose for the study stated in the Introduction and answer what the introduction asked. The interpretation of your key findings should also be repeated in the other power position of this section: the last paragraph.

Do not begin the Discussion with a second introduction, a summary of the results, or secondary information. Begin by directly stating the answer based on your findings in the opening sentence of the Discussion. If you feel this beginning is too abrupt, you can restate the purpose of the study or provide a brief context before stating the answer. Any statements placed before your answer should not exceed more than a few sentences.

If your answer is in the first paragraph of the discussion, the reader is sure not to miss it. Readers typically do not read the whole paper front to back. Rather, they skim over the Abstract, maybe read the Introduction, and then immediately jump to the Discussion. In the Discussion, readers instantly want to find the answer to the research question, that is, the interpretation of the key findings of the study. Readers do not usually take the time to read the whole Discussion. They will typically read the first paragraph and then go on to the last paragraph. Thus, your key findings and their interpretation(s) should be placed in these two power positions.

Example 13-1   **First paragraph of Discussion**

Question/Purpose:

Effectiveness of Peridomestic <u>Lyme Disease Protection Measures</u>

Answer/Interpretation of key findings:

Answer/Interpretation of key findings	*Our findings emphasize* the need to continue to promote personal <u>protection measures</u> to reduce the risk of <u>Lyme disease</u> infection. *We have identified* three reasonable personal measures that may be <u>protective</u> against <u>Lyme disease</u> when practiced: tick checks, bathing, and insect repellents. Performing tick checks within 36 hr after spending time in the yard may reduce one's risk by as much as 46%. In addition, bathing may reduce one's risk by up to 57%, and the use of insect repellent may be protective against the disease up to 75%.
Supporting evidence	

(With permission from Neeta Connally)

In Example 13-1, the interpretation of the key findings matches the purpose of the study. Note that the same key terms (*protection, measures, Lyme disease*) appear in the question/purpose as well as in the interpretation or answer to the question. This answer is immediately supported by key findings of the study.

Here is another example:

Example 13-2   **First paragraph of Discussion**

Question/Purpose:

Our goal was to determine <u>what part of the bindin polypeptide is responsible for the species-specific egg agglutination activities of the protein.</u>

Answer/Interpretation of key findings:

Answer to question	*Our results suggest that* <u>the part of bindin responsible for species-specific egg agglutination lies in the region of residues 75–121</u>. *We showed that* <u>residues 18–74 and 122–236 can be deleted without loss of egg agglutination activity</u>. All of the biologically active bindin deletion analogs were found to be <u>species-specific</u> by their ability to <u>agglutinate</u> exclusively *S. purpuratus* <u>eggs</u>. Deletion analogs that had any residues of region 75–121 deleted exhibited no significant activity above the bacterial control protein.
Supporting evidence	

(With permission from Elsevier)

In this example, the answer also matches the question/purpose posed. Note the *signals* used to guide the reader through the answer and supporting evidence in both of the previous examples. Note also the different verb tenses. To distinguish between results and the conclusions you draw from them, use the past tense for results ("showed," "were found," "exhibited") and present tense for general statements, interpretations, and conclusions ("suggest").

The exception to beginning the Discussion with an interpretation of the key findings is for papers that describe controversial topics and findings. For highly controversial topics, argue your case first by presenting your findings and explaining differences from other findings. This organization will help to logically prepare the reader for your upcoming argument. Present the controversial interpretation of your key findings at the end of the Discussion in these papers.

13.5   MIDDLE PARAGRAPHS

DISCUSSION GUIDELINE 5:
Organize the topics according to the science
or from most to least important.

DISCUSSION GUIDELINE 6:
Compare and contrast your findings with
those of other published results.

DISCUSSION GUIDELINE 7:
Explain any discrepancies, unexpected
findings, and limitations.

DISCUSSION GUIDELINE 8:
Provide generalizations where possible.

In this part of the discussion, it is not only important to put your findings in the context of your field but also to make connections to other important implications of the work or to other seemingly related or perhaps unrelated fields.

After stating and supporting your answer, mention other findings that were important. Tell your readers what you think your results mean and how strongly you believe in them. Organize these findings according to the science or from most to least important. To ensure that your

Discussion is organized rather than rambling, focus the story on the question/purpose of the paper that was stated in the Introduction.

Treat your secondary results as you did your main findings: Summarize and generalize them rather than simply repeating what you found. Do not discuss every single result you obtained. Rather, list and explain any general trends and tendencies and evaluate these.

The presentation of your arguments is a matter of personal style as is the order of the paragraphs between the first and last paragraph. To develop the middle paragraphs of a Discussion, organize the topics by proceeding from most to least important unless there is a reason for putting one topic before another. Explain any new findings and concepts obtained in your study, but do not present any new data that has not already been mentioned in the Results section. Also, do not repeat any information that has already been presented in other sections of your paper.

Mention any limitations of your study or unexpected findings, and present any new hypothesis or model based on your findings. If useful, include figures to illustrate complex models in the discussion. Compare and contrast your findings with those of previously published papers, but avoid the temptation to discuss every previous study in your subject area. Stick to the most relevant and most important studies. Explain any disagreements objectively, and credit and confirm the work of others. Give pro and contra arguments for your conclusion. Only if you mention both impartially will you sound convincing to the reader. Know that most of the time it is wise to present your opinion carefully rather than too strongly.

Comparisons and Contrasts

In addition to stating and supporting your answer, you need to explain how your findings fit in with existing knowledge on the topic. You can do so by comparing and contrasting your results with those found by others. One way to get started in comparing and contrasting your findings and interpretations is to prepare lists that contain your findings and those of others. Based on lists like these, you may be better able to see and discuss any similarities or differences between your work and that of previous reports.

When you mention any results that do not support your answer and conclusions, explain these findings as best as you can. If you can explain why a finding is conflicting, it is almost always worth doing so. If you cannot explain these findings, say so ("We cannot explain why...," "Although the reason for X is not obvious,...").

See the following example for an explanation of contrasting findings. In this example, the authors discuss a finding of another study that differs from the answer to their question ("In contrast to our observations ..."). The authors then discuss why previous findings cannot be directly compared with their results and go on to explain how previous studies had obtained these conflicting results.

Example 13-3	**Comparing and contrasting findings in the Discussion**
	We observed virtually no size classes of mtDNA mol-ecules. Since the undegraded circular mtDNA molecules were entirely of heterogeneous size, this observed size
Finding of paper	heterogeneity probably reflects the real situation within
Signal for contrasting finding	plant mitochondria.... *In contrast to our observations,* size classes of linear or circular molecules and species specific differences have been previously reported (24, 25). However, these studies were performed only with a fraction of supercoiled DNA (26), which most likely does not represent the complete set of molecules existing *in organello.* Supercoiled DNA isolated from a *C. album* sus-pension culture, for example, consisted exclusively of small circular plasmid mp1 DNA. Its oligomers were found in the open circular form, thus appearing indeed as a few
Explanation of conflict	size classes.
	(With permission from Springer)

Example 13-4 provides another example of comparing and contrasting findings and ideas in a Discussion:

Example 13-4	**Comparing and contrasting findings in the Discussion**
	The frequency of targeted events among integrative trans-formants was about 30% for transformation with a vector
Finding of paper	that shares 1kb of sequence homology with the genome.
	This targeting efficiency is comparable to that reported
Comparison to other study	for insertion vectors sharing more than 2kb of sequence homology with the moss genome (16). However, a tar-geting efficiency of 30% using 1kb of genomic sequence is considerably higher than that previously observed in
Contrasting finding	higher plants (0.13%) (18). The requirement for sequence homology for homologous recombination appears there-fore to be stringent and comparable to that reported for
Comparison to other study	mouse embryonic stem cells (37).
	(With permission from Springer Science and Business Media)

Limitations

Limitations of the study as well as assumptions should be explained in the Discussion, especially if their explanations are too long to be placed into the Materials and Methods section.

Example 13-5   **Explaining limitations in the Discussion**

Our data show that Aβ assemblies did not colocalize in drusen. It is important to note, however, that the epitope for Aβ may have been masked within the oligomeric structure, as is the case when Aβ monomers are transformed into amyloid fibrils (40). Therefore, we cannot preclude the possibility that the oligomeric cores in drusen are made up of Aβ.

Limitation

Unexpected Findings

Aside from comparing and contrasting your work with that of previous studies and describing limitations in your study, unexpected findings may also be mentioned in the Discussion section. Be alert to unexpected findings. Do not automatically assume that your experiment failed or that you made a mistake. Unexpected findings may be important—they may lead to new discoveries and alter the focus of your study.

When describing an unexpected finding, state at the beginning of the paragraph that the finding was unexpected (or surprising), and then explain it as best you can.

Example 13-6   **Describing unexpected findings**

Unexpected finding

To our surprise we discovered that the bindin fusion protein was being cleaved during isolation and purification. The proteolysis is remarkably efficient since only small amounts of the unprocessed form remain (Fig. 2, lane 3). We purified the cleaved bindin product to homogeneity by reverse phase HPLC and sequenced it to determine the site of cleavage. The predominant product is the mature bindin polypeptide containing an additional 4 amino acids of probindin and a minor product that corresponds to bindin containing a single additional amino acid. Both products contain arginine as the N-terminal amino acid. These results suggest that the fusion protein is cleaved at two sites: the Arg-Arg junction between the factor Xa linker and the probindin coding sequence and within the probindin segment at the Lys-Arg junction.

Description of unexpected finding

(With permission from Elsevier)

Generalizations

Sometimes you may be able to generalize your findings and formulate a hypothesis or propose a possible model. Explain how you arrived at your hypothesis or model. Consider illustrating complex models in figures. Describe how the hypothesis or the model works, incorporating a discussion of any figures if needed. If possible, also describe ways to validate your model.

Example 13-7   **Formulating hypotheses**

We found that the substrate ^{3}H-[9R]iP moves into the cells where it does not accumulate to concentrations higher than in the medium. However, the mechanism of ^{3}H-[9R] iP uptake is unclear. Because no extracellular activities for the deribolisation of ^{3}H-[9R]iP could be detected, we hypothesize that it is metabolized intracellularly to ^{3}H-iP and that the bidirectional transport of iP is based on passive

Hypothesis   diffusion.

Following is another example that explains a hypothetical model the authors came up with and points to a figure of this model:

Example 13-8   **Formulating a hypothetical model**

Repeated elements located in the amino- and carboxyl-terminus of the protein X vary in sequence for the different species tested. A hypothetical model of how these repeated elements might interact with the egg receptors is shown in Fig. 7. In this model, protein X is able to interact with complementary sites on its own receptor but not with those of a different species. Protein Y on the other hand contains ligands that interact not only with its own recep-

Hypothetical model
and figure   tor but also with that of another species. This model can explain the unidirectionality in cross fertilization.

13.6   LAST PARAGRAPH

At the end of the discussion, you should provide some closure by writing a one paragraph concluding summary. Readers typically expect to see two things in the summary of a scientific paper: an analysis of the most important results and the significance of the work. The analysis of the most important results is typically provided by the interpretation of your key findings, that is, the answer. Here too, the answer should match the question/purpose you posed in the Introduction and the answer presented in the first paragraph of the Discussion. Do not bring in new evidence for the summary. Rather, complete the "big picture" by restating your answer, that is, the interpretation of the key findings.

The significance of the work can be provided by including far-reaching interpretations and conclusions at the end of the Discussion section. Try to generalize your specific findings to other broader situations. Depending on your level of certainty, significance can range from the practical application to the theoretical proposition. Adding a practical application, giving advice, implying an action, or providing a proposition in the concluding paragraph gives the paper importance. Discuss any theoretical implications, possible applications, recommendations, or speculations based on your findings. If you pose any speculations or implications, base them on solid evidence and make sure that the reader understands that these are your speculations or implications.

		Level of certainty	Example
Practical	Application		"... can be used for ..."
	Advice		"X should be used to ..."
	Suggestion		"Our results imply ..."
Theoretical	Proposition		"We hypothesize that ..."

Because the conclusions are the major message of your paper, you should phrase them with great care. Possible ways to provide some closure of your work at the end of the Discussion section are shown in the next examples:

👍	**Example 13-9**	**Concluding paragraph with application**

Answer

In summary, our work reveals the functional interactions involved in the binding of antibiotics to the peptidyl transferase cavity of the bacterial ribosome. None of the antibiotics examined show any direct interaction with ribosomal proteins. Chloramphenicol targets mainly the A site, where it interferes directly with substrate binding. Clindamycin interferes with the A site and P site substrate binding and physically hinders the path of the growing peptide chain. Macrolides bind at the entrance to the tunnel where they

Key findings

sterically block the progression of the nascent chain. The structural model of the peptidyl transferase center in complex with the examined antibiotics *can* not only enable a

Significance indicated by a possible application

rational approach for antibiotic development and therapy strategies but *can* also be used to identify new target sites on the eubacterial ribosome.

(With permission from Macmillan Publishers Ltd.)

In this example, "In summary, ..." signals the conclusion. The first sentence of the paragraph is at the same time the topic sentence and the answer to the question. The word "can" in the last sentence indicates the importance of the work by signaling an application.

Other signals for a concluding paragraph are "Taken together, …" or "In conclusion, …" (see also Section 13.8), or the subheading "Conclusion." Sometimes, even an overall, summarizing question can serve as the signal of the concluding paragraph.

If you have drawn conclusions different from your original hypotheses, you might suggest ways in which these conclusions could be verified in future research. Do not merely say, however, that future research will be needed to clarify the issues without giving the reader any indication of what form this research might take.

The next example shows a conclusion that includes a speculation. This speculation is signaled by the words "believe" and "may."

👍   **Example 13-10**   **Conclusion with speculation and future direction**

| Signal of conclusion | In summary, we found no statistically significant associations between increased homocysteine (HCY) and age-related macular degeneration (AMD) after analyzing a large and well-characterized population of patients with and without maculopathy from two geographic areas in the United States. An analysis of smoking and HCY tertile subgroups did not show any association between smoking, increased HCY, and increased risk of intermediate or advanced AMD. An association between homocysteine levels and an increased risk of intermediate or advanced AMD *may* exist for patients for whom HCY is above the 90th percentile of HCY, as these patients were more likely to have intermediate or advanced AMD. When subjected to statistical analysis, this observation was found to be not significant however, and only a larger study cohort could determine whether there is any true association. |

Answer applies to the opening sentence; *Key findings and Conclusions* to the middle; *Future direction* to the final sentence.

13.7   IMPORTANT WRITING PRINCIPLES FOR THE DISCUSSION

Tone

The tone of your writing is important. "Beginners" often do very good scientific work but are intimidated by the knowledge and work of "experts." However, when you have collected enough data and are presenting your findings in a paper, you become an expert too. Make your writing convey confidence and authority. Show that you are knowledgeable about the subject, and take responsibility for your conclusions. Do not be afraid to take a stand.

Often ESL authors are also intimidated but mainly because English is not their native language. Some ESL authors do not realize when they sound too opinionated and their language use is too strong. Ask a native speaker, preferably a scientific editor, to read over your manuscript.

ESL advice

Native speakers will be much more aware about the fine nuances of their language.

Person and Voice

Do not just use third person and passive voice in the Discussion if it can be avoided. Instead, use first person and active voice to make your discussion more lively and interesting. The use of "we" is perfectly okay. If you are a single author, you may also consider using "we."

Tense

In the Discussion, you may find a mix of verb tenses. Remember that when you are referring to your findings and completed actions, use past tense. However, for statements of general validity and those whose information is still true, use present tense. Also use present tense for the answer to the question and for the statement of significance (see also Example 13-9).

Continuity

It is especially important to provide continuity in the Discussion such that the reader can read through the section without stumbling across unclear passages. To ensure such continuity, use topic sentences, transitions, and key terms. The topic sentence of each paragraph must indicate not only the topic or message of the paragraph but also the relation of the paragraph to the previous paragraph(s) and thus to the answer of the question/purpose. Readers who study your paper more carefully will mainly read the topic sentences to identify subtopics quickly.

13.8   SIGNALS FOR THE READER

DISCUSSION GUIDELINE 10:

Signal the elements of the Discussion.

Signal the different elements of the Discussion so that readers recognize immediately what they are reading about. Possible signals for unexpected findings, comparisons, conflicting results, limitations, and proposed hypothesis, are listed in Table 13-1 and Table 13-2.

ESL advice

Both of these tables provide great starting and reference points when you put together your first draft, and may be particularly useful for those authors whose native language is not English.

For unexpected findings, comparisons, conflicting results, and proposed hypothesis, you may use the signals shown in Table 13.2.

13.9   AN ALTERNATIVE: RESULTS AND DISCUSSION

The Results and Discussion sections are sometimes combined into one section. In a combined Results and Discussion section, there is no need for backward page turning. The results are discussed right after

Table 13–1 Signals for the Discussion

ANSWER	KEY FINDINGS	SUMMARY	SIGNIFICANCE	
In this study, we have shown that…	In our experiments…	In summary,…		**Level of certainty**
In this study, we found that…	…can be attributed to… We determined X by…	In conclusion,… Finally,… Taken together,…	Our findings **can/ will serve** to… …**can be used**…	
Our study shows that…	We found that…	To summarize our results,…	We **recommend** that X is… Y **should** be used for… …**is probably**…	
Our findings demonstrate that…	Our data shows that… …has been demonstrated by…	We conclude that… [overall question]	Y **indicates** that X **might**… These findings **imply** that X **may**…	
This paper describes…		Overall,…	Here we **propose** that… …we **hypothesize** that…	

Table 13-2 Signals for the Discussion

COMPARISONS	CONFLICTING RESULTS	LIMITATIONS	UNEXPECTED FINDINGS	PROPOSED HYPOTHESIS
…consistent with…(ref.) Similar to…(ref.) …has also been observed by…(ref.) X has been demonstrated …(ref.)	However, other studies found that…(ref.) …is controversial…(ref.) …does not agree with…(ref.) …has also been reported…(ref.)	…was not possible… …could not be measured… …was limited by… Further observations are needed to…	Surprisingly,… To our surprise… …was not expected.	Our results lead to the conclusion that… From these data we hypothesize that… We propose the following new principle…

ref. = reference

they are presented so that the reader can understand why they are important immediately. Note that in a combined Results and Discussion section, *all* results will have to be presented and discussed in a combined section and not just the most important summarized results of your paper.

13.10 COMMON PROBLEMS OF THE DISCUSSION

The most common problems of Discussion sections include

- The answer/interpretation of your key findings is not provided in the first paragraph (Section 13.4)
- No concluding paragraph is provided. The importance/significance of the study is not clear. (Section 13.6)
- Inclusion of irrelevant or peripheral information (Section 13.5)
- Results are repeated or summarized in the Discussion (Section 13.4)

Answer Not in First Paragraph

If you fail to provide the interpretation of your key findings or fail to answer your question in the first paragraph of the discussion, your readers will usually stop reading and jump directly to the concluding paragraph in hopes of finding the answer there.

Providing the answer in the first paragraph can be a problem if your Discussion is divided into subsections with independent headings. For such Discussions, consider providing an overall overview of the study before discussing individual key findings in subsections. Treat this overview paragraph like the most important power position of the Discussion, stating the main, overarching interpretation(s) based on your key findings.

Inclusion of Irrelevant Information

Often the importance of the results are not discussed or are not adequately discussed, which leaves the reader wondering about the significance of the paper. Indicate the significance of your findings at the end of the Discussion section in your conclusion. Take a position—you are the expert in your study now. Remember, the last paragraph(s) is the other power position within this section. Do not waste this position with anything other than the importance of the study (see also Section 13.6).

Although you should discuss how your research fits in with other existing findings in the field, the key is to concentrate on generalizing and summarizing your interpretations, comparisons, and contrasts by providing an overall picture.

Importance of Study Is Unclear

Beginning readers frequently include irrelevant information in the Discussion—often to fill in space and increase the length of the section. Such information may range from summarization of results; to secondary introductions; to discussion of minor experimental setups, limitations, or personal opinions. Do not be afraid to make your Discussion section shorter rather than longer. It is more important to discuss only immediately relevant findings than to fill in space. Irrelevant and repeated information is distracting to readers, as it does not allow them to distinguish clearly between what is important and what is less important.

13.11   SAMPLE DISCUSSIONS

Following are two complete discussion sections that show a good organization:

👍 **Example 13-11**   **Complete Discussion of an informative paper**

Interpretation/answer

Key findings

> Our results suggest that fertilization among sea urchins is intraspecies specific and that surface components of both gametes are involved in the fertilization specificity. Results of intraspecies specific fertilization between different gametes of several different individuals of the species *S. purpuratus* displayed varying degrees of fertilization success.

Comparison with previous findings

> Approximately one third of the examined individuals show significant differences in fertilization. The percent of individual crosses of gametes yielding significantly reduced fertilization specificity corresponds to the previously reported amount of "defective" gametes encountered during fertilization (5) and egg agglutination experiments (9). However, in our study, crosses for gametes of an individual that yield significantly reduced fertilization in one cross reached normal levels with other individuals of the opposite sex, suggesting that a low fertilization success is not due to immature or defective gametes.

Comparison with models of others

Suggested hypothesis

> It seems intuitive that the molecules that mediate sperm–egg interactions may play an important role in speciation since individuals are in separate gene pools if the sperm and egg cannot interact to form a zygote. How these systems evolve so that once common ancestors become reproductively isolated is not so obvious. Geographic isolation is believed to be the principal mechanism of speciation in marine animals (6). However, speciation does not have to be accompanied by major genomic reorganization (5, 7). A few mutations in bindin and its receptor can be sufficient to accomplish reproductive isolation because these proteins are major components of the fertilization mechanism. The hypothesis that bindins and their receptors contain multiple adhesive elements would allow mutations to occur within an individual element of bindin without catastrophic consequences for mutant individuals in the absence of a simultaneous compensating mutation in the receptor.

Summary

Significance

> In summary, fertilization among sea urchins appears intraspecies specific due to surface components of both gametes. It is conceivable that these surface components contain multiple adhesive elements. Based on our hypothesis, mutations in these elements may result in reproductive isolation and speciation.

Example 13-12   **Complete Discussion of a descriptive paper**

Discovery and implication

Our results suggest that the sparse matrix screen may be of general utility for establishing initial renaturation conditions for a wide variety of proteins. Eight of the nine proteins tested recovered significant amount of activity by this method. These proteins included an adhesive protein (bindin), proteins with disulfide bonds (lysozyme, β-gal, trypsin, acetylcholinesterase, HRP, BAP), proteins with multiple subunits (β-gal), and proteins with cofactors (HRP, BAP). The sparse matrix approach has the potential for the elimination of relatively tedious random searching for conditions that support renaturation. For example, we obtained bindin renaturation serendipitously with some of the buffers provided by the Crystal Screen Kit in an attempt to crystallize bindin.

Supporting evidence and explanations

The composition of the crystallization buffer resulting in renaturation differs widely from buffers previously known to be required for biological activity. As observed for bindin, conditions needed for refolding the other proteins tested are different from those required for their maximal enzyme activity. It appears that both crystallization and biological activity require a discrete tertiary conformation of the protein. Thus, solution conditions, which promote crystallization, will also promote the reformation of the native structure.

Further evidence supporting the discovery

The Crystal Screen kit contains three major components: a buffer, salts and a precipitating agent (see Table 1). In crystal growth the precipitant promotes the filling of an ordered lattice by favoring protein-protein contacts. Since the native protein structure in solution is frequently the same as that determined in the crystal (27, 28), it is perhaps not so surprising that there may be some commonality between conditions that promote protein renaturation and protein crystallization. This commonality might be the presence of precipitant in the solvent system, as most of the proteins tested do not renature in the absence of this component.

Description of discovery

Once initial renaturation conditions are identified, conditions for renaturation may be optimized by systematically altering other parameters which are known to be important for the refolding of a protein. The most common method for optimizing the yield of active protein is to perform a renaturation/reoxidation in the presence of low concentration of urea or guanidine hydrochloride (13). This situation can be mimicked in the sparse matrix screen by serial twofold dilution of the urea-denatured enzyme in the crystallization buffers.

Description of discovery

Extrinsic factors not encoded in the primary amino acid sequence may be of importance for the refolding of a protein (38). In the cell, accessory proteins are involved in regulating the rate of folding and association of a protein. Thus, the addition of molecular chaperones, cofactors, ions, and conjugate components such as carbohydrates, nucleic acids, and lipids to the buffer system may improve the rate and yield of renatured protein.

Renaturation of multiple subunit proteins or multiple domain proteins may be especially challenging and may require different conditions for each domain or subunit. Proteins with multiple domains are thought to refold slower than those with single domains (26). Since each subunit may have to refold independently, it may be necessary to screen each subunit for renaturation separately, and then test for the ability of the individually renatured subunits to reassociate in an active form. The screening for these conditions could be simplified by combining subunits from several renaturation systems and testing the mixture for activity. Optimum combination could be determined by testing mixtures containing progressively fewer combinations.

To summarize, here we report a sparse matrix method employing 50 different solvent systems for establishing initial solvent conditions that facilitate protein renaturation. Using this method, we tested nine different proteins for renaturation and observed significant amounts of renaturation for eight of the nine proteins tested. Once initial conditions are obtained, renaturation conditions may be systematically optimized by varying conditions such as pH, temperature, and protein concentration. This screening matrix is based on a set of solutions originally selected for protein crystallization. Our results suggest that this screening method may be widely applicable in identifying conditions that support renaturation and that the same conditions, which promote protein crystallization may also promote protein renaturation.

(With permission from Elsevier)

Description and application of finding

Limitations

Signal of conclusion

Key finding

Description of new finding

Speculation

13.12   REVISING THE DISCUSSION

When you have finished writing the Discussion (or if you are asked to edit a Discussion for a colleague), you can use the following checklist to systematically "dissect" the Discussion:

☐   1. Did you interpret the key findings/provide the answer to your research question?

☐ 2. Is the interpretation/answer to the research question provided in the first paragraph?

☐ 3. Is the answer stated precisely? (In present tense?)

☐ 4. Is the answer followed by supporting evidence?

☐ 5. Does the discussion follow a pyramid structure?

☐ 6. Is a summary paragraph placed at the end of the Discussion?

☐ 7. Is the significance of the work apparent?

☐ 8. Did you organize the topics according to the science or from most to least important in the middle of the discussion?

☐ 9. Did you compare and contrast your findings with those of other published results?

☐ 10. Did you explain any discrepancies, unexpected findings, and limitations?

☐ 11. Did you provide generalizations where possible?

☐ 12. Did you avoid restating or summarizing the results?

☐ 13. Are all elements signaled?

☐ 14. Revise for style and composition based on the writing principles of the book:

    ☐ a. Are paragraphs consistent? (Chapter 6, Section 6.2)

    ☐ b. Are paragraphs cohesive? (Chapter 6, Section 6.3)

    ☐ c. Are key terms consistent? (Chapter 6, Section 6.3)

    ☐ d. Are key terms linked? (Chapter 6, Section 6.3)

    ☐ e. Are transitions used and do they make sense? (Chapter 6, Section 6.3)

    ☐ f. Is the action in the verbs? Are nominalizations avoided? (Chapter 4, Section 4.6)

    ☐ g. Did you vary sentence length and use one idea per sentence? (Chapter 4, Section 4.5)

    ☐ h. Are lists parallel? (Chapter 4, Section 4.9)

    ☐ i. Are comparisons written correctly? (Chapter 4, Sections 4.9 and 4.10)

    ☐ j. Have noun clusters been resolved? (Chapter 4, Section 4.7)

    ☐ k. Has word location been considered? (Verb following subject immediately? Old, short information at the beginning of the sentence? New, long information at the end of the sentence?) (Chapter 3, Section 3.1)

    ☐ l. Have grammar and technical style been considered? (person, voice, tense, pronouns, prepositions, articles; Chapter 4, Sections 4.1–4.4)

    ☐ m. Is past tense used for results and present tense for descriptive papers?

    ☐ n. Are words and phrases precise? (Chapter 2, Sections 2.2 and 2.3)

    ☐ o. Are nontechnical words and phrases simple? (Chapter 2, Section 2.2)

> ☐   p. Have unnecessary terms (redundancies, jargon) been reduced? (Chapter 2, Section 2.4)
> ☐   q. Have spelling and punctuation been checked? (Chapter 4, Section 4.11)

SUMMARY

> DISCUSSION GUIDELINES:
> 1. State and interpret your key findings. Provide the answer to the research question.
> 2. Summarize and generalize.
> 3. Keep in mind who your potential readers will be.
> 4. Organize the Discussion in a pyramid structure:
>
> | First paragraph: | Answer based on key findings |
> | | Supporting evidence |
> | Subsequent paragraphs: | Comparisons/Contrasts to previous studies |
> | | Limitations of your study |
> | | Unexpected findings |
> | | Hypotheses or models |
> | Last paragraph: | Summary |
> | | Significance/Implication |
>
> 5. Organize the topics according to the science or from most to least important in the middle of the discussion.
> 6. Compare and contrast your findings with those of other published results.
> 7. Explain any discrepancies, unexpected findings, and limitations.
> 8. Provide generalizations where possible.
> 9. Signal the elements of the Discussion.

PROBLEMS

Problem 13-1
Evaluate the following statements found in different discussion sections. Use basic writing principles to improve the sentences:

1. Therefore, accuracy of Ebb's data is obviously doubtable and should be reevaluated.
2. A comprehensive pharmacological evaluation of different rhubarbs is therefore greatly necessitated for more rational and accurate clinical application of this important herbal medicine.
3. The influence of counter ions is obvious in Figure 3B.
4. Our future studies may reveal how certain single-point mutations affect the conformation of the protein and thus its interaction with the receptor.
5. The model proposed here however is currently an experimentally underdetermined system.

Problem 13-2

Consider the two different opening paragraphs of a discussion about a preventive measure against malaria. Which one is a better first paragraph for a Discussion and why?

Version A:

We trapped and counted the number of mosquitoes within the urban environment of the city of Caracas using conventional carbon dioxide traps. Nearly 70% higher numbers of adult *A. aegypti* were caught in settlements in the vicinity of irrigated urban agricultural sites compared to control areas without irrigated urban agriculture. When we evaluated malaria episode reports from people living in various parts of the city, we found that 18% of malaria cases were reported by people living in the vicinity of urban agricultural areas in the rainy as well as dry seasons, whereas only 2% of the control groups reported incidences of malaria per year.

Version B:

The results of this study show that open-space irrigated vegetable fields in Caracas can provide suitable breeding sites for *Aedes aegypti*. This is reflected in higher numbers of adult *A. aegypti* in settlements in the vicinity of irrigated urban agricultural sites compared to control areas without irrigated urban agriculture. In addition, people living in the vicinity of urban agricultural areas reported more malaria episodes than the control group in the rainy as well as dry seasons. Apparently, the informal irrigation sites of the urban agricultural locations create rural spots within the city of Caracas in terms of potential mosquito breeding sites.

Problem 13-3

Consider the two different concluding paragraphs of a discussion about desert frogs. Which one is a better conclusion for a Discussion and why?

Version A:

In conclusion, this study shows that desert frogs can avoid death by desiccation by maintaining a high body water content and water storage in their urinary bladder and by rapid hydration when water is available. These measures may be employed in combination with behavioral adaptations such as burrowing and change in pigmentation to minimize stresses tending to dehydrate the animals.

Version B:

A limitation of this study was the small number of animals, a single species of frogs, and the location of the study area, which took place in only one oasis in the Mohave Desert. Future studies should be extended to other species, a larger number of animals, and to a greater diversity of locations.

Problem 13-4

Given the following concluding paragraph of a Discussion about a simulation of a Sahel drought in the 20th and 21st centuries, explain why or why not this is a good final paragraph of an article.

We have described a global climate model (CM2) that generates a simulation of the 20th century rainfall record in the Sahel generally consistent with observations. The model suggests that there has been an anthropogenic drying trend in this region, due partly to increased aerosol loading and partly to increased greenhouse gases, and that the observed 20th-century record is a superposition of this drying trend and large internal variability. The same model projects dramatic drying in the Sahel in the 21st century, using the standard IPCC scenarios. No other model in the IPCC/AR4 archive generates as strong a drying trend in the future. Until we better understand which aspects of the models account for the different responses in this region to warming of SSTs, and devise more definitive observational tests, we advise against basing assessments of future climate change in the Sahel on the results from any single model in isolation. In the interim, given the quality of CM2's simulation of the spatial structure and time evolution of rainfall variations in the Sahel in the 20th century, we believe that its prediction of a dramatic 21st century drying trend should be considered seriously as a possible future scenario.

(With permission from the National Academy of Sciences, U.S.A.)

Problem 13-5

In the following paragraph, identify
—the signal for the conclusion
—overall results and their interpretation
—statement of significance

Conclusions

Habitat heterogeneity, as estimated by an advanced land cover classification, provides a stronger prediction of butterfly species richness in Canada than any previously measured factor. At large spatial scales, virtually all spatial variability (90%) in butterfly richness patterns is explained by habitat heterogeneity with secondary but significant contributions from climate (especially PET) and topography. Patterns of species turnover across the best sampled southern region of Canada are strongly related to differences in habitat composition, supporting species turnover as the mechanism through which land cover diversity may influence butterfly richness. Differences in climate are unrelated to butterfly community similarity at this scale, suggesting that the influences of energy on richness may be indirect or limited to within-habitat diversity. These results have significant conservation implications and indicate that the role of habitat heterogeneity may be considerably more important in determining large-scale species-richness patterns than previously assumed.

(With permission from the National Academy of Sciences, U.S.A.)

Problem 13-6

In this problem, you find the first and last paragraph of a Discussion for a manuscript with the title "Effectiveness of Peridomestic Lyme Disease Prevention Practices."

a. Evaluate the paragraphs (Does the first paragraph state the answer to the question? Does the last paragraph give a good conclusion and indicate the significance of the study?).
b. The opening/first paragraph of this Discussion is rather long. Suggest ways to condense it or to rearrange information.

DISCUSSION

Our findings emphasize the need to continue to promote personal protection measures to reduce the risk of Lyme disease infection. We have identified three reasonable personal measures that may be protective against Lyme disease when practiced. Performing tick checks within 36 hours after spending time in the yard may reduce one's risk by as much as 46 percent. Because studies have suggested it takes more than 24 hours for blacklegged ticks to transmit the etiologic agent[22, 23], prompt removal of ticks found attached to the body is a logical method of Lyme disease prevention. The effectiveness of performing tick checks has been suggested previously[8, 13], however this is the first time it has been demonstrated in a peridomestic setting. We also found that controls were more likely than cases to shower or bathe within two hours after spending time in the yard, and that bathing may reduce one's risk by up to 57 percent. Frequent bathing is not currently among the commonly recommended Lyme disease prevention measures. Although it is unlikely that bathing will remove ticks that have already attached to the body, taking a shower or bath soon after spending time outside may help to remove ticks that are yet unattached, or may create an opportunity to find ticks attached to the body. In addition, the act of bathing may indirectly prevent tick bites in that it necessitates the removal of clothing that may have blacklegged ticks upon them. Our data also suggest that the use of insect repellent may be protective against disease. This finding refers to all types of repellents, including those that contain N,N-diethyl-3-methyltoluamide (DEET), but not including permethrin insecticide. Responses varied greatly regarding the active ingredients of repellents used, and were often inconsistent with the brands and products named by the respondents. The effectiveness of insect repellents has been shown previously[8, 16], but not in the peridomestic environment. We found that wearing clothing treated with permethrin insecticide was rarely practiced. It may be that residents are unaware of permethrin products, or do not want to spend the time treating outdoor clothing (permethrin must be applied to clothing and allowed to dry for several hours before wearing). Clothing that is pre-treated with permethrin and sold in specialty stores may be too expensive for daily use in one's yard.

. . .

In summary, this study sought to evaluate the effectiveness of peridomestic Lyme disease prevention behaviors. Though studies suggest that Lyme disease risk in the Northeastern United States is largely peridomestic, members of the study population could have been exposed to ticks outside of their own yards. The analysis attempted to evaluate and control for this non-peridomestic risk by asking participants about their recreational, occupational, and travel exposures to ticks, although it is difficult to determine when and where people are exposed to ticks. Our findings emphasize the need to continue to promote personal protection measures to reduce the risk of Lyme disease infection. In the absence of a vaccine against Lyme disease, it is encouraging that the protective measures identified in our study—tick checks, showering or bathing after spending time in the yard, and use of repellent—are measures that anyone can take at limited expense to reduce their risk of infection. Clinicians and public health practitioners in Lyme disease endemic areas should continue to educate the public about these simple, practical methods for reducing Lyme disease risk after spending time outdoors.

(With permission from Neeta Connally)

Problem 13-7 Your Own Discussion
Write your own Discussion section based on the experimental results of your work.

Provide the overall question the paper is meant to ask. Use all basic writing principles studied, and follow the summary on how to write a Discussion section.

CHAPTER 14

Abstract

14.1 OVERALL

Most people (including editors and reviewers) will read your article only if your Abstract interests them. The Abstract is often also the only part of the paper—together with the title—that can be retrieved through a search, such as those done through Medline or Ovid. It is therefore essential that the Abstract interest your reader.

Knowing how to write an Abstract is one of the most important skills in science, as virtually all of a scientist's work will be judged first (and often last) based on an abstract. The ability to write a competitive abstract applies not only to research papers but also to grant proposals, progress reports, project summaries, and conference submissions and proceedings. Therefore, learning this critical skill cannot be underestimated.

14.2 CONTENT

The Abstract should fully summarize the contents of the paper in one paragraph. The Abstract must also be written such that it can stand on its own without the text. It must be concise, informative, and complete. Do not try to include every finding in your Abstract. Rather, include all the important details of the paper, but use as few words as possible. Write the Abstract with the nonspecialist in mind. Remember, you want to attract as wide an audience as possible.

The abstract selects the highlights from each section of the paper. Because the Abstract summarizes all of these highlights, it is easiest to write once your manuscript is complete. The Abstract covers all of the main information in the paper (Introduction, Material and Methods, Results, and Discussion) in a single paragraph.

In your Abstract, do not include any information or conclusion not covered in the paper. Avoid abbreviations, unfamiliar terms, and citations. Do not include or refer to tables or figures. Do not include any references, but be sure to include all the important key terms found in the title because the Abstract and the Title have to correspond to each other (see Chapter 6, Section 6.3 for more details on key terms, and distinguish between key terms and key words, also known as indexing terms — for key words see Chapter 15, Section 15.5.)

14.3   ORGANIZATION

ABSTRACT GUIDELINE 1:

Use an informative or structured abstract for research articles.

ABSTRACT GUIDELINE 2:

Abstracts of research papers include

    Question/Purpose (not required for descriptive papers)
    Experimental approach (not required for descriptive papers)
    Results/Description
    Conclusion (answer)/Implication
    Optional: short background
                  significance such as implication,
                  speculation, application or recommendation
                  (for investigative papers)

Abstracts for research papers (both investigative and descriptive papers) differ from those used for review articles or proposals. Research paper abstracts can be divided into informative and structured abstracts. The latter form is used mainly for clinical journals.

When you write your Abstract, follow the specific instructions from the journal to which you are planning to submit your manuscript. Although there will be some differences among journals, the content of the Abstract remains the same. An abstract for an investigative research paper should include the following:

    Question or purpose
    Experiments
    Results
    Conclusion (answer to the question) and implication

In addition to these basic parts, the Abstract may begin with a sentence or two of background information to help the reader understand the question and end with a sentence indicating the significance of the paper.

Although the Abstract is a miniversion of the paper, following its general organization, the Abstract does not give equal weight to all the parts of a paper. The Abstract may include a sentence or two of background information from the Introduction. It has to include the overall question or purpose of the paper found in the Introduction. It typically describes the experimental approach only generally and includes only the main results from the Results section. It also contains the answer to the research question/purpose, which is found in the Discussion. In addition, the Abstract may end with a sentence stating an implication, a speculation, or a recommendation based on the answer. However, avoid general descriptive statements that merely hint at your results or act like a rough table of contents.

👍 **Example 14-1**   **Abstract of an investigative research paper**

	1Interleukin 1 (IL-1), a cytokine produced by macrophages and various other cell types, plays a major role in the immune response and in inflammatory reactions. **2**IL-1 has been shown to be cytotoxic for tumor cells. **3**To determine the effect of macrophage-derived factors on epithelial tumor cells, **4**we cultured human colon carcinoma cells (T84) and an intestinal epithelial cell line (IEC 18) with purified human IL-1. **5**Microscopic and photometric analysis indicated that IL-1 has a cytotoxic effect on colon cancer cells as well as cytotoxic and growth inhibitory effects on intestinal epithelial cells. **6**As IL-1 is known to be released during inflammatory reactions, this factor may not only kill tumor cells but also affect normal intestinal cells and may play a role in inflammatory intestinal diseases.

Background · Question/Purpose · Experimental approach · Results · Answer/Conclusion/Implication

---

In Example 14-1, sentences 1 and 2 provide a short background. Sentence 3 signals the question, whereas sentence 4 states the experimental approach. Results are signaled in sentence 5. Sentence 6 provides the conclusion by stating an implication.

The format and length of the Abstract is generally specified in the *Instructions to Authors*. Usually, abstracts contain between 100 and 250 words. You should not exceed the maximum allowed, but you may certainly summarize your paper in fewer words.

Structured Abstracts

ABSTRACT GUIDELINE 3:

Use a structured abstract for clinical reports,
if requested.

ABSTRACT GUIDELINE 4:

Use subheadings for structured abstracts.

Structured abstracts are just like informative abstracts except that structured abstracts have subheadings (background, methods, results, and conclusions) and are often not written in complete sentences. These types of abstracts are most often found in clinical reports and clinical journals. They can sometimes be longer than informative abstracts—many contain a maximum of 400 words. Clinical journals in which these abstracts are required usually provide instructions for authors that state the length and subheadings for the structured abstract.

Example 14-2   **Structured abstract**

Background and Question/Purpose

Background In infants and children with maternally acquired human immunodeficiency virus type 1 (HIV-1) infection, treatment with a single antiretroviral agent has limited efficacy. We evaluated the safety and efficacy of a three-drug regimen in a small group of maternally infected infants.

Experimental approach

Methods Zidovudine, didanosine, and nevirapine were administered in combination orally to eight infants 2 to 16 months of age. The efficacy of antiretroviral treatment was evaluated by serial measurements of plasma HIV-1 RNA, quantitative plasma cultures, and quantitative cultures of peripheral-blood mono-nuclear cells.

Results

Results The three-drug regimen was well tolerated, without clinically important adverse events. Within four weeks, there were reductions in plasma levels of HIV-1 RNA of at least 96 percent (1.5 log) in seven of the eight study patients. Over the 6-month study period, replication of HIV-1 was controlled in two infants who began therapy at 2½ months of age. Plasma RNA levels were reduced by 0.5 to 1.5 log in five of the other six infants.

Answer/Conclusion

Conclusions Although further observations are needed, it appears that in infants with maternally acquired HIV-1 infection, combined treatment with zidovudine, didanosine, and nevirapine is well tolerated and has sustained efficacy against HIV-1.

(With permission from the Massachusetts Medical Society)

If you find it difficult to get started writing an Abstract, consider starting to write a structured abstract first, following the general setup of subsections. Then, convert the abstract to an informative Abstract by connecting the sentences using transitions and word location where needed.

Special Case: Abstracts for Descriptive Papers

Abstracts for descriptive research papers should include the following:

descriptive statement
description of the new findings
conclusion/significance/implication

 Example 14-3 a   **Abstract of a descriptive paper**

Descriptive statement	The "sexually deceptive" orchid *Chiloglottis trapeziformis* attracts males of its pollinator species, the thynnine wasp *Neozeleboria cryptoides*, by emitting a unique volatile
Description and implied significance	compound, 2-ethyl-5-propylcyclohexan-1,3-dione, which is also produced by female wasps as a male-attracting sex pheromone.

(With permission from the American Association for the Advancement of Science)

The preceding example is a very short abstract of just one sentence, but it contains all the necessary components of a descriptive abstract. The abstract relates the discovery of a "sexually deceptive" orchid in the first half of the sentence. In the second half of the sentence, the compound responsible for the deception is described ("a unique volatile compound, 2-ethyl-5-propylcyclohexan-1,3-dione"), and its significance is implied ("male-attracting sex pheromone").

Longer Abstracts of descriptive papers typically expand on the description of the discovery and on its significance. They may also contain additional background information. An example of a longer abstract for a descriptive paper is shown next:

 Example 14-3 b   **Abstract of a descriptive paper**

Background and problem	Indirect radiative forcing of atmospheric aerosols by modification of cloud processes poses the largest uncertainty in climate prediction. We show here a trend of increasing
Descriptive statement	deep convective clouds over the Pacific Ocean in winter from long-term satellite cloud measurements (1984–2005).
	Simulations with a cloud-resolving weather research and forecast model reveal that the increased deep convective clouds are reproduced when accounting for the aerosol effect from the Asian pollution outflow, which leads to large-scale enhanced convection and precipitation and hence an intensified storm track over the Pacific. We suggest that the wintertime Pacific is highly vulnerable to the aerosol-cloud interaction because of favorable cloud dynamical and microphysical conditions from the coupling between
Results/Description	the Pacific storm track and Asian pollution outflow. The

| Significance/ Implication | intensified Pacific storm track is climatically significant and represents possibly the first detected climate signal of the aerosol-cloud interaction associated with anthropogenic pollution. In addition to radiative forcing on climate, intensification of the Pacific storm track likely impacts the global general circulation due to its fundamental role in meridional heat transport and forcing of stationary waves. |

(With permission from the National Academy of Sciences)

14.4 APPLYING BASIC WRITING PRINCIPLES

Basic Principles

To write your Abstract, follow the basic principles covered in chapters 1 through 6. In the Abstract, pay particular attention to using simple words and avoiding jargon. Avoid noun clusters. Also, avoid abbreviations unless a long term occurs repeatedly in the Abstract. Here, it is even more important than in the rest of the paper to keep sentences short, dealing with just one topic each and excluding irrelevant points. To provide clear continuity throughout the Abstract, repeat key terms, use consistent order for details, keep the same point of view in the question and the answer, and use parallel form.

Verb Tense

The basic guideline is that if a statement is still true use present tense. For completed actions and observations, use past tense. Thus, when you write about the answer or describe a structure, use present tense because these statements are still true. However, when you describe the results of experiments, use past tense because these events are finished. When possible, use active verbs.

14.5 SIGNALS FOR THE READER

ABSTRACT GUIDELINE 5:

Signal the question, the experimental approach, the results, the answer, and the implication.

ESL advice

Because Abstracts are usually written as one paragraph, it helps the reader if you signal the different parts of the Abstract. Examples of signals for the abstract are shown in Table 14-1. Writers whose native language is not English may find this table particularly useful.

Table 14-1 Signals for the Abstract

QUESTION + EXPERIMENT	RESULTS	ANSWER/ CONCLUSION	IMPLICATION
To determine whether..., we...	We found...	We conclude that...	These results suggest that...
We asked whether......	Our results show...	Thus,...	These results may play a role in...
To answer this question, we...	Here we report...	These results indicate that...	Y can be used to...
X was studied by...			

Following is an abstract with components that have been signaled well.

👍	**Example 14-4**	**Abstract with signaled components**
		Human papillomavirus (HPV) vaccines provide an opportunity to reduce the incidence of cervical cancer. Optimization of cervical cancer prevention programs requires anticipation of the degree to which the public will
	Question	adhere to vaccination recommendations. ***To compare vac-***
	Experimental approach	***cination levels*** driven by public perceptions with levels that are optimal for maximizing the community's overall utility, ***we develop*** an epidemiological game-theoretic model of
	Results not stated clearly	HPV vaccination. The model is parameterized with survey data on actual perceptions regarding cervical cancer, genital warts, and HPV vaccination collected from parents
	Answer/Interpretation of results	of vaccine-eligible children in the United States. ***The results suggest*** that perceptions of survey respondents generate vaccination levels far lower than those that maximize over-
	Implication	all health-related utility for the population. Vaccination goals ***may be achieved*** by addressing concerns about vaccine risk, particularly those related to sexual activity
	Another implication	among adolescent vaccine recipients. ***In addition***, cost subsidizations and shifts in federal coverage plans ***may***
	Implication	***compensate*** for perceived and real costs of HPV vaccin- ation to achieve public health vaccination targets.

(With permission from the National Academy of Sciences)

In this example, the question "To compare vaccination levels...," leads the reader directly to the experimental approach, signaled by "we develop...". The complete description of the experiment in the sentence following describes the subjects of the study (parents of vaccine-eligible children) and sets the stage for the answer to the question. We know where the answer starts because it is signaled by "The results suggest...". The answer is followed by two implications, which are signaled by "...may be achieved..." and "In addition,...may compensate....". Note that the implications are written in similar form and have parallel signals, making it clear that both "…may be achieved …" and "…may compensate…" are of equal importance. To strengthen this Abstract, the results could be stated more clearly by, for example, providing the percentages and types of perceptions found in the study and signaling these.

14.6   COMMON PROBLEMS OF THE ABSTRACT

The most common problems of Abstracts include

- Omission of elements (Section 14.3)
- Excessive length (Section 14.4)
- Wrong type of Abstract (Section 14.6 following)

Omission of Parts

ABSTRACT GUIDELINE 9:
Do not omit any parts of the abstract.

If any of the parts (question/purpose, experimental approach, results, or answer/conclusion) are missing or obscured in an Abstract, the reader may have to reread the abstract several times because it is difficult to understand. The same problem arises when the parts of an Abstract are not signaled.

Example 14-5	Abstract without clearly stated question
Experimental approach	Using conventional violet red bile glucose agar as well as a new chromogenic Druggan–Forsythe–Iversen medium, which enables results to be obtained 2 days earlier than the conventional method, *Enterobacter sakazakii* was found in 5 out of 50 powdered infant milk formulas, 3 out of 30 dried infant foods, and in 4 out of 25 milk powders. Although both media detected *Ent. sakazakii*, the chromogenic Druggan–Forsythe–Iversen medium proved to be about twice as sensitive in the organism's detection than the conventional violet red bile glucose agar. Enrichment of the dry foods in Enterobacteriaceae enrichment broth increased the likelihood of detection almost an order of magnitude. *Salmonella* serovars, the standard organism used for testing food products for the presence of Enterobacteriaceae, were not detected even after enrichment, suggesting that monitoring dry milk and food products solely for *Salmonella* serovars is insufficient for the detection and control of *Ent. sakazakii*.
Results	
Conclusion/implication	

In Example 14-5, although some experimental approach, the results obtained, and the conclusion/implication of the work are stated, the author did not state what the question was. The description of the results lets the reader guess what the question of the paper is, but it should not be the goal to have the reader guess. Another problem with this abstract is that no context is given for the experiments done.

Revised Example 14-5	Rare forms of infant meningitis, necrotizing enterocolitis (NEC), bacteraemia, and neonate deaths have been associated with Enterobacter sakazakii. Although this organism has been isolated from a wide range of foods including cheese, meat, vegetables, grains, herbs and spices, its presence in powdered infant formula milk and other dried infant food products has not been investigated. In this study, we used conventional violet red bile glucose agar as well as a new chromogenic Druggan–Forsythe–Iversen medium, which enables results to be obtained 2 days earlier than the conventional method, to test 50 powdered infant milk formulas, 30 dried infant foods, and 25 milk powders for the presence of *Ent. sakazakii.* We detected *Ent. sakazakii* in 5 out of 50 powdered infant milk formulas, 3 out of 30 dried infant foods, and in 4 out of 25 milk powders. Although both media detected *Ent. Sakazakii,* the chromogenic Druggan–Forsythe–Iversen medium proved to be about twice as sensitive in the organism's detection than the conventional violet red bile glucose agar. Enrichment of the dry foods in Enterobacteriaceae enrichment broth increased the likelihood of detection almost an order of magnitude. *Salmonella* serovars, the standard organism used for testing food products for the presence of Enterobacteriaceae, were not detected even after enrichment, suggesting that monitoring dry milk and food products solely for *Salmonella* serovars is insufficient for the detection and control of *Ent. sakazakii.*
Background	
Experimental approach	
Question/Purpose	
Results	
Conclusion/ Implication	

In the revision, background information has been added in the first two sentences. The question is stated in the third sentence, following the experimental approach. Results are indicated, and an implication is provided. This Abstract is much easier to follow and to comprehend. Not only does it contain all the elements necessary for a well-written Abstract, it clearly signals the parts as well ("In this study, we used…to test…", "We detected…", "…suggesting that…").

Excessive Length

ABSTRACT GUIDELINE 10:
Keep the Abstract short.

One of the most common problems for abstracts is excessive length. When you write the Abstract, consider every word carefully. If you can write your Abstract in fewer words than the maximum allowed, do so. If you find yourself in a situation in which you have to condense your Abstract, follow the suggestions below (see also Chapter 4, Section 4.6).

To condense a long Abstract

- Omit unnecessary words and combine sentences
- Condense background
- Omit or subordinate less important information (definitions, experimental preparations, details of methods, exact data, confirmatory results, and comparisons with previous results).

An example of an unnecessarily long Abstract is shown next in Example 14-6:

Example 14-6   **Excessively long abstract**

Altered endothelial cell function appears to be an important part of the pathophysiology during a dengue virus infection. *The effect of such an infection on gene expression in primary human umbilical vein endothelial cells can be studied by Differential Display.* We confirmed altered gene expression detected by differential display by semi-quantitative and real-time fluorogenic RT-PCR. *Using these techniques,* we identified at least nine cDNAs with altered expression in virus-infected cells not previously reported. These cDNAs included the human inhibitor of apoptosis-1, 2'-5' oligoadenylate synthetase, a 2'-5' oligoadenylate synthetase–like cDNA, Galectin-9, Myxovirus protein A and 1, the regulator of G-protein signaling, and the endothelial and smooth muscle cell–derived neuropilin-like protein. *We also performed RT-PCR analyses on the nine cDNA and found* that dengue virus infection also appears to increase the expression of tumor necrosis factor-α (TNF-α), interleukin 1β (IL-1β), and the toll-like receptor 3 in HUVECs. These results point to the possibility of the activation of at least three signaling pathways during dengue virus infection of HUVECs. *These three signaling pathways include* the TNF-α pathway, the IL-1β pathway, and the type-I IFN (IFN-α/β) pathway.

(With permission from Irene Bosch)

Wordy

Wordy

Wordy

Wordy

This abstract is 184 words long and describes what the same abstract, when condensed, says in 133 words.

 **Revised
Example 14-6**

Altered endothelial cell function appears to be an important part of the pathophysiology during a dengue virus infection. We confirmed altered gene expression detected by Differential Display using semi-quantitative and real-time fluorogenic RT-PCR. The identified nine cDNAs included the human inhibitor of apoptosis-1, 2'-5' oligoadenylate synthetase, a 2'-5' oligoadenylate synthetase–like, Galectin-9, Myxovirus protein A and 1, the regulator of G-protein signaling, the endothelial and smooth muscle cell-derived neuropilin-like protein, and the phospholipid scramblase 1. RT-PCR analyses confirmed that dengue virus infection also increased the expression of the tumor necrosis factor-α (TNF-α), the interleukin 1β (IL-1β), and the toll-like receptor 3 in HUVECs. These results point to the activation of at least three signaling pathways during dengue virus infection of HUVECs: the TNF-α pathway, the IL-1β pathway, and the type-I IFN (IFN-α/β) pathway.

In the revised version, long wordy constructions have been replaced with more concise wording. Sentences have been combined, and background information has been condensed. Even though the original, longer abstract is quite readable, the shorter revision gets the overview across more clearly.

Wrong Type of Abstracts

Do not confuse or mix your abstract for a research paper with that for a review article or book chapter. Review articles or reports use different types of abstracts called *indicative abstracts* (see Chapter 18, Section 18.4.) Indicative abstracts provide the reader with a general idea of the contents of the paper and include little if any methods or results. Thus, an indicative abstract is essentially a table of contents in paragraph form. Unlike research paper abstracts, indicative abstracts usually are not self-contained and need to be read together with the text of the article. Often the final sentence of this type of abstract is an overview sentence, listing what will occur in the document, as shown in sentence 2 of Example 14-7 following:

 Example 14-7   **Indicative abstract**

This paper describes how plants use the structural diversity of oligosacchrides to regulate important cellular processes such as growth, development, and defense. We address the central remaining question of how cells perceive and transduce oligosaccharide signals and discuss current research aimed at providing the answer.

On occasion, authors blend research paper abstracts and indicative abstracts. In such mixed abstracts, most sentences are informative, but the last sentence is usually descriptive or indicative, stating that something will be described or discussed in the text. Such statements are not useful in an abstract for a research article, however, and should be avoided.

14.7   REASONS FOR REJECTION

The Abstract is usually the first part of a manuscript that the editor and the reviewers read. You are more likely to impress an editor or reviewer (not to mention your readers) if you master the skill of writing simple, clear, and concise Abstracts. You need to spike the interest of the reviewers with your Abstract. Very often reviewers will be tempted to judge your complete manuscript based on the Abstract alone. Several common reasons for quick rejection of an Abstract are

- **Lack of originality**—you must write about something new or better than what has been presented in previous work.
- **Lack of context**—you need to provide a background of your work as well as its implication. Never assume that the editor or reviewers will be sufficiently familiar.
- **Limited sample size**—few samples in a study may not convince a reviewer of the significance of the paper.
- **Lack of numbers**—the reviewers will not know whether there is anything of value in the paper.
- **Too many numbers/too much data**—you will leave reviewers with the sense that they have to figure out and interpret the results themselves.
- **Lack of conformity**—you should follow the instructions for authors very carefully so as not to antagonize a reviewer.
- **Wrong style of abstract**—differentiate between abstracts for research papers and other abstracts such as indicative abstracts for review papers.
- **Too many abbreviations**—this will turn off most reviewers (and readers).

14.8   REVISING THE ABSTRACT

When you have finished writing the Abstract (or if you are asked to edit an Abstract for a colleague), you can use the following checklist to "dissect" the Abstract systematically:

- ☐ 1. Did you distinguish between a research paper abstract and an indicative abstract?
- ☐ 2. Is the question/purpose stated?
- ☐ 3. Is the experimental approach stated?
- ☐ 4. Are the results indicated?
- ☐ 5. Is the answer/conclusion provided?
- ☐ 6. Are all elements signaled?

☐ 7. Is the length within the required limits?

☐ 8. Has the abstract been condensed as much as possible?

☐ 9. Is the significance of the work apparent?

☐ 10. Is the context clear?

☐ 11. Is the work original?

☐ 12. Revise for style and composition based on the writing principles of the book:

    ☐ a. Are paragraphs consistent? (Chapter 6, Section 6.2)

    ☐ b. Are paragraphs cohesive? (Chapter 6, Section 6.3)

    ☐ c. Are key terms consistent? (Chapter 6, Section 6.3)

    ☐ d. Are key terms linked? (Chapter 6, Section 6.3)

    ☐ e. Are transitions used and do they make sense? (Chapter 6, Section 6.3)

    ☐ f. Is the action in the verbs? Are nominalizations avoided? (Chapter 4, Section 4.6)

    ☐ g. Did you vary sentence length and use one idea per sentence? (Chapter 4, Section 4.5)

    ☐ h. Are lists and comparisons in sentences kept parallel? (Chapter 4, Section 4.9)

    ☐ i. Are comparisons written correctly? (Chapter 4, Sections 4.9 and 4.10)

    ☐ j. Have noun clusters been resolved? (Chapter 4, Section 4.7)

    ☐ k. Has word location been considered? (Verb following subject immediately? Old, short information at the beginning of the sentence? New, long information at the end of the sentence?) (Chapter 3, Section 3.1)

    ☐ l. Have grammar and technical style been considered (person, voice, tense, pronouns, prepositions, articles)? (Chapter 4, Sections 4.1–4.4)

    ☐ m. Is past tense used for results and present tense for descriptive papers?

    ☐ n. Are words and phrases precise? (Chapter 2, Sections 2.2 and 2.3)

    ☐ o. Are nontechnical words and phrases simple? (Chapter 2, Section 2.2)

    ☐ p. Have unnecessary terms (redundancies, jargon) been reduced? (Chapter 2, Section 2.4)

    ☐ q. Have spelling and punctuation been checked? (Chapter 4, Section 4.11)

SUMMARY

ABSTRACT GUIDELINES:

1. Use an informative or structured abstract for research articles.
2. Abstracts for research papers include

    The question/purpose

    The experimental approach

    The results

The conclusion
If useful, also include
            Background and Significance such as an implica-
            tion, recommendation, application, or speculation
3. Use a structured abstract for a clinical report if requested.
4. Use subheadings for a structured abstract.
5. Signal the question, the experimental approach, the results, the answer, and the implication.
6. Do not omit any parts of the Abstract.
7. Keep the Abstract short.

PROBLEMS

Problem 14-1

1. **In the following Abstract, check that all required parts are present and signaled correctly. Also check if basic writing principles have been considered.**
2. **Does the Abstract correspond to the Title?**

Structural Basis for the Interaction of Chloramphenicol, Clindamycin, and Macrolides With the Peptidyl Transferase Center in Eubacteria

Ribosomes, the site of protein synthesis, are a major target for natural and synthetic antibiotics. Detailed knowledge of antibiotic binding sites is the key to understand the mechanisms of drug action. Conversely, drugs are excellent tools for studying the ribosome function. To elucidate the structural basis of ribosome-antibiotic interactions, we determined the high-resolution X-ray structures of the 50S ribosomal subunit of the eubacterium *Deinococcus radiodurans* complexed with the clinically relevant antibiotics chloramphenicol, clindamycin, and the three macrolides: erythromycin, clarithromycin, and roxithromycin. We found that antibiotic binding sites are composed exclusively of segments of 23S rRNA at the peptidyl transferase cavity and do not involve any interaction of the drugs with ribosomal proteins. Here we report the details of antibiotic interactions with the components of their binding sites. Our results also show the importance of Mg^{2+} ions for the binding of some drugs. This structural analysis should facilitate rational drug design.

(With permission from Macmillan Publishers Ltd.)

Problem 14-2

1. **In the following Abstract, ensure that all required parts are present.**
2. **Is this an Abstract for an investigative research paper or for a descriptive paper?**

The cyanobacterial circadian pacemaker is an enzymatic oscillator, which orchestrates the metabolism of the bacteria to fit the day and night alternations of this planet. Interactions among KaiA, KaiB, and KaiC, the three components of this oscillator, result in many oscillatory properties *in vitro*, including an overall phosphorylation level of KaiC, an apparent segregation between synchronized phosphorylation and dephosphorylation reactions, and the size and the composition of this oscillator. To explain these properties, we propose here a molecular mechanism for this pacemaker within the framework of a cyclic catalysis scheme.

(With permission from Jimin Wang)

Problem 14-3
In the following Abstract, identify all essential components and their signals if provided:

Abstract

Many insects possess a sexual communication system that is vulnerable to chemical espionage by parasitic wasps. We recently discovered that a hitch-hiking (H) egg parasitoid exploits the antiaphrodisiac pheromone benzyl cyanide (BC) of the Large Cabbage White butterfly *Pieris brassicae*. This pheromone is passed from male butterflies to females during mating to render them less attractive to conspecific males. When the tiny parasitic wasp *Trichogramma brassicae* detects the antiaphrodisiac, it rides on a mated female butterfly to a host plant and then parasitizes her freshly laid eggs. The present study demonstrates that a closely related generalist wasp, *Trichogramma evanescens*, exploits BC in a similar way, but only after learning. Interestingly, the wasp learns to associate an H response to the odors of a mated female *P. brassicae* butterfly with reinforcement by parasitizing freshly laid butterfly eggs. Behavioral assays, before which we specifically inhibited long-term memory (LTM) formation with a translation inhibitor, reveal that the wasp has formed protein synthesis-dependent LTM at 24 h after learning. To our knowledge, the combination of associatively learning to exploit the sexual communication system of a host and the formation of protein synthesis-dependent LTM after a single learning event has not been documented before. We expect it to be widespread in nature, because it is highly adaptive in many species of egg parasitoids. Our finding of the exploitation of an antiaphrodisiac by multiple species of parasitic wasps suggests its use by *Pieris* butterflies to be under strong selective pressure.

(With permission from the National Academy of Sciences, U.S.A.)

Problem 14-4

1. **Write your own Abstract.**
2. **Revise your own Abstract: Check that all required parts of an Abstract are present and signaled correctly. Also check if writing principles are applied correctly.**
3. **Have a peer comment on your Abstract.**

CHAPTER 15

Titles

15.1   OVERALL

TITLE GUIDELINE 1:
Aim to attract readers.

TITLE GUIDELINE 2:
State the main topic of your study in the title.

TITLE GUIDELINE 3:
Your title should separate your article from other
articles in the field.

The title is the single most important phrase of an article. Many readers will discover your paper by seeing it listed in Current Contents or a similar secondary service. Most readers will read your abstract only if the title interests them. They will judge the paper's relevance on the title alone.

To identify the main message of the paper, state the main topic in the title and make the title interesting to attract readers. Because a title in a scientific research journal will be used in indexing systems and bibliographic databases, it should also be informative and accurate.

A title is typically stated as a phrase, but it may also be a complete sentence. However, a full sentence with an active verb is usually not a good title—neither is an overly long phrase.

You need to ensure that your title separates the article from all other articles in that field. Do not hesitate to try out several versions of a title on colleagues and friends. When you convey the message of your paper, be assertive, but do not brag. Be exact and clear or readers for whom you wrote the paper will never read it.

15.2   STRONG TITLES

TITLE GUIDELINE 4:

Use a strong title: Make it clear and complete but succinct.

The first thing readers, editors, and reviewers see is the title. It is therefore very important that your title is strong. A strong title should fulfill three criteria: It needs to be clear, complete, and succinct.

Clarity of a Title

Unclear titles confuse or mislead readers, and can give an annoying first impression. Some titles are not clear because word choice is too general as shown in Example 15-1.

 Example 15-1   Effect of hormones on tumor cells

The title in Example 15-1 is not clear because words in it are unspecific, such as the category terms "hormones" and "tumor cells." What hormones and which tumor cells? The specific hormones and tumor cells should be listed. Otherwise, the title is essentially meaningless. At the same time, this title is unclear because it does not state what specific effect has been observed on the study. The revised example is a much stronger title:

 Revised   Effect of **testosterone and estradiol on the growth and**
Example 15-1   **morphology of rat epithelial** tumor cells

Titles can also be ambiguous because of unspecific words such as *and* and *with*. Avoid these words as they tend to confuse readers.

 Example 15-2   Tracking long-distance migration of gray whales <u>with</u> geolocators

---

In Example 15-2, the relationship between "gray whales" and "geolocators" is not clear. It reads as if gray whales have geolocators. "With" needs to be replaced by a more specific word.

| Revised Example 15-2 | Tracking long-distance migration of gray whales **by using** geolocators |

Ambiguity can also arise because of noun clusters as is shown in the following example:

| Example 15-3 | Involvement of amygdala in the <u>oxotremorine memory enhancing effect</u> |

| Revised Example 15-3 | Involvement of amygdala in memory enhancement by oxotremorine |

The original title contains a confusing noun cluster. In the revised title, this noun cluster has been split up, and the word *effect* has been omitted, making the title much clearer.

Aside from unspecific word choices and confusing noun clusters, abbreviations, especially nonstandard abbreviations, can create unclear titles. The only abbreviations that are acceptable in titles are those that are better known than the words they stand for, such as DNA (deoxyribonucleic acid), and those for chemicals such as N_2O_5 (dinitrogen pentoxide). If you are unsure of whether an abbreviation will be clear, write the words.

Completeness of a Title

A title should not only be clear, it also has to be complete. To make a title complete, include and highlight the key items of your study (e.g., a specific disease seen in a certain group of people, a novel assay, the species studied). Keep in mind, however, that in the title, readers can only absorb three or four details and that the title cannot replace the Abstract. Concentrate on the most distinctive aspect of your work. Details of secondary importance can be presented in the Abstract or Introduction. Announcing the main variables of the paper is stronger than trying to fit all the variables into the title.

Consider the following title:

| Example 15-4 | Dengue virus activates human umbilical vein endothelial cells |

Although this title orients the reader to the area of research, it does not give any specifics as to how the activation takes place. Adding a few more specific words completes the title and sets it apart from others in the field.

| Revised | Dengue virus activates human umbilical vein endothelial cells |
| Example 15-4 | **via the tumor necrosis factor-α pathway** |

To check that your title is complete, compare your title with the question and answer and ensure that you use the same main key terms in the title as in the question and answer of the paper (see also Chapter 10 for more information on question and answer of a paper and Chapter 6 for more insights into key terms).

Succinctness of a Title

Although a title has to be clear and complete, you need to use the fewest words possible to describe your work adequately. Many titles are simply too long. Note that short titles have more impact than long titles.

Most journals also prefer short titles, typically 100 characters (including the spaces between the words) or 10 to 12 words. Limit your words by cutting out trivial words and phrases that contribute nothing to the information in the title. The aim is not to fill the space allowed but to state the message of your paper clearly and completely in as few words as possible.

To make a title succinct, omit unnecessary words and condense necessary words as much as possible. Unnecessary words and phrases that can be omitted include "Nature of," "Studies on," and the like.

 Example 15-5   Examination of the differential response properties of single units within neuronal clusters in the inferior colliculus

In this example, "Examination of" is unnecessary. Even "properties" could be omitted.

 Revised   Differential response of single units within neuronal clusters
Example 15-5   in the inferior colliculus

Aside from unnecessary words and phrases, articles at the *beginning* of the title can be omitted. However, do not omit articles before singular nouns later in the title.

 Example 15-6   <u>The</u> kinetic analysis of the Na-ATPase in the corneal epithelium

 Revised   Kinetic analysis of the Na-ATPase in the corneal epithelium
Example 15-6

Titles can also be shortened by condensing necessary words. One way to condense a title is to use a category term instead of details. Another way to condense a title is to use an adjective instead of a noun, for example, "reduced" instead of "reduction in" or "seizure-induced" instead of "induced by seizures." A third way to condense a title is to use noun clusters instead of prepositional phrases. These techniques must be used carefully, however, to avoid creating an ambiguous title. An example of a longer title that can be shortened using noun clusters is shown next.

☞	**Example 15-7**	Diabetes promotes cholesterol deposits in the corneas of the eyes of rabbits
☟	**Revised Example 15-7**	Diabetes promotes cholesterol deposits in **rabbit eye corneas**

Note that in the revised example the word *eye* can be dropped if the target journal is highly specific, such as the *Journal of Ophthalmology*, as most readers would know that eye corneas are being referred to.

15.3   THE TITLE PAGE

Titles are usually written on the first page of a manuscript, the Title page. This page contains not only the title of the paper but also the names and affiliations of the authors. The names are written in Western style name order: first name, middle initial, then last name. Superscript numbers after each name identify the footnote that contains the affiliation of the author and the corresponding author. Footnotes may also identify authors who contributed equally to the work. As some journals may have very specific requirements regarding the content and format of the title page, be sure to check the *Instruction to Authors* of your target journal.

Aside from the title and the authors, some journals may require you to provide a running title, key words, or a total word count on the Title page. Check the *Instructions to Authors* for any other rules and requirements.

ESL advice

15.4   RUNNING TITLE

TITLE GUIDELINE 5:
The running title should be recognizable as a short version of the title.

A running title is a short version of the complete title. This short version is placed either on the top as a header or on the bottom as a footer of every page (or alternating with the authors' names, on every other page) of a journal to identify the article.

 Example 15-8   **Title and Running Title**

Title: Immunocytochemical localization of seizure-induced increases in ornithine decarboxylase in the rat hippocampus

Running Title: Ornithine decarboxylase in the rat hippocampus

---

15.5   KEY WORDS

KEY WORDS GUIDELINE 1:
Select important and specific terms as key words.

KEY WORDS GUIDELINE 2:
Avoid words that appear in the title.

KEY WORDS GUIDELINE 3:
Avoid general single key words that may apply to a very large number of papers.

Many journals require an author to provide three to ten key words or phrases. These are placed either on the title page or after the Abstract and will help in cross-indexing the article.

Key words should name important topics in your paper. Select the most important and most specific terms of your paper. Choose words or phrases that you would look up if you were trying to find your own paper and that would attract the readers you hope to reach. Avoid using terms that appear in the title already. Terms from the title are used in cross-referencing as well. These terms should not overlap. Using terms different from those in the title will broaden the selection of words and terms that can be looked up.

If possible, use terms and phrases instead of individual words. For example, use "cholesterol degradation" rather than "cholesterol" as looking up papers with cholesterol as a search word will produce far too many hits.

If necessary, include a term as an indexing term even if the term does not appear in your paper. For biomedical articles, for example, it is best to select current specific terms from the medical subject headings (MeSH) list of *Index Medicus*. Science databases can then be searched by such key words (see also Chapter 8, Section 8.2 for a list of top science databases.)

Example 15-9   **Sample Title Page**

Seasonal variation of ibotenic acid and muscimol in *Amanita muscaria*

Natalya Motka[2], Klaus Malbeck[3], Leng-Fei Wu[1], Alicia Fernández[2]
and Samuel Girald[1*]

[1] *Department of Biol. Chemistry, Yale University, New Haven, CT, USA.*
[2] *Institute of Experimental Botany, Academy of Sciences of the Czech Republic, Rozvojová 263, CZ-16502 Prague, Czech Republic*
[3] *Max Planck Institute for Plant Biology, Dortmund, Germany*

* To whom correspondence should be addressed
    Telephone: +001 (203) 234 8006
    Fax: +001 (203) 785 3360
    Email: samuel.girald@yale.edu
    Address: Department of Biol. Chemistry
        Yale University
        New Haven, CT 06520

Running title: Ibotenic acid and muscimol in *Amanita muscaria*
Key words: neurotransmitter, GABA agonist, NMDA glutamate receptor, neurotoxin, psychoactive, mushroom poisoning, muscarine

---

15.6   REVISING THE TITLE

When you have decided on a title (or if you are asked to edit a paper for a colleague), you can use the following checklist for the title:

- [ ] 1. Does the title attract readers?
- [ ] 2. Does the title state the main topic of your study?
- [ ] 3. Is the title clearly and unambiguously stated?
- [ ] 4. Have abbreviations been avoided?
- [ ] 5. Is the title complete?
- [ ] 6. Did you use the same key terms as the question and the answer of your paper?
- [ ] 7. Is the title succinct? Have unnecessary words and phrases been omitted?
- [ ] 8. Does the title separate your article from others in the field?
- [ ] 9. Is the running title recognizable as a short version of the title?

SUMMARY

> TITLE GUIDELINES:
> 1. Aim to attract readers.
> 2. State the main topic of your study in the title.

3. Use a strong title: Make it clear and complete but succinct.
4. Your title should separate your article from other articles in the field.
5. The running title should be recognizable as a short version of the title.

KEY WORDS GUIDELINE:
1. Select important and specific terms as key words.
2. Avoid words that appear in the title.
3. Avoid general single key words that may apply to a very large number of papers.

PROBLEMS

Problem 15-1
The following titles are unclear or not specific. Make suggestions on how to improve these titles. Invent additions if needed.

1. Variation of fossil density with Triassic sedimentary deposits
2. Antibacterial, anti-inflammatory, and antimalarial activities of some African medicinal plants
3. Classification of Fowl Adenovirus Serotypes with genome mapping and sequence analysis of the hexon gene
4. Hemolymph-dependent and -independent responses with Drosophila immune tissue
5. Temperature dependence of fir tree carbon dioxide sequestration
6. Stable, immunogenic, and nasal-specific formulation of NoV vaccine using VLP and adjuvant components
7. Microscopic observation drug susceptibility assay for the diagnosis of TB

Problem 15-2
The following titles are not complete. Suggest information to add to make these titles complete.

1. Differences of old-world and new-world species of hazelnuts
2. Preparation of single nanocrystals of platinum
3. Widespread increase of bat mortality rates
4. Transmission of coccidioidomycosis

Problem 15-3
The following titles are too long. Condense them to make them more concise.

1. The adaptive value of cued seed dispersal in desert plants: Seed retention and release in *Mammillaria pectinifera* (Cactaceae), a small globose cactus

2. Analysis of historical data of Groundwater flow in the South Wales coalfield used to inform 3D modeling
3. The effect of negative mood on persistence in problem solving
4. The intensity of hurricanes is linked to increased atmospheric dust originating from the Sahara desert
5. A cost effectiveness analysis is used to determine how best to optimize flu virus vaccination strategies
6. New cellular antioxidant activity assay can be used to quantify antioxidant activity effectively
7. Analysis of temperature and light requirements for seed germination and seedling growth of *Sequoiadendron giganteum*

Problem 15-4

Ensure that the title for an article on the antioxidant effects of blueberries is specific, complete, concise, and unambiguous. The title should convey the following facts:

- Blueberries have an antioxidant effect
- Eating blueberries increases plasma antioxidant capacity
- If milk/milk protein is digested together with blueberries, no increase in plasma antioxidant capacity is observed

Choose the best title from the list:

1. Antioxidant effect of blueberries reduced with milk
2. Increase of plasma antioxidant capacity of blueberries prevented with milk protein
3. Consumption of blueberries with milk influences plasma antioxidant capacity
4. Milk consumption prevents antioxidant effect of blueberries
5. Milk consumption effects plasma antioxidant capacity
6. Influence of milk protein on antioxidant effect of blueberries

Problem 15-5

For the following titles, find a corresponding running title:

1. Transcoronary transplantation of progenitor cells after myocardial infarction
2. Analysis of the surface characteristics and mineralization status of feline teeth using scanning electron microscopy
3. Inflammatory mechanisms in chronic obstructive pulmonary disease
4. The roles of subsurface carbon and hydrogen in palladium-catalyzed alkyne hydrogenation
5. Picosecond coherent optical manipulation of a single electron spin in a quantum dot
6. Structural analysis of *E. coli* hsp90 reveals dramatic nucleotide-dependent conformational rearrangements

7. Ecological correlations to decline and extinction of the endemic Brazilian tree frog *Corythomantis greeningi*
8. Determination of chlorophyll density for corn from spectral reflectance data
9. Rare structural variants disrupt multiple genes in neurodevelopmental pathways in schizophrenia
10. Otoferlin, defective in a human deafness form, is essential for exocytosis at the auditory ribbon synapse
11. NMR imaging of catalytic hydrogenation in microreactors with the use of para-hydrogen
12. Generation and photonic guidance of multi-octave optical-frequency combs

CHAPTER 16

Revising the Manuscript

Revision is the key to successful writing. Once you have written down everything you could think of for a first rough draft, let your draft "incubate" for at least a day or two. Then it is time to revise it. You may also pass it on to coauthors for their comments and suggestions. Let your coauthors know that this is a working first draft. When you read your first draft again—with or without comments from coauthors—you will see passages that you would like to change, recognize portions that need work, think of points to include, and notice those to omit or condense.

16.1   REVISING THE FIRST DRAFT

Content and Placement

GUIDELINE:

FIRST REVISION

1. Check the first draft for content and content location.

When you revise your first draft, check it first for content and organization. Make sure that all the essential points you want to make have been included. Everything you say should contribute in some way to your question and answer (see also Chapter 10) and no steps should have been left out. You may have included unnecessary material, left out essential evidence, or discussed points in the wrong order. Any irrelevant points need to be removed, and any missing evidence should be included.

In revising, you essentially work your way from the outside in. Therefore, check the individual sections of the paper first, then the paragraphs, and then sentences. Check especially that your manuscript is logically organized. All the parts, paragraphs, and sentences must be in the right order before you revise the style.

The overall structure of your paper should conform to the following outline:

Title:	**3–4 important key terms**
Abstract:	**Content: Question/Purpose, Experimental approach, Results, Interpretation/Answer, Significance**
Introduction:	Organization: funnel shape (known, unknown, question, experimental approach) **First paragraphs: Background** **2nd to last paragraph: Unknown** **Last paragraph: Question/purpose and experimental approach. Optional are main results and significance.**
Materials and Methods:	Organize chronologically, most to least important, or by subsections
Results: 1. Paragraph(s):	**Overview of most important/interesting result(s)**
	Middle paragraphs: Describe other results. Organize chronological or most to least important—every result segment should contain purpose of experiment, experimental approach, results, and their interpretation

	Last paragraph:	**State interesting result(s) or summarize main findings if Results section is lengthy**
Discussion:	Organization: Pyramid shape (interpretation of findings, compare and contrast, models, conclusion, significance)	
	1. Paragraph	**Interpretation of most important results/answer to the question of the paper; support and defend interpretation**
	Middle paragraphs:	Chain of topics, compare and contrast findings, list limitations, etc.
	Last Paragraph:	**Conclusion: summary of main findings and significance (future directions)**

Note that in this overall outline, key power positions are written in bold, as they indicate the most crucial structural locations of an article. Pay particular attention to the content and location of these power positions throughout your revisions.

Once all of your structural components are in place, make sure that you have not missed any content component. Use the checklists provided for each section of a paper (see end of Chapters 10–15) to double check that you have included all relevant components in each section. It may help to clearly mark and label important components of each section on a print version of the manuscript to check their completeness using the summaries provided in Chapters 10 through 15. Identifying each essential component will make it clear to you if there is anything missing or not clearly signaled.

Ensure that you have phrased the purpose of the study/question and interpretation of results/answer the *same* in all key locations: Abstract, Introduction, and the first paragraph of the Discussion and Conclusion. Make certain that your question and answer make sense together.

GUIDELINE:

LOGICAL ORGANIZATION AND FLOW

2. Check logical organization and flow of sections
and subsections. Use the checklists at
the end of Chapters 10–15.

When you are happy with the structural organization and content of power positions, revise each section for logical organization and flow. Ensure that headings refer to the text they describe. Look at how the ideas are distributed among the paragraphs, and make sure that your arguments are logical. Is it clear how and why the evidence presented supports the interpretation of your findings? Is it clear why a particular experimental approach or technique is appropriate? Have the main concepts been clearly and logically connected?

To check for logical flow, verify that you have a chain of topic sentences running throughout the paper. When read by themselves, the topic sentences should be sufficient to provide a rough outline of the paper.

It may also help to make a reverse outline of your manuscript by going through it paragraph by paragraph. Check that this reverse outline is logically organized. You may even want to compare this reversed outline to your original outline to ensure that you have not missed any content (see also Chapter 7, Section 7.5 for a discussion of outlines).

GUIDELINE:
STYLISTIC REVISIONS

3. Revise for style only after you are satisfied with the content and organization.
   Use the basic writing principles to check for word choice, word location, sentence structure, and paragraph structure.
   Pay particular attention to key terms and transitions.
4. Condense where possible.
5. Proofread your manuscript.

Once you are satisfied with the content and organization of the first draft, revise it stylistically. You will probably see a lot to change. Here, too, start by working your way from the bigger structures toward the smaller ones.

For good flow between sections and between paragraphs, ensure that the transitions between paragraphs and sections are smooth and that they tie the pieces together. Pay particular attention to key terms and transitions within paragraphs as well. Add transition phrases and clauses to create the overview of the story (see also Chapter 6, Sections 6.1 and 6.3). Then, use the basic writing principles discussed in this book to check for paragraph structure, sentence structure, and word choice (Chapters 2–6). Inch through your manuscript sentence by sentence, word by word. Consider word location (see Chapter 3 and Chapter 6, Section 6.3). Check whether you have paid attention to either jumping word location or to a consistent point of view.

Look for all possible ways to condense your paper: Omit unnecessary details, unnecessary words, and unnecessary paragraphs (see Chapter 2, Section 2.4 and Chapter 6, Section 6.4). Most readers, editors, and reviewers prefer short, meaty, clear papers. Ensure that you have not repeated any information unnecessarily. Scientists have a tendency to repeat the same information in different sentences or sections using the same or different wording. Your writing will be more concise if you learn to recognize such repetitions. Finally, proofread the text for punctuation, spelling, and typographical errors.

You will not be able to do all this revising on one draft, so revise in stages. Do as much as you can on the first revision. When you no longer see anything to change, put the paper in a drawer again for a few days. Then you are ready to work on the second draft.

16.2   SUBSEQUENT DRAFTS

GUIDELINE:
SUBSEQUENT REVISIONS

- Let some time elapse between revisions. Count on ≥6 drafts.
- Recheck for content and logical organization.
- Check for style and revise if needed.
- Give complete copies of revised manuscript to your coauthors.
- Show your manuscript also to a colleague in a related field and a friend in a different discipline.
- Ask for comments and constructive criticism in writing.
- Be prepared to accept the criticism.

After you have waited a few days, you will be ready to look at your paper with fresh, critical eyes. Start anew by rechecking your draft for content and logical organization, and then recheck for style, especially word location.

When you have revised the paper structurally and stylistically as much as possible, give copies to any coauthors and/or your mentor. Show it to colleagues in the same or related fields of work as well as to a friend in a different discipline. Readers unfamiliar with the manuscript are more likely than you to spot inconsistencies, jargon, parts that are not logical, and other flaws. Be aware though that too many critics may result in chaos and confusion.

ESL advice

If you are not a native English speaker, ask someone who is fluent in English to review your paper as well. Reviewers and editors are more likely to reject a manuscript that does not read well or shows poor English.

Give everyone complete copies of the paper. Ask for comments in writing. Oral comments are easy to forget, especially if there are a lot of them. If you are only provided with oral comments, immediately sit down and write them out to ensure that you will remember as much as possible about what was said.

Ask for constructive criticism not only of style, spelling, and punctuation but especially for content and logic. Do not be defensive when you receive harsh criticism. However, you should be concerned if your critics are indifferent, afraid to hurt your feelings, or say very little.

Be prepared for more revisions. If you consider the criticism valid, incorporate it. If you do not think it is valid, at least think it over. You do not have to agree with the people reviewing your paper, but you should respect their opinions. Give particular consideration to passages questioned by more than one reviewer. Such passages usually need extra attention even if you as the author do not immediately see the problem.

Your critics may not be able to pin down what is wrong, but the fact that they are questioning it probably means it could be improved. In revising these questionable passages, check for poor sentence location, sentence structure, word location, and word choice as well as for noun clusters, unclear comparisons, lack of parallel form, change in key terms, lack of transitions, and use of nominalizations.

Remember, revision is the key to strong writing. Revising large sections in a one sitting will make your document smoother. You may not be able to do this in the first few revisions, but the more you revise, the more you will be able to read through the document in one go.

Most authors need at least six to ten drafts to get a paper ready for submission. Be aware, however, that the process of revision can be endless. There will inevitably be always something that you would like to change, but you should not spend forever writing one paper. At some point, you have to stop revising. Keep in mind that the writing does not need to be perfect, just clear.

The final draft should be approved by all authors, but know that the final version will not read exactly as each individual coauthor would like it to read.

16.3   REVIEWING A MANUSCRIPT

Evaluating the work of others is one of the best ways to reinforce familiarity with revising strategies. Such peer review can also give you a deeper understanding of how writing affects different readers.

How you review a peer's work depends on when in the writing process you are doing the reviewing. Early drafts should be evaluated primarily with respect to major components of the paper such as the research question, the answer of the question, and the logical overall organization of the paper. Subsequent drafts should be evaluated particularly for style and composition as well as flow; whereas a final draft should be reviewed for every aspect of the paper including its topic, question, logical approach, organization, style, composition, impact, grammar, and spelling.

The least helpful comment to receive from a peer reviewer is "It looks OK to me." To be an effective reviewer, you need to be as specific as possible and point out both strengths and weaknesses. Point out particular places in the paper where revision will be helpful. Do not hesitate to note when something is unclear to you, scientifically or in terms of the writing. If you disagree with the comments of another peer reviewer, say so. Not all readers react the same way, and divergent points of view can help writers see options for revising.

Always treat the author with respect. Avoid snippy comments such as "So what?" Instead, make suggestions and recommendations on how to improve and strengthen certain passages or on what else to add or omit from the document. If a passage reads well, point out this strength. If an argument is difficult to follow logically or does not make sense, raise objections politely or ask for explanations to clarify the argument. Write your comments either between the text lines or on the margins of the draft. Better yet, use the "Track Changes" option of MS Word as this will clearly show the author where and how to revise a document.

Read the paper you have been asked to review at least twice, once to get an overview of the paper and subsequent times to provide constructive criticism for the author. If you stumble across a phrase or sentence, this often indicates a spot that needs correcting. Mark such potentially faulty passages for the author. If possible, make suggestions on how to improve them. Use the following checklist as well as those provided at the end of Chapters 10 through 15 to ensure all important components have been included. When you have finished reviewing the document, you can then write up your overall impression of the paper by summarizing its strengths and weaknesses and listing specific areas of concern.

CHECKLIST FOR PEER REVIEW

CONTENT

Purpose and Interpretation

- ☐ Is the overall purpose of the paper and/or central question clear?
- ☐ Does the interpretation of the findings answer the overall question of the paper?

Support

- ☐ Is there sufficient evidence to support the answer?
- ☐ Is every paragraph and sentence in the paper relevant to the overall question?
    Are there portions of the text that could be omitted?

Overall

- ☐ Does the paper advance the field?
- ☐ Does it provide interesting and important insights into the topic of interest?
- ☐ Have power positions been considered (especially in the Introduction, Results, and Discussion)?

Individual sections

Title:
- ☐ Is the title strong?

Abstract:
- ☐ Does the abstract adequately summarize the paper?
- ☐ Have all necessary elements been included (question, experimental approach, results, conclusion)
- ☐ Is the abstract concise?

Introduction:

☐ Does the introduction clearly state the overall question of the paper?
☐ Does the question follow the unknown?
☐ Are all elements (known, unknown, question, experimental approach) clearly signaled?
Has the topic been reviewed? _____

Materials and Methods:
☐ Have all experiments been described adequately?

Results:
☐ Is the main finding presented?
☐ Are there errors in factual information, logic, analysis, statistics, or mathematics? _____
☐ Are all figures and tables explained sufficiently?

Discussion:
☐ Is the overall interpretation of the results clearly stated?
☐ Did the writer adequately summarize and discuss the topic?
☐ Has a clear conclusion been provided?

ORGANIZATION
Overall organization

☐ Is the overall organization of the paper clear and effective?
Are there unclear portions? _____
Could the clarity be improved by changes in the order of the paper? _____
☐ Does the language seem appropriate for its intended audience?

Individual sections

☐ Introduction: Does the introduction follow a funnel structure?
☐ Materials and Methods: Are experiments organized logically?
Results:
☐ Is the main finding presented in the first paragraph?
☐ Are all figures and tables labeled properly?
Discussion:
☐ Is the overall interpretation of the results clearly stated in the first paragraph?
☐ Is the significance clearly stated in the last/concluding paragraph?
☐ Is the discussion ordered in a way that is logical, clear, and easy to follow?

References

☐ Have references been cited where needed?
☐ Are sources cited adequately and appropriately?
☐ Are all the citations in the text listed in the References section?

STYLE AND COMPOSITION

- ☐ Are the transitions between sections and paragraphs logical?
- ☐ Are key words repeated exactly?
- ☐ Are the paragraphs and sentences cohesive?
- ☐ Has word location been considered?
  Are there any grammar, punctuation, or spelling problems?
  _____

- ☐ Is the style concise?
  Are there any wordy passages? _____
  What other problems exist? _____

OVERALL QUALITY

What are the paper's main strengths? _____
What are the paper's main weaknesses? _____
What specific recommendations can you make concerning the revision of this paper?_____

SUMMARY

FIRST REVISION:

1. Check the first draft for content and content location.
2. Check logical organization of sections and subsections. Use checklists at the end of chapters 10 through 15.
3. Revise for style only after you are satisfied with the content and organization.
   Use the basic writing principles to check for word choice, location, sentence structure, and paragraph structure. Pay particular attention to key terms and transitions.
4. Condense where possible.
5. Proofread your manuscript.

SUBSEQUENT REVISIONS:

- Let some time elapse between revisions. Count on ≥6 drafts.
- Recheck the first draft for content and logical organization.
- Recheck for style, and revise if needed.
- Give complete copies of revised manuscript to your coauthors.
- Show your manuscript also to a colleague in a related field and a friend in a different discipline.
- Ask for comments and constructive criticism in writing.
- Be prepared to accept the criticism.

CHAPTER 17

Final Version and Submission

17.1 GENERAL ADVICE ON THE FINAL VERSION

When you have finished revising the paper, make sure that the final versions of the tables and figures correspond to your manuscript. Recheck that you have followed the *Instruction to Authors* exactly (see also Chapter 7, Section 7.2.) Pay attention to detail such as font (Times Roman size 10 or 12 is the most preferred), margins (usually 1 in. all around), line spacing (usually double spaced), word count, page numbering, and line numbering if needed. Ensure that your sections are in the right order. Then make enough copies of the complete corrected version for each of your coauthors and any other person who may need one. If there are no more corrections, make sure that the version you submit is really the final version and that it is complete.

17.2 SUBMITTING THE MANUSCRIPT

GUIDELINE:
Submit to only one journal.
Remember that first impressions are important.

Send your manuscript to only one journal at a time. Your paper will only be considered for publication if not submitted elsewhere. For advice on how to select a journal for submission of your article, see Chapter 7, Section 7.2 on Audience and Journal Choice.

Electronic submissions are standard these days. When submitted electronically, an article can be edited, stored, and distributed easily by the publisher. Electronic submission is also the fastest way to send your manuscript to an editorial office. Instructions for electronic submissions differ from journal to journal. They are usually very detailed and specific and should be followed carefully. If you are unsure how to submit your paper electronically, ask for help from someone that has experience with electronic submissions, particularly when it comes to submitting figures in a specific format and resolution.

When you submit your files, ensure that you are submitting the latest version of the manuscript. Name the file clearly and according to the submission instructions. Some journals require you to submit your manuscript as separate sections—cover letter, abstract, text, figures, tables, and sometimes even a list of potential reviewers or of those you consider to have competing interests and would like to block from reviewing your manuscript. Check online instructions to determine whether this is the case, and prepare your final version accordingly.

The biggest problem in submitting papers electronically is getting figures into the correct format and resolution. Here, it is especially crucial to follow guidelines and suggestions explicitly from the start so that figures and tables do not have to be remade at the last minute to fit the journal specifications. Usually, when you submit figures, you will receive electronic notification that the figure is in the correct format and has been accepted for submission.

17.3 WRITING A COVER LETTER

GUIDELINE:
Send a brief cover letter to go with the manuscript.

Always send a brief cover letter to the editor to go with the manuscript. In most scientific fields, the primary goal of a cover letter is to let the editor know why your research is novel and why it belongs in that particular journal. Make your case convincingly and quickly. That is, keep the cover letter short and simple while at the same time stating your objectives clearly. Do not list all your achievements to date, and do not list any complicated background information or details of your study. Also, do not ask a well-known scientist to contact the editor on your behalf.

Your cover letter to the editor should be carefully written and well presented. It is typically submitted electronically with the manuscript. Find out the name of the editor if you can. Addressing the editor by name will make a good first impression.

Start your letter with an introductory sentence stating the title of your manuscript, and mention that you would like to submit it for publication to that particular scientific journal.

Tell the editor why your research makes an important contribution in the second paragraph. You may add a sentence that describes the results and their importance. You may also state there why your work will be of interest for your target journal. Keep this paragraph short. Do not include a complete summary of your research. If you are submitting to a high-profile journal covering a wide variety of fields, such as *Science* or *Nature*, then you should argue why your paper would be interesting and relevant for a wide variety of disciplines.

If applicable, mention any special features of your paper. Such features may include prior related publications or conference abstracts, permissions to cite others' work or figures, or large tables and graphs. Be prepared to send files of related publications or conference abstracts. Where relevant, assure the editor that informed consent was obtained in accordance with ethical guidelines or that experimental animals were well treated and cared for. Attach signed copies of any required copyright assignment forms of your target journal.

Some journals allow you to suggest names of possible reviewers and to mention anyone to whom the editor could turn for more information. If allowed to do so, list only names of people who are not close colleagues, students, family, or friends. Editors will automatically exclude these from being reviewers. Make sure you include all relevant contact information for each suggested reviewer. You may list reviewers either within the letter itself or you can refer the editor to a separate section you have submitted. You may also be allowed a request for certain people not to review your manuscript. Explain why you make a particular request if possible. For example, explain that it might lead to a conflict of interest if any direct competitors review your work. Although most editors will try to accommodate you, be aware that your request may be ignored. Some journals also limit the number of reviewers that you are allowed to block.

Check the *Instructions to Authors* of your target journal to see if you have to send any additional documentation in your letter to the editor. This documentation may include signed statements of all authors or copyright assignment forms.

Two sample cover letters are shown in Examples 17-1 and 17-2. The first letter is very brief but contains all important information including the title of the manuscript and the purpose of the letter. It also mentions the fact that the manuscript is an extension of a previous publication, which indirectly indicates that the manuscript advances the field, providing significance to the paper. The letter concludes with a general, positive closing statement.

👍 **Example 17-1** **Sample cover letter**

Dear Dr. Riccardo,

Introductory paragraph contains title and purpose	We would like to submit an article entitled "Characterization of two Herpes simplex virus type I transcripts" by A. Wagner and J. Klein, for consideration of publication in the *Journal of Virology*.
Special feature	The work reported in this article extends the work described in our earlier article, "Isolation and localization of HSV-1 mRNA abundant prior to viral DNA synthesis" (*J Virol*

Special feature	2002;10:45–51). For your reference, a copy of this work has been included with the supplemental material.
Concluding sentence	We look forward to hearing whether you can accept this article for publication.

Yours sincerely,

Alfred Wagner Ph.D.

The letter in Example 17-2 is a bit more detailed. Its first paragraph provides the purpose of the letter as well as the name of the manuscript. The second paragraph is a succinct statement about the contribution the work makes to the field and its expected impact. This letter also shows a sample request on who not to consider as reviewers (note again that requests for whom to include or exclude as a reviewer usually should be done by invitation only; check the *Instructions to Authors*).

Example 17-2	**Sample cover letter**
	Dear Mr. Moore,
Introductory paragraph states title and purpose	We would like you to consider the enclosed manuscript "XXX" by A. Piroletti. J. Klein, and U. Wettla for publication in your journal.
Overview of manuscript and impact	Our manuscript describes the development and testing of a novel vaccine against *Vibrio cholerae*. The study describes an innovative approach to vaccine development and would be of great interest in the medical field. Vaccines for this organism, which is the principle agent of cholera, have been available only on a limited basis. A new vaccine would therefore add much value to the treatment of this disease.
Special request	We would like to ask that Drs. D. Fuller and F. Gruppe, both from ABC University, not be selected as reviewers for this work, as they are in direct competition to our laboratory.
Concluding sentences	Thank you for considering our manuscript for publication in your journal. We are looking forward to hearing from you soon.

With best regards,

William Smith, Ph.D.

17.4   THE REVIEW PROCESS

GUIDELINE:

Check *instructions to authors* to know how soon to expect a decision on your paper.
Be courteous and professional at all times.

Most journals will send an acknowledgment that the paper has been received. The *Instructions to Authors* or the acknowledgment you receive may say how long the journal will take to tell you the fate of the paper. If you have heard nothing by a week or two after the promised time, inquire whether a decision has been made. If you still do not hear back from the editor when another 2 weeks have passed after your inquiry, a phone call is not out of place. Be courteous and professional at all times, and do not submit the paper to any other journal until you get a letter of rejection from your target journal.

When the managing editor has received your manuscript, the editor will assign it to two to three anonymous, qualified reviewers. These reviewers will get back to the editor with specific comments and recommendations on the article. The editor then reads your paper and the comments of the reviewers. The editor may also comment on your paper, summarize the major comments of the reviewers, and state which comments should be taken most seriously. Most important, the editor decides whether to accept, accept pending revision, reject with encouragement to resubmit, or outright reject your manuscript.

17.5   LETTER FROM THE EDITOR

GUIDELINE:

Be prepared to make editorial changes even in the
event of acceptance.
Do not waste time trying to figure out who your
reviewers were.

The answer from the journal of your choice usually includes a letter from the editor and the evaluation and comments from the reviewers. In the letter, the editor will tell you whether your manuscript has been accepted or not. Reply to any letters from the editor as quickly as possible.

Paper Accepted
Papers are seldom accepted without revision. If your paper is one of these rarities, thank the editor briefly. If the paper is returned to you for minor changes and corrections to be made, make the corrections carefully and by the requested date.

Accepted Pending Revision
If the editor requests revisions before the paper is accepted, and if the requested changes are few or minor ones, make the necessary changes if you agree. Make sure that your coauthors approve of the changes. Many journals expect revised manuscripts to be returned within a certain time frame; otherwise, they will be considered new submissions. It is not in your interest to pass the set time frame when returning a manuscript for publication.

In the cover letter you send with the revised version, thank the editor and reviewers for their advice, and enclose a list of the changes you have made in response to their suggestions. If you have rejected one or more of the recommendations, justify in the letter to the editor why, but keep rejections of recommendations to an absolute minimum. The editor will then decide whether the authors have addressed all concerns satisfactorily. The paper may be sent to one or both reviewers again.

Rejected with Encouragement to Resubmit

If the editor rejects your manuscript but encourages you to resubmit after revisions, the revisions requested are probably major ones. Now you have to decide whether you can and will make the changes requested by the reviewers. If you decide to resubmit your manuscript to the same journal, you need to include another letter to the editor explaining how you have addressed the concerns and recommendations of the reviewers (see Section 17.6 on Resubmission). When you resubmit your rejected manuscript, the editor will most likely have it reviewed again, and this review may include the same reviewers as the first one. Sometimes the authors are asked to revise a paper a second time before it is accepted for publication.

If the editor offers nothing more than "further consideration" and requests major changes in his or her answer letter, decide whether the effort of making the changes is worthwhile. If you disagree, submit your manuscript to another journal. A second editor may give you a different opinion.

Rejection

If the editor rejects your paper, read the reasons for rejection carefully. There are different ways you can respond now:

If the editor says the article is outside the scope of the journal for whatever reason, send the article to another journal.

If the editor says the article is too long and needs changes, decide whether to make the suggested changes—but, again, submit the revised article to a different journal. The editor would have included an offer to reconsider the article after revision if he or she had wanted to.

If the editor says the reviewers have found serious flaws in the paper, you should probably not resubmit the manuscript to the same journal. You should even consider obtaining more and better information before submitting it elsewhere. If you are sure that the editor and reviewers are wrong, send the paper to another journal or write a short but polite letter saying why you think the paper should be reconsidered. Reserve such appeal letters for extreme cases only, however, such as when a review is flawed. Do not phone the editor.

Realize you are not alone. Up to 50% of articles submitted receive an initial rejection. More prestigious journals such as *Science* and *Nature* have even higher rejection rates (80%–90%). Relax. Then go to work on a revised version for a new journal that same day. The longer you wait, the harder it will be to get started again. Follow the suggestions of the reviewers when you revise your paper. You may get the same reviewers again even when you submit elsewhere.

Overall, do not react too fiercely if you receive a rejection letter from the editor or reviewers. Reviewers and editors can be very helpful, and your paper will usually improve if you take their advice. Writing a furious letter or—worse—arguing on the phone will not get your paper published in that particular journal. It will be better to consider the recommendations you received and to make constructive use of them. Do not give up. Perseverance usually pays off.

Reviewers are often, but not always, anonymous. Their anonymity may annoy you, but do not waste time trying to guess who they are. Your guess is likely to be wrong, and you may feel resentful toward the wrong person for the rest of your career. Rather spend your energy on revising or rewriting your paper.

17.6 RESUBMISSION

GUIDELINE:

FOR RESUBMISSIONS

Address every comment raised by the editor
or by the reviewers.

If you have been invited to resubmit your paper after substantial revisions and have decided to do so, your second letter to the editor and any responses made to the criticism and comments of the reviewers must be written very carefully and clearly and must answer *all* of their concerns. This letter may be the last chance to get your manuscript accepted to this particular journal. You may or may not have the same reviewers as in your original submission. In any case, assume that the editor and the reviewers do not recall your paper in every detail, if at all.

Your response will consist of two parts: a letter to the editor and accompanying responses to the reviewers. Address the second letter to the same editor unless you know that the original editor is no longer there. In the first paragraph, thank the editor and the reviewers for their helpful comments and suggestions and reintroduce the title of your manuscript. Mention any manuscript number or identification number that has been assigned to your paper. In subsequent paragraphs, respond directly to the comments from the editor. In the last paragraph, introduce the more detailed list of your responses to the reviewers' comments, and shortly and politely state that you hope all concerns have been addressed and that you look forward to hearing about a final decision soon.

 Example 17-3 **Sample response letter for resubmission**

Dear Mr. Moore:

| Introductory paragraph; includes title and number as well as a statement of thanks | We are submitting a revised version of our manuscript "ABC" (03-7062). In this revised version, we have addressed the concerns of the editor and the reviewers. We thank you for the helpful comments and suggestions. |

In response to comments from the editor, in your letter dated August 10th, 2008, you suggested to redraw Figure 4 such that the intron of gene A is more easily identified. In the revised version of our paper, Figure 4 has been redrawn according to your recommendations. You also suggested increasing the number of patients tested for the vaccine by at least a factor of 10. We now have analyzed our data with more than 50 times the original number of patients. We have included this new set of results in our revised manuscript (pp. 11–12 and Fig. 4).

Response to editor comments

We have revised the manuscript based on the suggestions and advice of the reviewers. An item-by-item response to their comments is enclosed. We hope that these revisions successfully address their concerns and requirements and that this manuscript will be accepted for publication.

Mention detailed responses to reviewers

Concluding sentences—letter ends on a positive note

Looking forward to hearing from you soon.

With best regards,

William Smith, Ph.D.
Principal investigator

Attach the accompanying responses to the comments of the reviewers to this letter. Your responses should address *every* comment the reviewers have raised. Explain what specific changes you have made and why. If you disagree with a reviewer's suggestion, it is particularly important to explain why. You usually will have to have a solid reason to disagree and should do so very sparingly. Note that accompanying responses may be quite long.

Example 17-4   Sample response to reviewer comments

Responses to Reviewer 1:

We thank Reviewer 1 for the critical comments and helpful suggestions. We have taken all these comments and suggestions into account, and they have improved our manuscript considerably.

1. Reviewer 1 requested to map the cysteine residues involved in the disulfide linkages and to study the mechanistic significance. We have performed these experiments, and the results are now shown in Fig. 5. As mentioned in the manuscript, the full-length protein A contains 22 extracellular cysteines (p. 13, line 12). We did not attempt to do similar experiments with protein A because of the large number of possible disulfide linkages.
2. The reviewer suggested the citation of Norris and Manley with respect to the C-terminal truncation results. We have added this citation and compared the results (p. 12, line 6ff).
3. The reviewer was concerned about the clarity of describing the deletion series. We have improved this description and added the exact amino acid residues of the deletions in the Methods section (p. 5, line 5).
4. ...

After you have addressed each comment and suggestion of Reviewer 1, you need to address those of Reviewer 2 in a similar fashion, starting with a polite introduction and then moving on to each individual comment and response. Address every concern of Reviewer 2, even if some of the concerns are identical to those of Reviewer 1. Note that the style of the response letter in Example 17-5 has been changed to show another way to write such responses. Here, the comments of the reviewer are shown in italics, and the corresponding answers of the authors are shown below each reviewer comment. Answers clearly reference where in the document the corresponding change has been made.

👍 **Example 17-5**   **Sample response to reviewer comments**

Reviewer 2 also made many helpful comments and suggestions, and we thank this reviewer for them. We have taken all these comments and suggestions into account as follows:

Reviewer's comment ——

1. *p. 3, line 37. The GenBank accession numbers should be matched with the "isolate" designation in Table 2.*

Author's answer

We have revised the designation of the GenBank accession numbers in the table in the manuscript. Please refer to **Table 2**.

2. *The pathogenic role of X in acute gastroenteritis is uncertain, particularly when rotavirus and norovirus were detected in 18 out of 20 X positive samples. What was the prevalence of rotavirus infection in this sample pool?*

Reviewer's comment

Author's answer

The prevalence of rotavirus infection in this sample pool was 61.5%. We have included this information in the text. Please refer to **The Study** section **page 7, lines 98–99**.

3. ...

Before you resubmit your manuscript, recheck it as carefully as you did for its original version that was submitted. Make sure the *Instructions to Authors* have been followed meticulously. Recheck especially page numbers, number of words, numbering of figures and tables, and the reference order and list, as they may have changed during revision.

17.7   PAPER ACCEPTED

GUIDELINE:

When your manuscript has been accepted, celebrate!

When your manuscript is accepted, the text is set into type by the publisher. This process will take several weeks or months. In the meantime, you may be asked to assign copyright to the journal and to order any reprints.

It is likely that you will not hear from the journal for several weeks or months thereafter. Then the first proofs or galley proofs are sent out to the corresponding author. These proofs are a sample printing of your paper and will look very much the way the paper will appear in the journal. Examine the proofs or their electronic form closely and carefully. Look for errors and make corrections and last minute edits. At this point, you should not introduce any new material nor make any substantial changes. Also, do not make minor changes of wording or emphasis. Usually, only factual errors such as incorrect data in a table should be corrected now. If you need to make substantial changes, discuss this with the editor first. Return any future proofs or an electronic version thereof by the requested date as quickly as possible, usually within 48 hr. The next time you will see your paper will be in print.

SUMMARY

SUBMISSION GUIDELINES

1. Submit to only one journal.
2. Remember that first impressions are important.
3. Send a brief cover letter to go with the manuscript.
4. Check Instructions to Authors to know how soon to expect a decision on your paper.
5. Be courteous and professional at all times.
6. Be prepared to make editorial changes even in the event of acceptance.
7. Do not waste time trying to figure out who your reviewers were.
8. For resubmission, address every comment raised by the editor or by the reviewers.
9. When your manuscript has been accepted, celebrate!

B. REVIEW ARTICLES

CHAPTER 18

Review Articles

18.1 OVERALL

Review articles outline the overall picture of a particular topic as it is currently understood by scientists in that field. Their main emphasis is on knitting together theories and results from a number of studies to describe the "big picture" of a field of research. Some review papers also evaluate methods and results. Thus, review articles are not original articles with new data but secondary sources representing a well-balanced summary of a timely subject with reference to the literature. These articles are usually invited by an editor, although they can also be submitted independently. Most review articles, even if invited, are peer reviewed.

18.2 CONTENT

The emphasis in review articles is on interpretation and evaluation. By comparing and contrasting all of the key studies done in a certain research area, a review article analyzes how each line of research supports or fails to support a theory. Such evaluations provide scientists with the most up-to-date information as well as with the history and a critical evaluation of the topic. Review articles also provide a solid background for a research paper's investigation, as comprehensive knowledge of the literature of the field is essential to most research papers. In addition, review articles outline any problems that are currently being addressed and explain the basis of any conflicts that exist between experts in the field. As the author of such an article, you can suggest which side of the conflict seems to be presenting the better arguments. You can also suggest possible next steps or propose a new model.

It is particularly important that the information in a review article is understandable to scientists in other related fields. Thus, reviews should use simple words and avoid excessive jargon and technical detail. Figures and boxed material should also be of a more general nature, summarizing and generalizing primary source data or highlighting new ideas. As review articles usually reach a wider audience, they also have a higher impact than regular research papers.

Selecting Source Material

Some research topics are much easier to write about than others. In writing a review article, it is wise to choose a topic of current interest and to pick a research topic about which articles are continuing to be published. Subjects of well-defined and well-studied areas of research typically give more fruitful topics. Avoid defunct or little-known areas of research.

To find out what is "hot," it is often helpful to read a couple of review articles from a variety of journals. Other literature reviews in your area of interest may also give you a sense of the types of themes you might want to look for in your own research or may suggest ways to organize your final review. Other ways to identify hot areas of research are through reading editorials and letters to the editor.

There are two main approaches to choosing an area of research to write about in a review article. One approach is to choose a point that you want to make and then select your primary studies based on this area of interest. Another approach is to read all the relevant studies and organize them in a meaningful way. That is, research the topic starting at general sources and work your way to specific sources. Begin by looking at textbooks or internet sources that are vetted through peer review or have refereed references backing up statements. These are generally considered to be tertiary sources and can provide a good overview of a topic. Next, narrow your search by reading up on topics that have been summarized in secondary sources such as review articles. For both approaches, use secondary sources to come up with a general skeleton for your review paper. Then, use gathered, specific information from primary sources, which are first-hand accounts of investigations and include journal articles, theses, and reports, to fill in the skeleton of your outline.

For all your sources, keep good records from the beginning, particularly citation information. There is nothing more aggravating than having to rediscover where the source material originated from.

18.3   ORGANIZATION

REVIEW ARTICLE GUIDELINE 1:

Overall structure:

Title

Abstract (indicative)—not always required

Introduction

Main analysis section
Conclusion and/or recommendations
Acknowledgments
References

The standard organization of review articles does not follow the IMRAD (Introduction, Materials and Methods, Results, and Discussion) format. Instead, the organization of review articles usually includes an introduction to the topic, a main section with headings and subheadings, a conclusion with recommendations for further research, and a lengthy reference section. Some review articles also contain an abstract. Check the *Instructions to Authors* of your target journal. Depending on what type of review article you write, your paper could follow a slightly different organization.

When you write a review article, provide focus and direction. Focus your review on any trends, solutions, or unsolved issues. A good and critical review will develop new insights into the field and propose potential new research opportunities. Information you select from your references should be able to let you answer not only what is known about the topic but also why the topic is important, what gaps exist, and how they can be filled.

To compose a logically structured review article, create an outline. If you decide to compose your article without one, at least check the overall organization using a reverse outline during the revision stage. To create a clear outline for your review paper, create subsections based on the information you have gathered from the literature. Give these subsections individual headings and subheadings, and then sort the information you have collected into the various subsections. Use bullet points or whole sentences under each heading or subheading (see also Chapter 7, Section 7.5 for more guidelines on how to construct effective outlines). Subsequently, sort the information under each heading or subheading by similarities, contrasts, gaps in knowledge, and so forth.

As you are filling in your outline, reread the source articles to ensure that you have not missed anything. Identify additional papers if needed, and resort your material again if necessary. Writing a review article is an iterative process. When you are satisfied with your outline, start writing the review article by linking all the ideas under each subheading (see Chapter 7, Section 7.5 on how to compose a document.) The following sections provide more details on how to logically organize the Introduction, Main Analysis section, and Conclusion of a review article.

In Example 18-1, an outline of a review paper is presented:

 Example 18-1   Outline of review article

Title: Species-specific cell recognition: Gamete adhesion of sea urchins

Abstract—use an indicative abstract (see Section 18.4)

Introduction

      Cell adhesion—overview

      Cell adhesion model system—sea urchins

In this outline, the authors funnel from a general overview of cell adhesion to specific gamete interactions in sea urchin fertilization in the Introduction. For the Main Analysis section, the article is logically organized into discussing macromolecular interactions in sea urchin fertilization, the species specificity of these interactions, and the molecular interactions of sperm adhesion. The Main Analysis section ends with a proposed model for the interaction of the specific sperm and egg adhesive molecules based on the collective evidence from previous studies presented in prior subsections. This hypothesis fills the gap in knowledge of the exact molecular interactions during sea urchin fertilization. The review article concludes by summarizing the main findings and resulting hypothesis and by projecting the possibility of similar interactions during fertilization in other species.

18.4   ABSTRACT OF A REVIEW ARTICLE

REVIEW ARTICLE GUIDELINE 2:
If an abstract is required, use an indicative abstract.

Indicative abstract = Table of contents in paragraph form

REVIEW ARTICLE GUIDELINE 3:

Structure the indicative abstract as follows:
- Background (optional)
- Problem statement (optional)

- Purpose/topic of review
- Overview of content

Not all review articles require an abstract. If they do, their abstracts are usually written in form of indicative abstracts, providing the reader with a general idea of the contents of the paper. Indicative abstracts differ from informative abstracts. Unlike abstracts for research papers, indicative abstracts are essentially tables of contents in paragraph form. They usually are not self-contained and need to be read together with the text of the article. They may contain some background information and/or a problem statement and should state the purpose or topic of the review. They usually include little if any methods or results. Whereas some review article abstracts end with a statement of significance, interpreting the main findings of a topic for the readers, most end with an overview sentence, listing what will occur in the document as shown in sentence 2 of Example 18-2:

👍	**Example 18-2**	**Indicative abstract**
		This paper describes how plants use the structural diversity of oligosacchrides to regulate important cellular processes such as growth, development, and defense. **We address the central remaining question of how cells perceive and transduce oligosaccharide signals and discuss current research aimed at providing the**
	Overview of content	**answer.**

The next example shows a complete, longer abstract for a review article. This example contains all of the elements of an indicative abstract: background, problem statement, statement of topic, and overview of content.

👍	**Example 18-3**	**Indicative abstract**
	Background	Aerosols serve as cloud condensation nuclei (CCN) and thus have a substantial effect on cloud properties and the initiation of precipitation. Large concentrations of human-made aerosols have been reported to both decrease and increase rainfall as a result of their radiative and CCN
	Unknown/Problem	activities. At one extreme, pristine tropical clouds with low CCN concentrations rain out too quickly to mature into long-lived clouds. On the other hand, heavily polluted clouds evaporate much of their water before precipitation
	Purpose/Topic statement	can occur, if they can form at all given the reduced surface heating resulting from the aerosol haze layer. We propose
	Overview of content	a conceptual model that explains this apparent dichotomy.

(With permission from the American Association for the Advancement of Science)

18.5   INTRODUCTION OF A REVIEW ARTICLE

REVIEW ARTICLE GUIDELINE 4:

Organize the Introduction
- Background
- Unknown or problem
- Purpose/topic of review
- Overview of content

The Introduction of a review article should provide the big picture of the topic and grab the readers' attention. It should present some general background and state the central purpose/topic of the review. It should also make clear why the topic warrants a review.

After discussing the general background and aspects of existing research, present any recent developments and describe what the problems with the existing research are and/or what is unknown. Subsequently, explain the overall purpose of the review article. This statement should be followed by a description of its organizational pattern. Do not make the introduction longer than 1/5 of the review article.

The statement of purpose/topic of a review article is similar to the question or purpose of a research paper. Your statement of purpose/topic will not necessarily argue for a position or an opinion; rather, it will argue for a particular perspective on the topic. The statement tells the reader how you will interpret the significance of the subject matter under discussion and lets the reader know what to expect from the rest of the article. Thus, these sentences present the topic of the entire article, and all paragraphs and sentences in your article relate to them. Sample statements of purpose/topic for literature reviews include

---

Example 18-4   **Purpose/Topic statement for review papers**

a   The prevalence of implementation failure in HCOs is of great concern in the medical community.

b   The development of new antibiotics has become the main focus of several biotech companies.

c   Mathematical modeling of disease transmission is important to maximize the utility of limited resources.

In the Introduction of your review article, you may state the overall significance of the article. See Chapter 10 for a more detailed discussion of various components of introductions.

Following are two examples of introductions of review articles.

Example 18-5 a   **Introduction of a review paper**

Background

Global climate change is altering the geographic ranges, behaviors, and phenologies of terrestrial, freshwater, and marine species. A warming climate, therefore, appears destined to change the composition and function of marine communities in ways that are complex and not entirely predictable (1–5). Higher temperatures are expected to increase the introduction and establishment of exotic species, thereby changing trophic relationships and homogenizing biotas (6). Because organisms in polar regions are adapted to the coldest temperatures and most intense seasonality of resource supply on Earth (7), polar species and the communities they comprise are especially at risk from global warming and the concomitant invasion of species from lower latitudes (8–10).

Unknown/Problem

Background

Shallow-water, benthic communities in Antarctica (<100-m depth) are unique. Nowhere else do giant pycnogonids, nemerteans, and isopods occur in shallow marine environments, cohabiting with fish that have antifreeze glycoproteins in their blood. An emphasis on brooding and lecithotrophic reproductive strategies (11, 12) and a trend toward gigantism (13) are among the unusual features of the invertebrate fauna. Ecological and evolutionary responses to cold temperature underlie these peculiarities, making the Antarctic bottom fauna particularly vulnerable to climate change. The Antarctic benthos, living at the lower thermal limit to marine life, serves as a natural laboratory for understanding the impacts of climate change on marine systems in general.

Purpose/
Topic statement

Overview

Recent advances in the physiology, ecology, and evolutionary paleobiology of marine life in Antarctica make it possible to predict the nature of biological invasions facilitated by global warming and the likely responses of benthic communities to such invasions. This review draws on paleontology, biogeography, oceanography, physiology, molecular ecology, and community ecology. We explore the climatically driven origin of the peculiar community structure of modern benthic communities in Antarctica and the macroecological consequences of present and future global warming.

(With permission from Annual Reviews)

Example 18-5 b   **Introduction of a review paper**

	Mitochondrial genomes differ greatly in size, structural organization and expression both within and between and the kingdoms of eukaryotic organisms. The mito-chondrial genomes of higher plants are much larger (200–2400 kb) and more complex than those of animals
Background	(14–42 kb), fungi (18–176 kb) and plastids (120–200 kb)
Unknown/Problem	(Refs 1–4). Although there has been less molecular ana-lysis of the plant mitochondrial genome structure in com-parison with the equivalent animal or fungal genomes, the use of a variety of approaches—such as pulsed-field gel electrophoresis (PFGE), moving pictures (movies) dur-ing electrophoresis, restriction digestion by rare-cutting enzymes, two-dimensional gel electrophoresis (2DE) and
Topic statement	electron microscopy (EM)—has led to substantial recent progress. Here, the implications of these new studies on the understanding of *in vivo* organization and replication
Overview	of plant mitochondrial genomes is assessed.

(With permission from Elsevier)

18.6   MAIN ANALYSIS SECTION OF A REVIEW ARTICLE

REVIEW ARTICLE GUIDELINE 5:

Organize the Main Analysis section logically into subsections either

- chronologically
- thematically
- methodologically

REVIEW ARTICLE GUIDELINE 6:

Logically organize information within the Main Analysis subsections (similarities, contrasts, gaps in knowledge, etc.).

One of the most difficult tasks in writing a review article is finding the best structure of the Main Analysis section. This section should pre-sent any experimental evidence by describing important results from recent primary literature articles and explain how these results shape current understanding of the topic. Mention the types of experiments done and their corresponding data, but do not repeat the experimental procedure step for step. In addition, point out and address any controversies in the field. You may use figures and/or tables to present your interpretation of original data or show key data taken directly from the original papers.

The overall organization of the Main Analysis section should sequentially unfold ideas in a logical order. The best structure may not become obvious until several drafts have been written (see also Sections 18.2 and 18.3 for information on selecting sources and deciding on which information to use).

Generally, this section of a review article can be organized chronologically, thematically, or methodologically. If your review follows the chronological method, you could write about topics according to when they were published. Alternatively, you could examine the sources under the history of the topic. Such an organization would call for subsections according to eras within this history.

In contrast to the chronological presentation of topics, thematic reviews are organized around a topic or issue rather than around the progression of time. For example, as you deal with various levels of evidence pertaining to a question, your Main Analysis could move steadily downward in the level of inquiry from the organism, to the organ, to the cell, to the molecular mechanisms within the cell. However, progression of time may still be an important factor in a thematic review.

A methodological approach differs from the preceding two in that its focus usually does not have to do with the content of the material. Instead, it focuses on the "methods" of the researchers, and topics are organized accordingly by techniques or by methods or approaches.

Once you have decided on the organizational method for the main section of the review, the subsections you need for the review paper should arise out of your organizational strategy. In other words, a chronological review would have subsections for each vital time period, whereas a thematic review would have subtopics based on factors that relate to the theme or issue.

For some reviews, you might need to add additional sections that are necessary for your study but do not fit in the organizational strategy of the main section (e.g., "Current Status" or "Future Directions"). In some instances, you may also merge a section with another or omit one altogether. Similar to the overall structure, what subsections you include in the body may only become clear as the review evolves.

In the next example, a partial Main Analysis subsection of a review article is shown. This subsection has its own heading ("The Opposing Effects of Aerosols on Clouds and Precipitation"), indicating what the subsection deals with. Under this heading, the authors provide some context and state the problem. This statement is followed by (a) a proposed solution from the literature and (b) their own opinion, completing the logical structure of the subsection.

👍   **Example 18-6**   **Main analysis section of a review paper**

Subheading   **The Opposing Effects of Aerosols on Clouds and Precipitation**

With the advent of satellite measurements, it became possible to observe the larger picture of aerosol effects on clouds and precipitation. (We exclude the impacts of ice nuclei aerosols, which are much less understood than the effects of CCN aerosols.) Urban and industrial air pollution plumes were observed to completely suppress precipitation from 2.5-km-deep clouds over Australia (*20*). Heavy smoke from forest fires was observed to suppress rainfall from 5-km-deep tropical clouds (*21, 22*). The clouds appeared to regain their precipitation capability when ingesting giant (>1 μm diameter) CCN salt particles from sea spray (*23*) and salt playas (*24*). These observations were the impetus for the World Meteorological Organization and the International Union of Geodesy and Geophysics to mandate an assessment of aerosol impact on precipitation

Context

(*19*). This report concluded that "it is difficult to establish clear causal relationships between aerosols and precipitation and to determine the sign of the precipitation change in a climatological sense. Based on many observations and model simulations the effects of aerosols on clouds are more clearly understood (particularly in ice-free clouds);

Problem/Unknown   the effects on precipitation are less clear."

A recent National Research Council report that reviewed "radiative forcing of climate change" (*25*) concluded that the concept of radiative forcing "needs to be extended to account for (1) the vertical structure of radiative forcing, (2) regional variability in radiative forcing, and (3) nonradiative forcing." It recommended "to move beyond simple climate models based entirely on global mean top of the atmosphere radiative forcing and incorporate new global and regional radiative and nonradiative forcing metrics as they become

Proposed solution   available." We propose such a new metric below.

(With permission from the American Association for the Advancement of Science)

18.7   CONCLUSION OF A REVIEW ARTICLE

REVIEW ARTICLE GUIDELINE 7:

In the conclusion section, summarize your topic, generalize any interpretations, and provide some significance.

The conclusion section is one of the main highlights of a review paper. It recaps your review and your main conclusions, recommendations, and/or speculations.

You need to phrase this section with special care, summarizing and generalizing main lines of arguments and key findings. Discuss what conclusions you have drawn from reviewing the literature, and restate your interpretations. In addition, provide some general significance of the topic and results, and discuss the questions that remain in the area. Although this section is often longer than the conclusion section of a research paper, try to keep it brief.

Example 18-7 provides an example of a well-written Conclusion section for a review article. It first provides a summary of the main findings in the field, generalizing these findings for the readers ("All of these effects appear destined to ..."). In the second paragraph of this Conclusion section, the authors state their overall opinion/interpretation of the findings in the field and provide a general recommendation on how to solve the problem ("Global environmental policy must immediately be directed to ...").

Example 18-7   **Conclusion of a review paper**

Beginning in the late Eocene, global cooling reduced durophagous predation in Antarctica. Despite some climatic reversals, the post-Eocene cooling trend drove shallow-water, benthic communities to the retrograde, Paleozoic-type structure and function we see today. Now, global warming is facilitating the return of durophagous predators, which are poised to eliminate that anachronistic character and remodernize the Antarctic benthos in shallow-water habitats. Rising sea temperatures should in general act to reduce the mismatch between the development times of invasive larvae and the length of the growing season. Increased survivability of planktotrophic larvae will decrease the selective advantage of brooding and lecithotrophy, increasing the pool of potentially invasive species. Warming temperatures will also increase the scope for more rapid metabolism and should ultimately obviate the adaptive value of gigantism. All of these effects appear destined to amplify the ongoing, worldwide homogenization of marine biotas by reducing the endemic character of the Antarctic fauna.

Summary

The fact that benthic predators are already beginning to invade the Antarctic Peninsula should be taken as an urgent warning. Controlling the discharge of ballast water from ships will be difficult but not impossible. Whether or not humans are the proximal vectors, however, the

Interpretations and recommendations

long-term threat of invasion in Antarctica has its roots in climate change. The Antarctic Treaty cannot control global warming. Global environmental policy must immediately be directed to reducing and reversing anthropogenic emissions of greenhouse gases into the atmosphere if marine life in Antarctica is to survive in something resembling its present form.

Interpretations and recommendations

(With permission from Annual Reviews)

18.8   REFERENCES

Like in academic research papers, your interpretation of the available sources must be backed up with evidence. Therefore, cite primary and secondary sources where needed. The type of information you choose should relate directly to the review's focus. See Chapter 8 for more information on references.

18.9   SIGNALS FOR THE READER

REVIEW ARTICLE GUIDELINE 8:
Signal all the necessary elements in a review article.

As for research papers (investigative or descriptive), all the parts of the review article should be signaled to the reader. Signals are given at various levels and may consist of subheadings, topic sentences, as well as specific phrases and terms. Examples of such signals are listed in Chapter 10 (Introduction), Chapter 13 (Discussion), and Chapter 14 (Abstract).

18.10   COHERENCE

REVIEW ARTICLE GUIDELINE 9:
Use topic sentences and techniques of
continuity to tell the story.

To ensure that the overall story is clear, use topic sentences and consider the following:

- Word location (Chapter 3 and Chapter 6, Section 6.3)
- Key terms (Chapter 6, Section 6.3)
- Transitions (Chapter 6, Section 6.3)
- Sentence location (Chapter 6, Sections 6.1 and 6.2)

18.11   COMMON PROBLEMS OF REVIEW ARTICLES

The most common problems of Review articles include:

- Lack of analysis and commentary—Writing the review as a simple list of facts and dates without interpretation and inference.
- Not stating the unknown or problem—this leaves the reader hanging and wondering why the review is of interest
- Lack of logical organization of subtopics (Section 18.6)
- Review article is not objective—does not show conflicts between research "camps"
- Referencing errors (incorrect or missing citations, incorrect dates or volume; see also Chapter 8)

18.12   REVISING THE REVIEW ARTICLE

When you have finished writing the review article (or if you are asked to edit one for a colleague), you can use the following checklist to "dissect" the article systematically:

☐   1. Does the topic present something of interest to the field?
☐   2. In the Abstract (if present), is the purpose or topic stated precisely?
☐   3. Is an overview of the article provided in the Abstract?
☐   4. Does the Introduction have the following components?
      ☐   Background
      ☐   Problem or unknown
      ☐   Purpose/topic or review
      ☐   Overview of content
☐   5. Is your Main Analysis section logically organized?
☐   6. Did you analyze and interpret all information (rather than simply listing facts and dates)?
☐   7. Does your paper present information objectively, including contradictory data and ambiguities?
☐   8. Is the topic summarized and interpreted in the Conclusion section?
☐   9. Is the significance of your analysis clear?
☐   10. Do all the components logically follow each other? (Is the unknown what one would expect to hear after reading the background/known? Is the thesis statement really the statement one would expect to read after reading the unknown?)
☐   11. Did you double-check references and citations for errors?
☐   12. Are all the necessary elements in the review article signaled clearly?
☐   13. Revise for style and composition based on the writing principles of the book:
      ☐   a. Are paragraphs consistent? (Chapter 6, Section 6.2)
      ☐   b. Are paragraphs cohesive? (Chapter 6, Section 6.3)
      ☐   c. Are key terms consistent? (Chapter 6, Section 6.3)

☐   d. Are key terms linked? (Chapter6, Section 6.3)

☐   e. Are transitions used and do they make sense?
(Chapter 6, Section 6.3)

☐   f. Is the action in the verbs? Are nominalizations avoided?
(Chapter 4, Section 4.6)

☐   g. Did you vary sentence length and use one idea per sentence? (Chapter 4, Section 4.5)

☐   h. Are lists parallel? (Chapter 4, Section 4.9)

☐   i. Are comparisons written correctly? (Chapter 4, Sections 4.9 and 4.10)

☐   j. Have noun clusters been resolved? (Chapter 4, Section 4.7)

☐   k. Has word location been considered? (Verb following subject immediately? Old, short information at the beginning of the sentence? New, long information at the end of the sentence?) (Chapter 3, Section 3.1)

☐   l. Have grammar and technical style been considered? (person, voice, tense, pronouns, prepositions, articles) (Chapter 4, Sections 4.1–4.4)

☐   m. Is past tense used for results and present tense for descriptive papers?

☐   n. Are words and phrases precise? (Chapter 2, Sections 2.2 and 2.3)

☐   o. Are nontechnical words and phrases simple? (Chapter 2, Section 2.2)

☐   p. Have unnecessary terms (redundancies, jargon) been reduced? (Chapter 2, Section 2.4)

☐   q. Have spelling and punctuation been checked? (Chapter 4, Section 4.11)

SUMMARY

GUIDELINES:

1. Overall structure:
     Title
     Abstract (usually optional and indicative)
     Introduction
     Discussion
     Conclusion and/or recommendations
     Acknowledgments
     References
2. If an abstract is required, use an indicative abstract.
    Indicative abstract = Table of contents in paragraph form
3. Structure of the indicative abstract:
     • Background (optional)
     • Problem statement (optional)
     • Purpose/Topic of review
     • Overview of content

4. Organize the Introduction:
   - Background
   - Unknown or problem
   - Purpose/Topic of review
   - Overview of content
5. Organize the Main Analysis section logically into subsections either
   - chronologically
   - thematically
   - methodologically
6. Logically organize information within the Main Analysis subsections (similarities, contrasts, gaps in knowledge, etc.).
7. In the Conclusion section, summarize your topic, generalize any interpretations, and provide some significance.
8. Signal all the necessary elements in a Review article.
9. Use topic sentences and techniques of continuity to tell the story.

PART FOUR

Grant Proposals

CHAPTER 19

Proposal Writing

19.1   GENERAL

One of the greatest challenges for a scientist is finding sources of money to allow for research, salaries, supplies, travel, innovation, and technology. From graduate students and postdoctoral fellows to faculty members and investigators, competing for grants occurs at all stages of academic research careers. To be successful, scientists require funds and have to find sources for this money. Grant writing is as central to a scientific career as is writing papers and doing research at the bench.

Grant writing can make or break a research career no matter how good or innovative a scientist's ideas are. However, many candidates struggle with it and make needless mistakes that impair potentially successful applications. Common mistakes include overambitious research plans, incoherent or unfocused experimental designs, and poor write-up and presentation of the proposed study. If you learn to address these problems, your grant applications stand a good chance of receiving a favorable review. The best way to learn about what constitutes a successful proposal aside from practicing to write one is to read grant applications that were funded, noting how they are structured, how much detail to include, and which ideas sold.

Another major stumbling block associated with attaining grants is locating funders. Do your homework. Read up on the various funding agencies, and carefully review individual grant programs. Then decide which funders and programs align most closely with your research objectives. Funding agencies sponsor grant programs for various reasons. Before developing a grant proposal, it is vitally important to understand the goals of the funder and particular grant programs. Familiarize yourself with proposals the sponsor has funded previously to get to know the type of research and writing style that is expected.

If your proposed study does not fall within the grant maker's interest areas, the grant maker may not fund your work. Discuss your ideas with colleagues and with program officers at the funding agencies. You may find that for a particular project to be eligible for funding, the original concept may need to be modified to meet the criteria of the grant program. The success of grant proposals depends on six factors:

1. The sponsoring organization
2. The innovative nature or critical importance of the proposed project
3. The competition level
4. The skills of the grant writer in building a compelling case
5. Good luck
6. Good timing (in terms of whether the funder and/or society are ready for the idea or approach)

Thus, careful targeting of a funder and careful preparation of a proposal can distinguish success from failure.

Aside from these factors, know that in the last few years, interdisciplinary proposals have become of more interest to many funders. Thus, view your research in a broader context and consider such collaborations.

Before you submit a proposal, or start writing one, find out whether there is an internal competition for this application. Many funders accept only a limited number of applicants per institution and year. In these cases, organizations usually decide which projects have priority through internal competitions.

As reviewers are busy people who are faced with many more requests than they can grant, they have to find out quickly and easily the answers to the following main questions:

What do you want to do?
How much will it cost?
How much time will it take?
Why is the proposed work important?

19.2   TYPES OF PROPOSALS

There are various types of proposals:

- Solicited proposals—in response to a specific request for proposals (RFP) by a funder
- Unsolicited proposals—no solicitation, but you believe the funder will likely be interested
- Preproposals—based on an abbreviated version of your proposal, the funder decides if a full proposal is warranted
- Continuation or Noncompeting proposal—continued funding is contingent on whether initial progress is satisfactory
- Renewal (usually competitive)—request for continued support of an existing project

19.3 CHOOSING A SPONSORING AGENCY

When you are looking for funding for a project, there are various options. Broadly, funding sources can be divided into governmental and private sectors. The obvious funding agencies are federal agencies—private funders might not be so evident.

To have potential funders available at a glance, make a list and add onto it throughout your career. When looking for private funders, start with networking. Check the acknowledgment section of all those related research articles piling up on your desk to see who is funding those projects. Contact your institution's program officers at the Office of Sponsored Programs (also called Office of Development, Corporation and Foundation Relations, or Office of Research) for a list of funding sources. Colleagues whose work resembles yours might also be able to advise you about potential sources. Program officers and/or proposal coordinators have a list of private foundations and corporations that support scientific research. Program officers exist to help you secure funding, so use them. Make an appointment, and see what services they are able to offer.

In seeking a funder, you need to identify the agenda and priorities of the various options. You also need to find out the level of desired innovation and the respective risk tolerance (is the funder likely to fund more risky research in which outcome is uncertain?) Again, read previous proposals that the sponsor has funded. Take advantage of seminars or online guidance the agency offers, and, if possible, talk to a program officer at the funding agency. Most are more than happy to give feedback and advice. (See also Section 19.7, subsection "Research.")

19.4 FEDERAL AGENCIES

GRANT WRITING GUIDELINE 1:
Do not be afraid to contact funding agencies.

The most obvious sources for scientists to obtain funding are federal agencies. If you are a medical researcher, you usually apply to the National Institutes of Health (NIH). If you are a nonmedical scientist or are looking for funding for education in nonmedical fields, you apply to the National Science Foundation (NSF). Other federal agencies for engineering, for example, include the Department of Energy (DOE) and the National Aeronautics and Space Administration (NASA). If you expect to get tenure, you most likely will need to secure funding from one of these agencies.

The NIH consists of several institutes with different missions, whereas the NSF organizes its research and education support through several directorates and offices (Table 19-1).

Note that the NSF places substantial emphasis on education, women, underrepresented minorities, and persons with disabilities in their consideration process.

Table 19-1

NIH INSTITUTES

- National Cancer Institute
- National Eye Institute
- National Heart, Lung, and Blood Institute
- National Human Genome Research Institute
- National Institute on Aging
- National Institute on Alcohol Abuse and Alcoholism
- National Institute of Allergy and Infectious Diseases
- National Institute of Arthritis and Musculoskeletal and Skin Diseases
- National Institute of Biomedical Imaging and Bioengineering
- National Institute of Child Health and Human Development

- National Institute on Deafness and Other Communication Disorders
- National Institute of Dental and Craniofacial Research
- National Institute of Diabetes and Digestive and Kidney Diseases
- National Institute on Drug Abuse
- National Institute of Environmental Health Sciences
- National Institute of General Medical Sciences
- National Institute of Mental Health
- National Institute of Neurological Disorders and Stroke
- National Institute of Nursing Research

NSF DIRECTORATES AND OFFICES

- Biological Sciences
- Computer and Information Science and Engineering
- Cyberinfrastructure
- Education and Human Resources
- Engineering
- Geosciences

- Integrative Activities
- International Science and Engineering
- Mathematical and Physical Sciences
- Polar Programs
- Social, Behavioral and Economic Sciences

Detailed instructions and guidelines for applications to the NIH are described in the application for a Public Health Service grant PHS 398. For NSF solicitors, guidelines are found in the Grant Proposal Guide (see also https://www.fastlane.nsf.gov/fastlane.jsp for the NSF online Web site). Check these guidelines frequently, as they are subject to change.

Do not be afraid to contact these funding agencies. The NIH, for example, has program officers there to help. Tell them what you are interested in or send a one- to two-page white paper, an educational overview that addresses the problems of your topic and how to solve them. Do not view the NIH as a black box. Follow up with the designated federal contact person to determine the estimated ratio of applicants to grantees based on prior rounds, planned deadlines for future grant cycles, the constituents who will serve on the review panel, and the precise definitions of funding criteria. Many federal agencies also conduct workshops for applicants, and online sources and examples are available from them as well.

Securing any type of funding is difficult, time consuming, and unpredictable; and if you are considering federal agencies, you need to be aware of potential changes, especially in an election year. Presidential policies can impact the budgets of the NIH, NSF, and other federal funders. Government grant making can be political, so obtain support from elected officials, and recognize that geographic considerations might limit your competitiveness. Remember also that the application process and bureaucratic obstacles are far more time consuming and cost consuming for government grants than for private foundations.

Federal grant applications tend to be fairly long (often about 20 to 25 single-spaced pages for the narrative) and have very specific, lengthy instructions. Therefore, when tackling a government submission, begin preparing well in advance of the deadline, break the application into manageable sections, develop a time line for completing each section, delegate some of the preparation work to colleagues or staff if possible, and carefully review your draft against the selection criteria to ensure that no detail is overlooked.

19.5   PRIVATE FOUNDATIONS

GRANT WRITING GUIDELINE 2:

Write a letter of inquiry to a private foundation
or corporation before preparing or mailing out a proposal.

Although many scientists will apply primarily to government agencies, private foundations and corporations can be equally good choices. Foundations come in various sizes and types, but their grants can be important and substantial. In fact, in times of low federal support for research, private funders may be preferable. The notion that private sources fund at a much lower level or are more competitive is not correct. In terms of the success rate, you may be much more likely to get funded by a private funder than by government agencies at certain times and for particular projects.

Most foundations have specific areas of interest for which they award funds. It is essential that you identify those foundations whose interests match your proposed project. Seldom will a foundation fund a project outside of its stated field of interest. (See also Section 19.7, subsection "Research.")

Unsolicited proposals to foundations have a better chance of success if they are preceded by an informal contact. This contact is usually a brief (not more than two pages) letter outlining the proposed project, suggesting why it may be of interest to the foundation, and requesting an opportunity to discuss it in further detail. Such a letter permits an investigator to make inquiries to several foundations at once and gives an interested foundation the chance to offer suggestions before a formal proposal is finalized.

In general, foundations are interested in innovative projects that (a) address pressing national, international, or regional problems; (b) are relevant to new methods in education; (c) can serve as a model or stimulus for further or related work in its general area; (d) can be continued after the end of the funding period without further assistance from the foundation; (e) are highly innovative and/or promise to solve a question with broad ramifications; and (f) are not eligible for funding by governmental agencies or the investigator's own institution. The letter of inquiry should highlight whichever of these characteristics best fit the project at hand.

When you are considering approaching a private funder (foundation or corporation), it is important to get help and guidance from your university's administration such as the Corporations and Foundations Relations Staff or the Office of Research. In some cases, you may even be *required* to work through the university when approaching a potential funder.

19.6   CORPORATIONS AND OTHER FUNDERS

Aside from federal agencies and large private foundations, corporations, individuals, family foundations, and public charities also may sponsor projects.

Corporate giving may be handled by the corporation directly, or it may be done through a company foundation. Corporate foundations derive their grant-making funds primarily from the contributions of a profit-making business. These foundations often maintain close ties with the donor company but are legally separate organizations. Their boards are often made up of corporate officers, their endowment funds are separate from the corporation, and they have their own professional staff. They are subject to the same rules and regulations as other private foundations. Although corporations award grants, they often give to get. What they hope to receive in return is exposure, publicity, community respect, market share, and intellectual property rights or parts thereof. Corporate funding can be a good source of support for new initiatives, special programs, and special events, but often it may result in publication restrictions, as corporations are more interested in producing a product rather than publishing research results.

Individuals are the largest source of funding for nonprofit organizations. To receive funding from an individual, you usually must have direct ties and overlapping interests such as alumni giving to their alma mater.

Family foundations receive assets from individuals or families. Usually, at least one officer or board member of the foundation is a family member. The family member plays a significant role in governing and/or managing the foundation, often on a voluntary basis receiving no compensation. Family foundations often fund in their immediate geographic area, with out-of-state exceptions only made for the donor's alma mater or a hospital where a relative received medical treatment.

Public charities include churches or associates of churches, certain educational organizations, hospital and medical research organizations, endowment funds organized and operated in connection with state and municipal colleges and universities, and publicly supported organizations that normally receive a substantial part of their support from a governmental unit or from direct or indirect contributions from the general public. Some public charities make grants, although most provide other services. Note that private foundations usually derive their assets from a single source—such as an individual, family, or corporation—and not from the public. In the sciences and medicine, certain public charities, such as the American Chemical Society or the Polycystic Kidney Disease Foundation, can be important sources of funding, often providing scholarships, fellowships, and grant awards.

There are many useful fundraising resources available on the Internet including program guidelines, application materials, and reports issued by private foundations. In addition to those Internet sites mentioned under Section 19-8, you may also consult the following:

> **http://research.unc.edu/grantsource/private_agencies.php**
> **http://www2.lib.udel.edu/subj/foce/resguide/found.htm**—which contains a concise resource guide from the University of Delaware Library
> **http://www.mcmaster.ca/ors/sources/sources_additional.htm**— lists various funding agencies in alphabetical order
> **http://foundationcenter.org/findfunders/topfunders/top50 giving.html**—lists the 50 largest corporate foundations ranked by total giving
> **http://www.guidestar.org/**—is the best source of information on grant making public charities

19.7   PRELIMINARY STEPS TO WRITING A PROPOSAL

GRANT WRITING GUIDELINE 3:
Obtain and strictly follow the proposal guidelines.

GRANT WRITING GUIDELINE 4:
Consult administrators (department chair or dean and proposal coordinators).

Do your homework. Most funders provide detailed instructions or guidelines regarding the format, content, length, and font size of a proposal. These guidelines can be downloaded and should be studied carefully before you begin writing the draft. Follow the guidelines! Funders post these guidelines for a variety of reasons; sometimes, it is out of respect for the time of their reviewers who commonly review proposals without pay and only out of commitment to their field of study. A uniform format also makes it easier to compare proposals and provides an easier means for program officers to gauge whether a proposal contains the necessary information and detail needed for evaluation. Enforcement of the rules is commonly Draconian, and many proposals are discarded without review because they fail to adhere to guidelines.

Provide the required information in the order listed, and answer all questions completely. Also, plan to start early—submission deadlines are usually ABSOLUTE.

Format requirements are typically

- 8½″ by 11″ paper
- Single spaced with all margins measuring at least 1"
- At least 12-point font in Times New Roman

Most proposals are judged on content and presentation. Therefore, the presentation must be neat, professional, and organized.

Early discussion of potential problems will smooth the way for the proposal later. You will benefit by consulting two persons at an early stage in the planning of the proposal: your department chair (or dean) and a university representative who maintains a liaison with the funder you have in mind. These resources can provide valuable help and advice both in substantive and administrative matters. In addition, if the proposal includes a budget, you may be required to contact the office responsible for administering grants and contracts for your institution before you send out your proposal—the best route is usually through your department/program's business office.

The department chair should be informed of your intentions and especially of any aspect of the proposed project that might conceivably affect departmental administration or your departmental duties. You will eventually be asking the chair to approve the proposal and thereby endorse your plans for staff and facility commitments. Project officers and/or proposal coordinators are a general source of help for the whole process of planning and writing the proposal as are professional grant writers and editors. Proposal coordinators can review the proposal for completeness, ensure compliance with all requirements, and raise pertinent questions that must be resolved before the proposal will be approved for submission such as human subjects review, the use of animals, potential conflicts of interest, equipment purchase, biological hazards, proprietary material, cost sharing, intellectual property, and many other matters. If you are a student or postdoctoral fellow, work closely with faculty and advisors in developing proposals.

Proposal Organization

Note that some foundations and corporations have very few or even no guidelines on how to write a proposal when you apply there. In these cases, it will be up to you to decide which sections to include. As a general guideline, know that almost all grants will contain the following basic sections in some form or other:

1. Abstract
2. Specific Aims
3. Background/Introduction/Statement of Need
4. Preliminary Results
5. Research Design/Research Plan/Methodology/Implementation/ Proposed Methods and Procedures/Approach/Strategy
6. Alternate Strategy/Technical Problems and Alternate Routes
7. Summary/Impact Statement
8. References
9. Budget
10. Personnel/Biographical Sketches

Depending on the funder, your presentation preference, and the topic, some of these sections may be combined; others may be split into two or more subsections or may have differently named headings. All these basic sections are described in more detail in the following chapters. Some funders will ask for additional special sections, which are discussed in a separate chapter (Chapter 25.)

In general, federal agencies will require you to provide much more detailed proposals than private foundations or corporations. However, proposal details required by some private foundations may even exceed those requested by federal agencies.

Tips for Postdoctoral Fellows and Junior Faculty Members

Start early on to build your personal list of potential funding sources. Pay particular attention to those agencies that have special grant competitions for postdoctoral fellows or junior faculty. NSF, for example, has the CAREER (the Faculty Early Career Development) program, which provides funding for 4 to 5 years and ranges from $200,000 to $500,000.

Take advantage of pilot projects, projects that are planned as a test or trial. You may be able to get a smaller amount of funding to test out projects. Although these projects do not come with large amounts of money, they can make a huge difference in reviewers' reactions to a larger funding proposal. In addition, check out institutional sources of research funds. Find out whether junior faculty are eligible and what the deadlines are.

Familiarize yourself with how research grants are processed and administered. Inquire about resources in your new department or elsewhere on campus for helping with grant preparation and processing. Some departments and universities offer substantial assistance; but in others, you will be largely on your own.

Above all, give yourself plenty of time to search for funding sources and to write the proposal. Estimate how long it will take you to write a proposal, then double or triple the time to come close to the actual time you will need to spend.

Research

In-depth research of targeted funders is essential for success. Aside from preparing a list of potential funders and obtaining the instructions for writing the proposal, you need to discover other important details about the funder. Every hour invested in research increases your chance of success.

What do you need to look for when trying to identify potential funding agencies? Most important, you need to ensure a match between you and the agency. Study the grant maker's priorities to decide on the most competitive project for a given grant maker. Find out if the agency's interests match what you are proposing. Funding agencies give away money only for projects that further a well-defined goal that falls under the agency's statement of purpose or mission. Find out what kinds of projects they do and do not fund. For example, look at the agency's annual report to see the kinds of projects that have been funded. Determine whether geographical limitations are imposed.

Other details you need to know include the following:

- Have a pretty good idea how much money you will need to ask for to get your project done. If you need more than the sponsor offers on average, check to see whether the amount you are asking for will automatically eliminate you from consideration.
- Find out what the proposal deadlines are to ensure that you can meet them. Estimate that it will take you about 2 to 3 months to prepare a well-written proposal. Also take into account the start date of your project. It will take from 3 to 6 months for you to find out whether your proposal will be funded.
- Know who will review your proposal. Knowing who your reviewers will be, particularly their training and background, will help you gear your writing level to that specific audience.
- Inquire what the indirect costs (or "overhead") are. Institutionally administered grants more often than not require overhead or "indirect costs" to pay the institution back for the services it offers to the researcher. The amount funding agencies allow for indirect costs varies from one agency to another (from 0% to more than 70%). The amount you ask for from the granting agency is the sum of the direct and indirect costs.
- Find out the name and contact information for a program officer at the agency. Consider contacting the program officer to get a feel for how receptive the agency might be to your proposal and to get answers to questions that remain. The program officer is the single best source of information about a grant program, and most program officers are very willing to talk to potential applicants.

19.8   ONLINE RESOURCES

In addition to listings found using general Web search engines, the following specialized philanthropic Web sites contain profiles, articles, and links related to funders (last accessed October 2009):

www.grants.gov—Lists federal funding opportunities
http://grants.library.wisc.edu/federal.html
Contains information on grant programs from all kinds of U.S. government departments, agencies, and other federal entities
http://www.esri.com/grants/about/funding_resources.html—List of federal agencies
http://www.pitt.edu/~offres/proposal/propwriting/websites.html—Contains writing tips as well as a list of federal funding agencies and their instructions
http://foundationcenter.org—List of private foundations in the US
http://searchedu.com/—List of foundation and corporate grants of $75,000 or more awarded to colleges and universities since

1995. These grant listings are drawn from *The Chronicle Guide to Grants,* a compilation of grant announcements previously published in *The Chronicle of Higher Education* and *The Chronicle of Philanthropy.*

http://www.philanthropysearch.com—The *Chronicle of Philanthropy* is the leading fundraising news source

http://sciencecareers.sciencemag.org/funding—The journal *Science* and the American Association for the Advancement of Science list various funding search engines including those for international grants and fellowships.

Federal 990 tax forms: These forms are required of foundations and charities with gross receipts over $25,000 annually. The 990 tax forms provide information on funder assets, expenditures, grant making, and board members. Guidestar, Grantsmart, the Foundation Center, and many state government Web sites have begun developing online archives of 990 forms:

Guidestar: **www.guidestar.com**

Grantsmart: **www.grantsmart.com**

Additional sources include the following:

- Newsletters, annual reports, and Web sites of nonprofit organizations similar to your own
- Personal contacts who can help your organization approach a grant maker
- RFP announcements from large private or federal agencies detail competition instructions and funding guidelines
- Cultivated relationships between your institution and a funder. Many foundations are loyal to grantees; a modest initial gift can lead to an ongoing partnership, with larger grants awarded each year
- E-mail discussion groups for grant writers. Professionals frequently share successful proposals and exchange insight about funders through these e-mail lists.

19.9   STARTING TO WRITE A GRANT

GRANT WRITING GUIDELINE 5:

Make sure the first page is perfect.

There is no best way to write a successful proposal, but successful proposals share similar characteristics. Know that your proposal will, with luck, be read by one or two experts in your field. The program officer and many members of the panel or board that judges your proposal against others likely will not be experts. You have to write your proposal for their benefit as well.

When you are ready to write, sketch out a research "mission statement" or overall objective/goal. Then, identify the specific aims that define how you will accomplish this goal. Draft expected project outcomes in measurable terms, and draft a realistic time line. If you have not done so already, conduct appropriate preliminary studies. For young scientists, it is particularly important to see if and how this piece of research fits into their overall plans.

The first page of any proposal is the only page that may get read by the funder or reviewers. Thus, pay particular attention when outlining, writing, and revising this page. This page usually contains the executive summary or the abstract and specific aims. These have to be perfect. The summary should act as a stand-alone summary of the entire proposal. Every word in it will count and should therefore be weighed carefully.

Grant Writing

The actual writing part of a proposal is not easy. It requires skill and experience, especially in style and composition of technical documents. If you do not have these skills, educate yourself or existing staff in proposal preparation. You can also consider hiring a grant writing specialist or a fundraising consultant. If you hire a writer or consultant, look for a scientific grant writer or editor. General grant writers might be able to provide input in style and grammar but often lack the analytical mind and the scientific background to aid in overall composition, content, and structure of scientific grants.

Attending a grant writing class, using this book, or online Web sources can help beginners learn the basics of grant writing and allow more seasoned researchers to improve their writing styles. Of course, any designated staff person should already know the fundamentals of clear writing, have an analytical mind and scientific background, and be able to pay excruciating attention to detail. Above all, you need to be concerned that your proposal is well organized.

Online resources that discuss the standard components of proposals or offer courses on grant writing include the following:

> http://foundationcenter.org/getstarted/learnabout/proposalwriting.html—The Foundations Center site offers online advice and training courses on grant writing.
> http://www.jcdowning.org/resources/generalguide.htm—The Downing Foundation Guide for Grant seekers offers explanation of grant writing's basic principles.
> http://www.mcf.org/mcf/grant/writing.htm—Writing a Successful Proposal: These tips from the Minnesota Council on Foundations offer answers to common questions for proposal writing and submission.
> http://www.silcom.com/~paladin/promaster.html—Paladin Group on Writing Proposals outlines the standard components of a full proposal, including the budget and attachments.

http://www.nsf.gov/pubs/1998/nsf9891/nsf9891.htm—The
National Science Foundation (NSF) has a valuable guide on how
to write proposals for the NSF

19.10 INTERACTING WITH THE FUNDER

Just as it is imperative to create a strong case to sell programs, it is equally essential that you establish and maximize a close relationship with funders and potential funders. The cultivation process often begins when you call a grant maker to request current guidelines. Culture this relationship by making phone calls if warranted, sending reports, Christmas cards, and invitations to visit your facilities. The cultivation of a funder should never stop. Personal contact with a funder increases the likelihood of success. Be aware that it may take years of cultivation before some funding agencies make a significant contribution to your work.

Maintaining good relations involves good stewardship. Writing personalized cover letters and thank you letters or e-mails is an absolute must. Mail a thank you note, even if the application is not funded. Express your appreciation for the funder's time during the review process and your admiration for the funder's philanthropic priorities. If funding is not approved, consider a courteous phone call or send a polite letter requesting reviewer comments and suggestions for future submissions. Aside from good stewardship, report key results and provide timely progress reports, as these are often used to justify grant programs or to obtain renewal grants, particularly for federal funders. In addition, if you are an established researcher, you may also volunteer some time for consultation or for reviewing of other people's grants for a foundation.

When you receive a positive response to your letter of inquiry and an invitation to submit a full proposal, consider setting up a meeting or telephone conference with a program officer at the foundation. Discussion with the foundation can lead to excellent feedback on such points as time lines, evaluation expectations, and budget items. You may even get feedback on preliminary drafts of your proposal.

SUMMARY

GRANT WRITING GUIDELINES:
1. Do not be afraid to contact funding agencies.
2. Write a letter of inquiry to private a foundation or corporation before preparing or mailing out a proposal.
3. Obtain and strictly follow proposal guidelines.
4. Consult administrators (department chair or dean and proposal coordinators).
5. Make sure the first page is perfect.

CHAPTER 20

Letters of Inquiry and Preproposals

A letter of inquiry (LOI; sometimes also called letter of intent) is a short proposal in letter form. For foundations, LOIs are a quick way to screen potential candidates for funding. For you, an LOI is a way to get an invitation from the foundation to submit a complete proposal. LOIs may be used instead of full grant submissions by some funders and as preproposals or screening devices by others. You should carefully review the application instructions to find out if an LOI or preproposal is required and what form it should take. Your LOI can make or break your relationship with a foundation.

20.1   GENERAL

LOI GUIDELINE 1:

Scrutinize every sentence for detail, clarity,
and conciseness.

The letter of inquiry is crucially important to securing funding for your project. It is the most critical step to get one foot inside the door.

Before composing an LOI, you must research the foundation's priorities first. Your letter must establish a connection between your project's goals and the foundation's philanthropic interests. Respect the funding source's stated preferences for geographic region, type of grant, and program areas.

An LOI is much shorter than a full proposal—no more than two to three pages plus the budget. Yet an effective LOI is often more difficult to write than a full proposal. Although the LOI is a miniproposal, do not just chop down your proposal to fit onto three pages. Every single word in your LOI or preproposal needs to be weighed and should be important. There are two secrets to a successful letter of inquiry: condense, condense, condense and edit, edit, edit. With only two to three pages of text, each sentence must be scrutinized when editing. Focus on detail, clarity, and conciseness. The LOI must succinctly but thoroughly present the need or problem, the proposed solution, and your organization's qualifications for implementing that solution.

When you need to prepare an LOI or preproposal, consider involving the help of a university representative who maintains liaisons with the funder, particularly someone trained in dealing with foundations and corporations. These professionals are well qualified to provide advice in approaching and coordinating activities with these organizations. They can also assist with budgetary concerns and with the actual writing and editing of proposals and letters.

Check submission requirements. Be aware that many funders ask for electronic submission; others require you to send your LOI or preproposal by regular mail. To submit your LOI by regular mail, address it to the appropriate contact person at the funding agency. Know that a rejection notification is not always sent out to inform you that your project is unsuitable.

20.2   CONTENT AND ORGANIZATION

LOI GUIDELINE 2:

Follow the funder's guidelines EXACLTY.

LOI GUIDELINE 3:

Adjust the level of writing to the review board.

If the foundation has published guidelines for an LOI or preproposal, follow them EXACTLY. Sometimes, LOIs or proposals are rejected simply because they do not conform to the required format specified by the funder's guidelines. Although the content and organization of an LOI can vary considerably from funder to funder, the primary challenge is to write clearly and concisely.

If an outline for the LOI or preproposal is not provided by the foundation, the following structural elements offer a starting place:

Abstract
Introduction/Background
Statement of Need
Objective and Specific Aims
Strategy and Goals
Leadership and Organization
Budget
Significance/Impact

Note that each of these elements does not contain more than one or two paragraphs, as the LOI has to be very short. Also, the sequence of these elements is somewhat flexible depending on your project and organization. For example, instead of a summary or abstract, you may also consider stating your objective or hypothesis first. Follow the objective with specific aims (optional) and then provide a rationale, the approach, and an impact statement. Note that you often do not have to include citations or a list of references.

You may write the LOI separately from an accompanying cover letter or as part of the letter. If you are writing a preproposal, it should always be separate from the cover letter. Label the LOI and preproposal as such, and include a title. It is helpful if you make it plain that you are submitting an LOI or preproposal right from the start so the funder cannot miss it. In addition, in the header for each page, indicate your institution, your last name, and the date.

Adjust the level of writing to the foundation and the review board. Include facts, concrete verbs, and sentences that show action, but do not lecture the reader. Include an explanation of the issue you are addressing and how you will do it. Keep the foundation's interests in mind. You need to address these to sell your idea. Above all, convey confidence in the project and in your ability to carry out the work.

Pay special attention to verbs and their tense. Use the future tense *will* to describe anticipated project outcomes.

Consider the following example:

 Example 20-1   If we receive the grant, then we <u>could</u> double our research space.

The conditional sentence structure of Example 20–1 is less authoritative and less certain than the definitive statement shown in the revised example:

 Revised Example 20-1   An X Foundation grant **will** fund 5,000 square feet of a new DNA sequencing center.

Note that using first person (*I* or *we*) and referring to your program as "our program," or your aims as "our aims" rather than "the program" or "the aims" often seems more natural and compassionate.

Generally, your LOI should contain the answers to the following questions:

> Why this project?
> Why you?
> Why at your institution?
> Why this sponsor?
> Why now?

Format your LOI carefully. Rather than filling each page with long blocks of text, aim for section breaks that catch the eye. Consider offering at least one format break per page and perhaps a bold heading or a short list of bulleted items. However, a LOI should be recognizable as a letter—do not include any figures, tables, photographs, or diagrams. Also, do not use color printing, and do not include a cover sheet. Preproposals, on the other hand, may contain an illustration or two.

20.3 ABSTRACT/OVERVIEW

LOI GUIDELINE 4:
Include the following elements in the abstract:
> Background/General context
> Statement of need/Problem
> Objective (and specific aims—optional)
> Approach
> (Funding request)
> Impact/Significance

Summarize the proposal in the first paragraph. The abstract serves as the executive summary for the LOI and should include the name of your organization and a description of the project. The abstract may also contain the qualifications of project staff, a brief description of the experimental approach, and possibly a timetable and the amount requested. This summary is sometimes also called an *overview* and may be combined with the introduction.

Pay special attention to power positions in your abstract. Put the most effort into writing the first sentence. Write and rewrite it. Start by providing general context in a brief sentence. Then state the problem or need, followed by the overall objective of the project, the specific aims (optional), and your proposed experimental approach or strategy. You may request a specific dollar amount for the proposed project and justify it. Conclude by talking about the expected outcomes and the impact of the proposed work. Remember to focus on philanthropic needs rather than on your or

on institutional needs. Even if you seek funding for equipment or laboratory space, emphasize how these will impact services for people in need or
allow you to gain new knowledge in the field more rapidly.

A well-written abstract is shown in Example 20-2. This abstract contains all required elements and starts with a strong first sentence. It is
logically constructed and presents the individual elements in the order
most reviewers expect to find them:

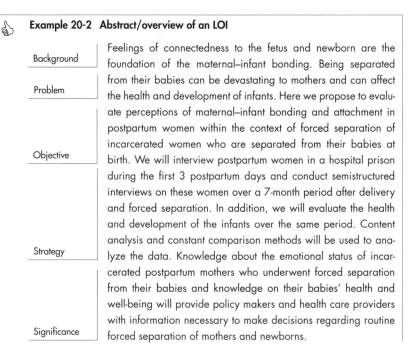

Example 20-2 Abstract/overview of an LOI

| Background | Feelings of connectedness to the fetus and newborn are the foundation of the maternal–infant bonding. Being separated |

| Problem | from their babies can be devastating to mothers and can affect the health and development of infants. Here we propose to evalu- |

| Objective | ate perceptions of maternal–infant bonding and attachment in postpartum women within the context of forced separation of incarcerated women who are separated from their babies at birth. We will interview postpartum women in a hospital prison during the first 3 postpartum days and conduct semistructured interviews on these women over a 7-month period after delivery and forced separation. In addition, we will evaluate the health and development of the infants over the same period. Content |

| Strategy | analysis and constant comparison methods will be used to ana- lyze the data. Knowledge about the emotional status of incar- cerated postpartum mothers who underwent forced separation from their babies and knowledge on their babies' health and well-being will provide policy makers and health care providers with information necessary to make decisions regarding routine |

| Significance | forced separation of mothers and newborns. |

20.4 INTRODUCTION/BACKGROUND

LOI GUIDELINE 5:

Focus on highlights in the Introduction/
Background portion.

Follow the abstract/overview paragraph with one to two paragraphs of
background information, which may also include important preliminary results. However, remember that a successful LOI or preproposal
results in an invitation to submit a longer proposal. Save the extensive
narrative history for that document and just focus on the highlights
here as shown in the following example. This example starts with general background to provide context and then lists the problem and offers
the hint of a possible solution by generally reporting on some preliminary results:

Example 20-3   Introduction of an LOI

	Multi-drug resistant (MDR) TB is of increasing concern in countries with a high burden of HIV (1). Recent reports of MDR TB isolates resistant to several second-line drugs have magnified concerns about the continued transmission and effectiveness of treatment of MDR TB. Furthermore, the worldwide emergence of TB resist-
Background	ant to several second-line drugs (2), coined "extensively" drug-resistant (XDR) TB (3) has hampered advances in HIV treatment.
Overall Problem	Together, MDR TB and XDR TB account for an increasing proportion of TB cases globally (3). To decrease morbidity and mortality from drug-resistant TB, the evaluation of new laboratory methods and clinical protocols has been prioritized (2,3). However, epi-
Specific Problem	demic models of TB transmission have yet to be integrated with operations research methods to determine the most effective con-
Preliminary data for a solution	trol strategies for drug-resistant TB in resource-limited settings.
	(Alison Galvani, proposal to a federal agency, modified)

20.5   STATEMENT OF NEED

LOI GUIDELINE 6:

Do not overuse statistics.

The statement of need is an essential element of the letter of inquiry and must convince the reviewers that there is an important need that can be met by your project. The statement of need may include, if appropriate, a description of the target population and geographical area, appropriate statistical data in abbreviated form, and examples.

Example 20-4   Statement of need

Salamander limb deformities detected in California are of concern to public health experts. It is important to explore the underlying causes of these deformities to establish preventive measures. An urgently needed next step in the protection of public health is the examination of the X hypothesis. Investigations of this hypothesis may not only determine the causative agent of the observed limb deformities but also forgo the danger of potential developmental problems in humans.

If you provide statistics to indicate the need for funding, do not overuse them. Most funders are well aware of such numbers if they have a particular interest in your topic. For foundations that fund projects throughout the United States, focus on how local statistics compare to national statistics.

20.6   OBJECTIVE AND SPECIFIC AIMS

LOI GUIDELINE 7:

Clearly identify the overall objective.

The overall objective of your LOI explains clearly what your long-term goal or mission is to accomplish your proposed work. It is the most important statement of your LOI. Ensure that funders can immediately identify this statement by either putting it in italics or boldface or by placing it into a separate section labeled, for example, "Objective" or "Goal." Signal this statement ("Our overall objective is ..."; "The goal of this study is to ..."). The overall objective can be subdivided into specific (short-term) aims, which can be listed within the abstract, with the objective, or as a separate section. The objective should follow logically from the statement of need or the problem. Two sample objective statements are shown in Example 20-5.

 Example 20-5   Objective

    a   The objective of this proposal is to develop an entirely new approach to X through Y and Z.

    b   We propose to use a novel class of experiments to test nearly every theoretical prediction of T-violation that can account for the matter–antimatter asymmetry in the universe.

The specific aims link the overall objective to your research plan. Between two to five specific aims are standard, and these are often listed just after the objective as shown in Example 20-6:

 Example 20-6   Objective and Specific Aims

The proposed study aims to develop treatment strategies to minimize the emergence, amplification, and transmission of drug-resistant tuberculosis in the high-burden setting of Uganda. Specifically, we will (a) perform a meta-analysis of treatment success from five hospitals in the capital and (b) predict best treatment options through mathematical modeling.

20.7   STRATEGY AND GOALS

LOI GUIDELINE 8:

Avoid excessive details in the experimental approach portion.

The strategy/approach paragraph should describe briefly how you plan to address your objective. Present a clear, logical, and achievable solution to the stated need. Describe the approach briefly, summarizing major activities. This section will be presented in far greater detail in a full proposal. A brief strategy/approach section is shown in the following example:

Example 20-7 Strategy/approach
Experimental Methodology.

To block parasite transmission in mosquitoes, we will express *Plasmodium* products in the *Anopheles* commensal symbionts using expression system A. This system enables secretion of the expressed effector molecules to interfere with parasite viability in mosquitoes. We will also raise antibodies against this protein to demonstrate increased transmission in its absence by providing it in the infectious blood meal of mosquitoes.

In addition, we will evaluate the role of *Anopheles* symbionts on host reproductive physiology. We will focus specifically on oocyte and larval development. Aside from microscopic observations of the status of progeny development in the fertilized normal and aposymbiotic females, we will evaluate the expression of molecular markers associated with important genes in these processes. Our goal is to understand at the molecular level the basis of reproductive sterility that arises in the absence of the symbiotic flora. If the basis of this sterility is nutritional, supplementation of the female diet can rescue fertility and allow for the paratransgenic mosquitoes developed in the first year of the project to be fertile.

20.8   LEADERSHIP AND ORGANIZATION

LOI GUIDELINE 9:

Describe any expertise briefly, including
location if needed.

Briefly describe the project's leadership and their expertise as well as that of other key personnel. The description of your team should be concise and focus on the ability to meet the stated need. Depending on the nature of your proposed project, you might provide a few biographical sentences about the principal investigator(s), research scientists, postdoctoral fellows, graduate students, or key board members to add credibility. You may also want to provide a very brief description of your organization, and explain why your institution is the perfect place to conduct the proposed experiments. In addition, consider listing partners and collaborators if needed.

Following are some examples of personnel descriptions:

Example 20-8   Short bio/expertise of investigator

Dr. Paul Doe, Arthus Foundation Investigator and Chairman of the Department of Pathology, is a renowned expert on infectious diseases and has been instrumental in advancing genomics at Albert

Einstein University. By coupling characterization of hundreds of pathogens from around the world with genomics, his group has mapped over 50 genes and has identified functional mutations in 28 of these. His findings have provided new insight into the mechanisms underlying HIV, malaria, West Nile virus, and the flu. These studies have identified new targets and pathways for development of novel therapeutic approaches to these diseases.

 Example 20-9   Description of key personnel

a. Key Personnel: The proposed project will be led by Martin Brown (Ph.D., Associate Professor of Geology and Geophysics) and Ron Robinson (Ph.D., Professor of Applied Physics). Martin Brown's contributions to the project will build on his groundbreaking work in XYZ. Ron Robinson brings expertise in the design of Y. Both principal investigators have been collaborating on Z for a number of years.

b. Key personnel: Key personnel on the project include John Smith, a postdoctoral fellow working with Dr. Meng. Dr. Smith received his Ph.D. and postdoctoral training in the study of the molecular systematics of medical microbes. He has expertise in a number of analytical methods including bioinformatics tools and will apply his skills to epidemiological modeling in the proposed research.

20.9   BUDGET

LOI GUIDELINE 10:
State what the funding will be used for.

Briefly summarize the proposed use of grant monies. Request a specific dollar amount for the proposed project, and justify what funds will be used for and how they will be distributed. Be explicit about committed funding and pending proposals if needed. Remember that foundations sometimes consult other philanthropic entities before making grant decisions. You may also want to point out the likelihood or difficulty of obtaining funding from other sources.

The next three examples provide sample wording for a budget section of an LOI or preproposal:

 Example 20-10   Funding request

The funding requested for this study will cover the salary of a full-time research scientist who will perform all the proposed experiments under the guidance of the Principal Investigator. The amount requested will also pay for all the reagents such as monoclonal antibodies, serum, plastic ware for tissue culture, use of the flow cytometry facility, and chemical reagents.

Example 20-11 Funding request

> We are requesting $150,000 to carry out the proposed study. This amount will cover the salary of a postdoctoral fellow, equipment, and travel to South America to collect data and specimens.

Example 20-12 Budget

> **Budget:** The total project cost is $xxx,xxx, of which $yyy,yyy is requested from the ABC Foundation and $zzz,zzz is institutional support. The Foundation funds requested are allocated as follows: Personnel, xx%; Equipment, xx%; and Operations, xx%.

20.10 IMPACT AND SIGNIFICANCE

LOI GUIDELINE 11:

End the LOI or preproposal with a broad impact statement.

In the last paragraph of your LOI, state the impact and significance of the project if funding is provided as requested. Focus on how human beings will benefit rather than on construction, staff, or institutional issues. You may also mention evaluation plans such as measurable objectives and reporting to stakeholders. Examples 20-13 and 20-14 show how typical impact statements are worded. Pay particular attention to the last sentence. It should relate the importance of your study as broadly as possible:

Example 20-13 Impact statement

> Understanding the interaction between HCV virions and the extracellular milieu is critical for combating this virus. These interactions can be illuminated by defining the biochemical composition of the virus particles, elucidating the mechanism of their assembly and maturation, and determining the differences in the entry of high-infectivity, low-density particles to those of low-infectivity, high-density particles. Insights into these mechanisms will contribute to developing new drugs and vaccines to combat this major human pathogen.

Example 20-14 Impact statement

> Novel, bioengineered crops will allow farmers to grow varieties of plants with enhanced productivity, quality, and improved ability to adapt and survive when faced with adverse environmental conditions. This need is particularly urgent in light of the global food crisis, which has put almost a billion people at risk of hunger and malnutrition. By designing new technology and developing varieties that allow more sustainable farming, the project aims to provide innovative and sustainable solutions to the problems faced by crop growers worldwide.

20.11 COVER LETTER

LOI GUIDELINE 12:

Send a cover letter together with your LOI
or preproposal.

In a separate cover letter to your LOI or preproposal, indicate that you are sending an LOI. Do so in the first paragraph. In the second paragraph, describe your project very briefly—no more than one paragraph—indicating the need, objective, strategy, and impact.

The final paragraph should indicate the next step, identify the contact person, and summarize the project's importance. Unless the foundation's guidelines state that attachments should be included with inquiry letters, resist the temptation to slip newsletters and other promotional material into the envelope. Such material is more likely to be appreciated during the full proposal phase or for a site visit. Trust that the presentation of your project and the clarity of your letter will entice the foundation into requesting additional information. Above all, be courteous and professional.

 Example 20-15   Cover letter

Dear Dr. Miller:

We would like to submit the accompanying proposal on *Speciation Progression due to Global Warming* for the ABC Foundation's consideration. The proposed project is highly innovative and very ambitious, and we are very confident that it will bring about far-reaching advances in our understanding of and approach to climate changes.

Under the leadership of Professor X, the project will be overseen by a 10-member scientific advisory board, which will.... Although Y University will provide expertise and allocate state-of-the-art equipment and laboratory facilities, additional support is vitally needed, and the boldness of this project precludes funding from government sources. We are therefore turning to the ABC Foundation and its distinguished record of investing in research that expands the frontiers of knowledge.

We look forward to hearing the results of your deliberations and to future opportunities to collaborate in pursuit of our mutual goals.

With best regards,

20.12 VERBAL PROPOSALS

LOI GUIDELINE 13:

Give verbal proposals a similar outline as
LOIs or preproposals.

For corporations, it is not uncommon to give a verbal proposal in the form of a PowerPoint presentation during a meeting or site visit. Such presentations provide great opportunities for questions and answers with the potential sponsor. Often a concept paper or LOI is prepared together with the verbal proposal, but a formal proposal may only be submitted if the sponsor is interested. Know that an invitation for a formal proposal does not guarantee funding.

Construct any verbal proposal using the same overall outline of an LOI or preproposal. Do not overcrowd your slides. See also Chapter 28 for more information on how to prepare and deliver a clear presentation.

20.13 LOI OUTLINES

Following are two general overall outlines for an LOI or preproposal. Use these only if the funder does not provide any guidelines.

Outline 1:   Abstract/Overview
  - Context/background (optional)
  - Statement of need
  - Long-term mission/objective (and specific aims—optional)
  - Strategy
  - Significance/impact of initiative

Introduction/Background
  - Context and Problem
  - Selected accomplishments/findings/activities/events to date

Statement of Need
  - Problem or unknown

Objective and Specific Aims/Goals
  - Planned activities

Strategy/Approach
  - Methodology
  - Timetable (optional)

Leadership and Organization
  - Key personnel profiles
  - Description of your organization (optional)

Budget (Operations, Personnel, Equipment, Travel, etc.)
  - Requests
  - Other commitments (other funding or institutional commitments)

Impact and Significance
  - Expected outcomes
  - Significance of outcomes within your field, outside your field, and to humanity

Outline 2:   Abstract (optional)
- Context/background (optional)
- Problem statement and/or statement of need
- Long-term mission/objective (specific aims—optional)
- Strategy
- Significance/impact of initiative

Goals/Overall Objective

Specific Aims

Rationale
- Context/background (optional)
- Preliminary findings
- Transition on how preliminary findings will be used in the proposal/LOI

Experimental Design
- Explanation of experimental approach—often subdivided by aims (Include: rationale, experimental approach, expected outcomes)

Leadership and Organization (as required)
- Key personnel profiles
- Description of your organization

Budget (Operations, Personnel, Equipment, Travel, etc.; as required)
- Requests
- Other commitments (other funding or institutional commitments)

Other (funding justification, relevance to foundations, etc.)

Impact and Significance (can also add expected outcome here)
- Within your field
- Outside your field
- To humanity

20.14 REVISING AN LOI/PREPROPOSAL

When you have finished writing the LOI/preproposal (or if you are asked to edit these sections for a colleague), you can use the following checklist to "dissect" the sections systematically:

☐ 1. Is the proposed topic original?

☐ 2. Are all the components there? To ensure that all necessary components are present, in the margins of the LOI clearly mark the following:

	Abstract/Overview
	Introduction/Background
	Statement of Need
	Objective and Specific Aims

	Strategy and Goals
	Leadership and Organization
	Budget/Fundraising
	Impact/Significance

☐ 3. Is the abstract kept short and within the set limits?

☐ 4. Does the abstract contain
  - Context/background (optional)
  - Statement of need
  - Objective
  - Strategy
  - Significance/impact of initiative

☐ 5. Is the overall objective stated precisely?

☐ 6. Does the Introduction contain the following elements:
  - Context
  - Selected accomplishments/findings/activities/preliminary studies
  - Organization (optional)

☐ 7. Did you follow the funder's instructions on how to write the LOI?

☐ 8. Is your proposed work doable in the time frame given?

☐ 9. Revise for style and composition using the writing principles of this book:

  ☐ a. Are paragraphs consistent? (Chapter 6, Section 6.2)
  ☐ b. Are paragraphs cohesive? (Chapter 6, Section 6.3)
  ☐ c. Are key terms consistent? (Chapter 6, Section 6.3)
  ☐ d. Are key terms linked? (Chapter 6, Section 6.3)
  ☐ e. Are transitions used and do they make sense? (Chapter 6, Section 6.3)
  ☐ f. Is the action in the verbs? Are nominalizations avoided? (Chapter 4, Section 4.6)
  ☐ g. Did you vary sentence length and use one idea per sentence? (Chapter 4, Section 4.5)
  ☐ h. Are lists parallel? (Chapter 4, Section 4.9)
  ☐ i. Are comparisons written correctly? (Chapter 4, Sections 4.9 and 4.10)
  ☐ j. Have noun clusters been resolved? (Chapter 4, Section 4.7)
  ☐ k. Has word location been considered? (Verb following subject immediately? Old, short information at the beginning of the sentence? New, long information at the end of the sentence?) (Chapter 3, Section 3.1)
  ☐ l. Have grammar and technical style been considered? (person, voice, tense, pronouns, prepositions, articles) (Chapter 4, Sections 4.1–4.4)
  ☐ m. Are words and phrases precise? (Chapter 2, Sections 2.2 and 2.3)
  ☐ n. Are nontechnical words and phrases simple? (Chapter 2, Section 2.2)
  ☐ o. Have unnecessary terms (redundancies, jargon) been reduced? (Chapter 2, Section 2.4)
  ☐ p. Have spelling and punctuation been checked? (Chapter 4, Section 4.11)

SUMMARY

LOI GUIDELINES

1. Scrutinize every sentence for detail, clarity, and conciseness.
2. Follow the funder's guidelines EXACLTY.
3. Adjust the level of writing to the review board.
4. Include the following elements in the abstract:
   - Background/General context
   - Statement of need/Problem
   - Objective (and specific aims—optional)
   - Approach
   - (Funding request)
   - Impact/Significance
5. Focus on highlights in the Introduction/Background portion.
6. Do not overuse statistics.
7. Clearly identify the overall objective.
8. Avoid excessive details in the experimental approach portion.
9. Describe any expertise briefly, including location if needed.
10. State what the funding will be used for.
11. End the LOI or preproposal with a broad impact statement.
12. Send a cover letter together with your LOI or preproposal.
13. Give verbal proposals a similar outline as LOIs or preproposals.

PROBLEMS

Problem 20-1

Write an abstract for an LOI on comprehensive measurements of CO_2 levels in ponds, lakes, rivers, and the ocean in the American Northeast. In this abstract, open with an important sentence or two and then state the problem/need, your overall objective, how you are proposing to solve it specifically, and the impact/significance of the proposed study. Feel free to invent, look up, or search for information on the Internet if needed.

Problem 20-2

Rewrite the following structured abstract to an abstract that would fit for an LOI:

Background: Cardiovirus is a common cause of gastroenteritis. For routine vaccination of Chinese infants, a new cardiovirus vaccine has been recommended.

Objective: To evaluate the impact and cost-effectiveness of the Chinese cardiovirus vaccine program using a dynamic model of cardiovirus transmission.

Expected outcome: Our analysis will indicate the impact and effectiveness of a rotavirus vaccination program.

Significance: Findings can be used to inform policy makers to prevent rotavirus infection.

Problem 20-3
Identify all components of the following LOI abstract/overview (background, need/problem, objective, specific aims, approach, impact):

Abstract

The role that soils play in mediating global biogeochemical processes is a significant area of uncertainty in ecosystem ecology. One of the main reasons for this uncertainty is that we have a limited understanding of belowground microbial community structure and how this structure is linked to soil processes. Building upon established theory in soil microbial ecology and ecosystem ecology, we predict that the structure of belowground microbial communities will be a key driver of carbon and nutrient dynamics in terrestrial ecosystems. We propose to test and develop the established theories by combining state-of-the-art DNA-based techniques for microbial community analysis together with stable isotope tracer techniques. By doing so, we expect to advance our conceptual and practical understanding of the fundamental linkages between soil microbial community structure and ecosystem-level carbon and nutrient dynamics.

    (Mark Bradford and Noah Frierer, proposal to private foundation)

Problem 20-4
Point out what the problems are with the following impact statement.

Impact/Significance.
Funding from the Foundation will not only provide for my postdoctoral fellow but also will result in the publication of two papers over the funding period.

Problem 20-5
Write a paragraph on personnel in which you will be the principal investigator leading the project.

CHAPTER 21

Abstract and Specific Aims

21.1 OVERALL

Most proposals open with this section. The section may be written as one paragraph or as two separate parts, which together are about one page. The Abstract and Specific Aims section should be self-contained, provide a broad overview of the proposal, and be general in nature.

Most reviewers will read your proposal only if your Abstract and Specific Aims section interests them. In fact, the Abstract and Specific Aims might be the only thing some reviewers read. It is therefore absolutely essential that this section is compelling and technically flawless.

21.2 ABSTRACT

Content

PROPOSAL ABSTRACT GUIDELINE 1:
In the Abstract and Specific Aims section include

Abstract:
Brief background
Unknown or problem
Objective
Preliminary results—if relevant

General strategy (may include specific aims if
not listed in separate part)
Expected Outcomes—optional
Significance/impact
Specific Aims

PROPOSAL ABSTRACT GUIDELINE 2:
The first sentences should be informative, short, interesting, and provide broad background.

The Abstract articulates the highlights from each section of the proposal. It includes all of the main information covered in the proposal (background, unknown or problem, overall objective, general strategy, and significance/impact) in a single paragraph. Open the Abstract with a short portion of background information that provides broad context for the reviewers. Follow this information with a brief statement of what is unknown or a problem and then with the overall objective of the proposal. After the overall objective, consider providing some preliminary results if needed and then describe your strategy in general terms. End the Abstract by adding a sentence or two about the overall impact of the proposed research. After the Abstract, add a lead-in sentence to introduce the specific aims.

Ideally, do not make the Abstract longer than one to two paragraphs. Together with the Specific aims, the Abstract is the most important section of your proposal and should therefore be placed in a power position on the first page of your proposal (or right after the title page if one is required). The first sentence of the Abstract deserves particular attention. Next to the title, it will be the first sentence the reviewers will read. Therefore, the first sentence needs to be perfect. It should provide general background, be informative and catchy, and be interesting to the reader at the same time. Do not state a cliché, however. Write and rewrite this sentence, and remember that shorter sentences are more powerful than long sentences.

The Abstract must be written such that it can stand on its own, without the detailed narrative. It must be concise, informative, and complete. Write the section with nonspecialists in mind, that is, provide a broad context for the reviewers. Avoid abbreviations, unfamiliar terms, and citations. Do not include or refer to tables or figures. Do not include any references, but be sure to include all the important key terms found in the title because the Abstract and the title have to correspond to each other. Include your specific aims in the Abstract unless they are repeated in a separate section shortly after the Abstract. In all cases, include your overall objective or goal of the project. Be aware of and address how your project furthers the goals of the sponsor.

The Central Point

PROPOSAL ABSTRACT GUIDELINE 3:
State the objective of your proposal precisely.

The most important statement in your proposal is the overall objective or goal of the work. To ensure that reviewers and funders immediately find this statement, consider highlighting it using boldface and/or italics. This objective provides an overview of the entire proposal, and every paragraph and sentence in your proposal relates to it. The overall objective can be subdivided into specific aims, which will be listed separately either as a separate section or within the Abstract or as a separate part of it.

Example 21-1   The proposed study will assess the benefits and costs of butterfly wing pattern designs to gain insight into broader questions of biodiversity in these and other species.

Example 21-2   Our objective is to treat rabies infections through vaccine vectors of recombinant adeno virus.

Example 21-3   The goal of our study is to map the fitness landscape of drug-resistant *Staphylococcus aureus* mutations in a high-burden area of the United States to facilitate the evolution of MRSA infections toward reduced transmissibility.

The overall objective and specific aims should follow logically from the previous statements of what is known or believed and what is still unknown or problematic. Thus, the objective and specific aims should state the purpose of the proposal one would expect after reading about what is unknown or problematic.

Types of Abstracts for Proposals

PROPOSAL ABSTRACT GUIDELINE 4:
Distinguish between technical abstracts and those written for a lay audience.

Proposals may contain two types of abstracts: technical abstracts and those written for a lay audience. Some institutions may require you to provide both. The format and length of the Abstract is generally specified by the organization to which you are applying. Usually, abstracts contain between 100 and 250 words. You should not exceed the maximum allowed, but you may certainly summarize your proposal in fewer words.

The content of the Abstract is the same for technical and lay abstracts. It is the level of sophistication of the writing that distinguishes the two. Use a technical abstract for the typical proposal to federal agencies. Abstracts for private foundations and corporations can be technical, but most are geared more toward an educated lay audience, especially if you know that the decision makers or members on the board are nonscientists or scientists outside your field of expertise. Lay abstracts will usually be presented to board members or may be posted on Web sites and therefore need to be widely understandable. An example of a technical versus a lay abstract is shown in Examples 21-4a and 21-4b.

 Example 21-4 a   Technical abstract with integrated specific aims

Background	Most people over the age of 35 years exhibit emphysema, a major manifestation of chronic obstructive pulmonary disease (COPD). Cigarette smoke, pollutants, and gender are thought to be important determinants of the severity of the disorder. Curative therapies or reliable diagnostic biomarkers do not yet exist for emphysema/COPD. **Our objective is to identify new diagnostic or therapeutic targets for emphysema by applying our recent discovery of novel molecules in mouse models to humans.** Aging or cigarette smoke-exposed mice exhibit lung changes that partially mimic human emphysema, and mice deficient in toll-like receptor Z, a canonical receptor for lipopolysaccharides, exhibit an accelerated form of spontaneous, age-induced emphysema. We hypothesize that the synergistic or additive effects of age and smoking on Z function in susceptible individuals may explain the pathogenesis and temporal characteristics of emphysema. We have identified two novel molecules regulated by Z, an oxidant-generating enzyme (X) and a protease (Y), and implicated both in the pathogenesis of emphysema in mice. This proposal will directly build on and expand our pilot findings. Specifically, we will first confirm the role of Z, X, and Y in the pathogenesis of age-induced and cigarette smoke-induced emphysema and validate their roles as therapeutic targets. Subsequently, we will analyze molecular interactions of these molecules in young and aged people in relation to cigarette smoke exposure, gender, and emphysema/COPD. These studies will provide important insights into the pathophysiologic mechanisms of emphysema, ultimately leading to the identification of novel targets for diagnostic or therapeutic interventions.
Problem	
Objective	
Preliminary results and hypothesis	
Strategy/ Specific Aims	
Significance	

(Patty Lee, proposal to private foundation; modified)

 Example 21-4 b   Lay abstract with integrated specific aims

	Emphysema is a major subset of chronic obstructive lung disease, predicted to reach epidemic proportions by 2020. The condition develops in most people over the age of 35 and can lead to the loss of oxygen exchange, lung enlargement, and, if severe, complete respiratory failure. Cigarette smoke, pollutants,
Background	and gender are thought to affect the severity of the disorder.

Problem

Objective

Preliminary
results

Strategy

Significance

Disease-altering treatments or reliable diagnostic features that can be used to measure the progress of the disease have not yet been determined. Therefore, **we propose to identify new diagnostic or therapeutic targets for emphysema by exploring its underlying mechanisms.** Using genetically altered mouse models, we have recently discovered two novel molecules involved in the development of lung emphysema, X and Y. We found that a substantial increase in these molecules destroys lung tissue, resulting in emphysema. Interestingly, both molecules are controlled by a specific cell wall structure (receptor Z). We believe that the synergistic or additive effects of age and cigarette smoke on Z's function may explain disease development and characteristics. Analysis of the role of receptor Z, as well as those of X and Y, in age-induced and cigarette smoke-induced emphysema will provide insights into the underlying mechanisms of the disorder and may ultimately lead to the identification of novel targets for diagnostic or therapeutic interventions.

(Patty Lee, proposal to private foundation; modified)

Note the much simpler word choices and more extensive definition of terms in the lay version of the abstract. Note also that in the technical abstract shown in Example 21-4a, specific aims have been integrated ("Specifically, we will ...").

Following are additional examples of a technical proposal abstract with and without integrated specific aims:

 Example 21-5   Technical abstract—with integrated specific aims

Background

Problem

Objective and
Specific Aims

Strategy

Significance

Plate tectonics distinguishes Earth from other terrestrial planets. Plate tectonics arise due to the convection in the mantle of the Earth; and for plate tectonics to occur, some mechanism must exist to compensate temperature-dependent viscosity. However, it is not well understood what this mechanism is. In this proposal, we aim **to approach this long-standing mystery through employing a MCMC algorithm that will (1) systematically explore various sampling strategies and (2) seek the fastest forward model calculation by benchmarking competing Stokes flow solvers, including the pseudo-compressibility method.** To lay a foundation for realistic 3-D applications, all computations will be conducted in the 2-D formulation. Microsoft Windows OS will be the main platform for developing the Monte Carlo code and the flow solver as well as for analyzing and visualizing the results of MCMC simulations. Stokes flow calculations will be done with the PI's ABC server. Understanding the physics of plate-tectonic convection in the Earth's mantle will have profound implications for our understanding of the habitability of a terrestrial planet and the evolution of life.

(Jun Korenaga, proposal to private foundation; modified)

In a combined abstract/specific aims section, the objective and specific aims should be highlighted by setting them off or by boldfacing them, as they are the most important portion of this section. Other passages whose signals are often placed in boldface or italics include the strategy of the proposal and the specific aims, as shown in Example 21-6:

Example 21-6   Combined technical abstract and aims

Background	Hepatitis B virus (HBV) causes acute and chronic hepatitis as well as hepatocellular carcinoma in infected individuals. Current therapies for chronic HBV infection are only moderately effective and are limited by severe side effects and viral resistance. Thus, there remains a need for new therapies for this serious disease. The most promising approach for a new therapy, and **the objective of this proposal,** is to treat HBV infections through vaccine vectors of recombinant vesicular stomatitis virus (VSV). These vectors induce strong protective CD8 T cell and antibody responses to a variety of pathogens and are showing great promise as therapeutic vaccines. We will examine the hypothesis that recombinant VSV vectors expressing the HBV structural proteins will make effective vaccines for prophylactic and therapeutic vaccination against HBV. **Our general strategy** is to develop an effective therapeutic vaccine and/or an improved prophylactic vaccine that provides long-term immunity in a single dose. Such a vaccine would have the potential to prevent millions of cases of HBV-associated carcinoma.

Background

Problem

Objective

Background

Strategy

Significance

Lead-in sentence | To evaluate this hypothesis, we will carry out three specific aims:

Aim 1. Generation of VSV vaccine vectors. We have previously produced ZZZ .... We will generate....

Aim 2. Characterization of the immune response to VSV/HBV vectors. Our preliminary data indicates that ... We will now comprehensively measure.... We will use...and expect to achieve....

Aim 3. Determination of the efficacy of VSV. We will determine

Specific Aims | if ...

(Michael Robek, proposal to federal agency; modified)

Following are two further examples of lay abstracts. Note that these generally do not contain detailed specific aims but rather present these as a general strategy:

Example 21-7   Lay abstract

Background	Infections caused by multidrug resistant bacteria have increased markedly. New and improved antibiotics are urgently needed to combat this ever increasing number of multidrug resistant bacteria. However, only two new classes of chemical antibiotics have

Background

Problem

Objective	been approved in the past 30 years. **We propose to develop a new family of designer ABC [antibiotics] that is effective against drug-resistant bacteria found in community and hospital settings.** We will design such antibiotics using ribosomes isolated from wild type *Staphylococcus aureus* and applying structural analysis, crystallography, as well as computational chemistry techniques. This new family of ABC [antibiotics] will offer a broader spectrum of application in the field of antibiotics and reduce ABC [antibiotics] resistance.
Strategy	
Significance	

 Example 21-8   Combined lay abstract and specific aims

Background	Dyslexia is the most common learning disability in children. Children with dyslexia have difficulty translating print into the sounds of spoken language. Otherwise, such children possess normal vision and intelligence and often demonstrate strengths in creative, visual, reasoning, and problem-solving abilities. Yet the needs and strengths of these students far too often remain unrecognized, underappreciated, and unsupported throughout their educational training. **We propose to organize a conference specifically to educate educators about dyslexia.** The conference will not only offer insights into the latest cutting-edge scientific research findings but also provide a forum for discussion and presentations of successful teaching practices for students who are dyslexic. Our goal is for new knowledge and perspectives disseminated through this conference to transform educational practices for students who are dyslexic, ultimately benefitting individuals with dyslexia at all educational levels.
Problem	
Objective	
Strategy	
Significance	

The proposed conference will

- Present the latest, cutting-edge scientific research findings
- Provide a discussion forum in which specific concerns and necessary accommodations for dyslexic students are addressed
- Demonstrate optimal strategies for teaching dyslexic students
- Introduce a film on the critical role of accommodations necessary for dyslexic students

Expected outcomes (optional)

- Present a panel of successful dyslexic college students discussing their individual experiences and successful strategies

21.3   SPECIFIC AIMS

PROPOSAL ABSTRACT GUIDELINE 5:
List your Specific Aims in active, precise language.

This section is where you need to identify clearly and concisely what you plan to accomplish. You need to describe the specific questions you intend to answer. These questions or specific aims should be thematically related and fit together to clarify the problem raised.

The Specific Aims section may be a separate section by itself or it may be combined with the Abstract. Reviewers will read this section very carefully, as it serves as an orientation for the rest of the proposal and provides a detailed approach to your overall goal.

The specific questions or aims link the overall objective to your research plan. Between two to five specific aims are standard. They should be written in a list rather than in paragraph form, and they should be placed in boldface to highlight them. It is best to formulate your Specific Aims in active, precise language ("To quantify …"). Each specific aim may be followed by a brief (one paragraph) description of the proposed approach. This description typically summarizes the preliminary results or rationale, the hypothesis, and the approach to be taken. If a hypothesis statement is included, consider placing it in italics to set it apart. As a final sentence in this narrative, you may also state expected outcomes.

 Example 21-9 Hypothesis and Specific Aims

HYPOTHESIS AND SPECIFIC AIMS:

We hypothesize that solar variability in energy output affects global temperature and thus climate. We plan to achieve our goal with two distinct specific aims:

Specific Aim 1 Model the interaction between turbulence and magnetic field. This interaction is a crucial element of the models and is currently represented through reasonable, but arbitrary assumptions. The 3D code will be thoroughly tested, and solar features required for the complex and time-consuming 3D approach will be determined.

Specific Aim 2 Optimize operating modes for the satellite and develop software for on-board calculations and data analysis. Findings obtained in Aim 1 will be incorporated in modeling the properties of solar variability. To extrapolate implications of this work for the problem of global warming, we will collaborate with climate modelers. We expect to gain further insights into …

21.4   SIGNIFICANCE AND IMPACT

PROPOSAL ABSTRACT GUIDELINE 6:
State the significance of the study at the end
of the Abstract.

You should state why your proposal and the expected outcomes are important. Do so by expressing what you expect to be the significance or impact of the study. Stating an impact is particularly important at the end of the Abstract, as most readers will scan only the first page and expect to find a statement of overall significance. This statement should be broad, explaining why your study is important in your field, for the scientific community, and society at large. However, do not overstate the significance of the study, as this would be as big a vote against your proposal as omitting that information.

👍 **Example 21-10**   These studies are important because they may result in new approaches to prevent and/or treat a disease that causes high mortality worldwide.

👍 **Example 21-11**   New knowledge and perspectives gained through this work will provide educators with the critical information to implement needed teaching techniques, ultimately transforming the lives of the next generation.

👍 **Example 21-12**   The envisioned biocomputing system would dramatically expand the frontiers of computation and multidimensional architectures as well as bioelectronic interfaces and could have wide-ranging applications in diagnostics, chemical and biological sensors, and security.

Although typically brief statements are used at the end of your Abstract and Specific Aims section (or at the end of the overall proposal), a longer impact statement is required for some proposals. In others (such as for certain NSF applications), you may also need to indicate educational components and/or outreach activities or even the inclusion of underrepresented minorities in the research or its impact. An example of a longer impact statement containing an educational component is shown in Example 21-13:

👍 **Example 21-13**

Intellectual Merit: Harnessing and controlling optical force on a chip will lead to the convergence of two important fields—nanophotonics and nanomechanics. Building silicon optomechanics and exploiting optical force on a silicon platform will also bring transformative advances in both photonics and nanoelectromechanical systems (NEMS).

Broader Impact: Translating this seemingly small yet fundamentally important phenomenon into an engineering reality will significantly impact not only the sciences and engineering but also society. Optically driven nanoscale machineries, as described in this proposal, could provide a prime example for the application of fundamental science in today's highly developed, technology-driven world, leading to diverse new applications of technology in a variety of fields.

The educational component of this career proposal will address new curriculum development in nanotechnology by creating hands-on nanoscience modules. The knowledge gained through this project will be disseminated through a plurality of widely accepted multimedia platforms designed for outreach to the local high schools, area community colleges, and further to the general public.

(Hong Tang, proposal to federal agency (NSF))

21.5   APPLYING BASIC WRITING PRINCIPLES

Basic Principles

To write your Abstract and Specific Aims, follow the basic writing principles (see Chapters. 2–6). Pay particular attention to using simple words and avoiding jargon. Also, avoid noun clusters and abbreviations unless a long term occurs repeatedly in the proposal. Consider waiting to introduce an abbreviation until the background section. If you choose to use an abbreviation that is not standard, show that you are doing so by including it first as a parenthetical. In the Abstract, it is even more important than in the rest of the paper to keep sentences short, dealing with just one topic each and excluding irrelevant points. To provide clear continuity, repeat key terms, use consistent order for details, keep the same point of view in the question and the answer, and use parallel form.

Verb Tense

The basic guideline is that if a statement is still true use present tense. For completed actions and observations, use past tense. For anything that has not been done yet and you are proposing to do in the future, use future tense ("will"). Prefer active verbs.

21.6   SIGNALS FOR THE READER

PROPOSAL ABSTRACT GUIDELINE 7:

Signal the unknown, the objective, the aims,
the strategy, and the significance.

Because Abstracts are usually written as one paragraph, it helps the reader if you signal the different parts of the Abstract. Examples of signals for the Abstract are shown in Table 21-1.

Table 21-1 Signals of the Abstract and Specific Aims section

PROBLEM OR UNKNOWN	OBJECTIVE	STRATEGY	SIGNIFICANCE OR IMPACT
…has not been determined	Our objective is…	We will achieve this goal by…	X will …
…is unclear	Specifically, we will …		…is important for…
X is limited by…	Our objective is to…	Specifically, we will…	These results may play a role in…
The question remains whether….	We propose to…	…by…	Y can be used to……ultimately
	We will examine the hypothesis that…	We will…	…resulting in…
		Our general strategy is to…	…will provide insights into…

21.7   COMMON PROBLEMS

The most common problems of Abstracts include

- Omission of parts
- Excessive length
- Unrealistic aims
- Excessive interdependence of aims

Omission of Parts

PROPOSAL ABSTRACT GUIDELINE 8:

Do not omit any parts of the Abstract.

If any of the parts (known, unknown, objective, strategy, or impact) are missing or obscured in the Abstract, the reader may have to reread the Abstract several times because it is difficult to understand. The same problem arises when parts are not signaled.

Consider the following example:

 Example 21-14   Technical abstract—separate aims section

> X is a major human pathogen, which infects over 100 million people per year, leading to high morbidity and mortality. Current therapies for X are expensive, poorly tolerated, and only partially effective in controlling the pathogen and in limiting disease. Recently, we and others succeeded in establishing a system to grow X in cell culture. These systems will allow us to completely dissect the life cycle of X. Our initial characterization of cell culture-produced X indicates unusual physical properties. Understanding of X's life cycle will aid in the development of improved pharmaceuticals.

Here, the most essential part of an abstract is not stated: the objective. This omission lets the reader come away wondering what the author is proposing to do. Including all essential components of the Abstract (with their signals) is a must. Otherwise, you stand little chance of being funded, as reviewers will not have the time to search for missing answers. The revised version of the same abstract makes clear what the author is proposing. Here all essential components have been included. The objective is stated, and a more specific problem statement has also been added:

 Revised   Technical abstract—separate aims section
Example 21-14

Background	X is a major human pathogen, which infects over 100 million people per year, leading to Y and Z. Current therapies for X are expensive, poorly tolerated, and only partially effective in controlling the pathogen and in limiting disease.

Specific Problem	<u>The development of improved pharmaceuticals requires a thorough understanding of X's life cycle. Yet, until now, only limited aspects of its life cycle have been studied in the laboratory because culture systems for X did not exist.</u> Recently, we and others succeeded in establishing systems to grow X in cell culture. These systems will allow us to completely
Preliminary Results	dissect the life cycle of X. Our initial characterization of cell culture-produced X indicates unusual physical proper-
Objective	ties. **Here we propose to determine the exact life cycle of X and to study its unusual physical properties in more detail.**
Significance	Understanding of X's life cycle will aid in the development of improved pharmaceuticals.

Excessive Length

PROPOSAL ABSTRACT GUIDELINE 9:
Keep the Abstract short.

One of the most common problems for abstracts is excessive length. When you write the Abstract, consider every word carefully. If you can write your Abstract in fewer words than the maximum allowed, do so. If you find yourself in the situation in which you have to condense your Abstract, follow the suggestions below:

To condense a long Abstract (see also Chapter 6, Section 6.4)

- Omit unnecessary words and combine sentences
- Condense background
- Omit or subordinate less important information (definitions, experimental preparations, details of methods, exact data, confirmatory results, and comparisons with previous results)

Unrealistic Aims

PROPOSAL ABSTRACT GUIDELINE 10:
Be realistic when listing aims.

Your aims should be specific. They shoud not be so broad as to be unsustainable. Be realistic in what you propose to achieve. Remember, to get your next grant, you will have to prove you followed through with the plan set forth in your specific aims.

Aims should also be feasable given your technical expertise, and it should be clear that the aims are relevant and important in the field. The aims should not appear to be merely a list of experiments but rather a targeted approach.

Excessive Interdependence of Aims

PROPOSAL ABSTRACT GUIDELINE 11:

Do not make aims too interdependent.

If an aim depends on the success of an earlier aim, it becomes dubious to the reviewers and may be a justified reason for rejection. Aims that are too interdependent appear risky because if just one aim fails, the success of the overall objective is in jeopardy.

21.8   REASONS FOR REJECTION

You need to spike the interest of the reviewers with your Abstract and Specific Aims section(s). You are more likely to impress reviewers if you master the skill of writing simply, clearly, and concisely.

Besides the reasons listed above, several other common reasons for quick rejection of a proposal are

- **Lack of originality**—you must propose something new or better than what has been proposed by others.
- **Lack of context**—you need to provide a background of your work as well as its impact. Never assume that the reviewers will be sufficiently familiar.
- **Limited sample size**—Few samples in a study may not convince a reviewer of the significance of the proposal.
- **Proposed work is too ambitious**—you will leave the reviewers with the sense that you are overstating what can be achieved within a given time frame.
- **Lack of conformity**—you should follow the proposal instructions very carefully so as not to antagonize a reviewer.
- **Too many abbreviations**—this will turn off most reviewers.

21.9   REVISING THE ABSTRACT AND SPECIFIC AIMS

When you have finished writing the Abstract and Specific Aims (or if you are asked to edit these sections for a colleague), you can use the following checklist to "dissect" the sections systematically:

☐ 1. Is the proposed topic original?
☐ 2. Are all the components there? To ensure that all necessary components are present, on the margins of the Abstract and Specific Aims clearly mark

background
unknown
objective
strategy

	(expected results)
	significance
	specific aims

☐ 3. Is the Abstract kept short and within the set limits (less than one page for Abstract and Specific Aims together)?

☐ 4. Is the objective stated precisely?

☐ 5. Do all the components logically follow each other? (Is the unknown what one would expect to hear after reading about what is known? Is the objective really the goal one would expect to read after reading the unknown? Does the significance of the project really come across?)

☐ 6. Did you follow instructions on how to write the proposal?

☐ 7. Is your proposed work doable in the time frame given? Are aims not too interdependent?

☐ 8. Revise for style and composition using the writing principles of this book:

    ☐ a. Are paragraphs consistent? (Chapter 6, Section 6.2)

    ☐ b. Are paragraphs cohesive? (Chapter 6, Section 6.3)

    ☐ c. Are key terms consistent? (Chapter 6, Section 6.3)

    ☐ d. Are key terms linked? (Chapter 6, Section 6.3)

    ☐ e. Are transitions used, and do they make sense? (Chapter 6, Section 6.3)

    ☐ f. Is the action in the verbs? Are nominalizations avoided? (Chapter 4, Section 4.6)

    ☐ g. Did you vary sentence length and use one idea per sentence? (Chapter 4, Section 4.5)

    ☐ h. Are lists parallel? (Chapter 4, Section 4.9)

    ☐ i. Are comparisons written correctly? (Chapter 4, Sections 4.9 and 4.10)

    ☐ j. Have noun clusters been resolved? (Chapter 4, Section 4.7)

    ☐ k. Has word location been considered? (Verb following subject immediately? Old, short information at the beginning of the sentence? New, long information at the end of the sentence?) (Chapter 3, Section 3.1)

    ☐ l. Have grammar and technical style been considered? (person, voice, tense, pronouns, prepositions, articles; Chapter 4, Sections 4.1–4.4)

    ☐ m. Are words and phrases precise? (Chapter 2, Sections 2.2 and 2.3)

    ☐ n. Are nontechnical words and phrases simple? (Chapter 2, Section 2.2)

    ☐ o. Have unnecessary terms (redundancies, jargon) been reduced? (Chapter 2, Section 2.4)

    ☐ p. Have spelling and punctuation been checked? (Chapter 4, Section 4.11)

SUMMARY

PROPOSAL GUIDELINES
1. For the Abstract and Specific Aims section, include
   Abstract
      Brief background
      Unknown or problem
      Objective
      Preliminary results
      General strategy (may include specific aims if not
      listed in separate part)
      Expected Outcomes—optional
      Significance/impact
   Specific Aims
2. The first sentences should be informative, short,
   interesting, and provide broad background.
3. State the objective of your proposal precisely.
4. Distinguish between technical abstracts and those
   written for a lay audience.
5. List your Specific Aims in active, precise language.
6. State the significance of the study at the end of the
   Abstract.
7. Signal the unknown, the objective, the strategy, the
   specific aims, and the significance.
8. Do not omit any parts of the Abstract.
9. Keep the Abstract short.
10. Be realistic when listing aims.
11. Do not make aims too interdependent.

PROBLEMS

Problem 21-1

For the following Abstract, ensure that the necessary elements (background, unknown, objective, strategy, significance) are present and clearly signaled.

Plants are our oldest source of medicines. Yet much of Earth's rich plant life remains unexplored. In recent decades, natural drug discovery has concentrated on tropical plants due to their great diversity. However, there is equally much diversity for the plants of our oceans, and this plant life has remained untapped. The overall goal of this proposal is to identify and purify natural chemicals of oceanic plants and to test their activity as potential medicines. We will apply new chemical fingerprinting technology for our screens of plant life in the Florida Keys and assess them for potential medicinal use using microbial techniques. Identification of plants with compounds active against important human diseases, and subsequent characterization of such compounds, will lay the foundation for new drug development and lead to novel treatment therapies and better outcomes for patients.

Problem 21-2

The following paragraph is an Abstract that has far exceeded its permissible length of 100 words. Shorten the abstract to 100 words or less by establishing importance. Omit unimportant information, and subordinate less important information.

Tourette syndrome (TS) is an inherited neurological disorder and is characterized by chronic motor and vocal tics. It has been reported recently that habit reversal therapy (HRT), a behavioral treatment for tics, may be effective in treating Tourette syndrome. The objective of this proposal is to compare the efficacy of HRT in reducing tics, improving life satisfaction and psychosocial functioning in comparison with supportive psychotherapy (SP) in outpatients with TS. We will assess 100 adult outpatients with TS and determine if HR but not SP reduces basic tic severity over the course of the treatment. The HRT as well as the SP group will show if life satisfaction and psychosocial functioning improves during active treatment. Reductions in tic severity and improvements in life satisfaction and psychosocial functioning should remain stable at the 6-month follow-up. Our results will indicate if HR has specific tic-reducing effects. In addition, our results will also show if SP is effective in improving life satisfaction and psychosocial functioning. Assessments of response inhibition may possibly be of some value for predicting treatment response to HR.

(167 words)

Problem 21-3

For the following Abstract, ensure that the necessary elements (background, unknown, objective, strategy, significance) are present and clearly signaled.

Global warming is arguably one of the most pressing concerns of our time. However, we lack an effective model to predict precisely by how much the temperature will rise as a consequence of the increased levels of CO_2 and other factors. The width of this range is due to several uncertainties in different elements of the climate models, including the variability in the Sun's rate of energy output. To gain greater insight into the relationship between solar energy output and global temperature, we propose to launch the internationally led ABC satellite in April 2012. Our aim is to collect for 2 years data on the solar diameter and shape, oscillations, and photospheric temperature variation. We will assess these data to model solar variability. Our findings will dramatically advance our understanding of solar activity and its climate effects.

Problem 21-4

1. The following Abstract is written in the format of one for a research paper. Rewrite this abstract to follow the format of a proposal abstract.

2. **Identify the individual elements of the proposal abstract (back-ground, unknown, objective, strategy, significance).**

Abstract

Background: The role of nurses continues to expand and shift in response to high societal demand for health care services. Within this landscape, it is important to more fully explore and understand the affect on employment status of the profession.

Methodology: This is a prospective, blinded descriptive study designed to collect lifestyle, employment, and demographic data on a sampling of nurses in the UK. A validated survey will be mailed to representative cohorts of nurses to determine gender, age, medical field, work hours per week, total time, and reasons for leave of absence from the workforce, and total leave time from the workforce that is planned. Standard descriptive statistics and multivariate analysis will be used to identify the impact of family and social responsibilities on work hours based on age and gender in the setting of different fields of medicine.

Results: Family and social responsibilities are expected to affect the number of work hours per week as well as the time and reason for work leaves. These responsibilities are anticipated to differ based on age and gender and may be more prevalent in some medical fields than others.

Conclusion: We study the effect of family and other social responsibilities on gender and age in diverse medical fields to better accommodate nurse professionals in relation to the health care delivery demands of our society. Identifying significant differences of employment patterns will enable policy makers to consider the effects of family and social responsibilities for the nursing profession.

Problem 21-5

Evaluate the following Intellectual Merit and Broader Impact statements written for a NSF proposal. Is the significance and impact of the proposed study clearly stated? Does the Broader Impact contain an educational component?

Intellectual Merit.

The entire research plan is composed of the following seven sub-themes: (1) the ambient state of stress in oceanic lithosphere, (2) the energetics of slab rollback, (3) the onset of convection with internal heating, (4) scaling laws for stagnant-lid convection with mantle melting, (5) the initiation of plate tectonics, (6) the history of ocean volume and global water cycle, and (7) the nature of core-mantle interaction. Collectively, they constitute a major step toward a better understanding of the long-term behavior of Earth, by tackling unresolved first-order issues all together. Moreover, each of them is designed to address a stand-alone, basic physics problem with potential applications beyond the scope of this proposal, reaching out to earthquake seismology, regional tectonics, planetary sciences, Precambrian geology, igneous petrology, plume dynamics, and geomagnetism.

Broader Impacts.

This proposal includes the education of one female Ph.D. student in theoretical geodynamics. The PI will also assimilate research results into three existing undergraduate/graduate courses he regularly teaches. In addition, a new 200-level undergraduate course will be developed on the physics of Earth's evolution, with hands-on experience in scientific computing. The PI will also conduct community outreach to assist K-12 teachers in developing public school curricula by showing how geophysical topics can be used as friendly examples in science classes.

(Jun Korenaga, proposal to federal agency; modified)

CHAPTER 22

Background and Significance

22.1 OVERALL

BACKGROUND AND SIGNIFICANCE GUIDELINE 1:
Aim to awaken interest.

The Background and Significance section has two main purposes: to awaken the readers' interest and to provide them with relevant background information to understand the proposal independently of other publications on the topic.

22.2 CONTENT AND ORGANIZATION

BACKGROUND AND SIGNIFICANCE GUIDELINE 2:
Follow a "funnel" structure.

Background:
    Within subsections:
        Background/known
        Unknown/problem/need
        Aim/hypothesis (optional)
    Summary (optional)
    Significance/Impact

The Background section describes what is already known about the problem including previous investigators' results and current theories.

Generally, readers expect the parts of the Background and Significance section to be arranged in a standard structure: a "funnel," starting broadly with background information and then narrowing to the unknown and a specific aim or hypothesis of the proposal. When you compose this section, consider, above all, the specific aspects of the topic that are of interest to the funder.

Do not simply describe what you already know but also what questions need to be answered to further understanding of the material. By identifying what needs to be answered, and why, you establish relevance and significance of the research. The rationale for your research should be clearly delineated. This section has to convince reviewers that specific aims, once achieved, will have significant impact on the topic in question. Do not just state that you aim to gain scientific knowledge in your field but rather tie the proposal in to some broader scientific or clinical picture.

When you write this section, be as objective as possible. Do not criticize the work of other investigators or possibly alienate reviewers with an opposing point of view. (Remember that you do not know who your reviewers might be!) Instead, provide clearly established facts but acknowledge controversy. You need to convince reviewers that you are thinking objectively about the topic. To do so, you need to present the pros and cons on the subject.

The length of this section can vary widely and depends largely on where the proposal will be submitted. Short proposals to private foundations may only require one to three paragraphs of background information, whereas federal grant applications such as R01s may need two to five single-spaced pages of detailed technical information. Often, pertinent guidelines state the maximum length or number of words allowed (note that reference lists are usually not included in this count).

The Background and Significance section may be divided into subsections according to topic. Such divisions make it easier for the reader to find key points at a glance and to recall information. Subsections may be arranged according to specific aims, but they may also follow some other logical arrangement under various headings.

The background information is often (but not always) followed by a short section on significance/impact. If no separate Significance/Impact section is required, at the end of the Background and Significance section, I recommend adding a very brief summary. Within this summary, you should state the unknown and the objective(s)/hypothesis, you may indicate your expected outcomes, and you should state the overall significance or impact. This impact should be very broad, addressing issues and topics of interest to the wider scientific field or even humanity as a whole.

22.3   ELEMENTS OF THE SECTION

Background

BACKGROUND AND SIGNIFICANCE GUIDELINE 3:

Provide pertinent background information,
but do not review the literature.

Start the section by providing context relevant to your proposed topic of interest. The amount of background information needed depends on how much the intended audience can be expected to know about the topic as well as on the guidelines of the organization to which you are applying.

For longer Background sections, compose the background portion similar to a literature review on the relevant topic of interest and focus on the proposed study. The review of the literature should demonstrate the applicant's knowledge of the research on the subject. Present the overall scope of the problem, but concentrate on aspects the proposal will address. Do not include an exhaustive literature review, however, and do not spend a lot of space on statistics if the topic is well known. After you provide some general context of your work, write about the existing research in the area, and discuss established scholarship. For shorter Background sections, a summary pertinent to the research you are presenting in the proposal should suffice.

Know that for federal grants you should always provide citations and indicate sources and references. However, for many private funders, sources do not need to be indicated. This is particularly important to know when your proposal length is very restricted (2–4 pages total), as cutting out all references and citations will allow you more space for text and figures. Do not worry about leaving out citations and references if needed in these cases. Your proposal is not intended for publication. Very few people will see it, and when they do, it will be solely for considering funding.

Following are some sample background sections in which the elements found in a well-written Background section are pointed out.

☝ **Example 22-1   Background funneling to unknown**

Context/ Background	Global warming is arguably one of the most pressing concerns of our time. It has been linked to the rapidity of observed climate change—the fact that the Earth's temperature rose by approximately 0.7°C over the last century (the most dramatic increase documented in historic times) and the attendant threat posed by melting polar icecaps, rising sea levels, and potentially, more severe weather patterns.
Specific Problem/Need	We do not know yet what proportion of this global warming is due to human activity and what is due to natural variations. More important, we lack an effective model to predict precisely by how much the temperature will rise as a consequence of the increase of the levels of CO_2 and other greenhouse gases in the atmosphere of
Specific Aim	the Earth. In this proposal we aim to...

Example 22-2   Background funneling to unknown

Context/ Background	Stripes on the wings of butterflies are a common element and have long been regarded as a defensive strategy. They often indicate that a butterfly is unpalatable or toxic and are recognized by avian predators (1–3). An additional, but not necessarily exclusive, hypothesis is that striped wing pattern elements are important in butterfly mate recognition. This has been documented in some butterfly species, specifically in the family Danainae (4 and references therein). However, the role stripes play in mate recog-
Specific Problem/Need	nition has not been evaluated in a butterfly species in which the stripes are thought to serve a defensive function and warrants
Aim	further investigation (5). We propose to test this hypothesis using the butterfly *Ituna ilione* (Danainae).

Example 22-3   Background funneling to unknown

Context/ Background	The long-term sustainability of forest productivity depends on the interactions between plants and soil microbes, and their effects on resource availability. Plants rely on microbes to transform nutrients to available forms and microbes rely on plants to provide reduced carbon (C) for metabolism. The strong interdependence of this interaction has led to concerns that human induced increases in atmospheric CO_2 and nitrogen (N) deposition may be decoupling the C and N cycles in forest ecosystems (Asner et al., 1997), resulting in unpredictable feedbacks to long term forest productivity (Zak et al., 2003; Reich et al., 2006). Forest productivity is generally increased by elevated CO_2 (Ceulemans, 1999), and the magnitude of this growth enhancement is strongly regulated by soil N availability (Oren et al., 2001; Magnani et al., 2007). Thus, the stimulatory effects of elevated CO_2 on productivity have been predicted to decrease over time (i.e. a negative feedback) as pools of available N in soil become depleted (Strain & Bazzaz, 1983). However, empirical support for such progressive
Specific Problem	N limitation (PNL) in forests has been lacking (Johnson, 2006) suggesting that our understanding of the mechanisms by which trees influence soil N cycling needs further refinement.

(Richard Phillips; proposal to federal agency, modified)

Unknown and Aim

BACKGROUND AND SIGNIFICANCE GUIDELINE 4:
State the unknown, problem, or need.

After providing general context and specific aspects of existing research, describe shortcomings of or gaps in the existing research or unanswered questions. The unknown is clearest if you signal it; for example, write "is unknown" or "is unclear." You can also use other phrases to state the unknown, problem, or need for funding request: "… is the underlying problem" or "…is needed." Alternatively, you can imply rather than state the unknown by using a suggestion or a possibility ("Previous findings suggest that …"). Remember to use an objective tone when criticizing any previous work. Avoid antagonistic or judgmental phrases.

Unknown/need statements should allude to the specific aims or objective of the proposal and set the stage for the subsequent section on research design and methods. Ideally, unknown statements are placed toward the end of subsections within the Background section and are immediately followed by 1 to 2 sentences of a corresponding aim of the proposal or by a hypothesis. Here are some specific examples of stating an unknown/need followed by the corresponding aim or hypothesis.

👍	**Example 22-4**	It is unknown to what Antarctic ice core depth microbial life exists. We hypothesize that microbial life forms can survive even in the deepest ice cores due to interconnected liquid veins in which bacteria can move and obtain energy and carbon from ions in solution.
👍	**Example 22-5**	Because the consequences of global warming for life on Earth greatly depend on where the actual warming lies within this predicted range, it is critical to narrow the range by improving our understanding of the uncertain components of the climate models with utmost urgency. We will investigate these key issues in this proposal by…
👍	**Example 22-6**	These treatments are only moderately effective and are often accompanied by severe side effects and viral resistance. Thus, there remains a need for new therapies for this serious disease. We propose to use a novel class of experiments to test…

Significance and Impact

BACKGROUND AND SIGNIFICANCE GUIDELINE 5:
State an implication at the end of the section.

When you are asked to include a section on significance or impact, you should state in this section why your proposal and the expected outcomes are important. Do so by expressing what you think the significance or impact of the study is. Stating the significance or impact is particularly important at the end of the Background and Significance section unless a separate section on impact and significance is requested in the guidelines for authors. (See also Chapter 21, Section 21.4.)

Describe the significance of the problem by relating it to one or more of the following criteria, but do not overstate its significance. Instead, place your work into the proper context:

- Timeliness
- Practical solution to a problem
- Wide population or critical population
- Fills a research gap
- Has many implications for a wide range of practical problems
- May create or improve an instrument for observing and analyzing data
- Provides possibility for a fruitful exploration with known techniques
- May decrease costs (e.g., cost-effectiveness, equity of resource allocation across populations, etc.)
- May improve quality of life
- May bridge theoretical and practical knowledge
- May provide a sustainable solution to a problem

Two examples of a statement of significance or impact are shown next. Note that in both examples, the impact of the proposed study is not limited to just the advancement of the topic or the field but is stated much more broadly in terms of society.

Example 22-7   Significance and Impact Statement

Presently, we have a time-sensitive opportunity to model and gain greater insight into the relationship between global temperature and tidal cycles: the internationally led TOPEX/Poseidon Jason-2 satellite scheduled for launch in summer 2011. The TOPEX/Poseidon Jason-2 mission was developed in close consultation with the National Center for Meteorological Studies to provide a better understanding of variability in tidal cycles. It will gather data over three or more years on the solar diameter and shape, oscillations, and photospheric temperature variations. When these data are modeled, they will yield information on the magnitude, depth, and shape of the internal magnetic field in the Sun (variations that contribute to pressure, internal energy, and energy transfer) and thus allow us to model the engine of tidal variability. This satellite is the best opportunity in decades to advance our understanding of the solar activity engine and its climate effects.

In Example 22-7, the significance of the study is related to timeliness, whereas in Example 22-8, the impact is related to a possible solution to a major problem.

Example 22-8 Significance and Impact Statement

> Optically driven nanoscale machineries, as described in this proposal, could provide a prime example for the application of fundamental science in today's highly developed, technology-driven world. Harnessing light force in integrated silicon photonics will not only allow us to bring transformative impact in the field of NEMS but will also have significant impacts in the sciences and engineering as well as for society as a whole.
>
> *(Hong Tang, proposal to federal agency (NSF))*

Summary

If your Background section is long, and a statement of significance is not asked for, consider adding a one-paragraph summary at the end of it to remind the reviewers once again what the most important aspects of this section are. Examples of two such summaries are shown next.

Example 22-9 Summary

> **Summary.** Hepatocellular carcinoma is a major cancer found throughout the world, and a large proportion of this disease is associated with chronic HBV infection. New approaches to prevent or treat HBV infection therefore have the potential to prevent the development of this cancer in a large number of individuals. We will examine the hypothesis that VSV-based vaccine vectors represent a powerful tool for improved prophylactic and novel therapeutic vaccinations. These studies are important because they may result in new approaches to prevent and/or treat a disease that causes high mortality worldwide.
>
> *(Michael Robek, proposal to federal agency)*

Example 22-10 Summary

> **Impact and Significance.**
>
> We have proposed a new approach to vehicular fuel in a post-oil future by proposing production of ethanol and virtual ethanol storage. Our proposed approach avoids production of undesirable byproducts, does not compete with food production, and is viable on a global scale. It would thus generate a valuable alternate vehicular transport fuel. The investigations proposed herein will provide the first insights into this fuel development. The understanding that will be gained from these investigations will be extremely far reaching. Therefore, this proposal is uniquely relevant to the goals of the Trust for Research of Alternate Fuels.

22.4   SIGNALS FOR THE READER

BACKGROUND AND SIGNIFICANCE GUIDELINE 6:
Signal all the elements of the section.

Generally, all the parts of the Background and Significance section should be signaled so the reader does not have to guess about the information provided. The signals vary depending on how the known/background, unknown/need, objective/aim, and significance/impact were phrased. Numerous variations on these signals are possible. Some examples of such signals are shown in Table 22-1.

Table 22-1 Potential Signals for the Background and Significance section

BACKGROUND	UNKNOWN/NEED	OBJECTIVE/AIM	SIGNIFICANCE/ IMPACT
X is…	…is unknown	We propose to…	…may result in
X affects…	…is unclear	Our objective is to…	…will contribute to
X is a component of Y	…has not been determined	We will examine the hypothesis that	…may provide insight into
	…is needed…	Our overall goal is…	…are important for…
	…is necessary		…may be used to…

22.5   COHERENCE

BACKGROUND AND SIGNIFICANCE GUIDELINE 7:
Use topic sentences and techniques of continuity to tell the story.

In long Background and Significance sections, the story line can be difficult to follow. To ensure that the overall story is clear when ministories are placed into the section, you should use topic sentences and all of the techniques of continuity presented in this book:

- Word location—Chapter 3
- Key terms—Chapter 6, Section 6.3
- Transitions—Chapter 6, Section 6.3
- Sentence location—Chapter 6, Section 6.3

22.6   COMMON PROBLEMS

The most common problems of the Background and Significance section include

- Poor organization
- Lack of objectivity
- Amount of detail

Poor Organization

BACKGROUND AND SIGNIFICANCE GUIDELINE 8:
Organize your background section logically.

The section may also be divided into different subsections, which may be listed from most to least important or chronologically. In each subsection, organize your topic sentences and paragraph. Each subheading and topic sentence within each subsection should make it easy for the reviewers to absorb the key points at a glance. Follow a logical order.

Lack of Objectivity

BACKGROUND AND SIGNIFICANCE GUIDELINE 9:
Present evidence objectively.

Do not omit opposing viewpoints when presenting background information. Only if you discuss both, pros and cons, will you sound convincing to reviewers. Also remember that no reviewer wants to see himself or herself left out should they be working on the same topic.

Amount of Detail

BACKGROUND AND SIGNIFICANCE GUIDELINE 10:
Provide the necessary amount of detail.

Adjust your writing to the needs of the reviewers. Too much or too little detail can work against you. If you are writing to a panel of reviewers within your field, you may not have to provide much detail on the topic. If you are writing to an educated lay audience, however, your background section will need to start quite broad and will have to explain more technical details or use simpler expressions.

22.7 REVISING THE BACKGROUND AND SIGNIFICANCE SECTION

When you have finished writing the Background and Significance section (or if you are asked to edit these sections for a colleague), you can use the following checklist to "dissect" the sections systematically:

☐ 1. Are all the components there? To ensure that all necessary components are present, on the margins of the sections clearly mark

	background
	unknown
	aim/goal
	significance

☐ 2. Is the section written objectively?

☐ 3. Did you adjust your writing to the needs of the reviewers? Not too much detail nor too little.

☐ 4. Do all the components logically follow each other? (Is the unknown what one would expect to hear after reading about what is known? Is the aim really the goal one would expect to read after reading the unknown? Does the significance of the project really come across?)

☐ 5. Revise for style and composition using the writing principles of this book.

    ☐ a. Are paragraphs consistent? (Chapter 6, Section 6.2)

    ☐ b. Are paragraphs cohesive? (Chapter 6, Section 6.3)

    ☐ c. Are key terms consistent? (Chapter 6, Section 6.3)

    ☐ d. Are key terms linked? (Chapter 6, Section 6.3)

    ☐ e. Are transitions used, and do they make sense? (Chapter 6, Section 6.3)

    ☐ f. Is the action in the verbs? Are nominalizations avoided? (Chapter 4, Section 4.6)

    ☐ g. Did you vary sentence length and use one idea per sentence? (Chapter 4, Section 4.5)

    ☐ h. Are lists parallel? (Chapter 4, Section 4.9)

    ☐ i. Are comparisons written correctly? (Chapter 4, Sections 4.9 and 4.10)

    ☐ j. Have noun clusters been resolved? (Chapter 4, Section 4.7)

    ☐ k. Has word location been considered? (Verb following subject immediately? Old, short information at the beginning of the sentence? New, long information at the end of the sentence?) (Chapter 3, Section 3.1)

    ☐ l. Have grammar and technical style been considered? (person, voice, tense, pronouns, prepositions, articles; Chapter 4, Sections 4.1–4.4)

    ☐ m. Are words and phrases precise? (Chapter 2, Sections 2.2 and 2.3)

    ☐ n. Are nontechnical words and phrases simple? (Chapter 2, Section 2.2)

    ☐ o. Have unnecessary terms (redundancies, jargon) been reduced? (Chapter 2, Section 2.4)

    ☐ p. Have spelling and punctuation been checked? (Chapter 4, Section 4.11)

SUMMARY

BACKGROUND AND SIGNIFICANCE GUIDELINES
    1. Aim to awaken interest.
    2. Follow a "funnel" structure.
    Background:
        Within subsections:
            Background/known

> Unknown/problem/need
> Aim/Hypothesis (optional)
> Summary (optional)
> Significance/Impact
>
> 3. Provide pertinent background information, but do not review the literature.
> 4. State the unknown, problem, or need.
> 5. State an implication at the end of the section.
> 6. Signal all the elements of the section.
> 7. Use topic sentences and techniques of continuity to tell the story.
> 8. Organize your background section logically.
> 9. Present evidence objectively.
> 10. Provide the necessary amount of detail.

PROBLEMS

Problem 22-1

For the following Background section, ensure that the necessary elements (background, unknown, objective) are present and clearly signaled:

Healthy older adults often experience mild decline in some areas of cognition. The most prominent cognitive deficits of normal aging include forgetfulness, vulnerability to distraction and other types of interference, as well as impairment in multitasking and mental flexibility. These cognitive functions are the domain of the most evolved part of the human brain known as the prefrontal cortex, the brain region that is the last to fully mature in children and the first to decline as we age. Indeed, prefrontal cortical cognitive abilities begin to weaken already in middle age and are especially impaired when we are stressed. Loss of these organizational abilities is a particular liability in this Information Age when our demanding lives require that we multitask and navigate through endless interferences. Thus, understanding how the prefrontal cortex changes with age is a top priority for rescuing the memory and attention functions we need to survive in our fast-paced, complex world.

Problem 22-2

For the following Background section, ensure that the necessary elements (background, unknown, objective) are present and clearly signaled:

Age-induced emphysema ("senile emphysema") is an under recognized and poorly understood phenomenon that occurs in the lungs of most people over the age of 35 years. It leads to the loss of effective oxygen exchange in the lungs, resulting in progressive shortness of breath and, if severe, complete respiratory failure. We postulate that cigarette smoke-induced emphysema represents a form of accelerated lung aging, akin to what is observed in the skin and cardiovascular system of smokers. Toll-like receptor (TLR) deficiency predisposes people to emphysema. In mice deficient in TLR4, the canonical receptor for lipopolysaccharide (a component of specific bacteria) is an accelerated form of age-induced lung enlargement that resembles human emphysema both histologically and functionally (1). In people, TLR4 function declines with age (2, 3), which may in part contribute to "senile emphysema" even in the absence of significant exposures. Human studies have also found that chronic smoking leads to decreased TLR4 function in the lung and that the level of TLR4 depression is correlated with the severity of COPD (4). The synergistic or additive effects of age and smoking on TLR4 function in susceptible individuals may explain the pathogenesis and temporal characteristics of smoke-induced emphysema, but this synergy has not been explored in detail. We aim to explore how TLR4 is affected by age and cigarette smoke.

(Patty Lee, proposal to private funding agency, modified)

Problem 22-3

For the following Background section, ensure that the necessary elements (background, unknown, objective) are present and clearly signaled:

One potential therapeutic approach for treating chronic HBV infection is through therapeutic vaccination to disrupt the immunological tolerance and to induce an immune response that is capable of controlling the virus. Despite the promise of therapeutic vaccination for treating chronic HBV, progress in this area has been limited (16, 29, 34, 64, 73, 83). A successful therapeutic vaccination strategy must accomplish two goals. First, immunological tolerance must be broken, and virus-specific T cells must be generated. Second, these T cells must efficiently perform their effector functions, including killing target cells and producing the antiviral cytokines such as IFN-γ and TNF-α, which can noncytopathically inhibit virus replication. We will test the hypothesis that recombinant VSV vectors are ideally suited for therapeutic vaccination for chronic HBV infection.

(Michael Robek, proposal to federal agency)

Problem 22-4

For the following Background section, ensure that the necessary elements (background, unknown, objective) are present and clearly signaled:

The Pliocene climate was significantly different from our contemporary one. Most important, it was significantly warmer—high-latitude temperatures, for instance, were 4–6°C higher than today (4,5). Surface temperatures in polar regions were in fact so much higher that continental glaciers were absent from the Northern hemisphere, and the sea level was approximately 25 m higher than today. A number of numerical simulations have previously been conducted using coupled and atmospheric GCM and the data from the PRISM projects (6). However, many of these studies have assumed either that the Pliocene tropical climate was similar to the modern one or that the models themselves produced climate conditions in the tropics not far different from the modern one. Only recently, a wealth of new evidence has accumulated indicating that not only high latitudes but also the tropics and the subtropics had a very different climate in the early Pliocene. In particular, the tropical Pacific was characterized by what is often referred to as "a permanent El Niño-like state" (7). This phenomenon implies that the mean state of the Pacific had a significantly reduced or absent zonal SST gradient along the equator. In addition, a number of other dramatic climatic changes occurred throughout the tropics and the subtropics, resulting in very different patterns of sea surface temperatures from what we typically observe today. Because recently discovered changes in the tropical climate of the early Pliocene are so significant, we propose to undertake a systematic study to understand the physical mechanisms responsible for such a different climate state.

(Alexey Federov, proposal to federal agency, modified)

CHAPTER 23

Preliminary Results

23.1 OVERALL

The Preliminary Results section describes the major scientific contribution of your work to date, thus laying the foundation for the aims of the grant. It also describes what the unknown or problematic areas are and thereby sets the stage for the Research Design and Methods. The main purpose of the Preliminary Results section is to establish credibility in your capability to carry out the specific aims of your proposal. If you do not yet have many publications, related prior work or training can help demonstrate your competence.

23.2 CONTENT

General Content and Organization

PRELIMINARY RESULTS GUIDELINE 1:
Report all important previous findings relevant to the topic.

PRELIMINARY RESULTS GUIDELINE 2:
Indicate logical next steps or what is unknown or problematic.

PRELIMINARY RESULTS GUIDELINE 3:
Use figures and tables if necessary to enhance findings.

The Preliminary Results section presents the results of initial experiments carried out by you or your mentor's laboratory and points the reader to the data shown in the figures and tables. The section should show how your work to date has revolved around the topic that is directly pertinent to the funding agency's objectives. For established, independent investigators, this section carries more weight than for first-time applicants or training grants.

For grants that emphasize successful research, you especially need to prove your expertise in the area and establish credibility in your ability to perform the work. You will have to prove not only that you have some data on which to base your future aims but also that you and your team are sufficiently equipped (laboratory, skills, resources) to accomplish the goals set forth in the proposal. Some proposal formats also ask for this as a separate section in which you have to provide a more detailed overview of expertise and resources (see Chapter 25, Section 25.2). Do not overstate your competence or capabilities. Good reviewers will know if you do.

Organize the Preliminary Results section much like the Results section of a paper, with subsections pertaining to specific findings and observations. Use meaningful headings for these subsections. Report only results that are pertinent to the topic of the proposal. Exclude results that are not relevant. If needed, explain the purpose of an experiment briefly. Include results whether they support your hypothesis or do not, and explain any contradicting results if necessary. Consider ending each subsection with a sentence that summarizes the observations and conclusions.

Adjust the level of writing according to the topic and to your potential readers. Typically, novel and unusual techniques and methods will require more detailed descriptions and explanations. If you are writing for a group of scientists, provide sufficient scientific and technical details and results. In general, federal grants, especially R01s, require more detailed data and technical feasibility. Include figures and tables to support your hypothesis and to show your ability to perform the proposed experiments.

If you are writing for a lay audience, you probably need to provide more background and generalizations and discuss much broader implications. Here, too, include figure(s) and table(s) if needed. Consider your readers knowledgeable in the general research area but not experts on your specific topic.

Data and Their Interpretation

PRELIMINARY RESULTS GUIDELINE 4:

Interpret your results for the reader. Distinguish between data and results.

Do not simply describe the conclusions of prior publications or work. Instead, include data and their interpretation. Adjust the level of describing data and results according to your reviewers' background (technical or lay). Young investigators can show competence and capability despite having few publications by, for example, reporting results of preliminary studies or pilot studies, unreported data, or data presented at conferences. You may also include results determined by others, but you should clearly identify them as such.

Distinguish between data and results. Data are values derived from scientific experiments (concentrations, absorbance, mean, percent increase). Results are interpretations of data (e.g., "Growth rate **decreased** when samples were incubated at 15°C instead of 25°C.") In the Preliminary Results section, do not just present data but summarize and interpret their meaning for the reader by presenting them as results. See also Chapter 12, Section 12.2.

 Example 23-1   The plants grew 2 cm in 24 hrs when auxin was added.

Unless your readers are plant hormone specialists, they may not be able to put "2 cm in 24 hrs" into any relation, especially if no comparative value is given. Is "2 cm in 24 hrs" more or less growth than normal? The data are not interpreted for the reader.

 Revised   Plant growth **increased 50% in 24 hrs** when auxin was added.
Example 23-1

In the revised example, the data have been interpreted and are presented as a result, which has been given a comparative value to put it into relation for nonspecialists in the field. Thus, the revised example is much more meaningful for the reader.

23.3   ORGANIZATION

Overall Organization

PRELIMINARY RESULTS GUIDELINE 5:

Summarize and generalize relevant previous results.

PRELIMINARY RESULTS GUIDELINE 6:
Organize the Preliminary Results sections chronologically or from most to least important.

PRELIMINARY RESULTS GUIDELINE 7:
Organize longer Preliminary Results sections by subtopics and under each subheading chronologically or from most to least important.

Divide your Preliminary Results section into specific subsections to provide the reader with a better overview of this section. The overall structure within subsections is normally either chronological or from most to least important. You may also consider providing an overview, rationale, or introduction of your preliminary results before going into the specifics of your subsections. The next two examples show two such overviews:

  Example 23-2 Overview of Preliminary Results subsections

Preliminary studies

To understand the multitude of potential functions of short tandem repeats in the human genome, we are employing a battery of molecular, cellular, genetic, functional genomic, and bioinformatic approaches to analyze short tandem repeat-binding sites. We summarize these findings below.

  Example 23-3 Overview of Preliminary Results subsections

Introduction. We are initially focusing our VSV vaccine production on two viral antigens HBMS and HBC. We will produce VSV vectors that encode HBMS. Expression of HBMS has the advantage of encoding the largest number of possible cell epitopes without potential toxicity. In addition to producing these vectors, we will produce recombinant VSV vaccine vectors that express the viral capsid structural protein HBV Core. The use of these two particular proteins has the technical advantage that (a) they are highly expressed, (b) the immune response to these antigens is very well characterized both in humans and in mice, and (c) a variety of reagents (antibodies, ELISA assays, etc.) are readily available for analyses. We have initiated studies in these directions as described below.

(Michael Robek, proposal to federal agency)

Organization Within Subsections

PRELIMINARY RESULTS GUIDELINE 8:

Organize each subsection:
>   Purpose or background
>   Experimental approach
>   Results
>   Interpretation of results
>   Optional: problem statement/unknown/logical next
>   steps/important results

After the overview (if provided), describe details of subtopics in the different subsections. Start each subsection by presenting context first. You can provide context by giving a short background or writing an overview of the main findings on this topic or by stating the purpose of the particular set of experiments right after the subheading as shown as in Example 23–4:

Example 23–4   Subsection of Preliminary Results

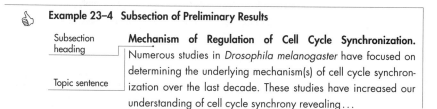

Subsection heading

Topic sentence

Mechanism of Regulation of Cell Cycle Synchronization. Numerous studies in *Drosophila melanogaster* have focused on determining the underlying mechanism(s) of cell cycle synchronization over the last decade. These studies have increased our understanding of cell cycle synchrony revealing...

After providing context, briefly state the experimental approach and the major findings on this topic in the field. Then give an interpretation of the results to make them meaningful for the reader.

When describing preliminary results, be sure to refer to each figure and table in the text, but do not repeat all the data shown in figures and tables. Instead, describe your results, and point the reader to a figure or table by citing this figure or table in parentheses after the description of the results:

Example 23-5   A total of 34 species was identified in this study, and they are listed in Table 3.

Revised Example 23-5   A total of 34 species were identified in this study (Table 3).

If appropriate, and the subsection's topic directly precedes a specific aim of the proposal, end the relevant subsection by stating what has not yet been determined or what logical next steps would be. Such statements will lead the reviewers into the next section describing your proposed research on this specific aim. You may also reemphasize important findings on which your specific aims are based. Consider highlighting such last thoughts or problem statements by placing them in italics or by underlining them. Examples of ending sentences within the Preliminary Results Section are shown in the next example; these sentences serve as a setup of the corresponding specific aim, which will be described in the Experimental Design section that usually follows the Preliminary Results section.

Example 23-6   Ending sentences of Preliminary Results subsections

(a) *This dramatic change indicates that further research into X may yield important results. However, this factor has only been investigated in one small observational study conducted in the absence of Y (19).*

(b) <u>At present, it is unclear what the function of A is.</u>

(c) *Together, these observations strongly suggest that X may increase Y by promoting Z.*

(d) **However, many educators are unaware of these findings and of the special accommodations required by children and adults who are dyslexic.**

(e) Therefore, we confirmed that <u>the specificity of the immunodominant A cell response in XX-immunized mice is directed to known epitopes in this antigen.</u>

If necessary, compare your data to that of other studies, speculate on possible mechanisms, or draw general conclusions as shown in Example 23-7:

Example 23-7   Specific conclusion at the end of a Preliminary Data section

Specific conclusion	The practicality of this approach is shown by the fact that Cooper and Pez at Air Products Corp have produced a practical H_2 storage material, N-ethyl carbazole, which holds ~5.5% w/w H_2 and is fully recyclable.

Following are three examples of subsections containing all necessary components:

Example 23-8   Preliminary Data subsection

Subsection heading	**The current Y vaccine.** Currently, a vaccine is available
Topic sentence	that prevents Y infection. This vaccine is a recombinant
Details of preliminary results	protein preparation consisting of the Y envelope protein expressed in yeast..... <u>Despite its success, the Y vaccine has</u>
Problem	<u>a number of characteristics that are suboptimal:</u>

1)...

2)...

Example 23-9   Preliminary Data Subsection

Subsection heading	**Downstream targets of X**
Topic sentence	Systematic and genome-wide analyses have been employed to identify putative downstream targets of X. Using cDNA
	microarrays, a number of targets were identified whose expression level is dependent on X (16, 18). Peders et al. (2003) found that the defined X consensus binding site is overrepresented in their set of target genes and identified a new potential binding site. Using bioinformatics, Lee et al. (2007) identified promoters with consensus X binding sites within 1kb of the ATG.... From these elegant studies (16–18), many of the "targets" could be linked to Y regulation because they included antioxidant and metabolic genes.
Details of preliminary results	
	These results generally agree with the concept that.... <u>These analyses suggested a large number of probable targets but did not</u>
Problem/Gap	<u>determine whether any are in fact direct *in vivo* targets of X.</u>

Example 23-10   Summary of Preliminary Data

	To search for slab-related heterogeneities beneath Central America we have performed P-wave seismic migration analysis using several intermediate and deep earthquakes in South America and several Californian seismic networks. Both conventional migration and the high-resolution migration method, Slowness Back azimuth Weighted Migration (SBWM) were tested.
Overview	
	Prior to migration, all of standard seismic arrivals (such as pP, sP, PcP, PP, etc.) were masked out to enhance weak but coherent energy from small-scale heterogeneities. A 3-D volume of migration grid points was defined in the study area (Figure 5a) from the surface to the lowermost mantle, with the vertical and horizontal grid intervals of 50 km. Theoretical travel time, back azimuth, and slowness were computed for each grid point and for each earthquake and receiver pair, using the IASP91 reference Earth model (25). We found that the migrated energy is concentrated at the topside of the high velocity anomaly imaged by tomography. Migrated energy is weaker in the SBWM image than that by conventional migration, but its distribution along the slab-like velocity anomaly remains similar, suggesting its robust nature. What these scatterers actually represent is of course an open question at present, and it will become clearer when we estimate the physical properties of these scatterers and compare them with mineral physics predictions for subducted oceanic crust.....These preliminary results support that our plan to investigate the fine-scale structure of the mantle beneath Central America is promising given the available source-receiver pairs for this region and the power of new-generation migration methods.
Details of preliminary results	
Specific conclusion/ Applicability for aim	

(Jun Korenaga, proposal to federal agency)

If your Preliminary Results section is long, consider adding a one paragraph summary at the end of it to remind the reviewers about the most important aspects of this section. Two examples of such summaries for a Preliminary Results section are shown next:

 Example 23-10 Summary of Preliminary Data

> Our preliminary data have started to unravel how IkappaB kinase alpha (Ikkα) regulates diverse functions. We have found that there is a new functional isoform of Ikkα that has a broad expression pattern. In addition, we have generated a polyclonal Ikkα antibody and successfully used this antibody for chromatin immunoprecipitation and for subsequent cloning to identify direct Ikkα targets. Our data also show that in addition to other phenotypes, reducing Ikkα signals affects cell survival. We will further examine the molecular relationship between Ikkα and cell survival, particularly examining how this affects progression of cancer.

 Example 23-11 Summary of Preliminary Data

> We produced recombinant WT VSV vaccine vectors that express the HBMS or HBC proteins and have preliminarily characterized the magnitude and specificity of the CD8 T cell response to HBMS. We now propose to (1) produce additional highly attenuated and boosting vaccine vectors, (2) test the ability of these vectors to induce an HBV-specific antibody and T cell response, and (3) determine if the VSV vaccine vectors can function as therapeutic vaccines using two mouse models for chronic HBV replication. These studies have the potential to result in new approaches to prevent and/or treat a chronic disease that is a major cause of cancer worldwide.
>
> *(Michael Robek, proposal to federal agency)*

23.4   IMPORTANT WRITING PRINCIPLES

Word Choice

PRELIMINARY RESULTS GUIDELINE 9:
Pay attention to word choice.

Words in the Preliminary Results section should be chosen carefully. Select the most precise and descriptive wording that reflects what you want to say. Use simple, precise words, and avoid jargon and repetition. (See Chapter 2 for more details on clear word choice.)

Pay particular attention to the following specific words and phrases in the section. These words are often used carelessly by authors, especially those for whom English is a second language (ESL authors), but they should be distinguished because of their implied meaning (see also Appendix):

Clearly/it is clear/obvious

Omit _clearly_ and similarly subjective terms and phrases. _Clearly_ makes authors seem arrogant and is an overt, but not necessarily persuasive, attempt to influence the reader:

Example 23-12	Figure 6 _clearly_ shows that Grizzly bears prefer south-facing slopes for hibernation.
Revised Example 23-12	Figure 6 shows that Grizzly bears prefer south-facing slopes for hibernation.

Significant

Significant in science refers to "statistically significant." If you write, for example, "Flow rate decreased significantly," the reader expects statistical details to follow this phrase. If you report results of statistical significance, specify the significance level.

If you do not plan to provide statistical details, use _markedly_ or _substantially_ instead of _significantly_. Also, remember you should quantify these qualitative words by using precise values or referring to data or figures such as in the following example:

Example 23-13	Flow rate decreased substantially (23%).

Tense

PRELIMINARY RESULTS GUIDELINE 10:
Use past tense for results but present tense for general conclusions.

Results are usually reported in past tense because they are events and observations that occurred in the past:

Example 23-14	a	Imidazole **inhibited** the increase in arterial pressure.
	b	Once nectar **was depleted** from the _D. wrightii_ flower, all of the moths **switched** to feeding from the _A. palmeri_ flowers.

Exceptions are results that are considered general knowledge and are still true. These findings and generalizations should be reported in present tense:

☞ **Example 23-15**   a   The *fgk* gene **has** several different introns.
                      b   These results **suggest** that learning the association between nectar reward and flower type is primarily olfactory mediated.

See also Chapter 4, Section 4.4 for more information and examples on verb tense.

23.5   SIGNALS FOR THE READER

To emphasize different portions of the Preliminary Results section, consider using signals, such as shown in Table 23-1:

Table 23-1

BACKGROUND/ PURPOSE	EXPERIMENTAL APPROACH	RESULTS	INTERPRETATION OF RESULTS	PROBLEM/ UNKNOWN
X is a member of…	…when X was subjected to…	Gyl et al. found…	…, indicating that…	…is unclear
Z was tested…	X was subjected to…	Experiment X showed…	…, consistent with…	X needs to be improved…
For the purpose of XYZ…	ABC was performed.	Ho reported that…	Thus, A is specific for…	…do not include…

ESL advice

These examples provide great starting points and may be particularly useful for those authors who have writer's block or whose native language is not English.

23.6   COMMON PROBLEMS

The most common problems of the Preliminary Results section include

- Findings not directly related to proposed work
- Experience of investigator not adequate
- Omission of important points or procedures
- Poor organization

Findings Not Directly Related to Proposed Work

PRELIMINARY RESULTS GUIDELINE 11:

Provide only findings directly relevant to your proposed work.

All findings and information in this section should be directly relevant to your proposed work. If you provide other, unnecessary information, it may confuse reviewers.

Experience of Investigator Not Adequate

PRELIMINARY RESULTS GUIDELINE 12:

Demonstrate your expertise.

You need to convince reviewers that you have sufficient expertise of the topic and/or preliminary data and collaborations to carry out the proposed Specific Aims. Expertise may be achieved through credentials, prior publications, preliminary data, training, or collaborations. In describing your preliminary data, you will also show indirectly that you have the necessary means and resources (laboratory, equipment, and team).

Omission of Important Points or Procedures

PRELIMINARY RESULTS GUIDELINE 13:

Do not omit important ideas or techniques.

If you neglect to describe an important idea or technique, you may not convince the reviewers that your project is feasible or that your expertise is sufficient.

Poor Organization

PRELIMINARY RESULTS GUIDELINE 14:

Organize subsections and the information within them.

If you organize your subsections and the information contained within them and provide logical transitions, the Preliminary Results section will be more interesting for the reviewers.

23.7   REVISING THE PRELIMINARY RESULTS

When you have finished writing the Preliminary Results (or if you are asked to edit this section for a colleague), you can use the following checklist to "dissect" the section systematically:

☐ 1. Are your findings directly related to the proposed work?

☐ 2. Are all important results and procedures described?

☐ 3. Did you indicate logical next steps or what is unknown or problematic?

☐ 4. Did you interpret your results for the reader and distinguish between data and results?

☐ 5. Is the Preliminary Results sections organized chronologically or from most to least important within subsections?

☐ 6. Do subsections state

Purpose/background
Experimental approach
Results
Interpretation of results
Problem statement/Unknown/logical next steps

☐ 7. Is the section organized well?

☐ 8. Do all subsections logically follow each other?

☐ 9. Did you follow instructions on how to write the section and what to include?

☐ 10. Revise for style and composition using the writing principles of this book:

☐ a. Are paragraphs consistent? (Chapter 6, Section 6.2)

☐ b. Are paragraphs cohesive? (Chapter 6, Section 6.3)

☐ c. Are key terms consistent? (Chapter 6, Section 6.3)

☐ d. Are key terms linked? (Chapter 6, Section 6.3)

☐ e. Are transitions used and do they make sense? (Chapter 6, Section 6.3)

☐ f. Is the action in the verbs? Are nominalizations avoided? (Chapter 4, Section 4.6)

☐ g. Did you vary sentence length and use one idea per sentence? (Chapter 4, Section 4.5)

☐ h. Are lists parallel? (Chapter 4, Section 4.9)

i. Are comparisons written correctly? (Chapter 4, Sections 4.9 and 4.10)

☐ j. Have noun clusters been resolved? (Chapter 4, Section 4.7)

☐ k. Has word location been considered? (Verb following subject immediately? Old, short information at the beginning of the sentence? New, long information at the end of the sentence?; Chapter 3, Section 3.1)

☐ l. Have grammar and technical style been considered? (person, voice, tense, pronouns, prepositions, articles; Chapter 4, Sections 4.1–4.4)

☐ m. Are words and phrases precise? (Chapter 2, Sections 2.2 and 2.3)

☐ n. Are nontechnical words and phrases simple? (Chapter 2, Section 2.2)

☐ o. Have unnecessary terms (redundancies, jargon) been reduced? (Chapter 2, Section 2.4)

☐ p. Have spelling and punctuation been checked? (Chapter 4, Section 4.11)

SUMMARY

PRELIMINARY RESULTS GUIDELINES

1. Report all important previous findings relevant to the topic.
2. Indicate logical next steps or what is unknown or problematic.
3. Use figures and tables if necessary to enhance findings.
4. Interpret your results for the reader. Distinguish between data and results.
5. Summarize and generalize relevant previous results.
6. Organize the Preliminary Results sections chronologically or from most to least important.
7. Organize longer Preliminary Results sections by subtopics and under each subheading chronologically or from most to least important.
8. Organize each subsection:
   Purpose or background
   Experimental approach
   Results
   Interpretation of results
   Optional: problem statement/unknown/logical next steps/ important results.
9. Pay attention to word choice.
10. Use past tense for results but present tense for general conclusions.
11. Provide only findings directly relevant to your proposed work.
12. Demonstrate your expertise.
13. Do not omit important ideas or techniques.
14. Organize subsections and the information within them.

PROBLEMS

Problem 23-1

Assess the following partial Preliminary Results section. Ensure that all the parts of a paragraph for the Preliminary Results section are provided and signaled:

ELISA for secreted HBMS. Because HBMS is a secreted protein (46), we determined whether the HBMS protein produced in VSV-infected cells is secreted into the culture media. We detected HBMS in the media of VSV-HBMS-infected cells but not in uninfected or recombinant WT

VSV-infected cells by both Western blot (data not shown) and qualitative ELISA (Table 2). We found by quantitative ELISA that secreted HBMS levels reached 35 ng/ml in the media of VSV-HBMS-infected BHK cells by 24 h post infection. <u>Therefore, the HBMS produced in VSV-infected cells is correctly processed for secretion.</u>

(Michael Robek, proposal to federal agency)

Problem 23-2
Assess the following partial Preliminary Results section. Ensure that all the parts of a paragraph for the Preliminary Results section are provided and signaled (purpose or background of the experiment, experimental approach, results interpretation of the results, problem statement/unknown/logical next steps):

Calculations with a General Circulation Model of the atmosphere (6) indicate that during idealized permanent El Niño-like conditions, the warming of the Eastern equatorial Pacific reduces the area covered by stratus clouds, thus decreasing the albedo of the planet. At the same time, the atmospheric concentration of the powerful greenhouse gas, water vapor, increases. This scenario may have happened during the early Pliocene, amplifying the warm conditions at that time. Consequently, projections of the effect of increasing concentration of greenhouse gases on climate should include a thorough consideration of the tropical climate conditions.

(Alexey Federov, proposal to federal agency, modified)

Problem 23-3
Assess the following partial Preliminary Results section. Ensure that all the parts of a paragraph for the Preliminary Results section are provided and signaled (purpose or background of the experiment, experimental approach, results interpretation of the results, problem statement/unknown/logical next steps):

There have been few measurements of root exudation in situ in forest ecosystems largely due to methodological challenges. As a first test to examine the role of CO_2 on exudation in loblolly pine, we grew seedlings for 150 days in controlled chambers under conditions designed to approximate the CO_2 and N gradients at the Duke FACTS-1 site. We found that pine seedlings increased mass specific exudation rates in response to elevated CO_2, and that the magnitude of the CO_2 effect was greatest in our lowest N treatment (Figure 5a). In contrast, higher N supply reduced exudation. Our preliminary data suggest that exudation may be increased by 60% in the elevated CO_2 plots, representing an increased flux of ~10 g C m-2 yr-1 to soil.

…

Our preliminary data also suggest that elevated CO_2 may increase the importance of rhizosphere processes by increasing the flux of soluble exudates from roots to soil. In addition, our results indicate that CO_2-

induced changes in the chemical composition of the exudates released can affect the magnitude of the microbial response and the quantity of N mineralized in the rhizosphere. Given the paucity of root exudation studies from trees in situ, and the absence of studies of the exudation response of field grown trees to elevated CO_2, there is a need to understand the effects of elevated CO_2 on both the quantity and chemical composition of the exudates released from roots.

(Richard Phillips, proposal to federal agency, modified)

CHAPTER 24

Research Design and Methods

24.1 OVERALL

The Research Design is usually the hardest section to write. It will receive the most scrutiny during review, as it is the heart of your proposal. Here you will need to convince the reviewers that your plans can achieve the stated specific aims. Many proposals are rejected by reviewers because of a bad Research Design section. Even though the preliminary data may be valid and interesting, the logical continuation and next steps may be obscure. Good style and clear, logical presentation are especially important for this section.

24.2 CONTENT

RESEARCH DESIGN GUIDELINE 1:
Address the problems stated in the Background and Preliminary Studies sections.

RESEARCH DESIGN GUIDELINE 2:
Adjust your level of writing to your audience.

RESEARCH DESIGN GUIDELINE 3:
Convey confidence—use future tense.

The main function of the Research Design section is to propose in detail what approach you will take to address each of your Specific Aims during the years of the proposed project. The section not only needs to clearly show the reviewers your plans to go about achieving the proposed aims, it also should explain alternate routes if you hit roadblocks.

To delineate your plans convincingly, the Research Design section should elucidate how you have arrived at your findings, what you expect to add to the topic, the specific experimental approaches, the expected outcomes of your experiments, what alternate approaches you could take, and what implications your project will have. It should also give the proposal significance by summarizing your specific aims and your expected outcomes.

Depending on the funder's requirements, the length and content of this section may vary. Other items that may be included in this section include a) potential limitations, such as possible technical problems that may arise and alternate plans you may implement; b) a time line, which shows that your experiments are doable in the proposed time and that you have a logical plan for carrying them out (see Chapter 25, Section 25.2); c) a description of your research team and what each member will contribute (see Chapter 25, Section 25.2); and d) a list of facilities and infrastructure you have available to you (see Chapter 25, Section 25.2).

Tone and Tense

The tone of your writing is important. Make your writing convey confidence and authority. Show that you are knowledgeable about the subject and take responsibility for your propositions. Do not be afraid to take a stand. The most important way to indicate that you are confident is to use future tense ("We will …"). Do not use "could," "should," "might," or "may," nor "we see an opportunity to."

24.3   ORGANIZATION

RESEARCH DESIGN GUIDELINE 4:
Organize the Research Design section into
subsections according to your Specific Aims.

RESEARCH DESIGN GUIDELINE 5:
Within the subsections, cover
Rationale/hypothesis (not always required)
Experimental design
Analysis
Expected results (outcomes and significance)
Alternative strategies (not always required)

Begin the Research Design section with a brief overview. Consider restating your overall rationale, objective, and approach as a reminder and transition for the reader. Then cover each Specific Aim in a separate subsection.

Give each subsection the exact heading as that of the corresponding Specific Aim. These subsections may be divided into further subsections. Consider following each Specific Aims heading by providing the rationale or hypothesis of the corresponding Aim. Then tell how you will go about finding the answer to the question of the Specific Aim. Explain your experimental design and your proposed analysis. The level of detail here depends on where you are planning to send the grant proposal. For federal grants, this section will have to be very detailed and technical. If you are applying to private foundations or corporations, read their individual instructions. Some require you to be as detailed as for federal proposals. Others simply ask for a short general overview of your planned approach.

An example of a partial Research Design and Methods section is shown next. This example starts off by relisting the Specific Aims as a title of the section. It then provides a brief overview followed by a subsection on experimental design and another on analysis. Such subsections may not always be labeled as such, but the content (experimental design and analysis) is usually provided in the section. Note that the analysis subsection is further divided into different components, which consist of an experimental approach, expected outcomes, and significance. These components may also be described under their own subheading. Expected outcomes and significance should be written in overview form. These latter two sections provide importance to your proposal.

Example 24-1 Research Design and Methods subsection

Heading of Subsection

Aim 1. Determining regulation of Y and Z by different XX isoforms

Overview

This aim focuses on examining how the different XX isoforms as well as XX-interacting proteins affect Y and Z.

Experimental Design

Experimental design

We have already established a function for at least one new isoform. To continue this analysis, we will generate peptide antibodies. Although peptide antibodies are slightly more expensive than traditional fusion constructs, they allow for the precise specificity required to delineate each isoform. We will use sequence analysis protocols as outlined that are specific for each antibody (Fig. 11).

Experimental Approach

Analysis

First, we will confirm the spatial and temporal expression patterns of the different isoforms to determine.... We will examine where the particular isoforms are expressed and.... Second, we will examine the expression of the different proteins by Western blot to see if we can detect distinct regulatory effects of the different isoforms. Finally,

Expected Outcome

if the antibodies function well, we will use them for additional analyses involving interacting proteins (Section 5.1.) These experi-

Significance

ments will ultimately define how the different isoforms are regulated.

Example 24-2 is another example of a Research Design and Methods section. This particular section starts by providing an overview of the experimental approach and methods that will be employed in the study. It does not contain a rationale. This is okay, as the rationale portion is optional. However, the section could be strengthened by stating the expected outcome.

 Example 24-2   Research Design and Methods subsection

Methods

Overview

<u>Research study components.</u> The four watershed sites will be studied as follows: We will establish 20 m × 20 m plots in each of the four high elevation watersheds. Transects will be established along the contour of each plot. Random locations along each transect will be used for each soil sample collection. Soil samples will be collected quarterly for two years for the eight, 3-monthly soil samplings.

Experimental design

(1) At the first of the eight 3-monthly samplings, the sites will be characterized for vegetation composition, soil carbon and nitrogen fraction sizes, bulk density, and aspect and slope.
(2) Eight, 3-monthly samplings for temporally responsive soil microbial and abiotic variables will be taken.
(3) In the second year of sampling, the remaining four 3-monthly samplings will be complemented by short-term, ^{15}N pulse-chase, field assays to follow the fate of inorganic and organic nitrogen inputs to the soils. In addition, ^{13}C-glucose will also be pulse-chase to follow its speed of movement through the soils.

Specific experimental approach

Soil sampling and processing: We will set up 20 m × 20 m plots in each of the four watersheds. Four transects will be established across each of these plots for selection of random sampling locations. Quarterly, ten 5 cm diameter and 10 cm deep cores will be taken and pooled, permitting microscale spatial heterogeneity in soil variables to be factored out of the watershed comparisons. Immediately following collection, composited soil samples will be sieved to <6 mm and extracted for NO_3 and NH_4-N determinations. Soils will be transported on dry ice (for later molecular analyses) or on wet ice and then transferred to −80°C or 5°C. At the first sampling date from each watershed, subsamples will be air dried and used for SOC fraction determinations. Additional cores will also be taken at the first sample date for bulk density determinations.

Expected outcome is not stated

(Theodore Gragson et al.; proposal to federal agency; modified)

Alternative Strategies

Often, the experimental design and analysis portion is followed by a subsection on alternative strategies, especially for federal grants and for proposals that are hypothesis based. As experiments frequently do not go as planned, reviewers often like to see that you have thought about potential pitfalls and alternatives strategies. If you do not foresee any pitfalls, consider stating that your experiments are straightforward, and no major roadblocks are expected.

Covering an alternative strategies subsection will indicate to the reviewers that you have thought through your experiments very carefully. It will also show them that the proposal has coherence even if some of the hypotheses or assumptions turn out different than anticipated. In this section, you may describe how you will continue your study if your outcome does not support a particular research hypothesis, that is, if one experiment determines the next. Think of the alternative approach section as a decision tree: if your experiment leads to A, you will do X; and if it leads to B, you will do Y; and explain how result A or B affects the interpretation of the second experiment or leads into a subsequent experiment. If you logically construct your alternative approach section, it shows reviewers that the study and experiments can be completed as proposed and will lead to useful results.

In the alternative approach section, you may also describe alternate routes to arrive at an expected finding in case one or the other experimental setup does not work. Know that all designs have limitations and that these limitations require either controls in the experiment or acknowledgment in interpretation and implications. Limitations may exist for interpretation of data such as those due to alternative explanations for observations. These limitations require a discussion of appropriate experimental controls. Other limitations arise due to generality of results and require qualifying interpretation of findings. Yet other limitations are due to a particular design. When you discuss limitations in your Research Design and Methods section, include only limitations of major concern to the purpose and potential conclusions. Do not list every possible item that "could go wrong," as this will not leave a positive impression on the reviewers.

An example of an alternative approach section is provided in Example 24-3:

 Example 24-3   Alternative Approach section

Expected Results and Alternative Approaches

Expected Outcome

We expect to successfully generate the highly attenuated, single cycle and G-protein exchange vectors for use in the vaccination studies. Based on the more than ten years experience using these techniques, and our successful recovery in collaboration with the X lab of recombinant WT VSV-HBMS and VSV-HBC, we do not anticipate any significant problems with this aim. However, if we encounter unanticipated problems, other alternative approaches exist. For example, if expression of HBMS or HBC does not induce a strong immune response in subsequent experiments, we can express the proteins from the first position within the VSV genome to increase protein expression (70). We chose the fifth

Alternative approach

| Alternative Approach | position because these recombinant viruses are better characterized and easier to recover (unpublished observations). However, expression from the first position in the genome typically leads to 3–4-fold greater protein expression and can therefore significantly enhance the immune response (63).

(Michael Robek, proposal to federal agency)

If you organize your subsections as described, reviewers are sure not to miss any important information. The following example shows a Research Design and Methods subsection that contains a subsection on expected results and alternative approaches:

 Example 24-4   Research Design and Methods subsection

SPECIFIC AIM 3. Determination of the efficacy of VSV as a therapeutic vaccine for HBV.

Rationale

HBV vaccine vectors elicit very strong protective antibody responses to virus-expressed proteins in antigen-naive animals (Table 1). However, relatively little is known about the ability of these vectors to induce an immune response in a therapeutic, rather than prophylactic, setting. One such ineffective immune response is associated with HBV infection. We will test the hypothesis that.... These experiments are important because they may identify therapeutic vaccination with VSV as a novel treatment for HBV infection.

Experimental Design and Methods

Mouse models of HBV replication. Studies on the immune response to HBV are limited by the relative lack of animal models in which to study virus infection. However, mouse models of HBV replication are available.... The use of mice for these experiments is advantageous in that.... Each model has a number of distinct advantages inherent to the specific system (Table 4). For example, model 1...and model 2.... The use of these two models is advantageous in that it allows us to perform our studies on therapeutic vaccinations in both contexts.

To measure virus replication and persistence, we.... Virus replication and persistence will be measured by ELISA. Based on a previously published study (30), we expect that....

Analysis

Animals replicating HBV will be immunized to generate a HBV-specific immune response. We will use five different immunization permutations, and analyze how.... Immunizations will be performed as described in Specific Aim 2a.

> **Expected Results and Alternative Approaches**
>
> If these experiments are successful as defined by the criteria outlined in 3c below, then we will test.... However, if these immunization procedures do not induce..., then we will determine if....
>
> *(Michael Robek, proposal to federal agency)*

To ensure continuity within the Research Design section, provide a chain of topic sentences. Use transitions and key terms and consider word locations and power positions. Organize the topics by proceeding from most to least important unless there is a reason for putting one topic before another. The topic sentence of each paragraph must indicate not only the topic or message of the paragraph but also the relation of the paragraph to the previous paragraph(s) and thus to the corresponding Specific Aim. Reviewers who study your proposal more carefully will read the chain of topic sentences to identify subtopics quickly. Note again that the subtitles for individual Specific Aims with the Research Design and Methods section should match the aims stated in the Specific Aims section of the proposal word for word.

24.4   CLOSING PARAGRAPH

RESEARCH DESIGN GUIDELINE 6:
Provide a concluding summary for the overall section.

RESEARCH DESIGN GUIDELINE 7:
Indicate the significance of your work.

At the end of the proposal, you should provide some closure by writing a one-paragraph concluding summary. This last paragraph needs to summarize your specific aims, generalize the expected outcomes, and give importance to the proposal.

Do not bring in new evidence for the summary. Rather, complete the "big picture" by restating your specific aims and the expected overall outcomes and significance.

The significance of the work can be provided by including far-reaching interpretations and conclusions. Try to generalize your expected findings to other broader situations. Depending on your level of certainty, significance can range from the practical application to the theoretical proposition. Adding a practical application, giving advice, implying an action, or providing a proposition in the concluding paragraph give the proposal importance.

		Level of certainty	*Example*
Practical	**Application**		" . . . can be used for . . ."
Theoretical	**Proposition**		" . . . will provide an under-standing of . . ."

You can highlight the significance of the problem by relating the problem to various general topics of urgency or importance such as an important point in time, solution to a practical problem, applicability to a wide or critical population, or filling an essential research gap.

Often the importance of the expected outcomes is not discussed or not adequately explained, which leaves the reviewer wondering about the significance of the proposal and the overall understanding of the researcher. It is therefore important that you indicate the significance of your expected findings at the end of the Research Design section.

Because the conclusions are the major message of your proposal, you should phrase them with great care. Possible ways to provide some closure of your work at the end of the Discussion section are shown in the next examples.

Example 24-5   Summary of Research Design and Methods section

Study Question | How does the one protein coordinate various neural cell functions to ultimately regulate migration and maturation? This grant application has three specific aims that are independent but closely related, and are all aimed at determining how *DCX* functions to modulate several different regulatory processes. We will find how the different *DCX* isoforms regulate Doublecortin outputs (5.1), we will perform genome wide location analysis to define direct outputs of Doublecortin under different conditions (5.2),

Specific Aims | and we will determine how changes in genomic stability affect Doublecortin-dependent phenotypes (5.3). At each step, we will examine how two *DCX* dependent processes are affected: neural migration and maturation. These findings will likely be relevant to cortical development and open up new areas to explore in the understanding of the connections between neural migration

Significance | and maturation.

The previous summary opens with the overall question of the study. A question such as this can be used instead of stating the overall goal of the project. The summary then lists the three specific aims and provides a brief, general overview of the experimental approach before it provides a significance/impact statement.

Following are two other examples geared toward a private foundation:

👍 **Example 24-6   Summary of Research Design and Methods section**

In summary, we propose to develop a new modeling approach to examine the intersection of behavioral, evolutionary, and ecological dynamics with the overall goal of developing novel theory on cooperation and conflict in real populations. This theory will be used to make specific predictions about cooperation and conflict during mating and parental care, allowing us to predict what factors determine when cooperation and conflict arise from social interactions. The ability to understand what factors shift an interaction from cooperative to conflict has broad relevance to our understanding of social behaviors in a wide variety of contexts, including human behavior.

👍 **Example 24-7   Summary of Research Design and Methods section**

Objective | Our research project will combine cutting edge imaging technique, genetic manipulation, and state-of-the art sequencing facilities to elucidate how X controls brain cancer cell invasiveness.

General Strategy | This innovative interdisciplinary approach will develop important new tools to examine how interactions with X affect neural migration and maturation. Our laboratory's ability to perform...coupled with our expertise in...provides us the unique opportunity to achieve these goals.

Specific Aims | We believe that aberrant neural migration can be blocked in one of two ways: 1) by targeting...; and/or 2) by inhibiting.... Both strategies represent novel therapeutic approaches to attenuate brain tumor invasiveness.

Expected Outcome | Our studies should identify the most critical targets for possible prophylactic treatment to block tumor cell invasiveness. The information we will gain from these studies will provide a solid foundation for better diagnosis and novel

Significance | strategies to treat invasive brain cancer and reduce brain cancer-related mortality.

24.5   SIGNALS FOR THE READER

RESEARCH DESIGN GUIDELINE 8:

Signal the elements of the Research Design and Methods section. Use subheadings.

Signal the different elements of the Research Design section so that the reviewers recognize immediately what they are reading. Possible signals for various elements include subheadings. Other possible signals are listed in Table 24-1.

Table 24–1

OBJECTIVE	EXPERIMENTAL DESIGN	EXPECTED OUTCOME	SUMMARY	SIGNIFICANCE
We will test the hypothesis that...	We will... Animals will be...	findings will define...	In summary,...	Our findings can/ will serve to...
We aim to determine if...	...will be analyzed...	The proposed study will...	In conclusion,...	...can be used...
the study has three specific aims...	To determine...	We expect to find...	Overall,... Taken together...	...will provide insight into...

ESL advice

This table may provide great starting and reference points when you put together your first draft and may be particularly useful for those authors whose native language is not English.

24.6 COMMON PROBLEMS

The most common problems of the Research Design and Methods section include

- Experimental plan is too ambitious
- Aims depend on previous aims excessively
- Amount of detail
- Expected results are not clear

Experimental Plan Too Ambitious

RESEARCH DESIGN GUIDELINE 9:

Your experimental plan should be feasible.

Your experimental plan cannot be too ambitious in scope or volume. In addition, you have to convince the reviewers that you have the required expertise to complete the proposed work. Remember that if you are looking to get the next round of research funded as well, you have to prove that you can follow through with this one.

Aims Depend on Previous Aim(s) Excessively

RESEARCH DESIGN GUIDELINE 10:

Do not use too many interrelated aims.

If the research design depends too much or solely on the first specific aim, failure of this aim will guarantee failure of any subsequent aims. As a consequence, your grant may not get funded unless it is a high-risk grant application with a promising high payoff. Including a section on alternative approaches may come in helpful to address failure issues, as such a section shows that the study will result in interpretable and important findings.

Amount of Detail

RESEARCH DESIGN GUIDELINE 11:
Do not use too much or too little detail.

Adjust your Research Design section according to who your potential readers will be. If you are writing for a very specific group of people, stay within that group's area of interest. If you are writing for a lay audience, you probably need to discuss much broader strategy and provide more generalizations and background. Generally, you want to keep the level of detail directly proportional to the novelty of the techniques and methods. Do not exaggerate, however. Note that if you make it sound too easy, the reviewers will think you do not understand the problem. If it sounds nearly impossible, they may think the risk is not worth funding.

Expected Results Are Not Clear

RESEARCH DESIGN GUIDELINE 12:
State expected results clearly and interpret them.

When your expected results are not clearly interpreted or when they are not listed, your reviewers may not follow your train of thought and may not understand how you will fulfill the proposed Specific Aims. Even if the results may seem obvious to you, the reviewers are not necessarily experts on this topic and will need to be told precisely what the expected outcomes are and what they mean; otherwise they are left hanging.

24.7   REVISING THE RESEARCH DESIGN AND METHODS SECTION

When you have finished writing the Research Design and Methods section (or if you are asked to edit these sections for a colleague), you can use the following checklist to "dissect" the sections systematically:

☐ 1. Is the proposed experimental plan doable in the time frame given?
☐ 2. Are most aims not too dependent on each other?
☐ 3. Did you convey confidence? (Use future tense?)
☐ 4. Has each of the Specific Aims been addressed under different subsections with about the same detail?

	rational (not always required) or purpose
	experimental design
	analysis

	expected results
	(alternative strategies)

☐ 5. Did you follow instructions on how to write this section?

☐ 6. Did you address the problems stated in the Background and Preliminary Studies sections?

☐ 7. Did you adjust your level of writing to your audience?

☐ 8. Did you organize the Research Design section into subsections according to your Specific Aims?

☐ 9. Did you provide a concluding summary?

☐ 10. Did you indicate the significance of your work?

☐ 11. Revise for style and composition based on the writing principles of the book:

    ☐ a. Are paragraphs consistent? (Chapter 6, Section 6.2)

    ☐ b. Are paragraphs cohesive? (Chapter 6, Section 6.3)

    ☐ c. Are key terms consistent? (Chapter 6, Section 6.3)

    ☐ d. Are key terms linked? (Chapter 6, Section 6.3)

    ☐ e. Are transitions used and do they make sense? (Chapter 6, Section 6.3)

    ☐ f. Is the action in the verbs? Are nominalizations avoided? (Chapter 4, Section 4.6)

    ☐ g. Did you vary sentence length and use one idea per sentence? (Chapter 4, Section 4.5)

    ☐ h. Are lists parallel? (Chapter 4, Section 4.9)

    ☐ i. Are comparisons written correctly? (Chapter 4, Sections 4.9 and 4.10)

    ☐ j. Have noun clusters been resolved? (Chapter 4, Section 4.7)

    ☐ k. Has word location been considered? (Verb following subject immediately? Old, short information at the beginning of the sentence? New, long information at the end of the sentence?; Chapter 3, Section 3.1)

    ☐ l. Has grammar and technical style been considered (person, voice, tense, pronouns, prepositions, articles?; Chapter 4, Sections 4.1–4.4)

    ☐ m. Is past tense used for results and present tense for descriptive papers?

    ☐ n. Are words and phrases precise? (Chapter 2, Sections 2.2 and 2.3)

    ☐ o. Are nontechnical words and phrases simple? (Chapter 2, Section 2.2)

    ☐ p. Have unnecessary terms (redundancies, jargon) been reduced? (Chapter 2, Section 2.4)

    ☐ q. Have spelling and punctuation been checked? (Chapter 4, Section 4.11)

SUMMARY

RESEARCH DESIGN GUIDELINES

1. Address the problems stated in the Background and Preliminary Studies sections.
2. Adjust your level of writing to your audience.
3. Convey confidence—use future tense.
4. Organize the Research Design section into subsections according to your Specific Aims.
5. Within the subsections, cover
   Rationale/hypothesis (not always required)
   Experimental design
   Analysis
   Expected results (outcomes and significance)
   Alternative strategies (not always required)
6. Provide a concluding summary for the overall section.
7. Indicate the significance of your work.
8. Signal the elements of the Research Design and Methods section; use subheadings.
9. Your experimental plan should be feasible.
10. Do not use too many interrelated aims.
11. Do not too much or too little detail.
12. State expected results clearly and interpret them.

PROBLEMS

Problem 24-1

On the left margin of this Research Design and Methods section, identify the components (rationale, methods, analysis, expected outcomes, alternate strategies, and significance).

Specific Aim:

Optimization of triage and treatment in the context of HIV and XDR TB.

In this model, we will improve upon preliminary studies to evaluate more precisely whether factors that can be assessed within a few days of a patient's admission can be used to determine which patients are most likely to be infected with drug resistant strains. We will thereby determine an optimal empirical treatment regimen for the interim period until their drug sensitivity results are available. Key clinical factors to be incorporated into the model include TB treatment history, hospitalization history, sputum smear results, lack of response to first-line treatment, chest X-ray deterioration, and rapid rifampin, kanamycin, and ciprofloxacin resistance testing results. Parameterizing our model with the data collected, we will calculate the likelihoods that a patient has non-MDR, non-XDR MDR, or XDR TB, and will determine what further studies are required to enhance confidence in the likelihood calculations. The model will also account for the benefits and disadvantages of different treatment regimens in terms of treatment outcomes, side effects, costs, pill burdens, and the probability of acquired or amplified resistance.

(Alison Galvani, proposal to private foundation, modified)

Problem 24-2

Assess the following, partial Research Design and Methods section. Ensure that all the parts for a Research Design and Methods section are provided and any unnecessary parts are omitted.

I. Laboratory and Field Behavior Study

<u>Objective:</u> To determine the function of butterfly wing patterns by characterizing the detailed behavioral responses of insect and predators in response to the Monarch butterfly, *Danaus plexippus* L. Laboratory experiments will incorporate information observed in the field.

<u>Approach and Analysis:</u> Field experiments will include a) tethering of live butterflies to vegetation to observe who predates them; b) observation of interactions of insect predators, such as mantids, with Monarch butterflies; and c) observation of interactions of avian predators, such as Blue Jays, with the same butterflies. For our experiments, artificial habitats will have representative irradiance, background color, and complexity based on ecological information we collect in the Mohave Desert. Our model insect predator will be *Mantis religiosa,* a common mantid species. Using high speed video, we will examine whether the width and number of stripes affect the target of mantid attacks.

Problem 24-3

Assess the following, partial Research Design and Methods section. Ensure that all the parts of a Research Design and Methods section are provided and any unnecessary parts are omitted.

Here we propose to study stress response behaviors in autistic children using both well-developed laboratory paradigms and dense array electroencephalographic assessments of maturing cortical coherence in response to inhibitory demands on standard neuropsychological tasks under non-stressful and stressful conditions. We will study children's response to stressors in a well-characterized cohort of boys and girls age 5 to 12 years old. We will follow the children over one year with detailed assessments of their social and scholarly development as well as their overall health. Laboratory stress procedures include asking the children to have a conversation with a trained professional and to complete simple math and social tasks in a controlled test setting. These procedures have been found to reliably induce changes in physiological indices of stress (heart rate, salivary cortisol, blood pressure) and to be a reliable index of individual differences in stress response.

CHAPTER 25

Budget and Other Special Proposal Sections

25.1   BUDGET

> **BUDGET GUIDELINE 1:**
> Follow specific instructions on how to prepare
> the budget if provided by the funder. Ask your
> administration for help.

> **BUDGET GUIDELINE 2:**
> Include direct and indirect costs.

> **BUDGET GUIDELINE 3:**
> Costs should be realistic and justified.

Aside from the Abstract, Specific Aims, Background, Preliminary Results, Research Design, and Summary/Impact, every proposal should contain a budget section. The budget is a detailed breakdown of the financial support requested from the sponsoring agency and should reflect the best estimate of the costs required to conduct the work outlined in the proposal.

Proposal budgets generally include two basic categories of costs: direct costs of the proposed project and indirect costs, which together equal the total costs. To prepare the budget, you will also need to know (1) the sponsor's likely upper limit and (2) whether cost sharing or matching funds are required. The grants and contracts administration office in a university usually helps in setting up a budget.

Depending on our proposed project as well as on the sponsor you are applying to, the following items may be included as direct costs in your budget:

Personnel:

If needed, include salaries and prorated fringe benefits (social security, retirement benefits, etc.). Also include an inflation factor and projected salary increases when applying for multiyear grants. For staff, identify the percentage of time that each individual will spend on this particular project and prorate the appropriate costs. If your case worker's salary is 100% covered by the County Department of Mental Health, then a supplemental 20% of this position's salary cannot be charged to a different funder. However, the 20% contribution to this project can be shown as a County contribution to the project. Calculate fringe benefits and payroll taxes on the exact cost percentage based on information from your human resources department.

Equipment: You may have to provide cost estimates for new equipment.

Operations:

Materials and supplies	List expenses for supplies, transportation, photocopying, or a similarly appropriate category.
Travel	Do not include nonessential travel in the budget for a project that only has local impact. If conference attendance is of vital importance for project effectiveness or results disseminations, identify the conference and provide ample justification in the budget narration.
Animal costs	Include costs for animal housing and care.
Facilities	Only list those relevant to the project and not other facilities costs.
Consultation	List any contracts if needed.
Other	If you include a "miscellaneous" or "other" budget category, be sure to itemize what this category refers to.

In addition to the direct costs, most budgets also include indirect costs, also known as overhead. These costs refer to ongoing expenses of operation and usually cover administration, utilities, taxes, repairs, and so forth. Rates can vary widely depending on what your organization requests and what the funder is willing to pay (range can be from 0%–70%).

Most funders provide detailed instructions for budget preparation; many provide budget forms or require a specific format. Always read an agency's guidelines before preparing a proposal budget. If the funding

Example 25-1

Table 25.1 Sample budget table for matching costs

CATEGORY	YEAR 1		YEAR 2		CUMULATIVE		TOTALS
	ABC Funding	Other Sources*	ABC Funding	Other Sources*	ABC Funding	Other Sources*	All Sources
Personnel (Salary + Fringe Benefits)							
Professor X, Principal Investigator (50%)	$ –	$xxxxxx	$ –	$xxxxxx	$xxxxxx	$xxxxxx	$xxxxxx
Associate Research Scientist (100%)	$xxxxxx	$xxxxxx	$xxxxxx	$xxxxxx	$xxxxxx	$xxxxxx	$xxxxxx
Personnel (Stipend + Fringe Benefits)							
Postdoctoral Fellow (100%)	$xxxxxx	$xxxxxx	$xxxxxx	$xxxxxx	$xxxxxx	$xxxxxx	$xxxxxx
Personnel Subtotal	$xxxxxx	$xxxxxx	$xxxxxx	$xxxxxx	$xxxxxx	$xxxxxx	$xxxxxx
Equipment							
		$xxxxxx			$xxxxxx	$xxxxxx	$xxxxxx
Equipment Subtotal	$ –	$xxxxxx	$ –	$ –	$xxxxxx	$xxxxxx	$xxxxxx
Operations							
Consumable supplies	$xxxxxx	$xxxxxx	$xxxxxx	$xxxxxx	$xxxxxx	$xxxxxx	$ xxxxxx
Animal costs	$xxxxxx	$xxxxxx	$xxxxxx	$xxxxxx	$xxxxxx	$xxxxxx	$xxxxxx
Travel/symposiums	$xxxxxx	$ –	$xxxxxx	$ –	$xxxxxx	$xxxxxx	$xxxxxx
Consultation	$ –	$xxxxxx	$ –	$xxxxxx	$xxxxxx	$xxxxxx	$xxxxxx
Facilities/Overhead (y%)		$xxxxxx		$xxxxxx	$xxxxxxx	$xxxxxxx	$xxxxxx
Other*	$ –	$ –			$xxxxxxx	$xxxxxx	$ xxxxxx
Operations Subtotal	$xxxxxx	$xxxxxx	$xxxxxx	$xxxxxx	$xxxxxx	$xxxxxx	$xxxxxx
Totals	$xxxxxx	$xxxxxx	$xxxxxx	$xxxxxx	$ xxxxxx	$xxxxxx	$xxxxxx

* Other includes institutional and external forms of support.

agency you are applying to does not have a set budget form, you may have to provide one yourself. Excel is a great program for setting up your budget, as it will calculate as you go. When you prepare your budget table, use only whole numbers (no cents) with proper formatting: $50,450 (not 50450). Do not overestimate or underestimate costs. Ensure that you conduct the project within the proposed budget, that the costs are realistic and justified, and that the budget matches the proposed goals and methods. Describe any cost sharing (what your institute/university will contribute). Once you have all numbers in place, check for their consistency between the project description, budget narrative, and budget line items. Be prepared to negotiate your budget. A sample budget is shown in Example 25-1.

For shorter proposals or if the full budget form is provided as an appendix, simpler tables may be shown within the text of the proposal. Examples of such tables are shown next.

 Example 25-2 Short sample budget table

Use of Funds

The total project cost is $xx,xxx. The requested funds are allocated as follows:

Personnel	PI, salary for 2 months	$xx,xxx including benefits
	Postdoctoral fellow	$xx,xxx including benefits
Equipment	laptop computer	$x,xxx
Travel	domestic and foreign	$xx,xxx
Publication costs		$x,xxx

 Example 25-3 Short sample budget table

PERSONNEL

Personnel	Title	Salary	%Effort	Requested	Fringe	Total
Peter Mayer	PI	$xxx,xxx	5%	$x,xxx	$x,xxx	$x,xxxx
Todd Blewett	Graduate Student	$xx,xxx	90%	$xx,xxx	$x,xxx	$xx,xxx
TBN	Postdoctoral Fellow	$xx,xxx	100%	$xx,xxx	$x,xxx	$xx,xxx

Subtotal for Personnel = $xx,xxx

MATERIALS AND SUPPLIES

Tissue Culture Supplies:	$x,xxx
Molecular Biology Reagents:	$x,xxx
Subtotal for Materials and Supplies =	**$xx,xxx**
Overhead (xx%) =	**$xx,xxx**
TOTAL REQUESTED =	**$xxx,xxx**

Budget Narrative or Budget Justification

In addition to the budget table described in the previous section, you may be required to provide a brief narrative or justification of each budget line item and to specify funding source(s). Such a section often consists of two parts: a budget justification/narrative and a list of other support. Examples on how to write such a section are shown following:

 Example 25-4   Sample budget narrative/justification

Budget Justification

Peter Warneke, Ph.D., Associate Professor and Principal Investigator, will devote 5% of his effort to this research project. He will provide guidance to the research staff and be involved in the design of the experiments.

Erin Toddl, Graduate Research Associate, will devote 90% of her effort to this research project. She will examine the localization of . . . .

Other Funding Sources

You may be required to list other active or pending funding sources:

 Example 25-5   Other funding sources

Other support

There is currently no other support available for the proposed project.

 Example 25-6   Other funding sources

<u>Current Support/Institutional Support</u>

Department of ABC, Y University, "Start-Up" Support (PI: John Doe) (9/1/2006-) $xxx,xxx

Roles of X in Z	Principal Investigator NIH/NINDS (R01) 02/01/04–01/31/08 $xxx,xxx direct costs/current year

Overlap: None

<u>PENDING</u>

Regulation of X	Principal Investigator Department of Defense (IDEA Award) 01/01/08–12/31/10 $xxx,xxx direct costs/year

Overlap: None

Investments of Organization

If your proposal is one that is meant to be matched or one that is meant to establish a partnership, you may also have to include a budget subsection about the investment your organization is making and that requested from the collaborating organization. Here, you should list any commitments that will be made toward the match or partnership such as facilities and equipment as shown in the next example:

 Example 25-7   Investments

Investments by ABC University

ABC University's investments in the field of chemistry constitute $xx million over x years. In addition to building institutional strength in Chemistry via a program of long-term investments in faculty recruitment and research infrastructure, ABC University has made specific new commitments to essential core facilities over the next 5 years. These include commitments for a clean room and for an Ion Cyclotron Resonance Fourier Transform Mass Spectrometer. The resulting cores will enrich the proposed collaboration and will be fully available to all participants.

These investments will be accompanied by continuing faculty recruitment for Chemistry at ABC University and are superimposed on a background of $xx million of research grants and contracts annually for Chemistry at ABC University (2009 data).

Summary of ABC University contributions to the collaboration:

	Years 1 through 5
Clean room	$xx,xxx
Ion cyclotron	$xx,xxx
Total	$xx,xxx

25.2   OTHER SPECIAL PROPOSAL SECTIONS

Overall

PROPOSAL GUIDELINE:

Check the proposal instructions carefully to ensure that your proposal contains all sections as required by the funder.

The overall content of a proposal varies depending on where you are planning to send it. Whereas many federal agencies have very defined requirements as to what sections to include in your proposal, many private funders do not. Even the names of various sections and their content may be different for the latter. In addition, you may be asked to include other special sections in your proposal. Check the proposal instructions carefully to ensure that your sections contain what is required by the funder.

The following elements are a list of special components that you may be asked to provide:

- Title Page/Cover Page
- Table of Contents
- Executive Summary
- Timeline
- Personnel
  - Organizational Chart
  - Expertise and Role
- Project Management
- Description of Organization
- Description of Facilities
- Expected Outcomes
- Evaluation Plan
- Future Directions
- Dissemination Plans/Data Sharing
- Impact Statement
- Relevance to Foundation
- Importance of Funding
- Recognition Statement
- Collaborative Arrangements
- Education and Outreach
- Biographical Sketches
- List of Publications
- References/Bibliography (not always required)
- Competing Interest Statement
- Endorsement Letters and Letters of Reference

Again, note that specific names for each section may vary and that some of these sections may be combined, and others may be split into two or more subsections depending on the funder.

Title Page/Cover Page

If required, provide a title page for the proposal. This page is the first page of the proposal and contains the title and the names and affiliations of the authors. You may also include the seal of your institution, the name of the funder, and the date of submission. Certain foundations, corporations, or federal agencies may have their own specifications for the title page. Check their instructions.

If you are coming from a country where names are usually written by listing last name before first name, know that in English the names are written in Western style name order: first name, middle initial, then last name. Take this into consideration when preparing the title page.

ESL advice

Table of Contents

For proposal narrations over five pages, a table of contents is often recommended. For a proposal over ten pages, a table of contents is vital. Even if the narration section is brief, a table of contents is helpful when multiple attachments or appendices are included. All of the attachments should be page numbered sequentially. Provide a title for appendices and other supplementary material. The table of contents and the sequence of attachments should follow the format specified by the funder.

Executive Summary

Some funders may require you to provide an Executive Summary. Usually, the Executive Summary does not exceed two pages and should be written for a well-educated lay audience. The components of the summary can be split into different subsections.

Aside from providing an overview and objective, the Executive Summary should contain methodology. Briefly describe the methods that will be employed, highlighting what is unique or distinctive in one to two paragraphs. You may also have to include a rationale, a summarized timeline, and a section on personnel. Identify the principal investigators and other key personnel, and summarize their expertise and roles in the project. In addition, the summary may include a brief version of the budget, stating the total cost of the project, the amount requested from the funder, and the amount of other sources of support such as from your institution. Describe how the percentage of funds requested from the funder will be allocated.

Time Line

In many proposals, you have to provide a time line for the research project, outlining the unfolding of the proposed research in reasonable detail. The research duration usually ranges between 1 and 5 years. Show that your experiments are doable in the proposed time and that you have a logical plan for carrying them out. It is a good idea to use a chart showing timing of activities.

If possible, indicate the desired start date or project period. Some foundation guidelines are explicit on when grant decisions are made and funds are dispersed. If this information is not available, allow between 9 months and 1 year after proposal submission for the start of your project.

 Example 25-8 Timeline

> The PI will be involved in all theoretical and empirical research. The postdoctoral fellow will focus on field research of X, whereas the graduate student will conduct fieldwork on Y. Theory and analyses will be conducted by all involved scientists. The first 2 years will focus on A. In Years 3 and 4, field data from the first 2 years will be used to test the general models and inform the development of theoretical predictions for the two study parts. Field experiments will then examine B and whether these responses can be predicted from the specific models examining the interactions between A and B.

Example 25-9   Timeline

We plan to complete the proposed project within 2 years, starting August 1, 2011. Specific milestones are depicted in the following time line:

AIM	Year 1	Year 2
Aim 1 Description of X	————————▶	
Aim 2 Generation of Y		————————▶

Personnel

Sponsors usually want to know who the team is. Provide information on your team in the personnel section. Allow insight into your team's background and respective areas of expertise. Sometimes you may even be asked to provide an organizational chart. This information may be combined with other sections under a project management section, but often funders request a separate section describing each member on the team. These descriptions can be done either in narrative form or by listing each team member's responsibilities and qualifications.

 Example 25-10   Personnel

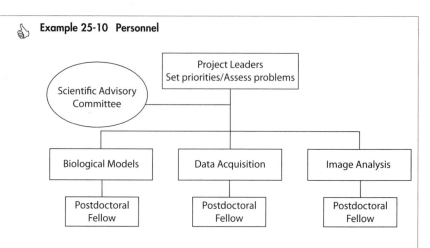

Personnel

The team is in a unique position to carry out the proposed project due to the complementary expertise of its members in three essential areas. Professors X, Y, and Z will together be responsible for directing the overall project, including setting priorities and assessing problem areas. Professor X will guide the biological parts of the work, whereas Professors Y and Z will manage the technical and mathematical aspects. The organizational chart of the project personnel illustrates the roles and responsibilities of key personnel. Work in each project area will be carried out by 4 to 5 postdoctoral fellows, 3 of whom will be paid by this grant; fellowships for the others will be funded from external sources. The entire project group will convene in biweekly group meetings. A Scientific Advisory Committee will be established consisting of experts in relevant areas.

 Example 25-11   Personnel

The PI is among the few scientists who is versatile in both A and B. She has worked on a variety of X problems in the last 12 years, including XX [9,15], XXX [17], and XXXX [20]. She has also made various contributions to AA, such as B [18,19], C [10,11,13], and D [12, 14]. She has been collaborating with experts of almost all disciplines, such as…. Before her faculty appointment at Y University, the PI was a Z Research Fellow at the University of California, Irvine. In 2004, she was awarded a 5-year NSF CAREER grant for her study on ZZ. In 2007, she was awarded the ABC prize from the AG Union, which is the highest honor bestowed to young scientists. Most recently, the PI was elected as a Frontiers Fellow by the National Academy of Sciences.

Collaborators:

An ambitious project of this nature requires varied expertise. The following investigators will contribute:

Dr. Dana Ghys, Associate Professor of X: Dr. Ghys is an ENT surgeon who will help us with injection and sampling of X.

Dr. Peter Stern, Associate Research Professor, Neuroscience, X University: Dr. Stern is an expert on immunohistochemistry and will consult on how best to perform injections.

 Example 25-12 Personnel

The funding for this project will support Mary Holmes, a talented scientist with expertise in X who recently joined my group as a postdoctoral associate. After her arrival, she developed Y, which will be used in this project.

For other examples on how to describe personnel, see also Chapter 20, Section 20.8.

Project Management

Often a funder who asks for an evaluation plan also asks for a management plan in which you have to describe the scientific advisory board, the external advisory board, the directors of the program, personnel, and the organizational structure. Whereas for smaller proposals, these sections may be individual sections, for larger proposals, a separate overarching project management section may be presented, which includes all these sections. A possible outline for a project management section is shown next:

Project Management
      Composition and Governance
            Overall Organizational Arrangement
                 Scientific Advisory Committee
                 Project Principal Investigator
                 Affiliated Investigators
      Staff and Facilities
      Budgeting and Accountability (funding, progress reports, etc.)
      Communication (among participating laboratories)

Description of Organization

It may be helpful to include the following passage or portions thereof on your organization in a proposal in which it is important to highlight the significance of your institution. Two such sample passages are shown in Example 25-13:

 Example 25-13 Why X University—Statement

    a   As an internationally recognized university, X attracts exceptional students and scholars. The University offers a remarkably open and flexible community that encourages working across disciplines and a transdisciplinary spirit of inquiry especially suited to broad, new, and innovative questions such as the one at the core of this proposal. Furthermore, the X community fosters and supports healthy risk taking by its investment in individuals with nontraditional career trajectories and in questions and programs that cross and/ or blur traditional academic disciplines and boundaries. The University supports structures to nurture creative thinking and innovative approaches even if these stand outside departmentally based organizational models.

    b   X University is one of the world's foremost research institutions. Comprised of the Graduate School of Arts & Sciences and a total of eleven professional schools, the University is diverse in many research areas. Its expert faculty and bright students are engaged in rigorous interdisciplinary collaborations, making substantial contributions across the academic spectrum in such fields as biomedical sciences, law, public policy, and the humanities. As such, the University maintains a prominent national and international presence, attracting a diverse student body of over 11,000 from countries around the globe. X's unique geographical location, together with its reputation for stellar academic scholarship and advancement, makes the University a leader among research institutions.

Description of Facilities

If you are asked for details on your facilities, describe those available to you, particularly those at your institution, as well as those required for the completion of your study.

 Example 25-14   Description of facilities

Professor X's laboratory occupies 3,900 ft² in the Center for Biochemistry. Professor Y's laboratory (1,000 ft²) is located in the immediately adjacent Center of Immunobiology. Professor Y also directs the BBB lab, which occupies 760 ft² of space in the basement level of the same building. Professor X's laboratory contains all the major equipment needed for modern biochemical research, including −70°C freezers, PCR machines, spectrophotometer, electrophoresis equipment, power supplies, protein sequencer, HPLC system, and three chemical fume hoods. The BBB lab is equipped with three advanced CCC setups, including DDD setups.

Expected Outcomes

Although sometimes Expected Outcomes will be included in other subsections, such as in the research plan and/or research design or in the significance section, other times you will be asked to place your expected outcomes into a separate section. State how your proposed study will advance the field(s) involved and mention also any broader applications or significance.

 Example 25-15   Expected outcomes

The proposed project represents a major step toward a self-consistent theory of X. Overall, the study will broaden our understanding of Y. For the field of physics, this project will produce a new model that can generate insights into X. For the computational field, the project will lay a knowledge base for the application of specific methods to large-scale Y.

Evaluation Plan

For certain funders, evaluation plans are very important and need to be addressed in your proposal. These usually list annual or final reports, meetings, and conferences. An evaluation plan (performed by external evaluators) may also be an important component of a federal proposal, especially for NSF grants. Such a plan is required particularly for large-scale, multi-investigator proposals. In the plan, you have to show how you can prove that you have been successful in meeting your objectives.

If the proposal is for a project that will be implemented over an extended period of time and that will have many complex, far-reaching effects, the evaluation will become a very complicated task. For such a project, it might be advantageous to budget for one or more professional evaluators (if it is not already necessitated by the granting organization).

 Example 25-16 Evaluation plan

> The project will be measured not only by publications but by its impact on the greater community. As an evaluation mechanism, we plan to a) establish a Scientific Advisory Board (SAB) composed of experts in A, B, and C and b) have two retreats to discuss and identify major scientific issues relevant to the project. The SAB will consist of Professors Peters, Falk, and Gusto. We will convene the scientific advisory board once a year to discuss and identify major scientific issues relevant to the project and to provide evaluation. We will also consult with the AAA foundation for the selection of external advisors. At the end of Year 1, 2, and 3, we will have an annual retreat where all involved parties will participate and present their work by posters and talks. The SAB will be present and will reconvene to advise and offer improvements. A final report will also be presented to the AAA Foundation.

Future Directions

Some sponsors ask for a section on future directions. Such a section should cover what will happen after funding is expended and should describe how the proposed project will continue.

 Example 25-17 Future directions

> The proposed research aims to develop X methodology for the analysis of desertification. We expect that development of this methodology will find utility in a broad range of application areas and will advance the field of geology in general. An initial commitment from the AAA Foundation will serve to launch the project, but we hope to leverage AAA support to obtain additional funds from other granting agencies. With this goal in mind, we plan to submit one or more such proposals by the end of the second year. Over time, we will seek to extend the X-based approach into complementary areas of geology with the addition of collaborators from our university and other institutions as appropriate.

Dissemination Plan/Data Sharing

If you need to include any dissemination plans in your proposal, describe how the findings of your study would be publicized. Include any meetings, conferences, publications, reports, newsletters, broadcasts, teaching materials, or Web sites.

 Example 25-18 a   **Dissemination plan**

We plan to present our research findings at two major conferences: the Fall Meeting of Y and the General Assembly Z. In addition, we aim to submit our research articles to either the Journal of A or B Journal International. These two journals are the top-class journals in the field, thus securing the most critical reviews for our research.

 Example 25-18 b   **Dissemination plan**

The Institute Web site will contain a regularly updated resource section on current understanding of X as well as a section on publications and materials from the Institute. We will make available information on the methods developed in phase one and also offer trainings in the use of Y. Results achieved from our studies will be published in peer-reviewed, high-profile journals and presented at national and international meetings. We will also develop instructional materials derived from X. These materials will be a critical part of our plan to disseminate the work of our community of scholars. These instructional materials are also a key step in developing prevention and intervention approaches for large groups of children and adults and are aimed to help more individuals access a capacity for healthy adaptation to X, a goal we anticipate will be a priority of the Institute in its later phases of development.

Impact Statement

An Impact statement is similar to a summary but expands more on the significance portion of it. This statement is often one of the most important paragraphs in a proposal. As described in earlier chapters, in the Impact statement, highlight the significance of the problem by relating it to one or more of the following criteria:

- important point in time
- solving a practical problem
- applicability to a wide or critical population
- filling a research gap
- having many implications for a wide range of practical problems
- advancing a specific field of research

Describe the impact your expected outcomes will have for future projects in your field, outside your field, and for society at large.

Example 25-19   Impact statement

X represents perhaps our best hope for tackling many of the seemingly intractable problems facing the world: from engineering tissues to expanding computing capacity; from responding more effectively to global climate change and pollution to meeting the threats from bioterrorism. The preceding requests will have both immediate and long-term impact on advancing X and its applications. The enhanced Seed Project Funding will nearly double the number of new projects that can be started, benefiting not only University students and faculty but also the new fields and discoveries that will emerge. Y will have both immediate and long-lasting positive impact on the kinds of research that may be conducted on campus, thereby encouraging faculty from every field to extend their research horizons and to think more broadly about new opportunities for A, B, and C than would otherwise be possible.

On a long-term scale, support for a new building would have the greatest and most lasting impact on X. Such a dedicated facility will meet evolving, long-term educational and research needs. In recognition of such a transformative gift, the new building will provide a wide range of highly visible naming opportunities such as an appropriate recognition of the AAA Foundation. It is our hope that the AAA Foundation will be a partner in this important endeavor.

Relevance to Foundation

Certain funders request a section that describes very directly in what way your proposal is relevant to the vision and goals of the funder. The content of such a section depends largely on the goals of the sponsor. In general, you need to point out and highlight any alignments of your work with the vision and goals of the funding agency. A sample section is provided in Example 25-20:

 Example 25-20   Relevance to foundation

Our proposed project aligns extremely well with the vision and goals of the AAA Foundation. The significance of gaining insight into why and how X occurs is monumental. We intend to approach such studies by joining Y and Z in a multifaceted, cross-disciplinary approach to gain critical and transformative knowledge in the field of A. In addition to Y and Z, the implications of this research touch other core themes of the AAA Foundation such as D, E, and F. Our findings could be a tremendous potential benefit to society. Our work will include inquiries into ABC to gain insights into X, promoting an overall new line of inquiry that will encompass the understanding of these different and yet connected core themes.

Importance of Funding

Some funders like to know why their funds are important for your project. Explain if your funding options are limited, and highlight the importance of your proposed research.

 Example 25-21   Importance of funding

> The goal of this proposal is to support an innovative program that brings together A, B, and C. An application to the NIH would not be appropriate at this time because this work is high risk, as it will be conducted at the interfaces of three radically different disciplines. However, the potential payoff in terms of advances in technology, theory, and not least our understanding of XXX requires forward-looking investment. Thus, the information that we expect to obtain from this proposal will be relevant to many different fields of biology and medicine. The methodology will be widely applicable beyond the field of XXX, for example, to studies of Q, O, and P. The AAA Foundation can provide the critical initial impetus to enable these advances.

Recognition Statement

You may have to include a Recognition Statement in your proposal, which should describe how your institution will recognize any extraordinary funding from your sponsor. Possible recognitions include press releases, naming opportunities, acknowledgments in publications, or conferences to which the funder will be invited.

 Example 25-22   Recognition statements

> 1. Z University proposes to name the program in recognition of the AAA Foundation's support. This program will consist of the faculty, postdoctoral fellows, and students associated with the research program over the 3 years of the grant.
>
> 2. Press releases will be issued by Z University regarding noteworthy accomplishments resulting from the AAA foundation sponsored research. Approval of these texts will be sought prior to release.
>
> 3. Scientific papers and presentations that arise from the research conducted within this project will acknowledge the support of the AAA Foundation.
>
> 4. At the end of Year 1, 2, and 3, the program will host a retreat for all involved parties (students, postdoctoral fellows, investigators, Scientific Advisory Board members). These events will honor the AAA Foundation for its support, and representatives of the AAA Foundation will be invited to attend.

Education and Outreach

Some funders, especially the NSF, ask for a section on education and out-reach. Subsections that may be included under this umbrella are

Workshops
Guided scientific tours
Seminar series
School programs
Scientist training

Collaborative Arrangements

Some proposals require you to include a Collaborative Arrangement. Such a section should state with whom you are planning to collaborate, what this collaboration would involve, and how the collaboration would be maintained.

 Example 25-23   Collaborative arrangements

Collaborative network

Our collaborative network will be maintained by visits, e-mail, conference calls, and meetings. We will have a minimum of two conference calls per month, and an annual meeting, as well as site visits to both the United States and country X to facil-itate greater collaboration between X scientists. The proposed research is based on a collaboration that has already been established and has worked well for several years. Our mod-eling study will be an expansion of the collaborative process that has already resulted in the detection and initial study of Z. Furthermore, several members of our collaborative network are instrumental in the implementation of Z treatment and control protocols. Thus, the proposed research will have a direct impact on public health policy designed to reduce Z.

Biographical Sketch

In your biographical sketch, include your professional information as well as selected publications. Typically, biographical sketches do not exceed one to two pages. Keep your biographical sketch short, listing only the most important details. Do not cram too much information into it.

 Example 25-24   Biographical sketch

John Doe

Current Address
Dept. of Biochemistry, Y University, XLML 332
3445 Froh Street, P.O. Box 3454345, Cincinnati, Ohio, USA
Phone: ####### Fax: #######; e-mail: john.doe@yuniversity.edu

Education

Ph.D. in Molecular Biology, 1983, University of California, Los Angeles, CA

B.S. in Biology, 1973, Lafayette College, Easton, PA

Professional Experience

Professor, Dept. of Biochemistry, 1997 – present, Yale University, New Haven, CT

Asst/Associate Professor, Dept. of Biochemistry, 1989–1997, Yale University, New Haven, CT

Other Appointments and Awards

- President, International Society of Biochemistry, 2001 – present

Other Activities:

- Co-editor-in-Chief, *XXX Journal,* Elsevier (1999 – present)
- Associate Editor, *YYY Journal* (1991 – present)
- Editorial Board, *Journal of ZZZ* (1997 – present)

Selected Publications (5 of 122 total)

1. XXX, YYY and **J. Doe,** *XXX Journal* 3749: 61–76, 2006.
2. AAA, BBB and **J. Doe**. *Medical Imaging,* 13(9): 40–48, 2002.
3. CCC, DDD, EEE and **J. Doe,** *Science,* 22(5): 86–100, 2002.

. . .

List of Publications

Investigators are sometimes asked to provide a list of their publications. Provide this list using a scientific format in presenting the references. If your list is too extensive, consider listing only selected publications, and label the list as such.

 Example 25-25   List of publications

Selected Publications

Cove, D.J. & Knight, C.D. The moss Physcomitrella patens, a model system with potential for the study of plant reproduction. Plant Cell, 5 (1993) 1483–1488.

Cove, D.J. Regulation of Development in the moss Physcomitrella patens. In: Russo, V.E.A., Brody, S., Cove, D. & Ottolenghi, S. Development. The molecular genetic approach. pp 179–193. Springer Verlag Berlin, 1992.

Cove, D.J., Kammerer, W., Knight, C.D., Leech, M.J., Martin, C.R. & Wang, T.L. Developmental genetic studies of the moss, Physcomitrella patens. Symp. Soc. Exp. Biol. 45 (1991) 31–43.

. . .

References

For the references, list in bibliography format what information sources were used to develop your proposal. List important journal articles and review papers, and place citations throughout the proposal narrative in which information is drawn from these works. Both the in-text citations and the reference list should adhere to a format common to scientific publications.

Competing Interest Statement

Some proposals ask for a statement about conflict of interests. A conflict of interest exists when an author (or the author's institution), reviewer, or editor has financial or personal relationships that inappropriately influence (bias) his or her actions or have the potential to do so. Such relationships are also known as dual commitments, competing interests, or competing loyalties. These relationships vary from those with negligible potential to those with great potential to influence judgment, but not all relationships represent true conflict of interest. Conflicts can occur for other reasons such as scientific judgment, financial relationships, personal relationships, academic competition, and intellectual passion. All participants in the peer review and publication process must disclose all relationships that could be viewed as presenting a potential conflict of interest. Disclosure of these relationships is also important in connection with editorials and review articles because it can be more difficult to detect bias in these types of publications than in reports of original research.

 Example 25-26   Competing interest statements

1. "The authors declare no competing financial interests."
2. "No potential conflict of interest relevant to this article was reported."
3. "Dr. A. reports serving as a consultant to C. Clinic, and serving as a consultant for a patent-infringement case involving U.S. patent xx/yyy, zzz, in the treatment of...."
4. "Dr. B reports receiving lecture fees from X and grant support from Y. No other potential conflict of interest relevant to this article was reported."
5. "Y University owns a patent, xx/yyy,zzz, that uses the approach outlined in this article and has been licensed to Z."
6. "The applicants have nothing to disclose."

Endorsement Letters and Letters of Reference

Many grant proposals require an endorsement letter or a letter of reference. Such letters are often written by a department head, a dean, or even the president of a university. If such a letter is required, give plenty of time to the person composing this letter so your application is not delayed because of it. You may even consider drafting this letter to help the writer with information about you and your project.

Example 25-27   Endorsement letter

Date

Dear Mr. Miller:

I am delighted to write this letter of support for Professors Laura South and Peter Plowe's application to the AAA Foundation. We are honored that these outstanding faculty have been invited to present a full proposal on their program entitled "Moon Phases and Tidal Variability."

Professors South and Plowe aim to answer important questions on if and how moon phases influence tides throughout the planet. Funding for their proposed program will allow them to bring to light the forces that drive tidal variability. I am convinced that a collaboration between the AAA Foundation and X University on this project will be very productive and advance our shared institutional values and aims in this and related fields.

We are extremely grateful for the previous awards provided by the Foundation and thankful for the opportunity to submit this current request. We look forward to hearing the results of your deliberations and to future opportunities to collaborate in pursuit of our mutual goals.

Sincerely,
Dean of X University

For examples of Letters of Reference/Recommendation, see Chapter 29, Section 29.8.

SUMMARY

BUDGET AND PROPOSAL GUIDELINES

1. Follow specific instructions on how to prepare the budget if provided by the funder. Ask your administration for help.
2. Include direct and indirect costs.
3. Costs should be realistic and justified.
4. Check the proposal instructions carefully to ensure that your proposal contains all sections as required by the funder.

CHAPTER 26

Revision and Submission

26.1 GENERAL

Writing a convincing proposal is a problem of persuasion. Do not send the same generic boilerplate proposal to a random list of funders. Always tailor the proposal and the specific budget request based on extensive research into the funder's priorities.

26.2 BEFORE SENDING OUT THE PROPOSAL

PROPOSAL REVISION GUIDELINE 1:
Edit, edit, and edit; then have other people edit.

Proposals range widely in their quality. You can improve your chances for obtaining grant dollars enormously by ruthlessly writing and rewriting. Therefore, edit, edit, and edit; then have some other people edit. Ensure that the proposal is clear and concise. Also, work with your administration and get all required signatures for supporting documents. Before mailing, ask a coworker to help you double-check that every required attachment is included.

26.3 REVISING THE PROPOSAL

PROPOSAL REVISION GUIDELINE 2:
Ask for help.

PROPOSAL REVISION GUIDELINE 3:

Be clear and concise.

PROPOSAL REVISION GUIDELINE 4:

Make sure the first page is perfect.

Know that program managers and board members are often inundated with proposals; therefore, you may have one minute or less to grab their attention.

Key ways to get their attention:

- Present a clear and concise proposal
- Make the proposal look neat and professional
- Follow all the requirements made by the funder

Above all

- *Ask colleagues inside and outside your field to help you improve your proposal.* Listen to what they say. Write and rewrite your proposal so the goals and intentions cannot be misunderstood and the proposal's significance is obvious. Be clear and concise.
- *Make sure that the first page is perfect.* Assume that many readers will get no further than the first page. It should therefore act as a stand-alone summary of the entire proposal.

When you are ready to revise the remaining text of the proposal, start by revising for content and logic first, and then revise the overall structure. Ensure that all necessary elements are contained in each section. For ease of reference, double-check the overall structure against a proposal outline such as the following:

 Example 26-1 Proposal Outline

Abstract/Aims:
background, problem/need, overall objective (specific aims), approach, impact

Specific Aims: list aims
Under each aim: rationale, approach, expected outcome

Background:
1. Paragraph: background/introduction, problem, objective
Last paragraph—hypothesis/approach; expected outcome

Preliminary Results (Review of Literature):
by Specific Aims, chronologically/most-to-least important
1. paragraph/sentence overview of literature
Last paragraph—summary and unknown

Research Design/Study Methods:
by Specific Aims and by subsections
within subsections: cover design, analysis, outcome, alternate
strategies, significance
   1. paragraph/sentence overview of aim/experiment
  Last paragraph — summary and significance

Conclusion: Expected outcomes and significance

Note that in this overall outline, key power positions are written in bold, as they indicate the most crucial structural locations of a proposal. Pay particular attention to the content and location of these power positions throughout your revisions.

When your structural components are in place, revise each section. Refer to summaries and checklists in the previous chapters for the proposal Abstract and Specific Aims, Background and Significance, Preliminary Results, and Research Design and Methods (Chapters 21–24). After you are satisfied with content, logical arrangement, and structure, revise for style and composition, using the basic scientific writing principles presented in Chapters 2 through 6. Pay particular attention to word, sentence, and paragraph locations. Then check for spelling and punctuation. Expect several revisions, and give yourself enough time to write and revise the proposal. Consider also time for colleagues and mentors to read and comment on any drafts, and do not forget that it will take you time to incorporate their suggestions into the proposal. Calculate several weeks of preparation and editing before sending out a proposal. For specific questions regarding the proposal, you may also consider contacting the program officer.

After you have revised the proposal a final time, recheck that you have followed the guidelines of the foundation if they were provided. Ensure that you have paid attention to detail such as font and margins and that your sections are in the right order. Double-check that you have included all required sections in the requested order. Get all required signatures from your administration or business office.

If there are no more corrections, and if you are required to send hard copies, make the required number of copies of the proposal for the funder. Also, make one copy for yourself. For electronic submission, make sure that the version you send is really the final version and that it is complete.

Factors That Kill Your Proposal

The following factors are considered the most common reasons for a proposal not getting funded:

1. Logical inconsistencies
2. Project feasibility not convincing
3. Failure to focus on the overall problem or need and the resulting payoff provided through your project
4. No persuasive structure: poorly organized, power positions and location of information has not been considered

5. Proposal does not distinguish itself from work of others
6. No compelling potential impact is offered
7. Key points are buried or not signaled
8. Proposal is difficult to read: full of jargon, too long, or too technical
9. Oversimplifying the problem at hand
10. Credibility killers: misspellings, grammatical errors, wrong technical terms, inconsistent format, and so forth

26.4   SUBMITTING THE PROPOSAL

PROPOSAL REVISION GUIDELINE 5:
Ensure that you have included all sections in the order requested.

Depending on the funder, your proposal may have to be submitted electronically or by hard copy. Electronic submission is the fastest way to send your proposal to the funder.

Instructions for electronic submissions are usually very detailed and specific and should be followed carefully. If needed, ask for help from someone that has experience with electronic submissions.

When you submit hard copies, package your final copies with care. Realize that first impressions are important. Submit only a neat proposal that follows all the requirements provided by the funder. Check that the copies to be submitted are complete. Consider placing the proposal pages in a folder or binder. Ensure that the proposal will reach the funder in time. Use a dependable delivery service, check schedules and delivery times, and address the package clearly. Keep a complete spare copy of everything in case the package is lost or damaged in the mail.

Include a brief cover letter to the sponsor (or an endorsement letter with the Dean's or President's signature) to go with the manuscript. Keep the cover letter short and simple. Introduce the project and your organization. Do not list all your achievements to date, and do not list any complicated background information or details of your proposed study. Any cover letter or endorsement letter should be carefully written and well presented. Find out the name of the program officer if you can.

Start the letter with an introductory sentence stating the title of your project and that you would like to submit it for funding. Add a sentence that describes the expected outcomes and their importance. You may also state why your proposal would be of interest to the funder.

Some funders allow you to suggest names of possible reviewers. Make sure you include all relevant contact information for each suggested reviewer. You may list reviewers either within the letter itself or you can refer the editor to a separate sheet.

A sample cover letter is shown in Example 26-2:

 Example 26-2 Cover letter

Name   date

ADDRESS

Dear Dr. Brunner:

I am pleased to submit the enclosed proposal, *Gene Targeting in the moss Pyscomitrella patens,* to the ABC Foundation. The project investigators, Professors Gary Jillter, who will lead the effort, as well as Professor Robert Johnson, are very enthusiastic about their project. The proposal's strong interdisciplinary approach and ambitious aims would seem to fit well within the Foundation's goals of supporting innovative research that lies outside the funding purview of conventional sources.

The proposed project aims to use recent breakthroughs in X to achieve Y. Support from the ABC Foundation for equipment and investigations will be critical to enable these scholars to address fundamental scientific questions, surmount a range of technical challenges to basic research, and point the way to the design of new technologies that promise to improve treatment methods in cancer therapy.

Thank you for your willingness to consider this proposal. We look forward to hearing the results of your deliberations and to a future opportunity to collaborate.

Yours sincerely,

Signature

26.5   BEING REVIEWED

Usually, when your proposal has passed the first screening, it will be discussed at panel meetings. Know that reviewers have many proposals to review but only a limited amount of time to do so. They also bring with them different experiences, which may affect the review process. Similarly, their level of knowledge in your area of expertise will influence their understanding and judgment.

If you are not familiar with the review process, do not hesitate to clarify the process with the program officer in charge of your proposal. A clear understanding of the process, such as who will be reading your proposal, for example, research staff, board of directors, and so forth, will also help you better tailor the technical details of the proposal to the appropriate audience.

Many foundations that do not have staff with scientific background will employ a scientific advisory committee to review the content of the grants. In this case, your proposal will have to appeal technically to them as well as to the foundation staff. A good rule of thumb is to integrate a *Scientific American* style of writing with that of a news and review section in *Science* or *Nature*.

26.6   SITE VISITS

PROPOSAL REVISION GUIDELINE 6:

Offer to meet the funder and put your strongest effort forward: yourself and your team.

Be prepared for a site visit, if requested. Many philanthropic organizations that make large grants will require a site visit along with proposal submission. When you get that call or e-mail announcing a visit, do not panic. First you must determine the expectations of the visitor. Knowing what they expect to get out of the site visit will be key to your success in garnering the grant. For example, some foundations use site visits only as a formality. This may be the case when you have already established a good solid relationship through past grants. Other funders may look to validate that all the staff are well versed in the goals and expectations of the proposal or that research space and equipment is adequate for completion of the project. Some foundations look for steps for sustainability and backing by the research institution. Many foundations do not like to make grants in which they are the sole funder—they like to know that other funding organizations also think that the project is worth investing in.

Try to pinpoint who is coming, what they want to get out of the meeting, and who in your organization they want to meet with. Consider involving a professional at your organization who is used to dealing with site visits from potential funders. Chances are, these professionals will be happy to help out or provide advice.

Do clean up your laboratory and office, but do not stress your staff with your nervousness. When money is involved, know that "guys wear ties" (and, of course, women should dress up, too.) So dress appropriately; but again, do not make your staff dress as if they are going to a high school prom.

Confirm the meeting and your expectations in advance. Prepare a packet to hand to the funder as part of your greeting. If a meal is involved, you might be asked to suggest a local restaurant. Be prepared with a few choices.

During the site visit, be objective and try to answer any questions they may have with respect. Do not get angry or complain. If you think they are wrong, stay objective and respond with justification.

Most funders will want to see what they might be funding in action. They also love the opportunity to ask questions. Thus, it may not be a bad idea to even make the offer to come by and meet the funder, or invite them to visit with your organization. In brief, put your strongest asset forward: yourself and your team.

26.7   IF YOUR PROPOSAL IS REJECTED

PROPOSAL REVISION GUIDELINE 7:
Be persistent. Try, try again.

Funding depends on many things, some are beyond your control. Every sponsor—whether federal, state, private foundation, or industry—has its own review and decision-making processes. Commonly, when proposals are reviewed, they will be scored or ranked. Rankings are initially determined by expert reviewers who usually are asked not only for written comments but also for a relative evaluation. This might be a verbal evaluation such as "excellent, very good, good, fair, poor"; or it might be a numerical scale such as that from the NIH, which provides a numerical score from 100 to 500 (lower scores are better). If a review panel is convened, it too will rank the proposals, taking the opinions of the expert reviewers into consideration.

It is rare, especially for novices and particularly when applying to federal agencies, that a proposal is successful on its first submission. It is best to approach a first-time submission with the attitude that it will be a learning experience. Know that rejection is devastating, particularly if accompanied by scathing commentary. Do not get frustrated if your proposal does not get funded. Nearly all researchers, sooner or later, have an application rejected. If possible, find out the reason for rejection. It may be something that you can adjust, giving you another chance to be considered for funding. Reapply if possible. Do not give up! Some funders encourage resubmission of worthy but unfunded proposals on the basis of reviewer comments.

Even if you will not reapply to the same agency, reevaluate your proposal/project, and make any necessary changes. Do not throw out your proposal! You can use many parts again, possibly in other proposals.

26.8   RESUBMISSION OF A PROPOSAL

If you revise a rejected proposal based on reviewers' comments, you may have to decide to resubmit the application to the same funding program or to a different one. If you have not received any reviewers' comments or other feedback from the funder, consider seeking feedback from the program officer. If the agency is encouraging about resubmission, your chances of success on a second go-round are good. If the agency discourages resubmission, consider finding an alternative funding source or modifying your project idea or approach. Reworking a proposal and resubmitting it, either to the original agency or to a different one, often results in a funded project.

Most reviews are meant to be helpful and intended to provide constructive criticism, pointing out weaknesses in any aspect of the application including the scientific ideas presented, the research methods, and the clarity of presentation. If you plan to resubmit a proposal, you have to decide how and to what degree you will respond to reviewers' comments. Consider discussing the reviews with senior faculty members in your department and possibly with a program officer. They may be able to offer insights on the interpretation of reviewers' comments and to help you decide how best to respond to the comments, including to seemingly unconstructive reviews. Comments that focus heavily on the style and organization of the proposal may be relatively easy to address. Comments that call for a basic scholarly or scientific revision, however, are likely to be more difficult to attend to. If possible, work closely with the program officer and be willing to rethink aspects of the project based on the agency's feedback.

If you are resubmitting a proposal to the same funding agency, clearly state when a proposal is a resubmission, and clearly identify revisions that have been made based on reviewers' comments in the proposal. If you are planning to resubmit your proposal to a different funding agency, do not make any revisions based on reviewers' comments for the first funding agency if those comments run contrary to the priorities of the new agency or program.

Some sponsors, such as the NIH or NSF, limit the number of resubmissions of a proposal to two. If resubmitted, these proposals have compulsory redress by the principal investigator. For the NIH, this is usually restricted to the first three pages of the proposal and is essential for increasing your chances of funding on resubmit. Here you need to show the reviewers that you have made efforts to address their concerns, and point out where the reviewers were mistaken in their assessment. It is common to get reviewers who do not know the proposal's field but make uninformed comments, which must be dealt with in the resubmission. However, do not antagonize reviewers by arguing about critiques. If you disagree with a reviewer's suggestion, explain why in a professional, objective manner. You will have to have a solid reason to disagree and should do so very sparingly.

Following is a sample redress of a resubmitted proposal:

Example 26-3   Resubmission—Addressing Reviewers' Comments

Introduction

This is a resubmission application. We are pleased that the panel members have acknowledged the significance of X. Our revised proposal has been strengthened by the incorporation of changes suggested by the reviewers and provides more detail about Y. Major changes are pointed out below. Critiques of the reviewers are italicized.

Critique 1
For Aim 1, reviewer 1 asked how procedures for development and validation of X will be addressed.

We have explained in detail how our models will be developed and validated in Section D.2.

For Aim 2, the reviewer asked for a detailed description of Y.
We have now added these details in Section D.3.

For Aim 3, reviewer 1 requested more details of X. Specifically, the reviewer wanted to see...
We have added these specifics in Section D.5. We apologize for the lack of clarification concerning these details. We have added a new detailed description of the data that are being used in this study (Section G)...

Critique 2
Reviewer 2 asked us to mention Z and to provide more details of its derivation.
This is a good point. For this reason, we are now comparing....

In regard to Aim 1, reviewer 2 asked us to define all terms.
We apologize for any previous unintended omission, and have defined these terms in the text.

In addition, reviewer 2 was unclear why it is necessary to develop a local database system to collect data instead of available standard database software. The reviewer also inquired....
It is necessary to develop a local database because this permits dynamic collection of data as the disease progresses. The system is a simple Microsoft access-based system used to collect....

Critique 3...

26.9   IF YOUR PROPOSAL IS FUNDED

PROPOSAL REVISION GUIDELINE 8:
Know who is negotiating your agreement.

When funding is offered, you may have to negotiate with the sponsor. Hopefully you will have involved your Grants and Contracts office for assistance prior to submitting the proposal. The Grants and Contract office has your interests at heart and will usually ensure that you are not being taken advantage of. Surprises can be fatal.

Know who is negotiating your contract. Stay involved and keep others involved as well.

SUMMARY

REVISION GUIDELINES

1. Edit, edit, and edit, then have other people edit.
2. Ask for help.
3. Be clear and concise.
4. Make sure the first page is perfect.
5. Ensure that you have included all sections in the order requested.
6. Offer to meet the funder and put your strongest effort forward: yourself and your team.
7. Be persistent. Try, try again.
8. Know who is negotiating your agreement.

PART FIVE

Posters and Presentations

CHAPTER 27

Posters

27.1 FUNCTION

Posters facilitate the rapid communication of scientific ideas. A poster is a visual presentation of your work and can serve as a concise communication tool for small groups of people. When presented well, it can be more effective than a talk in establishing a relationship with your audience because such presentations allow you to interact one on one with the people who are interested in your research. Not only can a poster attract viewers and provoke curiosity, it can also serve as an advertisement and as a summary of your work, which can be viewed in your absence. Although posters should look professional, the actual poster sessions are usually informal and interactive, unlike a talk in which people are generally more afraid to ask and answer questions. Aside from developing good scientific writing skills to convey your research results, mastering excellent presentation skills is arguably the most effective and most rapid way to disseminate new findings in science.

27.2 CONTENT

POSTER GUIDELINE 1:

Design the poster around your research question. Include

> Title
> Abstract
> Introduction
> Materials and Methods

Results

Conclusion

(References)

(Acknowledgements)

POSTER GUIDELINE 2:

Concentrate only on the main points in each section.

POSTER GUIDELINE 3:

Find visual ways to show your work—the
illustrations tell the story.

Posters primarily are visual presentations that contain text to support the graphics. Overall, posters follow the standard scientific format in that they contain all the sections found in a research paper except for the Discussion. Posters include: Title, Abstract, Introduction, Materials and Methods, Results, Conclusion, and optional are References and Acknowledgments.

The major differences are that poster sections concentrate only on the main points of each section and present these sections briefly and visually. The poster is not a publication of record, so excessive detail about methods or vast tables of data are not necessary. This material can be discussed with interested persons individually during or after the session or it can be presented in a handout.

Sometimes it can be helpful to think of each part of the poster as a slide that you would show to an audience. As done for a slide presentation, try to keep text to a minimum by using key words and phrases throughout the poster. Do not simply paste your journal articles, conference abstracts, or other manuscript portions onto poster board. Instead, present a well-designed and engaging display of scientific information:

- Design the poster around your research question. During your poster presentation time you can use the discussion time with individuals to expand on issues surrounding that central theme.
- Provide an explicit, take-home message. Summarize implications and conclusions briefly, and direct them toward an educated audience. Depending on the conference, you may also have to gear your poster toward experts in your field.
- Give credit where it is due. Include an Acknowledgments section if needed, in smaller size type (14–18 points), in which you acknowledge contributors and funding organizations.

Remember that the clarity of the presentation stems from the proper arrangement of information and that graphical elegance is often found in simplicity of design. Because a poster is a visual presentation, you need to find visual ways to show your work. Use schematic diagrams, arrows, and other strategies to direct the visual attention of the viewer rather than explaining it all in the text alone.

27.3   ORGANIZATION

POSTER GUIDELINE 4:

Vary the size and spacing of the poster sections to add visual interest.

POSTER GUIDELINE 5:

Aim for about 20% text, 40% graphics, and 40% empty space.

Overall

Scientific posters are judged by their content as well as by their presentation. Effective poster design is therefore important. One of the first things you should do when you find out that you have to present a poster is to find out how much space you are allowed, as this will determine the amount of detail you will be able to present. Unlike a manuscript, posters can adopt a variety of layouts (see, e.g., Figure 27-1). As long as you maintain sufficient white space, keep column alignments logical, and provide clear cues so your readers know how they should "travel" through your poster elements, you can get creative. However, consider power positions. In a poster, these are the central portion of the poster for illustrations and the top left corner and the bottom right corner for text. Your most important illustrations or results should therefore be placed in the middle of the poster. Your most important text sections (Abstract and Conclusion) should be placed in the corresponding text power positions, top left and bottom right, respectively (see also Chapter 6.2 for more detailed discussion of power positions). For clarity

- Present the information in a sequence that is easy to follow. You may wish to use numbers to help sequence sections of the poster and/ or choose your headings to show the flow of the information.
- Arrange the material into columns—most posters allow for three to four columns (see Figure 27-1).
- With the Abstract in the left upper corner and the Conclusions in the lower portion of your right column, consider demoting the unimportant text sections (References and Acknowledgments) to the very bottom portion of your poster.
- Vary the size and spacing of the poster sections to add visual interest, but do so in moderation.

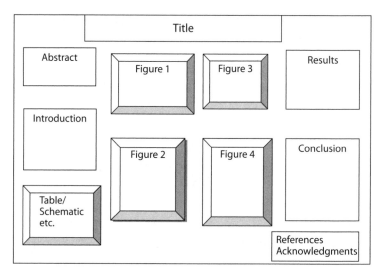

Figure 27-1 Horizontal poster layout

- For maximum visual layout and impact, aim for about 20% text, 40% graphics, and 40% empty space.
- Maintain a consistent style. Inconsistent styles give the impression of disharmony. They unnecessarily distract readers and can interrupt the fluency and flow of your messages. Ensure that headings on different pages of the poster appear in the same position on all pages.

Poster Background and Color

The background design is important in the presentation of your data. You can use mat board to make a solid background for the entire poster. You may also just frame the individual elements of the poster onto individual smaller pieces of poster board. Either approach works; the former gives a unified appearance and is easier to hang straight, whereas the latter is easier to carry to and from the meeting. It is also possible, but often expensive, to have your completed poster reproduced as a single large sheet of paper, which can then be rolled into a cylinder for transport.

For well designed poster backgrounds

Do

- *Use a colored background to unify your poster. The choice of a background color is up to you. You may also use a second background color to frame individual elements of the poster. If necessary or desired, add a single additional color by mounting the figure on thinner poster board first or by outlining the figure in colored tape.*

- *Use muted colors for the background. They are easiest to view for hours at a time and offer the best contrast for text, graphic, and photographic elements.*
- *Use a light background with darker photos; a dark background with lighter photos. Use a neutral background (gray) to emphasize color in photos and a white background to reduce the impact of colored photos.*
- *Use colors in a consistent pattern. Otherwise, your viewers will spend their time wondering what the pattern is rather than reading your poster.*
- *Also consider people who have problems differentiating colors, especially when designing graphics. One of the most common visual color weakness or blindness is an inability to distinguish green and red.*

Don't

- *Do not use more than two or three colors in the background—much more will overload and confuse viewers.*
- *Do not use overly bright colors. They will attract attention but then wear out readers' eyes. To add emphasis, you can use a more contrasting color for borders, but be conservative.*
- *Avoid using designs in the background.*

Text Format

Preparing a poster is very different from preparing a paper. Your main objective in preparing text for a poster presentation is to edit it down to very concise language. Know that people are attracted to posters that have good graphics, a clear title, and few words.

Suggested Text Fonts and Sizes

- Use a san serif font such as Arial rather than Times New Roman. San serif fonts are easier to read than serif fonts.
- Font sizes should be large enough to be read from 6 feet away.
- Use font sizes proportional to importance.
  - Title: 90 point, boldface
  - Subtitles: 72 point
  - Section headings (Introduction, etc.): 32–36 point
  - Other text: ideally 22–28 point
  - Boldface with 1½ to double spacing
- Pay attention to the text size in figures—it must also be large.
- Use lowercase lettering, with initial capitals where needed.
- Use italics to add emphasis.

Do

- *Use an interesting title in large font.*
- *Use only a small amount of text.*
- *Structure the poster such that you lead the reader through the panels.*
- *Use bullets and numbers to break text visually and make text more readily available.*
- *Double-space all text, using left justification; text with even left sides and jagged right sides is easiest to read.*
- *Use active voice when writing the text.*
- *Delete all redundant references and filler phrases (such as "see Figure 1").*
- *Spell check and proof text carefully before your final print out.*

Don't

- *Do not use too much text or information.*
- *Do not make font sizes too small.*
- *Do not arrange panels randomly.*
- *Do not use your conference abstract as text on the poster.*
- *Do not change the font in the poster. Stick with the same font type throughout.*
- *Avoid underlining or exclamation marks, as they are not considered effective.*

27.4   SECTIONS OF A POSTER

POSTER GUIDELINE 6:

Minimize text and use images and graphs instead.

As posters are primarily a visual medium, you should minimize text and use images and graphs as much as possible (see the section on Photos, Figures, and Tables following). Use short sentences, simple words, and bullets to illustrate discrete points. Do not use long paragraphs. Instead, use short, simple statements. Also consider basic writing principles such as use of active voice and avoiding jargon and redundancies.

Title
The title should convey the topic, the approach, and the system (organism). It needs to attract the reader and should be no longer than two lines. The title should be at least 2 in. (5 cm) tall.

Abstract
Some posters include an abstract. Others do not. If the abstract is published, there is no need to repeat it in the poster. If you are including an

abstract on your poster, keep the abstract to a minimum. Follow the guidelines of how to write a good abstract and ensure that all essential elements are present (see Chapter 14.) Include no more than 50 to 100 words.

 Example 27-1

Abstract

To examine the specificity of gene disruption in the haploid moss Physcomitrella patens, we have disrupted one member, ZLAB1, of the multigene Cab gene family. We found that the ZLAB1 gene is specifically targeted in three out of nine integrative transformants, indicating that gene disruption in P. patens is highly specific.

Introduction

Get your viewer *interested* in the issue or question while using the absolute minimum of background information and definitions. Quickly place your research in the context of published, primary literature; state your research question clearly; and provide a brief description and justification of the general experimental approach. Keep background information to a minimum—you are there to fill in details if needed. The introduction for a poster should be short (no more than 200 words):

 Example 27-2

Introduction

The landscape of the Middle East has been altered by human activity for most of the Holocene period. The rate of these modifications has accelerated in the last century, and today rapid population growth, political conflict, and water scarcity are common throughout the area. All of these factors increase the region's vulnerability to potentially negative impacts of climate change while decreasing the likelihood of successfully emerging region-wide adaptation strategies. In this study, we analyzed climate change in the Middle East during the 21st century as predicted by 18 Global Climate Models. The simulations were run as part of the Intergovernmental Panel on Climate Change Fourth Assessment Report (IPCC AR4) and used the Special Report on Emission Scenarios (SRES) A2 emission scenario, which is the scenario closest to a "business as usual" scenario in the SRES family.

(With permission from Roland Geerken, modified)

To reduce the length of a poster Introduction, you may also present the introduction in bullet point form as shown in Example 27-3:

 Example 27-3

Introduction

- Sea urchins are model systems for fertilization studies
- Fertilization among sea urchins is species specific
- Interaction of surface proteins (bindin and its receptor) on the egg and sperm are largely responsible for specificity

- **Differences in success rates of fertilization exists for individual urchins, but the reasons for these differences are unknown**
- **We analyzed fertilization specificity quantitatively for 10 male and 5 female Strongylocentrotus purpuratus urchins**

Materials and Methods

Briefly describe experimental equipment and methods. Do not describe them in the detail used for a manuscript. Employ figures and tables to illustrate experimental design wherever possible. Use flowcharts to summarize experimental procedures if needed. Include photographs or drawings, and use references where relevant—use only names and dates within the text. Again, stay well within a maximum of 200 words. In methods papers, this section will be longer.

A sample flowchart of a Materials and Methods section from a poster is shown in Example 27-4.

 Example 27-4

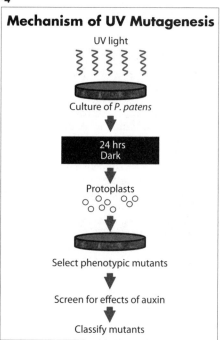

Results

Describe your most important, overall results. For basic research posters, this will be the largest portion of the poster. Most, if not all, of your findings should be presented in the form of figures and tables. Show your data analysis as it led up to your main findings. Provide engaging figure legends that can stand on their own and interpret your findings, particularly if you do not have any separate text on results. On posters, place legends with tables as well. Ensure a consistent order between your results and the conclusions.

 Example 27-5

Results

Table 2 shows the domain average multimodel ensemble mean change in annual temperature and precipitation. There is high agreement amongst the Global Climate Models for the predicted temperature change and significant disagreement for the predicted precipitation change. This is reflected in the magnitude of the change and standard deviation shown in Table 2.

Table 2: Multimodel ensemble mean change in annual temperature and precipitation

	TEMPERATURE (K)		PRECIPITATION (MM)	
	2050–2005	2095–2005	2050–2005	2095–2005
Mean Change	1.41	3.95	–8.42	–25.45
SD	0.32	0.73	16.08	28.66

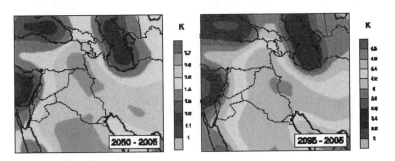

Figure 2: Multimodel ensemble mean change in annual temperature

(With permission from Roland Geerken, modified)

Sometimes the Results and Conclusions are displayed on the same panel, often in bullet-point format as shown in Example 27-6.

 Example 27-6

Results

- 30% of S. purpuratus gametes displayed twofold to ninefold differences in intraspecies fertilization efficacy
- Surface components of both egg and sperm are involved in intraspecies fertilization success rates

Conclusions

- We hypothesize that different alleles of the surface proteins are responsible for variable fertilization success rates

- • These alleles may contribute to the process of reproductive isolation and, ultimately, speciation
- • Sequences of individual surface proteins from different individuals of the same species may shed light onto this hypothesis

Conclusions

The last section of the poster is usually the Conclusion section. The discussion is left for you to present to your viewers and for publication. Assign importance to your results, and summarize them accordingly. Use the same key terms consistently throughout the poster, and make use of bullets, arrows, italics, or colored text to emphasize major points. Usually, conclusions are very brief, as the poster is indicative of only a portion of the research. Concentrate on your main findings and their interpretation. Do not list all of your research findings. Instead mention only two to four main points in your conclusion or summary. If written as bullet points, these findings will be more visually pleasing than a whole paragraph of text.

 Example 27-7

Conclusions

- • Mean annual temperatures will increase by ~4K by the late 21st century.
- • Changes in precipitation are more variable; the largest change, however, is a precipitation decrease that occurs over an area covering the Eastern Mediterranean, Turkey, Syria, Northern Iraq, Northeastern Iran, and the Caucuses.
- • Changes in precipitation will have a significant impact on fresh water resources.

(With permission from Roland Geerken, modified)

As part of a conclusion, you may also consider depicting a model or other figure to highlight a hypothesis or proposed model or mechanism such as the one shown in the following example:

 Example 27-8

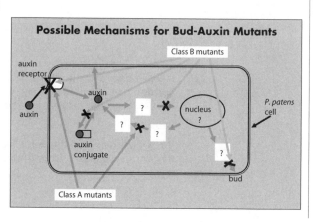

Possible Mechanisms for Bud-Auxin Mutants

Sometimes, a brief section titled "Future Research" follows the conclusion section. In this section, explain briefly how you are planning to extend the presented work.

References

References should be included if the technique is someone else's, but keep references to a minimum and keep them brief. Usually, names, dates, and journal information is enough for the Reference List. To cite information within poster text, use names and years only. If possible, do not use more than five citations.

Example 27-9   Text

Following SCI, a modified Basso-Beattie-Bresnahan (BBB) locomotor rating scale (Joshi & Fehlings, 2002), with 21 as normal and 0 as complete hind limb paralysis, was used to score locomotion in the open field after SCI (Basso et al., 1996).

References

Basso, D. M., Beattie, M. S., & Bresnahan, J. C. (1996). *Exp Neurol* 139, 244–256.
Joshi, M., & Fehlings, M. G. (2002). *J Neurotrauma* 19, 175–190.

Acknowledgments

If applicable, thank individuals for specific contributions to the project (e.g., equipment donation, statistical advice, laboratory assistance, comments on earlier versions of the poster), and mention who has provided funding. Also include in this section disclosures for any conflicts of interest and conflicts of commitment.

27.5   PHOTOS, FIGURES, AND TABLES

POSTER GUIDELINE 7:
Prepare illustrations well ahead of time.

POSTER GUIDELINE 8:
Keep exhibits simple.

POSTER GUIDELINE 9:
Make exhibits look attractive.

General Content and Format of Visual Aids

The success of a poster directly relates to the clarity of the illustrations and tables. Self-explanatory graphics should dominate the poster. Make illustrations well ahead of the meeting to give yourself time to check, replace, or improve them. Like text, graphic materials should be visible easily from a minimum distance of 6 feet.

Figures and Tables

All illustrations and tables should be comprehensible on their own. Keep exhibits simple and text to a minimum. Your exhibits should be attractive but professional, and individual figures should be recognizable as being part of a set (same colors, style of font, and emphasis techniques).

Most viewers look at the illustrations first. Not only do these need to be visually attractive, they also should be simpler than the ones in published papers such as the conference proceedings. You may also need illustrations such as flow diagrams, which are typically not found in papers.

For the best possible presentation, follow these guidelines:

Do

- *Make the illustrations tell the story*
- *Provide a title for each figure and table*
- *Provide a legend for each figure and table*
- *Use names for tables rather than numbers or letters*
- *Choose graphs rather than tables*
- *If you use tables, make them very simple*
- *Use contrast and colors for emphasis, especially to distinguish different data groups in graphs*
- *Make annotations and lines in illustrations larger than normal and symbols easy to tell apart*
- *Write all text horizontally*
- *Explain each variable and its significance*
- *Keep the scale consistent for all figures and graphs; keep axes consistent in all graphs*
- *Write equations large enough to be read from 6 feet away*
- *Remove all nonessential information from graphs and tables, and label data lines in graphs directly, using a large font type and color.*

Don't

- *Do not use more than six bars in bar graphs and no more than three to four lines in a line graph*
- *Do not use more than four columns and seven or eight rows in tables, including the title and column headings*
- *Avoid nonstandard colors, fonts, graphs, and abbreviations*
- *Avoid using patterns or open bars in histograms*
- *Do not present unnecessary or unimportant equations*

Examples 27-10 and 27-11 show two poster panels with well-constructed illustrations:

 Example 27-10

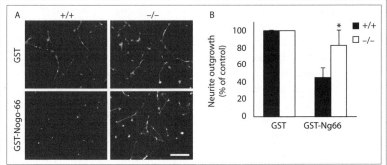

Neurite outgrowth assays using NgR KO P6 cerebellar neurons

(A) Dissociated cerebellar granule neurons isolated from *ngr* +/+ or *ngr* –/– P6 pups were plated on dried spots of GST or GST-Nogo-66 (45 ng) and incubated for 12 hr. Scale bar equals 50 μm.

(B) Quantitation of neurite outgrowth shown as percentage of GST control for *ngr* +/+ or *ngr* –/– P6 cerebellar neurons.

All data are represented as mean ± SEM. *, significantly different from wild type, *P* < .05 (Student's t test).

(With permission from Betty Lui)

 Example 27-11

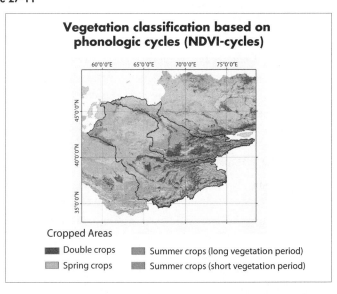

Vegetation classification based on phonologic cycles (NDVI-cycles)

Cropped Areas

- ■ Double crops
- ▨ Spring crops
- ▨ Summer crops (long vegetation period)
- ▨ Summer crops (short vegetation period)

(With permission from Roland Geerken, modified)

27.6   PREPARING A POSTER

If you are planning to present a poster at a conference, you usually have to write an abstract first. This abstract will be reviewed by a conference committee (see also Chapter 28, Section 28.2, as well as Chapter 14 for more details on writing an abstract.) If you are invited for the conference, you will be told whether your presentation is oral or a poster. Your abstract is made available to the conference participants and will often be published as submitted. You may also be invited for a paper, which gets published in the conference proceedings. Your abstract (and poster) may present preliminary results or results close to publication. The abstract has to fit the topic of the conference and should be interesting enough to attract an audience when displayed as a poster.

When your abstract has been accepted for poster submission, review the poster guidelines before starting to make the poster. Note especially whether the conference requires portrait or landscape poster format as well as the specific time and place you can display it.

Software and Hardware Options

Although some posters are hand created, posters generated electronically using a layout program usually look much more professional. Electronically generated posters can be printed as one large document using a variety of software packages such as Microsoft PowerPoint®, Adobe Photoshop®, Canvas®, CorelDraw®, Illustrator®, PaintShopPro®, Adobe FrameMaker®, Adobe InDesign®, Keynote®, Pages®, or Sun/Solaris®. Large-format printers come in various sizes. Some department printers can handle posters up to 42 in. wide (length is flexible). For more detailed instructions or supplements on how to use diverse programs to create a poster, see also the following Web sites (last accessed October 2009):

> **http://ssrl.brown.edu/support/design/large_posters**
> Creating a Large-Format Poster in PowerPoint

> **http://depts.washington.edu/mphpract/ppposter_3c.html**
> Creating a Poster Using MS PowerPoint

> **http://www.geo.mtu.edu/department/classes/ge511cpu/posters.html**
> Poster Presentations using FrameMaker (Adobe FrameMaker)

Other Useful Links

> **http://www.asp.org/Education/Howto_onPosters.html**
> Expanded Guidelines for Giving a Poster Presentation

> **http://lorien.ncl.ac.uk/ming/Dept/Tips/present/posters.htm**
> Poster Presentation of Research Work

> **http://www.osti.gov/em52/workshop/tips-exhibits.html**
> Tips for Effective Poster Presentations

> **http://www.bio.miami.edu/ktosney/file/PosterHome.html**
> How to Create a Poster That Graphically Communicates Your Message

http://www.swarthmore.edu/NatSci/cpurrin1/posteradvice.htm
Guidelines for Designing and Writing a Scientific Poster

http://wic.library.upenn.edu/multimedia/tutorials/posterPPT.html
How to Make a Great Poster Using PowerPoint

http://www.aspb.org/EDUCATION/poster.cfm
How to Make a Great Poster

Review, Review, and Review

POSTER GUIDELINE 10:

Be ruthless when you edit.

After you have drafted the poster sections, check them for mistakes, legibility, and inconsistency in style. Try different layout arrangements if necessary. Test out the layout and content on other people such as friends, colleagues, and your supervisor. Be ruthless when editing.

27.7   PRESENTING A POSTER

POSTER GUIDELINE 11:

Be at your poster during the assigned poster
session to answer questions and to tell viewers
about your work.

Arrive early at the display site. Unless you are confident the organizers will have proper supplies, bring a poster hanging kit with you. Hang your poster straight and neat, and do not encroach on your neighbor's space. You may consider bringing copies of a handout or miniature version of the poster for your readers. Such handouts can easily be created when the poster is designed using a layout program. Put handouts, business cards, and reprints nearby—on a table or in an envelope hung close to the poster so people can take them as they are passing by. Do not forget to restock supplies periodically if your poster is up for a long time.

Although the material you are presenting should convey the essence of your message, make sure you are at your poster during your assigned presentation time to be available for discussion. In addition, ensure that you have prepared a 5- to 10-min talk, highlighting the key points of your poster. Your task as the presenter is also to answer questions and provide further details and to convince others that what you have done is excellent and worthwhile. During the actual presentation, focus on your graphics. Use your poster as a visual aid—do not read it! Tell viewers the context of your research problem and why it is important, the objectives and how you achieved them, as well as the data and its significance.

27.8   SAMPLE POSTER

Example 27-12 shows a well designed poster with annotations.

(With permission from Alexey Federov and Jaclyn Brown)

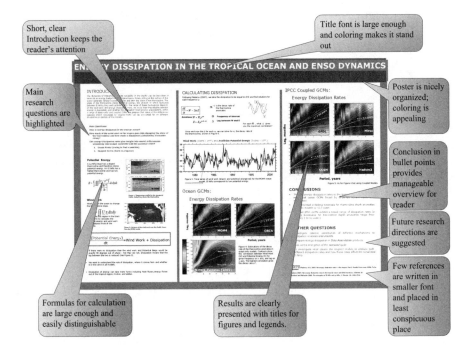

Short, clear Introduction keeps the reader's attention

Title font is large enough and coloring makes it stand out

Main research questions are highlighted

Poster is nicely organized; coloring is appealing

Conclusion in bullet points provides manageable overview for reader

Future research directions are suggested

Few references are written in smaller font and placed in least conspicuous place

Formulas for calculation are large enough and easily distinguishable

Results are clearly presented with titles for figures and legends.

27.9   CHECKLIST FOR A POSTER

Use the following checklist to ensure that you have addressed all important elements for a poster:

- ☐ 1. Do the illustrations tell the story?
- ☐ 2. Is the purpose of the research or topic stated precisely?
- ☐ 3. Did you avoid attaching the conference abstract?
- ☐ 4. Does the Introduction have the following components?
    - ☐ Background
    - ☐ Problem or unknown
    - ☐ Purpose/topic or review
    - ☐ Overview of content
- ☐ 5. Did you concentrate on the main points in each section?
- ☐ 6. Is the flow of the panels self-evident to the viewers?
- ☐ 7. Is the topic summarized and interpreted in the Conclusion section?
- ☐ 8. Do all figures and tables have a title and a legend?
- ☐ 9. Is your poster layout uncluttered?
- ☐ 10. Did you use visuals where possible rather than text?
- ☐ 11. Did you keep text to a minimum?
- ☐ 12. Is text written in sans serif font, and is the font large enough?

☐ 13. Are exhibits kept simple?
☐ 14. Are exhibits attractive? Is color used well?
☐ 15. Did you use active voice in the text?
☐ 16. Have all jargon and redundancies been omitted?
☐ 17. Did you proofread your text?

SUMMARY

POSTER GUIDELINES

1. Design the poster around your research question. Include
   Title
   Abstract
   Introduction
   Materials and Methods
   Results
   Conclusion
   (References)
   (Acknowledgments)
2. Concentrate only on the main points in each section.
3. Find visual ways to show your work—let the illustrations tell the story.
4. Vary the size and spacing of the poster sections to add visual interest.
5. Aim for about 20% text, 40% graphics, and 40% empty space.
6. Minimize text, and use images and graphs instead.
7. Prepare illustrations well ahead of time.
8. Keep exhibits simple.
9. Make exhibits look attractive.
10. Be ruthless when you edit.
11. Be at your poster during the assigned poster session to answer questions and tell viewers about your work.

CHAPTER 28

Oral Presentations

Every scientist should be able to prepare and deliver a good oral presentation. Although most scientists desire to present their work at international conferences, many of them also fear having to present a talk, especially if they are nonnative speakers. Unfortunately, as with scientific writing, most scientists are not formally trained in this art. However, being a good presenter is something that can be learned. For many people, this art is much easier to master than the art of writing a paper.

28.1   BEFORE THE TALK

ORAL PRESENTATION GUIDELINE 1:
Prepare your talk well ahead of time.

ORAL PRESENTATION GUIDELINE 2:
Practice, practice, practice.

ORAL PRESENTATION GUIDELINE 3:
Get to know your audience.

The two most important points in becoming a good presenter are being prepared and practicing. As soon as you know that you will speak, begin preparing your slides. Preparing your slides will already make your subconscious mind work on the words for the actual talk. To plan the best possible talk, you have to know your audience, however. Find out to how many people you will be presenting as well as their level of expertise. Design your talk accordingly in terms of its direction and necessary background information.

To deliver your talk well, it is of utmost importance that you practice and practice and practice. Without practicing a talk, you will not know whether you stay within the given time limit. Nor will you know if your flow of words is smooth or if your voice has the right pitch. Practicing will make you realize at least some of these potential problem areas. Consider videotaping yourself, or ask someone else to do this for you. Review the video and note any areas that may need improvement. In addition, take advantage of opportunities to give and practice presentations such as in departmental talks.

28.2   CONFERENCE TALKS AND ABSTRACTS

Getting invited to speak at a conference is an important recognition of your work and may not only help you to gather new ideas but also increase your visibility in the field. If you would like to get invited for a talk at a conference, you usually have to submit an abstract. You can submit abstracts for talks for preliminary work and pilot studies as well as for more polished work.

Sending in an abstract paper to a conference is not just a requirement to be considered for a talk, it also allows you to get initial feedback for work that you have not yet tried to publish. Thus, your abstract has to accurately summarize your work. As this abstract also has to convince the conference committee to invite you for a talk and the conference audience to attend your talk, it needs to take all these readers into consideration.

A conference abstract is usually longer than an abstract for a research paper. It may be 350 to 500 words long. The underlying format for a conference abstract is similar to that of a research paper but includes a title, a longer background/context portion, and sometimes a few references. The overall organization consists of

Background/context
Question/purpose
Experimental approach
Results
Conclusion (answer)/implication
(References)

See Chapter 14 for a more detailed description of these elements.

It is important to make a good first impression when submitting your conference abstract. Follow guidelines exactly or you may risk being eliminated from the start. Conference abstracts are usually considered a publication of a sort and many are published as a supplement to the association's journal.

Submit your abstract on or before the due date and in the required way, usually electronically or by e-mail. Ensure computer compatibility of documents, and include your name, title, organization, and contact details.

Abstracts are typically reviewed anonymously. A few conferences will send comments from reviewers about your abstract; this is very valuable information, and you should request it if available. When you present the same research at more than one conference, frame each paper a bit differently to match the focus of each conference.

28.3   CONTENT AND ORGANIZATION OF A SCIENTIFIC TALK

Content

To present a good talk within a given timeframe, you need to:

Choose what is most important
Display it in clear, uncluttered, visually attractive slides
Explain each slide in slow, simple, easy-to-understand English

The content of a talk is similar to that of a journal article with an introduction, results (combined with overview of experimental approach), discussion, and conclusion. However, an oral presentation needs to be structured and worded differently from a written paper. If you want to deliver your presentation well, do not just read it off a written text. Understand the differences of how information is extracted by readers versus listeners. Whereas text can be read and reread at your own speed, listeners get a chance to listen to a piece of information only once and at the speed the presenter sets. Similarly, listeners have no control over the order or type of information they will see and hear, whereas readers can easily scan headings and subheadings and skip ahead when they want. In addition, when presenting, the emotions of a speaker can easily be conveyed; this is much harder to do in writing.

If you have to present your talk in a language you do not know well, consider persuading a native speaker to listen to your talk and comment on pronunciation. Alternatively, ask a native speaker to make a tape recording of the talk for you. Listen to the tapes a few times. Note the pronunciation of difficult words and the intonation of sentences. If a translation service is provided for the talk, it's imperative to speak slowly so the translators do not get rushed in the translation—this cuts time off of your talk, as well. Practice the presentation as often as possible.

Layout of Talks of Various Lengths

Whether you are presenting a short, 10-min conference talk or a full hour seminar, follow the overall format of introduction, results, and discussion for your talk. Depending on your target audience, different numbers of introductory slides may have to be included. For a nonscientific audience, it is particularly important to include more background slides, as most speakers lose their audience in the first few minutes by failing to give an appropriate introduction to the problem.

Shorter talks are usually more difficult to prepare and to present than longer talks. For shorter talks, you need to be extremely selective of what is the overall most important information for each of the sections, and you need to prepare slides having this focus in mind. Generally, you will have to reduce the number of background slides for shorter talks and concentrate primarily on the main findings and their interpretations in the rest of the slides. In fact, depending on your topic and audience, you may have time to present only one or two main findings.

When you are time limited, consider skipping the title slide as well as the overview slide. Instead, just verbally inform the audience about your talk's overall title and about the outline for your talk. Do provide a concluding/summary slide in all cases.

General Organization

ORAL PRESENTATION GUIDELINE 4:

As an overview for your presentation:

Tell the audience what you are going to tell them

Tell them

Tell them what you have told them

ORAL PRESENTATION GUIDELINE 5:

Organize your slides and include:

Optional:	Title slide
First slide:	Overview of talk
Second slide:	Introduction and background
Subsequent slides:	Present what you studied and how you studied it. Present your results.

Final slide: Present your conclusions and the
                      main points that support it.

Optional final slide: Use a credit slide in which you
                      acknowledge those who worked
                      with you or financed your research.

Start your presentation with an overview of the talk by telling the audience what you will speak about: introduce the overall topic, mention how you will present your findings and that you will then summarize what you have told them in the talk.

Follow this overview with a presentation on the background of the overall topic that funnels down to your work, the unknown, and then the overall purpose/question of your project. State your experimental approach briefly when you present your results, and discuss findings in the general scientific context. In the final slide, present your interpretations and conclusions. When you are finished talking, thank the audience.

28.4   VISUAL AIDS

ORAL PRESENTATION GUIDELINE 6:

Know how to use visual aids.

Competent speakers must know how to use visual aids to clarify and reinforce their talk. Visual aids come in various shapes and forms. PowerPoint is currently the most powerful and impressive way of presentations. PowerPoint slides make a professional and effective presentation and can easily be changed at the last minute if needed. In PowerPoint presentations, you can also include short cartoons and jump back and forth between exhibits easily. You can either show the whole slide at once or build up an illustration using animations. Appropriate spots on the slide can be indicated using a laser pointer.

To present these slides, you need to know how to use a projector and be aware of potential problems arising when using different computer programs and systems. Particularly, when your presentation has to be converted from a Macintosh to a PC or vice versa, you need to check that figures and font types convert properly. Also check conversion between different programs or program versions on the same systems. Do not wait until shortly before your talk to check whether your slides will work correctly, however. Conversion problems may take time and expertise to resolve. Try out programs, equipment, and set-ups ahead of time, and bring your computer as a back-up. Be prepared to give the talk even if the slides fail.

As the speaker, you should take care not to obtrude the projection with your head, hands, or other parts of your body. In addition, ensure that you are not obstructing anyone's view.

Format of Visual Aids

ORAL PRESENTATION GUIDELINE 7:
Prepare visual aids well ahead of your talk.

ORAL PRESENTATION GUIDELINE 8:
Make exhibits look attractive.

ORAL PRESENTATION GUIDELINE 9:
Keep exhibits simple.

ORAL PRESENTATION GUIDELINE 10:
Think graphically.

Prepare visual aids well ahead of the meeting to give yourself time to check, replace, or improve them. Visual aids, regardless of whether you are using overheads, slides, or PowerPoint presentations, should be comprehensible on their own. Keep them simple. Your slides must be clear, legible, and easy-to-understand. Avoid non-standard colors, fonts, graphs, and abbreviations. You need to create exhibits that add to, not distract from your work. Your exhibits should be visually pleasing but professional. Moreover, individual slides should be recognizable as being part of a set (same colors, style of fonts and emphasis techniques).

Before you can prepare your slides, you need to decide on what and how much to include in your talk. A good rule of thumb is to communicate one main idea per slide, and emphasize this central message when speaking. Prepare slides with the audience in mind. The total number of slides you will be able to display in your talk depends on the time you are allotted to speak (see also Sections 28.2 and 28.3 for more information on content for various talks and time limits).

Format and Color

When presenting a talk, there is no substitute for clear, graphical visual aids. To create visually attractive slides, pay attention to how other people present colors, fonts, and graphics on slides that are easy on your eyes. Avoid bright colors, and avoid red/green (or blue/orange) color contrasts, as some people are color blind to these. Look for high contrast between background and writing or figures. Choose dark text against a light background or vice versa. Medium to dark blue background with white or yellow writing, for example, is commonly used and easy to read because of the high contrast between the background and writing.

Not only do your slides need to be visually attractive, they also should be simple. Your slides should be informative, but discipline yourself to use as few words as possible to convey this information. Use your voice to fill in the rest.

To create effective slides using as few words as possible, it is essential to know how to write clear but brief bullet points. Slides that are text heavy are usually constructed for the benefit of the presenter rather than for the benefit of the audience. Avoid such slides. To make your slides more attractive for your audience, use a maximum of 40 words per slide, 40 characters per line, and no more than 14 lines per slide for "word slides." Use key words and phrases—they are more effective than whole sentences. Avoid numbering items in a list. Use bullet points instead. Try to stick to three to five words per bullet point and to no more than seven bullet points per slide. At the same time, keep punctuation to a minimum, and start text at the left rather than in the center.

Visual fatigue is the biggest enemy of presentations, especially if you ask your audience to sit in a dimly lit room for 30 to 60 min. With just three to five words per bullet point, you will have to fill in the rest of the information, and the audience will have to pay attention to you, the speaker. This results in a more animated presentation, which is more interesting for the audience.

If your slides contain too much text, your audience will concentrate on reading the text and not listen to what you have to say, or they will be listening to you and not pay attention to what is on the slide. Neither case is what you as the presenter really desire. Here is an example of a slide that contains too much text.

 Example 28-1   Text slide

Overview of the Yale School of Medicine

- Founded in 1810, the Yale School of Medicine is a world-renowned center for biomedical research, education and advanced health care.
- The School is viewed internationally as a leader in biological and medical research.
- The Yale School of Medicine has over 900 faculty members and consists of 9 basic science departments and 17 clinical departments.
- The School of Medicine consistently ranks among the handful of leading recipients of research funding from the National Institutes of Health and other organizations supporting the biomedical sciences

To improve a text-heavy slide such as the one shown in Example 28-1, you need to decide on what information in each bullet point is really important and/or which bullet points can be omitted. Then list the important information by itself in a bullet point and omit the rest. A possible revision of Example 28-1 follows.

Revised
Example 28-1 A

Text slide

Yale School of Medicine Overview

- Founded in 1810
- Leader in biomedical research
- Over 900 faculty members
- 9 basic science departments
- 17 clinical departments
- Top biomedical research funding

In the revised version, the bullet points have been reduced to their main piece of information and are visually distributed better on the slide, all of which is preferred by the audience.

Note that lettering should be large enough. Ensure that the minimum font size is larger than or equal to 18 points. Use a sans serif font type such as Arial. Sans serif font types are easier to read than serif ones, such as Times New Roman. Lowercase lettering, with initial capitals where needed, is also easier to read than lettering that is all in capitals. Use maximum contrast and boldface for maximum legibility. All lettering on exhibits should be in the audience's language. Add emphasis by using colors and/or visuals such as photographs or figures where appropriate. Put the most important information in a larger print size. Italics can also be used to add emphasis, but avoid underlining or exclamation marks, as they are not considered effective.

If we further revise Example 28-1A, by adding a different background as well as a picture and using different styled and shaded text for

 Revised
Example 28-1 B

Text slide with background and graphic

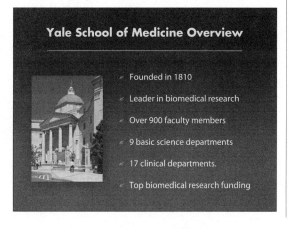

Yale School of Medicine Overview

- Founded in 1810
- Leader in biomedical research
- Over 900 faculty members
- 9 basic science departments
- 17 clinical departments
- Top biomedical research funding

the heading, a more visually attractive slide results, as shown in a further revision.

Revised Example 28-1B makes for a much more attractive and interesting slide than Example 28-1A. This is because Example 28-1B not only has a shaded background but also a pleasing picture that directly relates to the information on the slide.

Figures and Tables

Figures and tables on each slide should have a title, placed at the top and separate from the rest of the material by extra space. Use names for different groups rather than numbers or letters. However, do not simply transfer published figures or tables onto your slides. Their lettering is often too small, they do not have a heading, and more often than not, they contain additional information that is not needed for your slide.

Choose graphs rather than tables. Bar graphs are often preferable because their message can be quickly understood. In bar graphs, keep the number of bars to a minimum. If you can avoid using tables, do so. If you have to use a table, keep it to a maximum of four columns and seven or eight rows, including the title and column headings. Write all text horizontally, including labels for vertical axes if possible. For graphs, curves should be smooth and the lettering clear (i.e., Arial or Helvetica). Symbols must be easy to tell apart. Do not include a figure legend, but provide a key for figures if needed (see also Chapter 9 for advice on good lay-out of figures.) Make sure that exhibits are aligned well within the slide and with respect to each other and possible text.

An example of a well constructed slide containing figures is Example 28-2.

 Example 28-2 Slide with figures

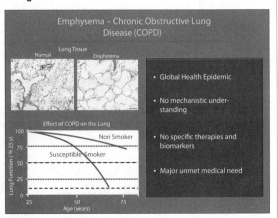

(With permission from Patty Lee, modified slide; and Robert Homer, EM images)
The figures and the text in Example 28-2 have been well placed. They are aligned nicely, visually pleasing, and easy to grasp. They are also clearly labeled, making for a very balanced slide.

Similarly, the schematic displayed in the slide for Example 28-3 is well balanced and easily graspable. The author could have used a text slide to explain the concept, but the schematic brings the message across much more effectively and memorably.

 Example 28-3 Slide with schematic

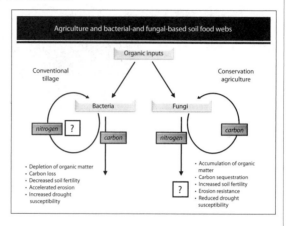

(With permission from Mark Bradford, modified)

Here is another slide in which a well-constructed bar graph is shown:

 Example 28-4 Slide with bar graph

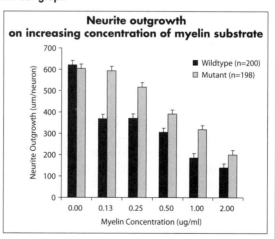

(With permission from Betty Liu, modified)

Note that in this slide, the title for the graph is at the same time the title for the slide. Depending on how your presentation is structured, it is possible to compose such titles if your slide contains a single figure. Note also that the slide contains no figure legend, but a key is shown for the graph. The presenter's words will fill in any additional necessary information and summarize the slide.

In other slides, the title may serve as a summary or overview of a finding depicted in the slide. This use of a title will reduce the number of words that need to be placed on the slide, as shown in the next example.

 Example 28-5 Slide with result in title

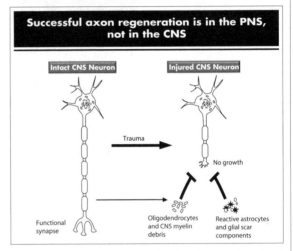

(With permission from SAGE publications)

28.5   PREPARING FOR A TALK

Time Limit

ORAL PRESENTATION GUIDELINE 11:

Stick to the time limit.

It is customary to assign a limit to the length of a speech. Adhere to this time limit. Find out whether you tend to speed up or slow down your talk during delivery. Allow for it during rehearsal. On average, you will be going through one to two slides per minute. Note that this is an average number, which very much depends on your slides. For some slides, you may spend considerably more time. For other slides, you may be able to go through them in substantially less time than the average. The only way to know how long it will take you is to time yourself when practicing your talk. Know that practicing in front of a small group of colleagues is more helpful than practicing in front of a mirror.

Nothing angers an audience or the organizers more than a speaker who goes overtime. To help you better keep track of the time, place a timer on the lectern. If you find that you need to leave out some important slides to stay within the time limit, do so. Do not speed through your talk. You run the risk of losing your audience if you present your material too fast.

If the chairperson signals to you that your time is up, summarize any remaining material, and give your conclusions immediately. You can prepare for this eventuality by inserting a hyperlink to the last slide strategically on some previous slides. This hyperlink may be disguised in a design feature or text of the slide.

Notes

ORAL PRESENTATION GUIDELINE 12:

Prepare notes.

Notes are the road maps for speakers. Too many notes can cause problems, however, and so can too few. To record notes, use index cards (3 in. × 5 in.) or write into the Notes section of your PowerPoint slides and display them using the presenter's view during your presentation.

Index cards are a convenient size to handle. They easily fit into a pocket and are not visible to the audience. Do not memorize a speech. Instead, strive to maintain an image of spontaneity. Most important, do not ever read a speech unless you are forced to by legal requirements or time restrictions.

Your notes on the index cards or on the computer should be easily readable—use extra-large print if necessary. Notes should be in outline form and should not be written in sentences. There are, however, a few exceptions when notes should be written out in full:

- Write out the opening of your speech.
- Write out the closing section of your speech.
- Write out transitions on note cards.
- Write out quotations in full.

Conquering Nervousness

No human has ever or will ever conquer nervousness. Only experience teaches a person to control his or her nerves and to appear confident to the audience.

Know that an audience usually will watch the body language of a speaker even more than listen to the speech. Thus, you need to use your body to communicate the message you want to present to your audience.

Confidence is the key: The most powerful way to appear confident is to look directly at the audience. You should also check your appearance and dress. Clothes should be immaculate and hair well groomed.

Study your surroundings. The more familiar you are with your surroundings, the more comfortable you will feel. Take a number of deep breaths while being introduced. If you have prepared and rehearsed the speech adequately, you may feel more comfortable. If your notes are well written, you are assured that a glance at them will put you back on track.

In your notes on the cards, or in your notes on screen, write out the opening sentence in full. Having it available in full will give you comfort in knowing where to find it if you need it. Deliver the opening sentence firmly and accurately—such delivery will give you confidence to continue. The early sentences of a speech should have been practiced over and over again, as confidence builds on itself. Nervousness normally abates when you realize that you are off to a good start.

Practice is the only way not to act in peculiar ways. Practicing in private helps only to a certain extent, however. Actually speaking before an audience is the only effective answer.

Self-Improvement Suggestions

To give yourself the opportunity to present the best possible talk, consider doing the following: Read aloud. Talk to yourself. Read to your children. Address your dog, and above all, practice in front of your colleagues. Hearing your voice is important as is finding the right intonation. Poetry makes the best material for practicing reading and speaking. Get away from the monotone. Learn to add excitement to your voice, and learn to add pauses for emphasis. Have fun while you are reading. Learn to read without gluing your eyes to the printed text. Use a recording device to play back your performance. Check quality of voice, word flow, delivery speed, vocabulary, grammar, and body movement. Become aware if you are uttering appalling sounds such as *"ums," "ahs," "uhs,"* and *"You know."* If so, break this habit. Videotaping yourself or recording your voice and playing back the tape is usually the best remedy for this annoying habit. Go over what you want to say often enough that the proper word will not be difficult to conjure. Moreover, actively seek speaking opportunities. <u>In short: Practice, practice, practice.</u>

Other important points

Be yourself. Being natural is the most valuable asset of a speaker.

28.6   GIVING THE TALK

Vocabulary to Know

> Dais or rostrum—a raised platform
> Podium—specialized dais used by orchestra conductors

Pulpit—specialized dais used by clergymen
Lectern—reading desk

Most speakers perform standing on a dais behind a lectern.

Setup

Before you present any talk, you should check the setup. Familiarize your-self with the room as well as with the equipment. Check lighting, plugs, chalkboard, and presentation equipment.

If you are planning to give a PowerPoint presentation without using your own computer, make sure your PowerPoint file is compatible with the program of the computer and the projector that you are planning to use. Make sure your notes are in the proper order. Last, but not least, get the correct pronunciation of the name of the person introducing you.

Going to the Lectern

Your movements should be unhurried and dignified. Do not begin your speech before you reach the lectern. Also, do not begin as soon as you reach the lectern. First, place your notes on the lectern. Place any other material you may need next to your notes, such as a pointer, a watch, or some water. Once you have your materials in place, compose yourself, and look out at the audience. Then thank your introducer and begin your speech.

28.7   VOICE AND DELIVERY

ORAL PRESENTATION GUIDELINE 13:
Make sure you can be heard by the entire audience.

ORAL PRESENTATION GUIDELINE 14:
Speak neither too fast nor too slow.

ORAL PRESENTATION GUIDELINE 15:
Avoid appalling sounds.

ESL advice

English is a language in which stress is crucial. Although pronunciation is important, it is less important than using the correct stress. If you have the stress right, you should not worry about having some kind of accent. If you have trouble pronouncing words correctly, it may help to put accent marks on syllables to be stressed in your notes and to mark places where your voice should pause. You may also want to underline phrases to emphasize.

Make sure you can be heard by the entire audience. Do not speak too softly. Soft speech signals that the speaker is uncertain. However, do not blast the audience out of its seat by the volume of your delivery either.

Pay attention to the pitch of your voice. Higher tones of pitch may lead the audience to assume that a speaker is less professional and more childlike. Speaking in deeper, fuller tones makes your voice more pleasant to listen to and can be achieved by using the diaphragm or lower throat to control the pitch of your voice rather than the upper throat or the nasal passages.

Speak neither too fast nor too slow. A good talk requires speech that is slower and clearer than in normal conversation. Speakers who are nervous often speak too fast. Try using a deliberate pace in speaking. Listen to yourself as you talk, and talk only as fast as you can comprehend it.

The Most Important Do's and Don'ts for an Oral Presentation

Do

- *Memorize the first few sentences*
- *Use a deliberate pace in speaking*
- *Look at the audience as much as possible*
- *Simultaneously show on your slides the information you are providing verbally*
- *Explain everything that is on a slide*
- *Show all key points on the screen*
- *Proofread your visuals for spelling*
- *Use spoken English*
- *Use simple words but technical terms*
- *Use uncluttered, visually attractive slides*
- *Use informative headings*
- *Be yourself—it is the most valuable asset of a speaker*
- *Pay attention to your body, arms, hands, and legs*
- *Dress appropriately*

Don't

- *Do not look only at the screen*
- *Do not look only at your notes*
- *Do not read word for word from your PowerPoint slides or from your notes*
- *Do not pace across the front*
- *Do not utter appalling sounds such as "ah" and "um"*
- *Do not speak in written English*
- *Do not fiddle with objects or play with your hair*
- *Do not read subheadings*
- *Do not use uninformative headings or text*
- *Do not skip over information on a slide*
- *Do not argue with a questioner in the audience*

28.8   VOCABULARY AND STYLE

Word Choice

ORAL PRESENTATION GUIDELINE 16:

Use smoothers and transitions.

Words used verbally are different from those used in written communications. When presenting a talk, you need the "smoothers" and soft transitions that are typically edited out in final drafts of published research articles. You need these soft transitions in your voice but not on the slides.

ESL speakers need to pay particular attention to the difference between written and spoken language. When you listen to native speakers talk, consider preparing a list for yourself that contains soft transitions. Do not overuse such smoothers, however. These transitions could include, for example

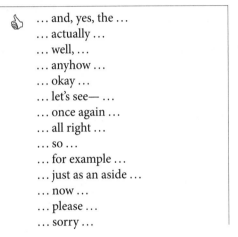

... and, yes, the ...
... actually ...
... well, ...
... anyhow ...
... okay ...
... let's see— ...
... once again ...
... all right ...
... so ...
... for example ...
... just as an aside ...
... now ...
... please ...
... sorry ...

As with soft transitions and smoothers, overview words and phrases are used commonly in talks but not in written presentations. Examples of overview words and phrases include the following:

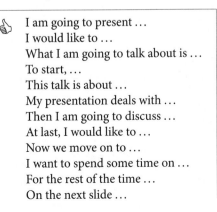

I am going to present ...
I would like to ...
What I am going to talk about is ...
To start, ...
This talk is about ...
My presentation deals with ...
Then I am going to discuss ...
At last, I would like to ...
Now we move on to ...
I want to spend some time on ...
For the rest of the time ...
On the next slide ...

ESL advice (side note)

Unlike overview words and smoothers, jargon is not preferred or accepted in a talk. Thus, get rid of the jargon (see Chapter 2, Section 2.4.) Listen to a recording of yourself. Do not use coarse words or profanity in a speech. Know your audience: Adjust the use of technical words accordingly, but never talk down to an audience, and do not use sexism or racism.

Grammar and Sentence Structure

Make a conscious effort to speak in reasonably short sentences. Do not let occasional slips of the tongue bother you. Correct yourself calmly and keep going.

Anecdotes, Jokes, and Personal Experiences

Do not forget that the message is the important factor. A joke is purely supplemental. Therefore, do not feel that your talk absolutely has to have a joke or funny cartoon. Many successful, clear presentations do not. If you do include humor in your presentation, ensure that it is not offensive to anyone in the audience and that your audience will understand the joke.

Personal experiences enliven and reinforce points made—provided, of course, that they fit logically into the speech. If you feel like telling your audience about a personal experience, I encourage you to do so.

28.9   BODY ACTIONS AND MOTIONS

ORAL PRESENTATION GUIDELINE 17:
Be conscious of body movement.

ORAL PRESENTATION GUIDELINE 18:
Keep eye contact.

ORAL PRESENTATION GUIDELINE 19:
Face the audience.

ORAL PRESENTATION GUIDELINE 20:
Stay within the presenter's triangle.

ORAL PRESENTATION GUIDELINE 21:
Use gestures.

Hands and Eyes

Be very conscious of head and eye movements. Eye contact with the audience is essential. Look at individuals in the audience, but do not exclude sections of listeners. Look at different sections of the audience at least 5 to

10 seconds at a time. If it distracts you to look at individual faces, you can look in between faces but at the level of faces for a large audience.

Do not look at your notes excessively. PowerPoint slides especially tempt speakers to glue their eyes on the computer screen. Resist this temptation; keep your eyes on the audience. Also, do not fix your eyes on a spot way beyond the EXIT sign at the back of the room.

Body

Keep the front of your body facing the audience as much as possible. Avoid hiding behind the lectern. Consider stepping out next to the lectern occasionally or moving closer to the audience at times, but avoid turning your back to the audience and speaking at the same time.

The best position for a presenter is on the left side of the room or screen as seen from the audience when facing the screen (Figure 28-1). This position will not only ensure that you do not block the projection of your slides but also that you will not block the view of the audience. It is also better to work your bullet slide from the left side, as text is written from left to right.

Arms and Hands

Use gestures to reinforce and complement your talk. Gestures should be smooth and not jerky and not a wild, windmill style. When you are not using your hands and arms, let them hang naturally. Do not stick your hands into trouser pockets. Do not grip the sides of the lectern. Do not cover the front of your body with your arms or fold your hands. Do not use your hands to straighten your clothing, rub your nose, explore your ears, smooth down your hair, or play with keys, bracelets, and so forth. Do not stand rigidly.

If you are using a laser pointer, learn how to use it correctly. Above all, learn where the "Off" button is. To use a laser pointer correctly, learn

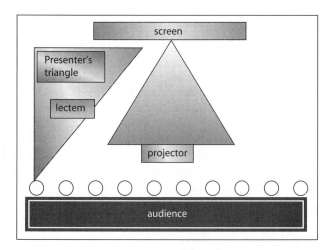

Figure 28-1 Ideal presenter's location

to employ a single, steady spot of light to show the audience where to look. Keep the light pointed at the spot 2 to 3 seconds and then turn it off. Pointing to items is most effective if you do not talk while you are pointing the laser to the point of interest. Talking will distract the audience from where to look. Exceptions exist such as when you are explaining a flowchart or comparing items on a slide. Hold the laser pointer in the hand closest to the slide presentation so you do not have to cross your body with the pointer or turn away from the audience when pointing out things on the screen. If your hands tend to shake while you are pointing, consider supporting the pointing hand with your other hand, or resting it on the lectern.

Feet

Your feet should be securely on the floor, each leg carrying an equal share of your body weight. Balance may be shifted occasionally but only for body comfort. Do not teeter back and forth or pace. Stay at the lectern unless you need to point out data on a slide or exhibit. However, when you are engaging the audience directly, outside the formal talk (during question and answer session for example), step away from the lectern to its side or move closer to the audience. When you show slides, turn halfway toward the screen rather than turning your back on the audience. Face the audience again after pointing out relevant parts of each slide. In general, your feet should neither be seen nor heard.

28.10 AT THE END OF THE PRESENTATION

ORAL PRESENTATION GUIDELINE 22:

Make sure that you are in charge.

ORAL PRESENTATION GUIDELINE 23:

Stay calm and polite.

When your speech is ended, stand for a short moment doing nothing. If your words and the tone of your voice do not make it clear that you have finished, you can thank the chairperson, or just say "Thank you" and stop. Do not worry about ending a bit early. No one has ever been upset if a speaker ended early, but the audience is easily upset if a speaker ends late. If questions are to follow immediately, stay at the lectern. Do not ask for questions yourself—this is the chairperson's job. Acknowledge any applause or "Thank yous." Gather your notes deliberately; also gather the remainder of your items. Then walk back to your seat dignified and unhurriedly.

28.11 QUESTIONS AND ANSWERS

How to Answer Questions

Often an oral scientific presentation is followed by a brief question-and-answer period during which anyone in the audience can ask a question of the presenter. Many, if not most, presenters are nervous about this period, especially about being asked questions to which they might not know the answer. The question-and-answer period is particularly frightening for nonnative speakers. The following are a few pointers to help ease you through your concerns.

Anticipate questions before the talk already. Try to look at your talk from the audience's point of view and envision what questions you would be asking. Practice receiving and answering questions with your peers, your Principal Investigator, or a colleague. The best way to practice with a colleague is to go to a quiet conference room. Give the talk standing up, using your slides (even if not quite finished), a projector, and laser pointer; in short, pretend to be in front of your audience. Be prepared to accept criticism from your colleagues.

To deal with questions after a talk in the best possible way

Do

- *Be courteous in your answers at all times*
- *Tell the audience that if there are questions, you will be happy to answer them at the conclusion of your talk*
- *Direct the answer to the entire audience*
- *Admit when you do not know the answer*

Don't

- *Do not give your audience an opportunity to interrupt your presentation*
- *Do not maintain eye contact with the questioner when you give your answer*
- *Do not make up an answer if you do not know*

If no questions are forthcoming during the question period

At many conference meetings, the chairperson is instructed to think up a question in case no one else from the audience asks one. Alternatively, before you begin your speech, plant the first question with the chairperson or with a friend in the audience. Once the ice is broken, other questions should flow spontaneously. Sometimes, you can start the session by asking a question yourself: "Many of you may have been wondering how ... ," and then go on to give the answer. Alternatively, you may ask the audience a question. Note, however, that the latter two options should be employed rarely, as in many fields they may be considered inappropriate.

If there is no chairperson

You must exercise firm control so that not too many people compete for questions at the same time. You must also be prepared to deal with a member of the audience who wishes to make a speech of his or her own. If time is running out for the question period, announce that you will be able to accept only one more question.

If questions are not relevant
Sidestep these questions gracefully. Be especially gracious when you duck a question.

If a single individual in an audience digs in and will not give up trying to turn the question period into an argument
Handle this person politely but resolutely. Isolate the opposition. Smile at the person. Say "It looks as if we do not agree on this point. Rather than take the time of the whole group, why don't we meet in the bar this evening and discuss it further." The odds are 10 to 1 that person will never show.

If a questioner is never satisfied with an answer but counters with a further question
Never look at such a questioner when you complete your answer. Look somewhere else and pick a new question as quickly as possible from a different sector of the audience.

Difficult questions
When a question is a tough one, always repeat it, ostensibly, to make sure all of the audience has heard it. This gives you a few seconds to think about the answer. You have one of several options:

- Ask if someone in the audience will help you answer
- Ask the questioner to rephrase the question
- Ask the questioner to come talk to you after the session
- Say: "I do not understand your question, please explain."
         "That is a good question, I will think about it."
         "I wish I could answer that."

28.12 OTHER SPEECH FORMS

SPEECH GUIDELINE 1:
Know how to make a proper introduction.

SPEECH GUIDELINE 2:
Know how to give an impromptu talk.

If you want to become a well-rounded speaker, you will not only have to learn to master the art of preparing and delivering presentations, but you will also have to learn to handle other speaking assignments confidently.
You must learn to

- make a proper introduction, and
- talk intelligently when called on unexpectedly

Making an Introduction

As an introducer, you should assume a secondary supportive role. The person being introduced is the star of the show.

The speaker who is being introduced has a name. Use it. Ask your speaker the exact form of his or her name you should use. Write it down, and get the pronunciation right. Ask the speaker what he or she would like you to say in your introduction. Do as the speaker wishes.

When the introduction has been completed, you say, for example, "Mary Peters will talk about … (Ladies and gentlemen), Mary Peters." When you have introduced the speaker, step away from the lectern and sit down.

At the end of the presentation, move back to the lectern and stand at the side of the speaker. You say, for example, "Thank you Mary Peters," and introduce the question and answer session if it is to take place immediately following the presentation.

Impromptu Talk

As a scientist you should be able to reply to a question and make spontaneous commentary when called upon unexpectedly. Such impromptu talks require background and a pool of knowledge to draw upon.

If you are called to the dais unexpectedly, keep your cool even though somebody called on you without warning. If you remain calm despite adverse feelings, you will win the audience on your side.

Walk SLOWLY. This gives you a few seconds to think. What should you be thinking about? Your opening sentence—and nothing else.

When you reach the lectern, politely acknowledge the chairman, and look calmly at the audience. Say, for example, "I am delighted to have this opportunity to tell you about…". Talk for a few minutes after delivering your opening sentence until you have had time to work out your closing sentence. Try to work out a good closing sentence: For example, you could shortly summarize again what you have talked about. Then smile for the last time to the audience, nod to the chairman, and leave the dais.

Remember

A good impromptu talk should never be more than 3 to 5 min long. Be conscious of elapsed time. Usually, 5 min is the outside limit for impromptu remarks.

A strong opening is essential. A strong close is even more important. What comes between should be short and concise. When you have delivered your close, stop and keep silent from then on. Remain polite at all times.

Things to Avoid

Omit afterthoughts. Stop when you come to your first close. Do not open your mouth again no matter what beautiful thoughts float into your mind.

28.13 CHECKLIST FOR AN ORAL PRESENTATION

Use the following checklist to ensure that you have addressed all suggestions in preparing for your talk:

- ☐ 1. Did you practice your talk—a lot?
- ☐ 2. Are your slides informative?
- ☐ 3. Is your talk within the time limit?
- ☐ 4. Are you aware of any appalling sounds or habits you show when presenting?
- ☐ 5. Did you find out who your audience will be?
- ☐ 6. Are your slides logically organized?
- ☐ 7. Do you have an overview slide?
- ☐ 8. Do you have a summary slide?
- ☐ 9. Did you prepare notes?
- ☐ 10. Did you write down
  - ☐ i. Your opening statement?
  - ☐ ii. Important transitions?
  - ☐ iii. Concluding remarks?
- ☐ 11. When preparing your slides, did you concentrate on the main points in each portion of your talk?
- ☐ 12. Do all figures and tables have a title and a legend?
- ☐ 13. Are visuals and text aligned well in each slide?
- ☐ 14. Is each slide logically organized and uncluttered?
- ☐ 15. Did you use visuals where possible rather than text?
- ☐ 16. Is text used sparingly but informatively?
- ☐ 17. Is the font large enough?
- ☐ 18. Are slides/figures/tables kept simple?
- ☐ 19. Are exhibits attractive? Is color used well?
- ☐ 20. Did you proofread your text?
- ☐ 21. Have you familiarized yourself with the setup?
- ☐ 22. Did you ensure that there will be no ugly compatibility problems in between computers or versions of computer programs?
- ☐ 23. Did you pack a (laser) pointer?

SUMMARY

ORAL PRESENTATION GUIDELINES:
1. Prepare your talk well ahead of time.
2. Practice, practice, practice.
3. Get to know your audience.
4. As an overview for your presentation
   Tell the audience what you are going to tell them
   Tell them
   Tell them what you have told them
5. Organize your slides and include

Optional:	Title slide
First slide:	Overview of talk
Second slide:	Introduction and background
Subsequent slides:	Present what you studied and how you studied it. Present your results.

Final slide:	Present your conclusions and the main points that support it.
Optional final slide:	Use a credit slide in which you acknowledge those who have worked with you or financed your research.

6. Know how to use visual aids.
7. Prepare visual aids well ahead of your talk.
8. Make exhibits look attractive.
9. Keep exhibits simple.
10. Think graphically.
11. Stick to the time limit.
12. Prepare notes.
13. Make sure you can be heard by the entire audience.
14. Speak neither too fast nor too slow.
15. Avoid appalling sounds.
16. Use smoothers and transitions.
17. Be conscious of body movement.
18. Keep eye contact.
19. Face the audience.
20. Stay within the presenter's triangle.
21. Use gestures.
22. Make sure that you are in charge.
23. Stay calm and polite.

SPEECH GUIDELINES:

1. Know how to make a proper introduction.
2. Know how to give an impromptu talk.

PROBLEMS

Problem 28-1

The following slide is text heavy. Reduce the amount of text to maximum five words per bullet point.

2007 Focus Group

Patients emphasized the value of:
- The welcome environment of the clinic
- Being able to communicate clearly in their native language
- The comprehensiveness of care offered during a visit
- Being seen by professional and caring clinical teams
- Feeling respected as human beings

Problem 28-2
Evaluate the following slide. How could it be improved?

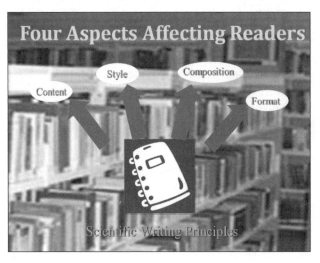

Problem 28-3
The following slide is text heavy. Reduce the amount of text per bullet point. Suggest what the presenter could do to highlight on the slide what part of the talk he or she is currently presenting.

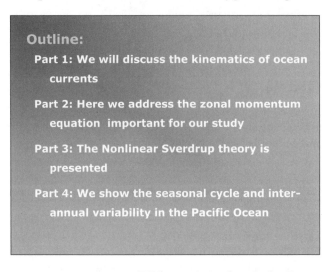

(With permission from Jaclyn Brown, modified)

Problem 28-4
Assess the following slide. Explain why this is not a good slide.

TABLE 2
List of Proteins Renatured by the Sparse Matrix Approach and Their Optimum Renaturation Conditions

Protein	Molecular mass (Da)	Structure	Standard assay buffer	Optimum renaturation buffer system	Max activity recovered (percentage of initial value)
BAP	80,000 (29)	Homodimer Zn²⁺, Mg²⁺ cofactor	100 mM NaCl, 5 mM MgCl₂, 100 mM Tris, pH 9.5	0.2 M Na acetate, 0.1 M Tris–HCl, pH 8.5, 30% (w/v) PEG 4000	138
HRP	40,000 (30)	Monomer, heme group, Ca²⁺, carbohydrate	100 mM CH₃COONa, pH 4.2	0.2 M Mg acetate, 0.1 M Na cacodylate, pH 6, 30% (v/v) 2-methyl-2,4-pentanediol	33
β-gal	540,000 (31)	Tetramer, with independent active sites (32)	Z-buffer (see Materials and Methods)	30% (v/v) PEG 1500	81
Lysozyme	14,388 (33)	Monomer	0.1 M K phosphate buffer, pH 7.0	0.2 M Mg acetate, 0.1 M Na cacodylate, pH 6, 30% (v/v) 2-methyl-2,4-pentanediol	333
Sperm bindin	24,000 (23)	Unknown	Seawater	0.2 M Na citrate, 0.1 M Na Hepes, pH 7.5, 30% (v/v) 2-methyl-2,4-pentanediol	100
Recombinant biodin	24,000 (34)	Unknown	Seawater	0.1 M Na Hepes, pH 7.5, 1.6 M Na, K phosphate	100
Trypsin	23,800 (35)	Monomer (α) dimer (β)	10 mM Tris, pH 8.0	0.2 M Mg chloride, 0.1 M Na Hepes, pH 7.5, 30% (w/v) PEG 400	11
Acetylcholinesterase	260,000 (36)	Aggregates, monomer	0.2 M Na phosphate buffer, pH 7.0	0.1 M Na Hepes, pH 7.5, 0.8 M Na phosphate, 0.8 M K phosphate	9

(With permission from Elsevier)

Problem 28-5
Assess the following slide. Explain why this is not a good slide.

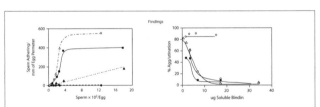

Fig. 4. Species specificity of sperm adhesion in *S. franciscanus* and *S. purpuratus* gametes. The number of adherent sperm was scored as a function of sperm concentration for all possible combinations of *S. franciscanus* and *S. purpuratus* gametes. ●, *S. purpuratus* sperm × *S. purpuratus* eggs; ♦, *S. purpuratus* sperm × *S. franciscanus* eggs; ▲, *S. franciscanus* sperm × *S. purpuratus* eggs; ■, *S. franciscanus* sperm × *S. franciscanus* eggs. Significant numbers of *S. franciscanus* sperm adhere to the surface of *S. purpuratus* eggs. The number of adherent sperm is normalized to account for the larger surface area of *S. franciscanus* eggs.

Fig. 2. Inhibition of egg agglutination by soluble bindin. *S. purpuratus* soluble sperm bindin inhibits egg agglutination non-species specifically. (○) Soluble *Stronglyocentrotus purpuratus* sperm bindin added to *S. purpuratus* eggs in the presence of particulate *S. purpuratus* bindin; (△) soluable *S. purpuratus* sperm bindin added to *S. purpuratus* eggs in the presence of particulate *S. franciscanus* bindin; (■) soluable *S. purpuratus* sperm bindin added to *S. franciscanus* eggs in the presence of particulate *S. franciscanus* bindin; (◇) soluable synthetic peptide corresponding to residues 69-130 of *S. purpuratus* bindin added to *S. purpuratus* eggs in the presence of particulate *S. purpuratus* bindin (see text)

(With permission from Elsevier)

Problem 28-6

Explain why the following statements are not good choices for an oral presentation:

1. "Thank you for listening to my talk. I hope it was not too confusing or boring."
2. "On this slide, please focus only on part D, and ignore parts A, B, and C."
3. "First, I will provide an overview, then tell you about the methods, show some specific results, and last, I will summarize my talk."
4. "Western blot analysis. Our Western blot analysis showed that . . . Sequencing. When we sequenced the insert . . ."
5. "This finding is in agreement with that of a previously published result reported by Lopez et al. in 2001 where it was shown that emergence of seedlings is temperature and humidity dependent."
6. "It was determined that frogs can hibernate under water for up to six months."

PART SIX

Job Applications

CHAPTER 29

Writing for Job Applications

29.1   OVERALL

Many students seek careers in science hoping to improve people's lives and to find knowledge from which all people can benefit. In fact, a background in science and engineering has become fundamental to more and more careers, as science and technology have become more central to modern society. Contributions of scientists extend well beyond research and development into teaching, business, industry, and government. People with bachelor's, master's, and doctoral degrees in science or engineering are involved in establishing and running companies, practicing law, fundraising, formulating policy, and consulting. Their careers reach far across businesses and institutions and cross national boundaries. The experience of doing scientific or technical work is supremely exhilarating for those with sufficient interest and determination.

To enter into a scientific career, or to change jobs and even careers, you need to be familiar with documents needed for job applications. In the next sections, I outline the most important documents required for job applications in the scientific and technical fields.

29.2   CURRICULA VITAE (CVs) AND RÉSUMÉS

JOB APPLICATION GUIDELINE 1:
Tailor your CV or résumé specifically to the position
and to the organization.

JOB APPLICATION GUIDELINE 2:

Your CV should be well presented and flawless.

The CV or résumé is typically the first item that a potential employer sees. A *curriculum vitae,* commonly referred to as a CV, is a summary of your educational and academic backgrounds, which is used primarily in academic and medical settings. A résumé is used for seeking employment in the private sector and contains a summary or listing of your relevant job experience and education.

For your CV or résumé, use a common font such as Times, Palatino, or Arial in 12 point. Use underlines, boldface, and italics sparingly, avoid fancy bullets and other ornaments, and keep left indents to a minimum.

International applicants often write the type of personal information on a *curriculum vitae* that would not be included on an American CV or résumé. When applying for a job position within the United States, provide only your full name along with your address and e-mail. Do not include birthday, marital status, or pictures.

ESL advice

CVs

CVs are usually very comprehensive and elaborate on professional history including every term of employment, academic credential, publication, contribution, or significant achievement. Typically, CVs have no length restrictions, but it is essential that your CV is clear, concise, and honest. Your CV should also be well presented and flawless.

The CV should start with your basic details including your contact information. After your contact details, include your credits and training or vice versa, depending on your personal preference. As your life is constantly changing and your career developing, you must consistently update this information. The most effective CVs are those that are customized to a specific application. Most potential employers will only glance at a CV; therefore, you need to highlight your main qualities on the first page (or two). To make your CV as effective as possible, think about what skills and qualities your potential employer may wish to invest in and why. Then organize and present your information based on the interests of the employer. Ensure that your CV is addressed to the organization to which you are currently applying—not the one to which you applied previously! Check that the address on the letter matches the one on the envelope.

CVs differ widely, depending on personal preference, where you are applying to, and your experience. The following list will give you an example of what categories you may include in your CV and approximately in what order:

Name & address (no personal information)
Education, with degrees & dates (most recent dates first)

Clinical certifications, with dates
Employment history, brief description & dates; most recent
    dates first (& brief description)—may also be placed in first
    position
Honors and awards (predoctoral and postdoctoral)
Grant funding
Leadership and service
Teaching experience
Laboratory skills
Publications—name bold faced, co-first authors easily identified
Invited presentations and seminars
Professional qualifications
Certifications and accreditations
Computer skills
Language proficiency
Unique technical abilities
Professional memberships

Note that for a CV for an academic job, the list of peer-reviewed publications is very important. List only those that are published or accepted for publication and not those submitted or in preparation. Often the applicant's name is in bold in the list of publications to make it immediately visible. List also any currently funded grants (amount, type, length, and agency) that you can bring with you. Such active grants may be extremely important for securing a new position.

The next example shows a sample CV that serves as an example of the appropriate format for a *curriculum vitae*. Note that educational qualifications and work experience are listed in reverse chronological order and account for the job seeker's entire career history.

 Example 29-1 Sample CV

JANET MILLER

Address
Telephone and e-mail

——————— **EDUCATION** ———————

Ph.D. Chemistry; *University of Utah*

      Area of Specialization: biological chemistry   2006

B.S. Chemistry, *cum laude*; *Washington University of California*   2001

————— **PROFESSIONAL EXPERIENCE** —————

Assistant Professor Yale University, Department of Biological Chemistry,   2008–present

———————— RESEARCH EXPERIENCE ————————

Postdoctoral Fellow: University of Pennsylvania; laboratory of
Dr. Enzo Russo,   2006 – 2008

———————— PUBLICATIONS ————————

Books

Miller, J. H. "Chemical synthesis—the new era," Oxford
  University Press, accepted for publication in 2008.

Articles and Book Chapters

Schulz, P.A., **Miller, J. H.,** Hartmann, E. (2008). Title. *Journal of
  Chemistry* 126: 1–8.
Miller, J. H., Knight, C., Cove, D. (2007). Title. *Science* 261:
  92–99
. . .

———————— ABSTRACTS ————————

Miller, J. H. and Kabe, C.G. (2008). Title. Protein Society
abstracts.
. . .

———————— COURSES ————————

. . .

———————— TALKS ————————

Synthesis of Y   VIIth International Conference on
                         Developmental Biology of the Sea Urchin.
                         Woods Hole, Massachusetts   2007
. . . .

———————— REFERENCES ————————

available on request

Résumés

In contrast to a CV, a résumé contains only experience directly relevant to a particular position. Your résumé should be tailor-made according to the position for which you intend to apply. It should be job oriented, goal specific, and very concise. Typically, résumés for industry are much shorter than those for academia, between 2 and 4 pages.

As the search for employment has become more electronic, résumés have followed suit. It is common for employers to accept résumés electronically and for job seekers to send out résumés electronically. However, beware that if you mass mail your résumé, it may no longer be tailored to a specific job position, which can have a negative effect on your chances of securing employment. Also know that using a required file format can create problems in maintaining formatting of your document.

Example 29-2

RÉSUMÉ

Peter Jones

CONTACT ADDRESS   Address; Telephone; Fax; e-mail

PROFILE

- Highly motivated organic chemist with extensive experience in organic synthesis, process chemistry, formulation, and chemical analysis
- Expert in GMP processing/production
- Excellent skills in synthesis of natural products with biological activity, heterocyclic chemistry for cancer, HIV
- Familiar with operation/data analysis of NMR, HPLC, IR, UV
- Excellent team player with good verbal and written communication skills

PROFESSIONAL EXPERIENCE

Jul 2006 – to present   **Sr. Scientist I/Sr. Investigator II; Bayer Pharmaceuticals**

- Led a research group specialized in biologically active molecule synthesis and custom synthesis
- Managed a process lab for preparation and processing of X
- Investigated quality and stability of drug products for preclinical and clinical trials

Nov 2003 – Jul 2006   **Chemist, Abbott Laboratories**

- Responsible for designing and conducting complex, multistep synthesis
- Synthesized carcinogenic substances for chemical toxicology research

Nov 2000 – Nov 2003   **Postdoctoral Fellow, ABC University**

- Studied biology of carcinogens
- Synthesis of carcinogenic compounds for cancer research

EDUCATION

- 1996: *Ph.D. in Biochemistry*, The University of Chicago
- 1992: *MS in Chemistry*, New York State University
- 1984: *BS in Chemistry*, Summa Cum Laude, Concordia College, Minnesota

AWARDS

2008   Nominee for Concordia College Alumni Achievement Award

2006   Outstanding Achievement Award, Abbott Global Medical Affairs

PROFESSIONAL ACTIVITIES

GRANT SUPPORT

- Ongoing Grants
- Complete Grants

SELECTED PUBLICATIONS

29.3   COVER LETTERS

JOB APPLICATION GUIDELINE 3:

Tailor the cover letter specifically to the position.

When you are applying for a job, you will not only have to send a CV or résumé but also a cover letter. Your cover letter creates a professional impression from the outset. Like the CV, it should be flawless, as it can make or break your application. In your cover letter, highlight the most relevant parts of your CV that can show your potential employer what you can do for them. Wherever possible, address your cover letter to a specific person, even if that means you have to call the organization. Check (and double-check) the letter for spelling and grammatical errors. If you are sending a hard copy, print it onto good quality paper.

The cover letter should be tailored specifically to the position and should be 1 to 1.5 pages long. After introducing yourself, highlight your accomplishments and state your research goal. Then state why you believe your expertise would add to the department to which you are applying. Here is an outline for the general organization of a cover letter:

Opening paragraph
    State the purpose of your letter
    Mention how you heard about the job
Middle paragraph(s)
    Highlight your past accomplishments
    Describe your research goals
    Explain why you believe you are a good fit for the position
Closing paragraph
    Mention any enclosures (CV, publication samples, Teaching
        Statement, etc.)
    Make positive closing remarks

Example 29-3 shows a sample cover letter that follows the outline given previously:

 Example 29-3 Cover Letter

Date

Dear Prof. Ying:

Purpose and name of position

I am writing to apply for the Assistant Professor position in your Department as posted in the October 10th issue of *Science*. I am convinced that my solid training in X-ray crystallography, my strong track record in cutting-edge research, and my long-standing passion for teaching make me well-suited for the position outlined in your advertisement.

Accomplishments and goals	In 2006, I completed my graduate studies at the University of California, San Francisco with a Ph.D. degree in biochemistry. Subsequently, I spent five years in Dr. Hannes Kari's group at the ABC University School of Medicine as a postdoctoral fellow, focusing mainly on structure function studies of the ribosomal protein Y. During this period, I was part of a team actively involved in solving several crystal structures of protein complexes involving this protein for several different organisms (see enclosed publications.) My current research is focused on ...
Reason for fit	The research and mission of your institute highly complements my own goals and objectives, and my expertise would add to your department's strength in molecular and structural mechanisms of biological processes. It would be an honor to join the University of Tennessee as an assistant professor, and I believe that my skills and talents will be a valuable asset for the University. In addition to leading my own laboratory, I am very enthusiastic about teaching, in which I have been actively engaged for the past five years. Classes that I have taught and which may be of particular interest to your department include ...
Enclosures and ending on positive note	Please find enclosed my *curriculum vitae*, Research and Teaching Statements, as well as three of my recent publications. The contact information for my references is included in my *curriculum vitae*, and letters of recommendation will be arriving under separate cover. Thank you for your consideration. I look forward to hearing from you.

Sincerely yours,

John Smith, Ph.D.

Enclosed: CV, Research Statement, Teaching Statement, Publications (3)

29.4   ACCOMPANYING DOCUMENTS

Aside from a cover letter and CV or résumé, you may be asked to send a Research Statement, a Teaching Statement, and letters of recommendation. These documents together will tell your potential employer about your education, skills, and expertise.

Research and Teaching Statements are components of many academic job applications for the life sciences and social sciences. A Research Statement summarizes your research achievements and proposes future studies. A Teaching Statement summarizes your teaching achievements, describes your teaching philosophy, and proposes planned courses or other educational components. These statements are usually requested for postdoctoral and faculty positions and are announced in academic job postings.

Even if you are not requested to write a Research or Teaching Statement, consider doing so, as writing them will help you focus your professional goals and improve your interview performance. For your potential employer, these statements, along with your CV, cover letter, and letters of recommendation, are important indications of your job readiness, your areas of specialty, your potential to get grants, your academic ability and research needs, as well as your compatibility with the department or school to which you are applying.

Each scientific field has different expectations for Research and Teaching Statements. It is therefore a good idea to obtain examples of statements. Such examples may be hard to come by, as many researchers consider them confidential information, especially in a competitive field. Try searching online for examples, but also do not hesitate to ask your advisor or other people in your department.

As with all other types of written communication, revise repeatedly, and have someone else read and comment on your statement(s) before you apply. These statements are very important, as they can make or break a job application. They should be written with great care and revised several times before you send them out.

29.5   RESEARCH STATEMENTS

RESEARCH STATEMENT GUIDELINE 1:
State your research achievements, current aims, and future goals.

RESEARCH STATEMENT GUIDELINE 2:
General Content
(Abstract)
Background
Current research
Research agenda
Relevance

RESEARCH STATEMENT GUIDELINE 3:
Tailor your Research Statement to a particular job posting and institution.

A Research Statement summarizes your research achievements to date, indicates your current aims, and states your future goals. It also describes how your research contributes to the field. The Research Statement should convince potential employers that you are knowledgeable and expert enough to carry out the proposed research. It should also show that your research will be different, important, and innovative. Like the CV and cover letter, your research statement should be tailored to a particular job posting and institution.

Content and Organization

The amount of detail and length of Research Statements vary among disciplines. Usually, Research Statements include an overview of your research, background information to show you are on top of your field, key issues remaining in the field, what directions you intend to take to contribute to the field, and why you are motivated to pursue these studies. Your statement may also include a brief abstract, although its inclusion is not required; and mention how you will incorporate graduate and undergraduate students in your research projects.

On average, Research Statements are about two to five pages long, and content is logically divided using headings and subheadings. Bullets and figures can also be found in these statements. The following list provides an outline for the different sections of a Research Statement:

Abstract (optional)	sometimes useful, but not required; provide an overview of your research and proposed plans; include background, overall objective, focus of work, approach, and significance.
Background	give context to your research efforts and describe relevant past research important in your future professional plans; keep the "big picture" in mind.
Current research	describe key findings and their importance as well as current promising lines of inquiry
Research agenda (3–5 year)	
	state your short- and long-term goals, approaches, and expected outcomes; provide your top 2 or 3 research questions/specific aims/hypotheses; proposed plans should be specific, credible, and realistic.
Relevance	indicate your studies' relevance to your field, potential employer, and to society

The following example shows a Research Statement that has been constructed based on an outline similar to the one shown previously:

 Example 29-4

Research Statement

John Smith

Previous Research

I have a long-standing interest in the molecular determinants for axon outgrowth in the central nervous system. While my Ph.D. thesis and earlier postdoctoral work focused on cell motility and cell-matrix adhesion, my years as an Associate Research Scientist centered on the mechanism by which CNS myelin inhibits axon outgrowth.

Research focus

Previous research	During my graduate studies in the Lineberger Cancer Center at the University of North Carolina, Chapel Hill, I studied the regulation of Rho family GTPases in response to extracellular matrix and growth factor signals. As a postdoctoral fellow at Yale University, I studied the molecular determinants for axon outgrowth in the central nervous system (CNS).
Current research findings	In my current postdoctoral work at Harvard University, I have focused on determining the mechanism by which CNS myelin inhibit axon outgrowth. . . . We have elucidated NgR-ligand interactions at the molecular level, and this will allow us to develop peptide antagonists to disrupt this receptor-ligand interaction and thereby promote axon regeneration after injury.

Future Research

I. Identify mutations in NgR that increase risk for schizophrenia.
These specific NgR mutants fail to transduce myelin signals and function as dominant negatives for endogenous NgR (Budel et al., submitted). I aim to determine the underlying mechanism for NgR signaling in humans, particularly in the case of schizophrenia. Our findings may aid in the development of NgR antagonists, thereby allowing axon regeneration after injury (Li et al., 2004).

Specific aims and expected outcome

II. Determine NgR interactions and effects on other cellular pathways . . . .

Overall objective and significance	Overall, my ultimate goal is to determine the roles of these mechanisms in distinct neuropsychiatric disorders and work toward the application of these findings to the recovery of a damaged neural system . . . . I take a strong experimental and collaborative approach to addressing these questions . . . .
Collaborations	Both of the above projects are collaborative, which I find the most productive (and enjoyable!) manner in which to further scientific understanding. In my current position at Harvard, I also initiated a collaboration with a geneticist to investigate . . . . I would bring this approach to ABC University and use it to address questions related to . . . .
Relevance	As a young faculty member, I will successfully synergize my neuroanatomical, developmental, molecular, cellular, biochemical, and functional neuroscience background with new multidisciplinary approaches to tackle novel mechanisms in the formation of circuits of distinct subpopulations neurons.
Cited literature	**Literature**

(Betty Liu, research statement, modified)

An alternate possible outline is the following:

Overview/Abstract
    Background/Context
    Overall objective
    Focus of work/Specific aims
    Overall approach
    Significance

Individual research projects
List individual projects/aims and for each:
    rationale
    research aim and approach
    expected outcome
    relevance to institution/department
Overall significance of research

An example of a Research Statement that is composed based on the alternate outline is shown next in Example 29-5. The individual components for this Research Statement are indicated.

 Example 29-5

Research Statement

Jane Smith

Overview

Background	Air pollution changes our planet's climate, but different types of air pollution have different effects. Urban air pollution and smoke from fires have been modeled to reduce cloud formation by absorbing sunlight, thereby cooling the surface and heating the atmosphere. My research aims to advance our knowledge of cloud formation in response to pollutants based on mathematical modeling. In particular, I am interested in gaining insight into general effects of atmospheric pollutants on precipitation and climate change. The study of cloud fraction in response to pollutants may provide important insights to explain how weather is affected and why Earth has warmed substantially in the last century.

Background

Overall objective and experimental approach

Specific focus

Significance

Individual Research Projects

1. Cumulus cloud formation and atmospheric pollutants
Collaborators: names
Funding: name

Rationale — Clouds are sensitive to land and water surface properties. Analysis of satellite data has documented the relationship between heavy air pollution and the formation of scattered cumulus clouds. Stable meteorological conditions and regular behavior of the clouds over the Amazon create an ideal case to study the impact of pollutants from biomass burning on cloud formation. . . .

Specific aim and expected outcome — My research aims to determine . . . Findings will allows us to gain insight into . . .

Collaboration and funding

2. Atmospheric pollutant effect in evaluating climate forcing
Collaborators: names
Funding: name

Rationale — Atmospheric pollution causes cloud formation with more numerous but smaller droplets, leading to less precipitation and longer cloud lifetime. Atmospheric pollutants also absorbed incoming solar radiation, however, which can reduce the cloud cover.

Specific aim and expected outcome	My research applies these ideas to a combination of different models and data assimilations to understand .... The energetics will provide a diagnostic tool for assessing ....

Overall Significance of Research

Significance	Mathematical modeling will advance our understanding of the effects of atmospheric pollutants on precipitation and climate change. Through the study of cloud formation and climate forcing, this interdisciplinary research will have important impacts on ....

Discuss and coordinate future research themes and strategies with your mentor. Be sure to represent yourself as an independent researcher, with different goals and achievements than your dissertation advisor. Include anything else that might set you apart from your peers (e.g., publications in top journals, important funding partners or collaborators, or breakthrough studies in a particular area of your interests). To be competitive in research today, you generally need a network. If applicable, indicate that you have established collaborations with various researchers across diverse disciplines.

It may also be helpful to point out potential collaborations that could be established with faculty at the department/university to which you are applying. In addition, and if applicable, mention funding organizations likely to support your research agenda and alternative projects showing the breadth of your interests. For candidates who are more senior, future goals are an expanded Specific Aims page as required for grant application.

What Do Job Committees Look for?

To create a strong and compelling Research Statement, consider the goals and facilities of the school you are applying to. Overly ambitious proposals, lack of a clear direction, and unclear significance of study usually result in a weak statement. Write as clearly and precisely as you can, adhering to the basic writing principles presented in earlier chapters (see Chapter 1–6). Job committees typically look for the following:

- A clear research focus and direction
- Your independence as an investigator
- Potential for funding
- Summary of accomplishments
- Your fit with the unit
- The resources you will need to be successful in your new job
- Your communication skills; how well you present your research

29.6 TEACHING STATEMENTS

TEACHING STATEMENT GUIDELINE 1:
Describe your teaching philosophy and approach.

TEACHING STATEMENT GUIDELINE 2:

General Content

Teaching philosophy

Teaching experience

Teaching goals

Relevance

TEACHING STATEMENT GUIDELINE 3:

Tailor your Teaching Statement to a particular job
posting and institution.

Aside from a Research Statement, a Teaching Statement is also often
requested for job applications. The Teaching Statement is similar to the
Research Statement in form. It is a one- to two-page essay written in
the first person that describes your teaching philosophy and approach.
The Teaching Statement should cover your teaching experience as well as
your teaching goals. In addition, consider including a brief statement on
how your teaching would add to the department or school to which you
are applying.

Usually, teaching institutions have a list of one to three courses that a
candidate must be able to teach. Your Teaching Statement should clearly
state that you have taught or have the competency to teach these course.
In addition, list areas outside of your research where you would feel comfortable in teaching a formal course, and indicate your willingness to
learn about new topics. If relevant, state how you would go about teaching
medical students—from clinical presentation to pathophysiology, genetics, and biochemistry, to treatment and long-term prognosis. The next
two examples provide sample Teaching Statements:

Example 29-6 Teaching Statement

TEACHING STATEMENT

Name

Teaching goal	As a teacher, my primary goal is to develop a student's enthusiasm for, and knowledge and questioning of, ecological understanding.

Classes

Teaching experience	Since arriving at UGA, I have developed or codeveloped three of the required classes for our undergraduate and graduate programs in ecology. General Ecology, where the class size is greater than 100 and includes students from across a range of scientific majors, has been the most challenging; but I feel I've made significant strides, which are reflected in my teaching evaluations. I have introduced a focus on data interpretation, hypothesis testing, and concepts into formal lectures by interactively working

through case studies taken from primary research papers. I take students outside to develop observational skills ....

In undergraduate seminar classes, both at freshman (Duke) and senior (UGA) levels, I've focused on both the contemporary topics within the field and on topics of societal relevance. For the former, this might involve discussion of niche and neutral theory and for the latter, evaluation of "Policy Forum" articles from *Science.* ...

In the undergraduate seminar classes, I place a lot of emphasis on professional skill development. As well as developing their writing, formal debate, and discussion, students are expected to select a research question and develop it into a class presentation. At the graduate level, I take this emphasis on skills development further. For example, an essential element for the required Ecosystems Class that I codeveloped with an aquatic ecologist at UGA is proposal preparation and evaluation. All students are expected to generate a proposal following NSF guidelines for ecosystem research and then go a few steps further by peer reviewing others within small groups, responding to peer-review .... Discussion is supported by informal lecturing to provide the background knowledge to facilitate effective discussion of the primary material.

Teaching experience *(margin note)*

Advising

The part of teaching I find most rewarding is in the one-to-one development of individuals. In my current position, this role takes a number of guises: I am the academic advisor for a cohort of undergraduate ecology majors, was nominated this year to be a Mentor for freshman students, advise three Ph.D. students ....

Graduate   Where I am the main advisor, my primary objective is to develop a student so that he/she can achieve and begin effectively his/her next career step. At the Ph.D. level, this involves generating scientists capable of independent research. I aim to produce well-rounded Ph.D. students who are conversant in topics across ecology, are independent researchers, effective presenters and proposal writers, experienced in teaching, and also in the process of publishing and reviewing manuscripts. To achieve this, students are required to participate in regular lab meetings, meet with me formally at a regular time every two weeks, present at local and national meetings .... I also work with them to submit proposals. My approach to postdoctoral advising is similar but with more emphasis on preparation for a faculty position, and so it includes providing opportunities such as involvement as a coinstructor on undergraduate classes.

Undergraduate   The requirement for all undergraduate science students in my laboratory is an independent-research project. I usually identify a question for a student to work on and then take them through all the steps, from literature review to the writing-up of the experimental results as papers. Students that have conducted such research with me have all entered graduate programs and published their results either as lead or coauthors.

Specific teaching goals *(margin note)*

Future teaching goals *(margin note)*

At ABC University, I would be keen to work with both undergraduate and graduate research students interested in soil

| Future teaching goals | ecology, ... I would welcome the opportunity to develop further graduate-level classes, such as the ecosystem ecology class I currently teach and codeveloped, and to offer advanced seminars on more specialized topics such as the philosophy of ecosystem ecology, ... I would certainly be interested, if desired, in developing classes that contribute to the undergraduate program as well as in facilitating independent research at this level. |

(Mark Bradford, teaching statement, modified)

Example 29-7 Teaching Statement

Teaching philosophy	One of the most rewarding aspects of an academic position is the opportunity to teach and interact with students, supporting their curiosity and helping them along in their career paths. My goals as a teacher are to help the students learn about the importance of science and in particular, ecology.
Overall teaching goal	The education plan of my work has several components, some of which are interconnected with my research, others falling outside my research topic. Specifically, the plan includes a comprehensive program designed to encourage students of different levels, from undergraduates to Ph.D. applicants, to consider careers in biomedical sciences. The central goal in teaching biomedical sciences to my students is to have them acquire critical assets, including knowledge, skills, and interest.
Teaching experience	My formal teaching career began as a teaching assistant at X University while I was an undergraduate and graduate student. I was a teaching assistant for a variety of courses including .... In addition to teaching the standard undergraduate courses as a postdoctoral fellow, I gave a number of lectures on selected topics in evolutionary biology and scientific writing. ...

Undergraduate education

I believe that students learn best by doing rather than by passively listening or watching. Thus, I am a strong proponent of interactive learning, asking students for suggestions, encouraging questions, and promoting discussions during lectures. Moreover, I have students directly apply taught skills to practical examples that I provide inside as well as outside class. To motivate students, I use clear, understandable, and exciting examples from a broad range of applications. For instance, ...

Aside from fostering critical thinking in my students, my additional goal as an educator is to promote good communications skills, as these are ever more important in today's interdisciplinary, and often international, research community. To that end, for the past three years, I have offered a special class on scientific communication once a year in my role as a postdoctoral fellow at X University. I plan to introduce this class to your department on my acceptance as well. In this class, students are presented with basic writing principles, which they learn to apply to writing for publication. Many practical examples set the stage for ....

| Specific teaching goal | Finally, I believe that a good teacher should always look for ways to improve teaching skills. For this reason, I plan to refine my teaching strategies by .... |

Graduate level education

At the graduate level, I will supply more specialized theoretical biology courses overlapping my research. These courses cover ... In addition, I am planning to offer specific communication courses on grant writing and oral presentation to graduate students. I have developed a manual for use in such classes based on ...

Aside from teaching graduate classes, I am also enthusiastic in supervising graduate student research. My objective is to have my graduate students participate in weekly group meetings during which updates for research projects are presented and critiqued. Students will have the opportunity to consult with all members of the group and will be exposed to different expertise in addition to receiving advice about their research projects.

Outside the weekly group meetings, graduate students will participate in seminar series.... In addition, I will encourage my graduate students to attend scientific meetings and prepare presentations for these. As well, I will encourage them to participate actively in writing publications and proposals through drafting content and revising documents.

My supervision and mentorship will take place in form of individual meetings, e-mails, and phone conversations, as well as group meetings. I will provide advice about scientific questions to be addressed, about research approaches, as well as on various forms of communication in the field. In addition, I will provide career advice ranging from how to respond to sensitive e-mails to securing a strong CV for the application to a faculty position.

Specific teaching goal

I plan to encourage them not only to work on their research project but also to develop their own ideas of research to pursue as independent investigators. In this regard, I will ask students to present quarterly research plans to the group and ... Furthermore, I will require them to gain teaching experience through presenting a lecture or two in a course attended by both undergraduate and graduate students....

Summary

In summary, I am committed to becoming an effective teacher with the goal of preparing my students with a solid foundation of knowledge, skills, and enthusiasm they will need to succeed professionally.

Teaching Portfolio

A Teaching Statement may be tied to a Teaching Portfolio. Such a portfolio is a collection of materials that illustrate your teaching strengths and accomplishments. This portfolio may or may not be required in the initial job application package. If it is not required initially, you may be asked to provide one later in the screening process. Be prepared for this request. It takes time to put a good portfolio together.

The contents and presentation of teaching portfolios vary widely from individual to individual and from field to field. A teaching portfolio may include, for example,

- Teaching Statement
- List of courses taught
- List of teaching awards and certificates
- Sample course material including a sample syllabus for a course
- Teaching evaluations by students and faculty members
- List of professional development in teaching
- Graded papers
- Teaching video

29.7   RESOURCES

Resources for Composing Research Statements

Electronic resources for composing a Research Statement include the following (last accessed March 2009):

> http://career.studentaffairs.duke.edu/graduate/find_job/apply/
> research_statements.html
> http://sciencecareers.sciencemag.org/career_development/
> previous_issues/articles/1820/writing_a_research_plan
> http://www.engin.umich.edu/students/graduate/aces/research.html
> http://www.otal.umd.edu/~sies/jo`bchecklist.html

Resources for Composing Teaching Statements

> http://chronicle.com/jobs/news/2003/03/2003032702c.htm
> http://ftad.osu.edu/portfolio/philosophy/Philosophy.html
> http://sunconference.utep.edu/CETaL/resources/portfolios/
> writetps.htm
> http://depts.washington.edu/cidrweb/resources/writingtips.html
> http://sciencecareers.sciencemag.org/career_development/
> previous_issues/articles/1820/writing_a_research_plan
> http://www.londonmet.ac.uk/deliberations/portfolios/
> iced-workshop/seldin-book.cfm
> http://www.utexas.edu/academic/cte/teachfolio.html

Printed Resources

> Mary Morris Heiberger and Julia Miller Vick, 2002. *The Academic Job Search Handbook* (Third Edition), University of Pennsylvania Press
>
> Richard Reis, 1997. *Tomorrow's Professor: Preparing for Academic Career in Science and Engineering.* IEEE Press, New York
>
> Christina Boufis and Victoria C. Olsen, Eds., 1997. *On the Market: Surviving the Academic Job Search.* Riverhead Books, New York
>
> Peter Feibelman, 1993. *A Ph.D. Is Not Enough!* Addison Wesley, Reading, Massachusetts

29.8 LETTERS OF RECOMMENDATION

LETTER OF RECOMMENDATION GUIDELINE 1:

In the letter of recommendation, highlight only
positive qualities.

LETTER OF RECOMMENDATION GUIDELINE 2:

Pay particular attention to the first sentence of the letter
of recommendation.

LETTER OF RECOMMENDATION GUIDELINE 3:

Make a letter of recommendation three to four
paragraphs long.

Requesting a Letter of Recommendation

Along with your CV, cover letter, and Research and Teaching Statements, you will probably need one or more letters of recommendation to secure a good position. When you have to ask someone for a letter of recommendation, explain exactly why the letter is needed and how important it is to you. Always offer to provide information that makes the writing task easier (CV, list of accomplishments, publication list, the due date, means of transmission [mail or e-mail], correct address of recipient). Also provide the recommender with a brief statement explaining *why* this opportunity is important to you and *how* it fits into your overall research or teaching plan. If the writer cannot or will not provide you with a letter, accept this decline gracefully.

Plan your request. Ask someone who is familiar with your work and knows you well enough to include details about you as a person on the letter. The writer should ideally write well, have experience composing letters of recommendation, and have the highest or most relevant job title. As a general rule, request your letter at least a month or two in advance. Also send with your request your CV and other documentation of work or service relevant to the position to which you plan to apply so that your letter writer has some material with which to work. If the recommender agrees to writing the letter in general but has a busy schedule, consider offering to write a first draft yourself, which you can subsequently pass on to the recommender.

Writing a Letter of Recommendation

There may be times when you are asked to write a letter of recommendation. Writing such a letter can be a privilege or a task. Agree to write a letter only for someone you know well enough and only if you can really write a supportive letter. It is important that you have a good understanding of the person's academic or professional history and goals. This letter will be important for the individual who has asked you to provide this recommendation. If you are honored to be asked, ask the person for a copy of their résumé or CV and a list of accomplishments to give you guidelines to use in composing the letter. This information is especially important if you are not sure what to say. Also ask for information on the job position to tailor the recommendation to that position. Describe the person truthfully. Highlight only positive qualities. The more personalized the letter of recommendation is, the more effective it will be.

If possible, use formal letterhead and include your title on the signature line to reinforce your credibility.

Content

Letters of recommendation provide valuable records of the candidate's previous experience, skills, and abilities. An effective letter of recommendation not only verifies experience but also builds credibility.

The information contained in a letter of recommendation depends on the type of letter and its intended audience. In writing letters of recommendation, it is extremely important to impress on the reader that you know this person very well. Avoid a long list of general praises. Instead, comment on the unique personality/traits that the individual exhibits.

A good letter of recommendation can take a substantial time to write. It should be a couple paragraphs in length and contain some or all of the following:

For a student
- Academic performance
- Honors and awards
- Initiative, dedication, integrity, reliability, etc.
- Willingness to follow school policy
- Ability to work with others
- Ability to work independently

For a researcher
- Previous position
- Summary of job responsibilities
- Strengths, skills, and talents
- Initiative, dedication, integrity, reliability, etc.
- Ability to work with a team
- Ability to work independently

Format

Here are some general guidelines for the format of a letter of recommendation:

1st Paragraph	Introduce yourself as the recommender and state the purpose of the letter. State how long

	you have known the person and in what capacity.
2nd/3rd Paragraph	Start by describing the person in general terms, and then mention specific traits of the person. Give specific examples, and make it relevant to the position being pursued. Include two or three outstanding attributes. Address specific qualities in order of importance.
4th Paragraph	Express your specific recommendation and confidence in the individual. Encourage the reader to contact you for additional information or with any questions.

You may also want to provide a phone number or e-mail address so employers can follow up if they have questions or want more information.

Pay particular attention to the first sentence of the letter. Its wording will communicate your overall opinion of the individual (often unconsciously). Here are some examples of opening sentences and their respecvtive strengths:

Strength of opening sentence

This letter pertains to ...   weakest

I am pleased to recommend ...

It is a pleasure to write this letter of recommendation for ...

It is a genuine pleasure and honor for me to recommend ...   strongest

---

 Example 29-8

> ### Letterhead
> Name and address
>
> Date
>
> Dear Dr. Smith:
>
> It is with the greatest pleasure that I write this letter of recommendation for Maria Miller. Maria has been working as a postdoctoral fellow in my laboratory at the ABC University since July 2007.
>
> I have been the laboratory head and chair of the department of epidemiology since 2001. Without hesitation, I can say that the expertise, efficiency, and professionalism shown by Maria in her research were surely among the best. These qualities are reflected in her successful publication record in top scientific journals. Her postdoctoral years at ABC University have been very productive. She has published excellent articles on various worldwide epidemics such as TB, HIV, influenza, and rubella. Maria also produced the first model on WNV transmission. Her research is often methodologically innovative and has important implications for public policy.
>
> Maria developed several new techniques in my laboratory and was sought out by her lab mates to help them when their assays

did not work. She was always able to ask insightful questions in Departmental Seminars covering a very wide range of topics and was never afraid of making a mistake, plunging right into difficult experiments. Moreover, Maria is gifted with superb written and oral communication skills. She kept a well-organized lab book, wrote up her research promptly and clearly, and produced great figures and posters. She also presented outstanding departmental and conference talks.

While working in my laboratory, Maria advanced her knowledge wherever possible by attending lectures whenever her schedule permitted her to do so. For example, she participated in courses at the School of Medicine on virology and microbiology and collaborated with a number of individuals within these fields. Maria was usually the first in my laboratory every morning, and we all enjoyed the collegiality and enthusiasm that she brought to my laboratory and department. She was always cheerful and ready to help, no matter the task.

In short, I recommend Maria without hesitation and have no doubt that this talented young investigator will succeed in the field of epidemiology. I am confident that her research will result in important contributions to your department. Please do not hesitate to contact me if I can provide any additional information that may be helpful.

Sincerely,

29.9   CHECKLIST FOR THE JOB APPLICATION

Use the following checklist to "dissect" the Research and Teaching Statements systematically:

OVERALL

☐ 1. Did you tailor your CV or résumé specifically to the position and to the organization?
☐ 2. Is your CV well presented and flawless?
☐ 3. Did you tailor the cover letter specifically to the position?

RESEARCH STATEMENT

☐ 1. Did you state your research achievements, current aims, and future goals?
☐ 2. Does your Research Statement follow one of the two following outlines?

    ☐ **A** Background
            Current research
            Research agenda
            Relevance
    ☐ **B** Overview/Abstract
            Individual research projects
            Overall significance of research

TEACHING STATEMENT

☐ 1. Does your Teaching Statement describe your teaching philosophy and approach?

2. Does your Teaching Statement contain the following?

☐ Teaching philosophy
☐ Teaching experience
☐ Teaching goals
☐ Sample course outline or syllabus (if a specific course is required)
☐ Relevance

LETTER OF RECOMMENDATION

☐ 1. Did you highlight only positive qualities in the letter of recommendation?

☐ 2. Did you pay particular attention to the first sentence?

☐ 3. Is your letter of recommendation three to four paragraphs long?

OVERALL STYLE AND COMPOSITION

Revise for style and composition based on the writing principles of the book:

☐ a. Are paragraphs consistent? (Chapter 6, Section 6.2)
☐ b. Are paragraphs cohesive? (Chapter 6, Section 6.3)
☐ c. Are key terms consistent? (Chapter 6, Section 6.3)
☐ d. Are key terms linked? (Chapter 6, Section 6.3)
☐ e. Are transitions used, and do they make sense? (Chapter 6, Section 6.3)
☐ f. Is the action in the verbs? Are nominalizations avoided? (Chapter 4, Section 4.6)
☐ g. Did you vary sentence length and use one idea per sentence? (Chapter 4, Section 4.5)
☐ h. Are lists parallel? (Chapter 4, Section 4.9)
☐ i. Are comparisons written correctly? (Chapter 4, Sections 4.9 and 4.10)
☐ j. Have noun clusters been resolved? (Chapter 4, Section 4.7)
☐ k. Has word location been considered? (Verb following subject immediately? Old, short information at the beginning of the sentence? New, long information at the end of the sentence?; Chapter 3, Section 3.1)
☐ l. Have grammar and technical style been considered? (person, voice, tense, pronouns, prepositions, articles; Chapter 4, Sections 4.1–4.4)
☐ m. Is past tense used for results and present tense for descriptive papers?
☐ n. Are words and phrases precise? (Chapter 2, Sections 2.2 and 2.3)
☐ o. Are nontechnical words and phrases simple? (Chapter 2, Section 2.2)

☐ p. Have unnecessary terms (redundancies, jargon) been reduced? (Chapter 2, Section 2.4)

☐ q. Have spelling and punctuation been checked? (Chapter 4, Section 4.11)

SUMMARY

JOB APPLICATION GUIDELINES:

1. Tailor your CV or résumé specifically to the position and to the organization.
2. Your CV should be well presented and flawless.
3. Tailor the cover letter specifically to the position.

RESEARCH STATEMENT GUIDELINES:

1. State your research achievements, current aims, and future goals.
2. Research Statement – General Content
   (Abstract)
   Background
   Current research
   Research agenda
   Relevance
3. Tailor your Research Statement to a particular job posting and institution.

TEACHING STATEMENT GUIDELINES:

1. Describe your teaching philosophy and approach.
2. Teaching Statement – General Content
   Teaching philosophy
   Teaching experience
   Teaching goals
   Relevance
3. Tailor your Teaching Statement to a particular job posting and institution.

LETTER OF RECOMMENDATION GUIDELINES:

1. In the letter of recommendation, highlight only positive qualities.
2. Pay particular attention to the first sentence of the letter of recommendation.
3. Make a letter of recommendation three to four paragraphs long.

Appendix

COMMONLY CONFUSED AND MISUSED WORDS

Learn to distinguish between words that more casual writers carelessly interchange. This list explains the meaning and use of the most commonly misused words scientific editors encounter. Strunk and White, Fowler, and Perelman have similar lists.

ABILITY, CAPACITY

Ability   The mental or physical power to do something or the skill in doing it. Ability can be measured; capacity cannot. (Some microorganisms have the *ability* to fix nitrogen.)

Capacity   The full amount that something can hold, contain, or receive. (The *capacity* of the beaker was 500 ml.)

ACCEPT, EXCEPT

Accept   *Accept* means to answer affirmatively, to receive something offered with gladness, or to regard something as right or true. (The scientists *accepted* his new theory.)

Except   *Except* is generally construed as a preposition meaning with the exclusion of or other than. It can also be a verb meaning to leave out. (Of the bacteria described, all are gram negative *except* Streptococci.)

ADMINISTER, ADMINISTRATE

Administer   *Administer* (The drug was *administered* orally.) Administration, however, is the noun of *administer*, which may lead to this common confusion between *administrate* and *administer*.

Administrate   *Administrate* means to manage or organize.

ACCURATE, PRECISE, REPRODUCIBLE

Accurate
: *Accurate* means errorless or exact. It is often used in the sense of providing a correct reading or measurement. (The readings obtained with the spectrophotometer were *accurate*.)

Precise
: *Precise* means to conform strictly to rule or proper form. (The value 5.26 is more *precise* than the value 5.3.)

Reproducible
: *Reproducible* means something can be copied or repeated with the same results. (The experiment described by William et al. was *reproducible*.)

ADAPT, ADOPT

Adapt
: *Adapt* is a verb and means adjust and make suitable to. (Most organisms *adapt* easily to minor changes in the environment.)

Adopt
: *Adopt* is also a verb and means to accept or make one's own. (We *adopted* a new protocol for DNA isolation.)

ADVICE, ADVISE

Advice
: *Advice* is a noun meaning suggestion or recommendation. (To measure dP/dt, we followed the *advice* of J. R. Boyd.)

Advise
: *Advise* is a verb and means to suggest or to give advice. (Dr. Boyd *advised* us to follow his protocol.)

AFFECT, EFFECT

Affect
: *Affect* is a verb and means to act on or influence. (The addition of Kl-3 to MZ1 cells *affected* their growth rate, i.e. it could have increased or decreased or induced.)

  Affect can also be a noun with a specialized meaning in medicine and psychology: an emotion. (People can experience a positive or negative *affect* as a result of their actions.)

Effect
: As a noun, it means a result or resultant condition. (We examined the *effect* of Kl-3 on MZ1 cells.)

  As a verb it means to cause or bring about. (The addition of Kl-3 to MZ1 cells *effected* their motility (i.e., it had caused or brought about). She was able to *effect* a change in his attitude.)

AGGRAVATE, IRRITATE

Aggravate
: When an existing condition is made worse, it is *aggravated*.

Irritate
: When tissue is caused to be inflamed or sore, it is *irritated*.

ALLUDE, ELUDE, REFER, REFERENCE

Allude
: *Allude* means to mention indirectly. (The authors of "Basic Medical Microbiology" only *allude* to brain abscesses in Chapter 10.)

Elude	*Elude* means to escape from or to escape the understanding or grasp of something. (The cause of the brain abscesses *eluded* us.)
Refer	*Refer* means to pertain or to direct to a source. (*Refer* to Barett et al. for more detailed listings. Questions *referring* to brain abscesses should be directed to a specialist in the field.)
Reference	*Reference* can be used as a noun, meaning a note in a publication referring the reader to another source. (Do not forget to include *references* in the paper.) The term can also be used as a verb, meaning refer to. (They *referenced* his work.)

ALTERNATELY, ALTERNATIVELY

Alternately	*Alternate* means every other one in a series. (Students *alternately* attended lectures on molecular biology and biochemistry every Saturday.)
Alternatively	A choice between two or more mutually exclusive possibilities. (Students can major in biology or, *alternatively*, in chemistry.)

AMONG, BETWEEN

Among	*Among* means in a group of or the entire number of. It is used to express the relation of one thing to a group. (We discovered one black sheep *among* the white ones.)
Between	Use *between* with two items or more than two items that are considered as distinct individuals. (We found no marked differences *between* our results and those reported in Nature.)

AMOUNT, CONCENTRATION, CONTENT, LEVEL, NUMBER

Amount	Quantity that can be measured but not counted. (The total *amount* of yeast extract required for the medium was 12 g.)
Concentration	The density of a solution or the amount of a specified substance in the unit amount of another substance. (The *concentration* of protein in the blood was 2.6 mg/ml.)
Content	A portion of a specified substance. (Soybeans have a high protein *content*.)
Level	Relative position or rank on a scale, often used as a general term for amount, concentration, or content. (Heart rate was at normal *level*. Protein *levels* (that is concentration) remained stable.)
Number	A quantity that can be counted. (The *number* of proteins in the aggregates varied.)

ANYBODY, ANY BODY, ANYMORE, ANY MORE, ANYONE, ANY ONE

Anybody	Refers to an unspecified person. Often used in place of everyone. (*Anybody* can get sick.)
Any body	A noun phrase referring to an arbitrary corpse or human form. (We found the head, but we did not see *any body*.) This rule also applies to everybody, nobody, and somebody.
Anymore	An adverb denoting time. (Doctors in the Western world don't prescribe chloramphenicol *anymore*.)
Any more	Used with a noun or as an indefinite pronoun. (We don't need *any more* measurements.)
Anyone	Refers to any person. Used like anybody and everyone. (*Anyone* can get sick.)
Any one	The two word form is used to mean whatever one person or thing of a group. (I would like *any one* of these apples.)

ASSAY, ESSAY

Assay	A test to discover the quality of something. (In this *assay*, a Pt electrode was used.)
Essay	A piece of writing, not poetry or a short story. (In this *essay*, she discussed her work in physical chemistry.)

AS, LIKE

As	*As* is a conjunction and is used before phrases and clauses. Rather than "*like* we just mentioned," say "*as* we just mentioned."
Like	*Like* is a preposition with the meaning of "in the same way as." (He was *like* a son to me.) Note the differences in the uses of *like* and *as*: *Let me speak to you as a father* (= I am your father and I am speaking to you in that character). *Let me speak to you like a father* (= I am not your father but I am speaking to you as your father might).

ASSUME, PRESUME

Assume	*Assume* means to take for granted or to suppose. It is usually associated with a hypothesis in scientific writing. (*Assuming* this hypothesis is correct, food should be supplemented with folic acid and vitamin B.)
Presume	*Presume* means to believe without justification. (Scientists *presume* a common ancestor for the two species.)

ASSURE, ENSURE, INSURE

Assure	*Assure* is to state positively and to give confidence and is used with reference to a person. (Let me *assure* you that we did not forget to add any enzyme.)

Ensure   *Ensure* means to make certain. (When you ligate DNA, you have to *ensure* that you do not forget to add the enzyme.)

Insure   *Insure* also means to make certain and is interchangeable with *ensure*. In American English, *insure* is widely used in the commercial sense of "to guarantee financially against risk." (We *insured* our car.)

BECAUSE, SINCE

See *SINCE, BECAUSE*

BUT, AND, BECAUSE

But, and, and *because* can be used to start a sentence as long as the sentence is complete. (Some bacteria can form endospores. *But E. coli* and *E. sakazakii* do not.)

CAN, MAY

Can   Means to be able to, to have the ability or capacity to do something. (Tetracyclin *can* be used to treat urinary tract infections.)

May   Indicates a certain measure of likelihood or possibility to do something. It also refers to permission. (Tetracycline *may* act by binding to a certain site of the bacterial ribosome.)

CAN'T, DIDN'T, HAVEN'T

These contractions should not be used in scientific writing at all. Write them out instead: *cannot, did not, have not*

COMPLIMENT, COMPLEMENT

Compliment   Compliment means praise. (John *complimented* Jean on her new dress.)

Complement   To *complement* means to mutually complete each other. (The protein bindin and its receptor *complement* each other.)

COMPRISE, COMPOSE, CONSTITUTE

Comprise   The conservative definition of comprise is to include or to contain. (The United States *comprises* many different states). Avoid the phrase *is comprised of.*

Compose   *Compose* means to make up or to create something. Frequently used in the passive voice. (Water *is composed of* hydrogen and oxygen.)

Constitute   *Constitute* means to make a whole out of its parts, to equal or amount to. (Many different organisms *constitute* a habitat.)

CONSERVATIVE, CONSERVED

Conservative   *Conservative* implies a medical treatment that is limited or treatment with well-established procedures. (Due to complications, we selected *conservative* treatment for this case.)

Conserved   *Conserved* means to keep constant through physical or evolutionary changes. (DNA sequences may be *conserved* between species.)

CONTINUAL, CONTINUOUS

Continual   Repeatedly, occurring at repeated intervals, possibly with interruptions. (Measurements were hampered by *continual* interruptions.)

Continuous   Without interruption, unbroken continuity. (Our experiments were *continuously* interrupted means the experiments were started and then interrupted, and this interruption never stopped. The light spectrum is *continuous.*)

CONTRARY TO, ON THE CONTRARY, ON THE OTHER HAND, IN CONTRAST

Contrary to   Preposition meaning in opposition to. (*Contrary to* our expectations, addition of Mg^{++} did not alter our results.)

On the contrary   *On the contrary* is used when one says a statement is not true. It is a subjective statement that indicates opposition and is usually used only in spoken English. ("It's exciting!" "*On the contrary*, it's boring!")

On the other hand   Use *on the other hand* when adding a new and different fact to a statement. Rarely used in scientific English. (It's cold, but *on the other hand*, it's not raining.)

In contrast   *In contrast* is also used for two different facts that are both true, but it points out the surprising difference between them. It is an objective statement for a marked difference and can be used in scientific writing. (pH values increased for procaryotes. *In contrast,* no pH difference was observed in eucaryotes).

DATUM, DATA

Datum   *Datum* is singular and means one result.

Data   *Data* is the plural form of *datum*. It should be used with the plural form of the verb. (Our *data* indicate that many experiments have to be repeated)
The same applies for *criterion, criteria*, and *medium, media.*

DESCRIBE, REPORT

Describe   Patients, persons, or cases are *described*. (We *describe* a patient with gynecomastia induced by omeprazole.)

Report   Cases and diseases are *reported*. (We *report* a case of omeprazole-induced gynecomastia.)

DIFFERENT FROM, DIFFERENT THAN

Different from   Correct expression. (The binding site of amoxicillin is *different from* that of penicillin.)

Different than   Incorrect.

DIE OF, DIE FROM

die of   Persons and animals *die of,* not *from,* specific diseases.

die from   Incorrect.

DOSE, DOSAGE

Dose   A *dose* is a specific amount administered at one time or the total quantity administered. (Patients received an initial *dose* of 10 mg.)

Dosage   *Dosage*, the regulated administration of doses, is usually expressed as a quantity per unit of time. (Give a *dosage* of 0. 5 mg every 2 hours.)

ENHANCE, INCREASE, AUGMENT

Enhance   Means to make greater in value or effectiveness. In scientific writing, terms such as *increase* or *decrease* are preferred over *enhance* because they are much more precise.

Increase   A general term that means to become or to make greater. (Although we *increased* the amount of nutrients, the number of bacteria decreased.)

Augment   Means to make something already developed greater. (Addition of A greatly *augmented* the effect of B on C.)

ETC/SO ON, SO FORTH, AND THE LIKE

Etc.   Can only be used when the contents of a noninclusive list are obvious to the reader. It is an imprecise expression and should generally be avoided in scientific writing. Instead, use *such as* or *including* at the start of the list, and put nothing at the end of the list.
Also avoid *and so on, and so forth*, and *and the like*.

EXAMINE, EVALUATE

Examine   Patients are *examined*.

Evaluate   Conditions and diseases are *evaluated*.

FARTHER, FURTHER

Farther   Refers to a physical distance. (You need to drive much *farther* than you think.)

Further	Refers to quantity or time. (The binding of X to Z needs to be examined *further*.)

FEWER, LESS

Fewer	Use *fewer* for items that can be counted. (*fewer* cells, *fewer* patients).
Less	Use *less* for items that cannot be counted (*less* medication, *less* water). Exceptions include time and money. (*less* than two years ago) (Jack has *less* money than John.)

FOLLOW, OBSERVE

Follow	A case is *followed*.
Observe	A patient is *observed*.
	To "follow up" on either approaches jargon, as does the term "follow-up study." However, in scientific writing, the use of the term *follow-up* is increasingly common, as is "follow-up study."

IMPLY, INFER

Imply	To *imply* is to suggest, indicate, or express indirectly. (These results *imply* that there is a key-lock mechanism between the enzyme and its receptor.)
Infer	To *infer* is to conclude or to deduce. (Looking at the interaction between the enzyme and its receptor, we can *infer* that a key-lock mechanism is used.)

INCIDENCE, PREVALENCE

Incidence	Means occurrence or frequency of occurrence. (The *incidence* of macular degeneration is high in the elderly.)
Prevalence	Means total number of cases of a disease in a given population at a specific time. (The *prevalence* of macular degeneration in the Western world in the year 2000 was 1 in 1,000 individuals.)

INCLUDE, CONSIST OF

Include	Means partial; to be made up of, at least in part; to contain. *Include* often implies partial listing. (Various antibiotics, *including* streptomycin and tetracycline, bind to the bacterial ribosome.)
Consist of	Means to be made up of, to be composed of. Usually used when a full list is provided. (Macrolites *consist of* ...)

INFECT, INFEST

Infect	Endoparasites *infect* or produce an infection.
Infest	Ectoparasites *infest* or produce an infestation.

INTERVAL, PERIOD

Interval   The amount of time between two specified instants, events, or states. (Optical density was measured at 10-min *intervals*.)

Period   An interval of time characterized by certain conditions and events. (Pterodactylus lived during the Jurassic *period*.)

IRREGARDLESS, REGARDLESS

Irregardless   Does not exist in English. The correct form is *regardless*.

IT'S, ITS

It's   Is a contraction of *it is* or *it has*. Preferably spell it out in scientific writing.

Its   Means belonging to. (Bindin is an adhesive protein found at the tip of the sperm cell. *Its* structure is unknown.)

LATER, LATTER, FORMER, LAST, LATEST

Later   Refers to time. (The fertilization membrane was not observed until *later* in the assay.)

Latter   Means the second of two items. (*Hpa*II and *Msp*I both cut the recognition sequence CCGG, the *latter* also when the sequence is methylated.)

Former   Means the first of two items mentioned. (*Hpa*II and *Msp*I both cut the recognition sequence CCGG, the *former* only when the sequence is not methylated.)

Last   *Last* refers to the last item of a series. (The samples were pooled, incubated at room temperature for 10 min, and centrifuged at 300 x g for 20 min. For the *last* step, 10 ml 80% ethanol was added.)

Latest   *Latest* refers to the most recent item in a chronological series. (The *latest* book on molecular biology was published about one month ago.)

LAY, LIE

Lay   To place or put something (Lay is a transitive verb; that is, it needs an object). Past tense is *laid,* past participle is *laid.* (Birds *lay* eggs.)

Lie   To rest on a surface. (Lie is an intransitive verb, that is, it cannot take a direct object.) Past tense is *lay,* past participle is *lain.* (The tree line ends where the lake *lies.*)

LOCATE, LOCALIZE

Locate   To determine or specify the position of something. (Next, we *located* the vagus nerve.)

Localize   To confine or to restrict to a particular place. (YT-31 was *localized* in the mitochondria.)

MEDIUM, MEDIA

Medium	Singular noun. Needs a singular verb. (LB *medium* was prepared at room temperature.)
Media	Plural form of medium. Needs a plural verb. (For the refined analysis, we used six different types of *media*.)

MILLIMOLE, MILLIMOLAR, MILLIMOLAL

Millimole	(mmol) An amount, not a concentration. (A 1 *millimolar* solution contains 1 *millimole* of a solute in 1 liter of solution (or a 1 *mM* solution contains 1 *mmol*/l of solution.)
Millimolar	(mM) A concentration, not an amount. (A 0.5 *millimolal* solution contains 0.5 *mmol* of a solute in 100 g of solvent. The final volume may be more or less than 1 l.)
Millimolal	A concentration, not an amount.

MUCUS, MUCOUS

Mucus	The noun. (*Mucus* is a viscous substance secreted by the *mucous* membranes.)
Mucous	The adjective meaning containing, producing, or secreting mucus.

MUTANT, MUTATION

Mutation	A *mutation* is an alteration in the primary sequence of DNA. (A *mutation* can be mapped, but a *mutant* cannot.)
Mutant	Refers to a strain of organism, population, allele, or gene that carry one or more mutations. (A *mutant* has no genetic locus, only a phenotype.)

NECESSITATE, REQUIRE

Necessitate	*Necessitate* means to make necessary or unavoidable. (The treatment may *necessitate* certain procedures.)
Require	*Require* means to have a need for or to demand. (A patient *requires* treatment.)

NEGATIVE, NORMAL

Negative	Tests for microorganisms and reactions may be *negative* or positive.
Normal	Observations, results, or findings are *normal* or *abnormal*. (The patient's behavior was *normal*.)

OPTIMAL, OPTIMUM

Optimal	Adjective meaning most favorable; never used as a noun. (The *optimal* (or *optimum*) temperature for X was 28°C.)
Optimum	Noun meaning the most favorable condition for growth and reproduction; often used as an adjective. (X reached its *optimum* 30 hours after induction.)

OVER, MORE THAN

Over   *Over* can be ambiguous. (The cases were followed up *over* two years). Instead, use "more than." (The cases were followed for *more than* two years.)

More than   Preferred.

PERCENT, PERCENTAGE

Percent   Means per hundred. Written as % together with number in scientific writing. (The yield of the gene product was less than 5%.)

Percentage Is a general part or a portion of the whole. Usually with a singular verb. (A small *percentage* of the crop was spoiled.)

PARAMETER, VARIABLE, CONSTANT

Parameter   Means a constant (of an equation) that varies in other settings of the same general form. Parameters are a set of measurable factors, such as temperature, that define a system and determine its behavior. (Changing the *parameters* of the system will result in a different outcome.)

Variable   A *variable* is a quantity that can change in a given system. (In the equation $y = ax + b$, x is the *variable*.)

Constant   A *constant* is a quantity that is fixed. (π is a *constant*).

PERTINENT, RELEVANT

Pertinent   Means having logical, precise relevance to the matter at hand. (These assignments are *pertinent* to understand the class material.)

Relevant   What relates to the matter or subject at hand. (He performed experiments *relevant* to is research.)

PRINCIPAL, PRINCIPLE

Principal   *Principal* can be either a noun meaning a person or thing of importance or an adjective meaning main or dominant. (The *principal* reason for not observing any gas formation was low temperature.)

Principle   *Principle* can only be a noun and means law or general truth. (This manual presents many writing *principles*.)

QUOTATION, QUOTE

Quotation   *Quotation* is a noun meaning a passage quoted. (Indicate a *quotation* with quotation marks.)

Quote   *Quote* is the verb and means to repeat or to cite. (Sentences copied from other sources should be *quoted*.)

QUANTIFY, QUANTITATE

Quantify   Often interchanged and a matter of personal preference. Both are used with the meaning to determine or measure the quantity of something.

Quantitate   *Quantitate* is preferred if one wants to emphasize that something was measured precisely.

RATIONAL, RATIONALE

Rational   An adjective meaning able to reason. (This is not a very *rational* thing to do.)

Rationale   A noun meaning basis or fundamental reason. (The *rationale* behind our theory is that cancer incidences are ever increasing.)

REGIME, REGIMEN

Regime   A *regime* is a form of government.
Regimen means a regulated system intended to achieve a beneficial effect. When a system of therapy is meant, *regimen* is the correct term.

Regimen   See *Regime*

REMARKABLE, MARKED

Remarkable   Is often incorrectly used to indicate a change that is notable but not significant. The correct word is *marked*. (There was a *marked* increase in binding.)

Marked   See *remarkable*.

REPRESENT, BE

Represent   Means to stand for or to symbolize. (Each data point *represents* the average of five measurements.)

Be   Equal, constitute. (Mercury *is* a toxic substance.)

SINCE, BECAUSE

Since   Use this word only in its temporal sense and not as a substitute for *because*. (We have had nothing but trouble *since* we moved here.)

Because   If you want to indicate causality, use *because*. (The reaction rate dropped *because* temperature dropped.)

SYMPTOMS, SIGNS

Symptoms   *Symptoms* apply to people. (The patient displayed no *symptoms* of the disease.)

Signs   *Signs* apply to animals. (A *sign* that the dog was sick was that she did not eat anymore.)

THAN, THEN

Than   *Than* is a conjugation that introduces an unequal comparison. (Birds are closer related to dinosaurs *than* to mammals.)

Then   Means next in time, space, or order. (The samples were centrifuged at 100 *x g. Then* the pellet was resuspended.)

TOXICITY, TOXIC

Toxicity   *Toxicity* is the quality, state, or degree of being poisonous.

Toxic   *Toxic* means poisonous. (Mercury is *toxic* for most organisms.)

VARYING, VARIOUS

Varying   *Varying* means changing. (Precise measurements could not be taken because of *varying* levels of humidity.)

Various   *Various* means different. (The crystals were of *various* sizes.)

VIA, USING

Via   *Via* means by way of. (The students went to the lab *via* the ice cream parlor.)

Using   *Using* means to put into service and by means of. (We isolated the plasmid DNA *using* a commercially available kit.)

WHICH, THAT

      Sometimes these words can be used interchangeably. More often, they cannot.

Which   Use *which* with commas for nondefining (nonessential) phrases or clauses. (One dog, which was 12 years old, survived.)

That   Use *that* without commas for essential phrases or clauses. A phrase or clause introduced by *that* cannot be omitted without changing the meaning of the sentence. Such essential material should not be set off with commas. (Dogs *that* were treated with antidote recovered.)

WHILE, WHEREAS, ALTHOUGH

While   *While* indicates time and temporal relationship. It means at the same time that. It is often incorrectly used instead of *although* or *whereas*. (Experiments were performed *while* patients were sleeping.)

Whereas   *Whereas,* often the word the writer intended, means "when in fact" and "in view of the fact that". (*Whereas* X increased, Y decreased.)

Although   *Although* is a conjunction meaning despite the fact that or even though. (*Although* the association between breast cancer risk and variant alleles varied slightly, the p values were not statistically significant.)

USE, UTILIZE

Use	Generally, *use* is the preferred term.
Utilize	*Utilize* means to find a new, profitable, or practical use for something. Viewed as an unnecessary and pretentious substitute for use.

Answer Key

SUGGESTED ANSWERS TO PROBLEMS

Chapter 2

Problem 2-1 Precise Words

1. All OVE mutants showed [**threefold; 30% increased**] iP concentrations.
   POINT: "Enhanced" is imprecise as well as the wrong word choice. "Increased" is the correct quantitative term for concentration. How much was the increase? Give a quantitative value.

2. Plants were kept [**at 0 °C? or –20 °C**] overnight.
   POINT: "In the cold" is imprecise. How is cold is cold? State the temperature or a range.

3. [**10%, 12 %, 6**] of the discovered exoplanets have an orbital period of less than 5 days.
   POINT: "Some" is imprecise. How much is some? Use a quantitative term.

4. In general, retinal explants treated with 0.5 µg/ml BSA exhibited [**30%, 10%, twofold increased**] growth rates compared with retinal explants treated with 0.5 µg/ml BSA.
   POINT: "Increased" is imprecise. How much increased? Use a quantitative term.

5. We field ionized those atoms not ionized by the HCP and detected the electrons produced [**using**] the rapidly rising field ionization pulse shown in Fig. 2B.
   POINT: "With" is the one of the vaguest terms in English. Because "with" can mean so many things, it is clearer to use a precise term whenever possible.

6. Briefly, cells were incubated with Mytomycin C for 3.5 hours, centrifuged [**HOW?**], harvested with trypsin [**HOW?**], and frozen in aliquots [**HOW?**].
   POINT: How exactly were the experiments done? Give details, such as $5,000 \times g$, 10 ug/ul trypsin, frozen at –20 °C in 1 ml aliquots.

7. To provide proof of concept for our hypothesis, we studied [**the human rotavirus VP7 serotype 1**] in **CV-1** cells.
   POINT: "A virus" is imprecise. What virus? Name it. Similarly for "host cells"; it is more precise to name the host cells.

8. Apart from the discussed main band, [**20% reduced, fourfold lower**] emissions were observed.
   POINT: Weaker than what and by how much weaker? "Weaker" is imprecise. State it quantitatively.

9. The current was [**reduced by 80%, increased 100-fold**] when temperature was increased.
   POINT: "Drastically" and "affected" are imprecise. How was the current affected? It can be increased or reduced. "Affected" is imprecise. "Drastically" is also imprecise. Science is quantitative. Thus, a quantitative detail such as "80%" is much clearer than a qualitative term such as "drastically affected."

10. (Last sentence in an Introduction) The present paper reports on [**XXX**] experiments that were performed to clarify this surprising effect.

POINT: What were the experiments and how were they done?

The way this introduction stands, "continuing" does not define anything. "Continuing" is imprecise. Experiments need to be defined; otherwise, there is no point to this sentence.

11. Only [**about 6%, 2%**] of the region under study exhibits larger reddening.

POINT: "Some" is imprecise. How much is some? Use a quantitative term.

12. Heating arises after recapture and subsequent equilibration following from the lowest $\bar{T} = 0.04$, which is obtained by imaging the gas [**1 ns, 5 ns**] after release from the trap.

POINT: "Shortly after" is imprecise. How long is shortly? Use a quantitative term.

13. A second calculated transition state that places the water molecule above the TFA ring is [**3% higher, twofold higher**] higher in energy than the first transition state.

POINT: "Little higher" is imprecise. How much is a little higher? Use a quantitative term.

14. The band showing vibrational splitting of 192/cm in Ne [**displaying**] the most intense peak at 444 nm can be identified [**using**] the A-> X transition of the dimer Ag_2.

POINT: "With" is the one of the vaguest terms in English. Because "with" can mean so many things, it is clearer to use a precise term whenever possible.

15. We studied the performance of different functionals [**in the presence of OR within, according to, as related to, by comparing, with respect to**] the size of the cluster.

POINT: "With" is the one of the vaguest terms in English. Because "with" can mean so many things, it is clearer to use a precise term whenever possible.

16. The afterglow of the blast wave was [**50%, 120%**] brighter than we expected.

POINT: "Markedly" is imprecise. How much is markedly? Use a quantitative term.

17. To determine the molecular events that ensue from this initial epithelial cell contact, we performed an analysis of proteins recruited to lipid rafts generated during [**15 min**] of infection with *P. aeruginosa*.

POINT: "Rapid" is imprecise. How long is rapid? Use a quantitative term.

18. It is also possible that, due to the high expression levels of MVP in airway cells, the residual [**5 to 10% of**] MVP present after siRNA treatment was sufficient for these other signaling and response pathways.

POINT: "Residual" is imprecise. How much is residual? Use a quantitative term.

Problem 2-2 Simple Words

1. support
2. used
3. are due to/ are caused by different conditions.
4. OMIT OR we believe that...
5. To ...
6. fit; agree
7. We studied ... systematically
8. For example,...
9. ground
10. 10-fold
11. cleaves

Problem 2-3 Commonly Confused/ Misused Words

1. **like, as:**
   Plasmids were isolated _**as**_ described by Beates (17).
   Our observations for C1P1-GpP localization were _**like**_ those of Andrews et al. (1989).
   Tropospheric ozone (O3) is a naturally occurring greenhouse gas formed _**as**_ a product of photochemical reactions
   The energy is transferred to the lattice before the electrons heat up to temperatures, _**like**_ they do in Cu.

2. **enhance, increase:**
   Metal atoms _ **increase** _ the relative intensity of the band at 476 nm.
   Soluble silicon in plants also has an active function in _**enhancing**_ host resistance to plant diseases.

3. **while, whereas**:
   Colonies of DH5 alpha cells transformed with the AB construct were able to degrade naphthalene, _**whereas**_negative control cells were not.
   The first enzyme was added _**while**_ the DNA mixtures were incubating at 37 °C.
   Tropical forests growing on highly weathered soils exhibit conservative P-cycling processes, _**whereas**_ conservative N-cycling properties are more common on younger soils.
   Higher temperatures favor the production of the 619 nm species, _**whereas**_ at 15 °C, an increase of intensity of the 476 nm band is observed.

4. **varying, various:**
   **Varying** water levels in a pond are often the result of climate conditions.
   Each student received _**various**_ concentrations of NaCl solution for the experiment.
   **Various** animals rely on darkness—to hide, to catch prey, to mate, to interact.
   Different varieties of semiconductors layered in solar cells respond to photons of _**varying**_ energies to produce electricity.
   Electrodes can be of _**various**__ sizes.

5. **effect, affect**:

Nutrition concentration was the most important factor _ **affecting** _ population size.

Although the cows were given steroids, the drugs had little _ **effect** _.

Ozone causes cellular damage inside leaves that adversely _affect_ plant production.

Energy supplies by electrons and ions or chemical reaction (e.g., oxidation) with impurities are just some of the _effects_ that might be responsible for luminescence via matrix.

6. **include, consist of**:

Her research interests __ **include** __ all areas of biochemistry and structural biology.

Components of Hyperion's crust _include_ solid H_2O and CO_2.

7. **that, which**:

Fish _ **that**_ live in caves show many adaptations to living in darkness.

At one electrode, hydrogen molecules are stripped of their electrons, __ which_ are then sent through an external circuit to do work.

Adaptive immune responses recognize novel viral antigens_that_ are not invariant but nevertheless are foreign to the infected organism.

8. **represents, is**:

25 mg of ketamine _is_ an overdose of anesthetic for mice. (Note: This sentence can be started with a number because the number refers to an actual scientific value.)

The Born–Oppenheimer approximation of uncoupled electronic and nuclear motion _is_a standard tool of the computational chemist.

9. **infers, implies**:

Both curves are of an identical shape, which _ **implies** _ a constant front profile as well as a constant velocity.

The Intergovernmental Panel on Climate Change has been criticized for _implying_ that climate-envelope models are more precise than they actually are.

10. **can, may**:

It _ **may**_ appear that Table 1 contains an essentially complete summary of patterns that occur in electrochemical systems.

Huge numbers of species _may_ be at risk of extinction from climate change.

Problem 2-4 Redundancies and Jargon

essential	some
many	like
ten times	because
despite	because, due to
during, while	to
cause	can
can, may	close to, near
about, concerning	to
if	because

(OMIT)   note that
by   square
most

Problem 2-5 Redundancies and Jargon

1. The doubling rate appeared to be **short**.
2. The following strains **were courteously provided by** or **obtained from** Dr. U. Miller (University of Minnesota).
3. The data of the analysis on cell cycle parameters have revealed that the cell cycle is advanced by factor X **(Fig. 1).**
4. Cytokinin overproducing mutants revealed **comparable rates to those of** the thiA1 auxotroph.
5. **After 2 hr**, we ended the incubation of CO_2 on a Ag(110) surface.
6. **Homologous recombination is the preferred mechanism of DNA repair in yeast.**
7. The effect of temperature on conductivity **was found not to** change dramatically. OR The effect of temperature on conductivity **did not** change dramatically.
8. Often, jewel weed **grows close to** poison ivy.
9. Transduction efficiencies *in vivo* were much higher than **those** *in vitro.*
10. Upon heat activation, filament size increased, and the number of buds decreased. Both **of these changes** were only seen for cytokinin mutants.
11. **To** test the adsorption performance under various near-field conditions of a waste repository, experiments on the retention of radioiodide by different organo-clays were carried out **at** elevated temperatures and high-molar saline solutions.
12. Although transition metals **can** form bonds with six shared electron pairs, only quadruply bonded compounds can be isolated as stable species at room temperature.
13. After infecting a host cell, a herpes viral DNA genome enters the nucleus where it is **transcribed** to both messenger RNAs (mRNAs) and noncoding RNAs (ncRNAs).
14. Large size has often been linked to elevated extinction risk in mammals **because** larger species tend to exist at lower average population densities and are disproportionately exploited by humans.
15. Aerobic phototrophic bacteria that **use** bacteriochlorophyll for light harvesting and charge separation were at first thought to be limited to selected, nutrient-rich environments, but similar organisms were later found to be ubiquitously distributed in the upper ocean.

Problem 2-6 Redundancies and Jargon

1. [OMIT] Collagen synthesis returned to normal 3 days postinjury.
2. **Although** our present knowledge on the subject is far from complete, this macromolecular structure can aid in the design of new antibiotics.
3. **Many** persons in whom severe neuroinvasive WNV disease develops have long-term disability or die as a result of their infection.

4. After 3 hr, the old medium was (removed and)/**replaced** with the same amount of fresh medium (was added).
5. The data in Table 1 **are consistent** with Brokl's (1999) model.
6. This indicates (suggests) that factor A interacts with factor B.
7. We conducted nearly identical seismic experiments adjacent to the north walls of both the HDR and BTF scarps (Fig. 1) **to** directly **correlate** the seismic layer 2A/2B boundary with mapped geologic units in young oceanic crusts.
8. In **many** cases, degradation leads to topsoil loss and a reduction in soil fertility.
9. **When** hydrogen nuclei fuse to form helium, they release energy.
10. **A** variety of newer reforestation methodologies are becoming available for areas that are currently deforested.
11. Long-range atmospheric transport of pollutants is generally assumed to be the main vector for arctic contamination **because** local pollution sources are rare.
12. Helper T cells perform critical functions in the immune system **through** the production of distinct cytokine profiles.

Problem 2-7 Abbreviations

Too many abbreviations make it hard for the reader to follow the paragraph. Abbreviations should be limited to standard abbreviations and to four to five nonstandard ones per 10-page manuscript. Nonstandard abbreviations should only be used if the term occurs more than 10 times in the paper or several times in quick succession.

Problem 2-8 Mixed Word Choice

1. A typical scientist spends many long hours, even on the weekend, in **the** laboratory.
2. We studied the **effect** of erythromycin on 5 **boys** and 3 **girls** in three different **assay**.
3. A graph displaying this data is shown in the slide **after next/in the next slide.**
4. We **used** a Sorvail centrifuge to obtain a sucrose gradient.
5. We **took** a picture of the gel we **ran**.
6. We observed a change in cluster size after **10 min**.
7. It was more difficult than expected to isolate the Nnkla –1 protein. OR To isolate the Nnkla –1 protein was more difficult than expected.
8. Absorbance was measured at **various** time points.
9. To **take** perfect slides of DNA nicks, an electron microscope is **essential**.
10. Postdocs should be taught how to improve **their** writing. OR Every postdoc should be taught how to improve **scientific** writing.
11. Onchocerciasis or river blindness is a chronic debilitating disease of **humans** caused by infection with the filarial nematode *Onchocerca volvulus*. OR A chronic debilitating **human** disease is onchocerciasis or river blindness, which is caused by infection with the filarial nematode *Onchocerca volvulus*.
12. To reduce the amount of data points, we **omitted** every alternative test point.

13. **The** distortion of the thiophene ring in quarterthiophene is affected by the medium (crystal or solvent) as well as by the intrinsic properties of molecules.

14. Smaller species should, in general, benefit more from the conservation of important threatened areas, **whereas** larger species will tend to benefit most from a conservation approach that also singles out individual species for particular attention.

15. Local nutrient enrichment from guano has been documented. **In contrast/However**, the possibility of contamination in areas near seabird nesting sites has been largely overlooked.

16. Mathematical tools **that** are not yet common in the field of cell biology need to be developed to perform image analysis.

17. The fate of endosomes, **which** eventually disappear as new synaptic vesicles reform, remains unclear.

Problem 2-9 Mixed Word Choice

a)

Sulfonamides were among the first **synthetic** agents used successfully to treat diseases. **Because of** their broad antibacterial activity, these drugs were **formerly** used almost exclusively in the treatment of **many different** diseases. **Fortunately**, other drugs have supplanted the sulfonamides as antimicrobial agents because all pathogenic bacteria are capable of developing resistance to the sulfonamides. Sulfonamides prevent the synthesis of folic acid, **which** is a coenzyme important in amino acid metabolism. Although sulfonamides are for the most part readily tolerated (OMIT), they do have some side **effects**.

b)

Butterfly wing patterns are amazingly diverse. One of the patterns **commonly** found on butterfly wings are eyespots. **Although** much is known about how these eyespots are produced on the developing wings, very little is known **about** why such patterns exist in butterflies. **The** function of eyespots for some species may be for mate recognition or attraction. In other species, eyespots may function as an anti-predator defense (14). Overall, eyespot patterns reveal high heritability (15).

Chapter 3

Problem 3-1 Sentence Interpretation

The end position in a sentence is more emphasized than the beginning position, and the independent clause is more stressed than any dependent clause. Thus, statement 2 is the one most likely to result in the paper being accepted. Statement 1 is the one least likely to result in acceptance.

Reason:

sentence	news in main clause	news in end position
1)	– negative	– negative
2)	+ positive	+ positive
3)	+ positive	– negative (dependent clause)
4)	– negative(+)	+positive (dependent clause)

Problem 3-2   Word Placement and Flow

a)

Vegetative cells are cells that are engaged in active growth and reproduction. Vegetative growth can be ceased by bacteria that can
produce endospores. Unlike vegetative cells, endospores are highly
resistant to heat, chemicals, and radiation. Resistance may be associated
in some way with the unique structure of the peptidoglycan layer of the
spore.

b)

Mangrove plantations attenuated tsunami-induced waves and protected
shorelines against damage. On the shorelines most damaged by the great
2006 tsunami, the area of mangroves had been reduced by 26% through
human activities. By conserving or replanting coastal mangroves and
greenbelts, communities can be buffered from future tsunami events.
The same buffer could be fulfilled elsewhere by the conservation of dune
ecosystems or green belts of other tree species.

c)

A quantum dot is a semiconductor nanostructure that confines the motion
of conduction band electrons, valence band holes, or excitons in all three
spatial directions. It has a discrete quantized energy spectrum. A quantum dot
contains a small number (on the order of 1–100) of conduction band electrons, valence band holes, or excitons. Small quantum dots, such as colloidal
semiconductor nanocrystals, can be as small as 2 to 10 nanometers.

Problem 3-3 Word Placement and Flow

Fleas often carry organisms that cause diseases. An example of an organism that is transmitted by fleas to humans is the plague *bacillus*. These *bacilli* migrate from the bite site to the lymph nodes. Enlarged lymph nodes are called "buboes," giving rise to the name "bubonic plague."

Problem 3-4 Word Placement and Flow

Thermophiles are microorganisms that grow at a temperature range between 45 °C and 70 °C. They are found in hot sulfur springs. Because thermophiles cannot grow at body temperature, they are not involved in infectious diseases of humans. However, it is unclear how they resist elevated temperatures.

After reading the last sentence, the reader expects to hear about possible mechanisms of resistance.

Problem 3-5 Word Placement and Flow

Extrasolar planets are common but possess a large variety of properties. Their orbital periods can range from as short as 1.2 days to as long as ~10 years. These planets are easy to detect because they require less observational time. Some of the planets are on eccentric orbits more typical of some comets in the solar system, whereas others are in multiple planet systems. Although the most recently discovered exoplanets have masses only one order of magnitude larger than Earth, some behemoths have more than 15 times the mass of Jupiter.

Problem 3-6 Subject-Verb-Object Placement

1. **Onchocerciasis is** now recognized as one of the major public health and socioeconomic problems in many tropical countries (Murdoch et al., 1996; OEPA, 1998). **There are approximately 18 million cases of onchocerciasis worldwide, and 80 million more people are at risk of infection.**

2. Since 1995, more than 150 **extrasolar planets have been discovered, most of them in orbits quite different from those of the giant planets in our own solar system.**

3. Early experiments revealed the extreme fragility of the superconductivity in the ruthenate superconductor Sr_2RuO_4 (SRO) **as demonstrated by a strong suppression of the transition temperature with impurities (4).**

4. Aside from these RNA structures found in bacteria, plants, and fungi, **a viral RNA has been found** to be able to bind ATP. **This viral RNA has a sequence very similar to an ATP binding RNA aptamer.**

5. **We recorded** the in-plane coordinates x and y and the time of detection t **for each detected atom released from the trap.**

6. After a warm spring, **female passerines often breed** earlier than they do after a cold spring (13, 14). **This behavior is a result of phenotypic plasticity (14, 15), an individual-level response to temperature.**

7. Allosteric ribozymes have been shown to perform logical functions (3). **These ribozymes take small molecules as input; their output makes cascading difficult, as it is a different form than the input.**

8. Cytoplasmic **incompatibility (CI) occurs** when a *Wolbachia*-infected male mates with an uninfected female, resulting in karyogamy failure and early developmental arrest of the mosquito embryo (5). **This incompatibility has attracted scientific attention as a potential vehicle for gene drive.**

9. **COF-1 [(C3H2BO)6·(C9H12)1] and COF-5 (C9H4BO2) have rigid structures, exceptional thermal stabilities (to temperatures up to 600 °C), and low densities. They also exhibit permanent porosity with specific surface areas surpassing those of well-known zeolites and porous silicates.** COF-1 and COF-5 can be synthesized using a simple "one-pot" procedure under mild reaction conditions that are efficient and high-yielding.

10. A superconducting quantum interference device (SQUID) is used to measure the phase difference between different real-space tunneling directions. **This device allows us to map the phase anisotropy of the superconducting order parameter.**

11. **To investigate asymmetric thermal propagation in a suitable 1D inhomogeneous medium,** we modified carbon nanotubes and boron nitride nanotubes so that they assumed a non-uniform axial mass distribution (Fig. 1).

12. We reported previously that a 50 **kDa protein is involved** in the mechanism in addition to poly(A) binding protein [23]. **This 50 kDa protein is also strongly associated with polysomal mRNA from *Saccharomyces cerevisiae*.**

Chapter 4

Problem 4-1 First Person

1. **We studied** the performance of different functionals with the size of clusters.
2. **We** thank Dr T. J. Guiermo for useful discussions.
3. **We/I** recommend that a triple regimen of antibiotics is given to infants with HIV.
4. Based on our results, **we conclude** that the measurements on translational cooling in desorption contain information about the dynamics of vibrational energy transfer.
5. **We observed** no delay in replication. (In a Materials and Methods section, authors may prefer third person or passive voice for this example.)
6. To remove PAH, **we** constructed transgenic *E. coli*. (In a Materials and Methods section, authors may prefer third person or passive voice for this example.)
7. We also studied the effect of Mg^{2+} and K^+ concentrations on [^{14}C] ATP-pmRNA complex formation. **We** found that binding is dependent on magnesium concentration showing an optimum at 6 mM and an apparent linear decrease at higher concentrations (Figure 2D).
8. **We are investigating** genetic drive mechanisms in the laboratory and in the field, including the incompatibility mediated by Wolbachia symbionts, which will allow us to apply this technology for disease control.
9. **We predict that** the twisted conformation **has** only one C–F bond in an anomeric position.
10. **We propose** a mechanism of electrocyclic rearrangements initiated by N-silylation.
11. J. D. Beckerle and colleagues studied vibrational energy relaxation on metals; but in this report, **we** establish a new class of nonequilibrium surface reactions.

Problem 4-2 Active and Passive Voice

1. **We obtained** the following results:...
2. **We examined** the isolated tissues for parasites.
3. **The team analyzed** the collected specimens and videotape recorded during 25 dives.
4. **Plants gauge** variations in day length.
5. **We recovered** several cytosine deaminases after 2 days.
6. **Two-pulse correlation measurements provide** insight into the dynamics of the excitation process.
7. **We saw** no colonies on blood agar.
8. **The addition of KA-1 to the reaction stimulated** phosphorylation.
9. **We will provide** additional arguments.... **We will discuss** the electronic structure....
10. **We refined** the structure at a resolution of 1.15 Å.
11. **Expedition members unveiled** their preliminary findings.
12. **We administered** the same dosage of antitumor drugs to patients for 60 days. OR **Patients received** the same dosage...

13. **The advent of giant magnetoresistance (GMR) helped launch** the emerging field of spintronics.

14. **The principle of detailed balance provides** another way of treating the translational cooling.

15. **We carried out** computational studies and searches in the CSD (Fig. S1) and the PDB to investigate the conformational preferences of the aromatic OCH3, OCH2F, OCHF2, and OCF3 groups.

Problem 4-3 Active Versus Passive Voice

1. Deep Water Coral, also known as cold water coral, are most often stony corals belonging to the phylum Cnidaria. One of the most common cold water corals is *Lophelia pertusa*. ***Lophelia pertusa* are affected negatively by fishing methods such as deep water trawling.** These methods tend to break corals apart and destroy reefs.
   To optimize word location, use the passive voice.

2. The photon is the basic unit of light and all other forms of electromagnetic radiation. It is also the force carrier for the electromagnetic force. The photon has no mass and thus can produce interactions at long distances. **It exhibits both wave and particle properties.** For example, a single photon may undergo refraction by a lens or exhibit wave interference but also act as a particle giving a definite result when its location is measured.
   Active voice is preferred. It allows a consistent point of view.

3. The bulk of inorganic compounds occur as salts. Important classes of inorganic compounds are the oxides, the carbonates, the sulfates, and the halides. **Many inorganic compounds are characterized by high melting points.** Inorganic salts typically are poor conductors in the solid state, and their solubility in, e.g., water and ease of crystallization varies.
   To keep a consistent point of view, passive voice is preferred.

Problem 4-4 Use of Tense

1. In our study, tree size **increased** with reduction of pesticides.

2. The larva of the Monarch butterfly **feed** on milkweed.

3. In our study, we **found** that there are significant flattenings of the PES along the reaction Pth with increasing cluster size, as Fig. 5 **indicates.**

4. In a study by Fermi, the vibrational levels **diverged** into the so-called Fermi dyad.

5. Females of this species **have** a unique projection from the dorsal part of the thorax.

6. Recent work by deValt (2001) **showed** that nematodes **reproduce** readily in the laboratory.

7. A phagemid construct encoding 2 transgenes in tandem **can** be used to transform *E. coli* cells.

8. Table 3 **shows** that in our study polychaetes **were** most abundant at depths of 10 to 16 m.

9. Baumann (1997) **found** that this bacterium **is/was** highly sensitive to pH.

10. We **observed** chemoattraction within 30 min of Shh addition.

11. Identification of restriction factors **gives** us more knowledge of the host cell restriction system.

12. Our results **suggest** that the Asn5 glycosylation site **plays** a critical role in the function of the KCEN 1 protein.

13. We **investigated** high-quality single crystals of vanadium oxide.

14. We **did** not observe rods in the diffraction patterns but instead well-defined Bragg spots, suggesting that at least 10 interatomic layers in the surface-normal direction **are** contributing to the interferences.

15. Coupled climate-carbon cycle models **suggest** that Amazon forests **are** vulnerable to droughts, but satellite observations **showed** a green-up in intact evergreen forests of the Amazon in response to a short, intense drought in 2005.

16. The K_a value for ß-cellobiosyl (approaching 10^3 M^{-1}) **is** comparable to that for many carbohydrate/protein interactions.

17. Hundreds of species of reef-building corals **spawn** synchronously over a few nights each year, and moonlight **regulates** this spawning event.

18. Arctic wind anomalies **were** part of a global-scale pattern of highly unusual circulation in 2007, the causes of which **are** as yet unclear.

19. To map pathways of motion, all observed Bragg diffractions of different planes and zone axes **were examined** on the femtosecond to nanosecond time scale.

20. ABSTRACT: The average translational energy **was** significantly lower than expected for the calculated surface temperature.

21. SUMMARY: In conclusion, we **found** significant deviations from the equilibrium behavior in fs-laser induced desorption of CO/Ru.

Problem 4-5 Sentence Length

Example 1

These results show that ATP and GTP are capable of binding to the same site in the mRNP. Higher specificity is shown for ATP than for GTP. The other nucleotides tested (CTP, UTP, AMP) did not stimulate the initiation of the translation. On the contrary, they produced some inhibition being most pronounced for UTP and CTP, suggesting that these nucleotides compete with ATP and GTP for the same site in the mRNP.   *(Average of 18 words per sentence)*

Example 2

The Dengue virus is a *Flavivirus* and belongs/belonging to the family Flaviviridae. This family includes over 60 known human pathogens such as those causing yellow fever, Japanese encephalitis, tick-borne encephalitis, Saint Louis encephalitis, and West Nile encephalitis. The Dengue virus is classified into four different serotypes, Types 1, 2, 3, and 4. Infection with any of these four serotypes can result in dengue fever (DF) or dengue hemorrhagic fever (DHF). The latter is particularly a concern during heterologous secondary dengue virus infections.   *(Average of 16.4 words per sentence)*

Example 3

Overall, in the TNF signaling pathway, h-IAP1 seems to enable its own expression. Once expressed, h-IAP1 exerts its antiapoptotic effect downstream in the TNF α signaling pathway by stimulating proteolytic cleavage of terminal caspase-3, -7 and caspase-9. Thus, h-IAP1 inhibits protein degradation and caspase-mediated apoptosis.   *(Average of 15 words per sentence)*

Example 4

When thiamine derivatives bind to *E. coli* mRNA coding for the enzymes that catalyze the thiamine biosynthesis, an mRNA conformation is stabilized. This conformation inhibits translation (22b).   *(Average of 13 words per sentence)*

Note that "Reports show that" can be omitted, as it adds nothing of importance to the sentence.

Example 5

Together with on/off photodetectors in the reflected channel, two combinations of half-wave plates and polarizing beam splitters form the photon-subtraction modules. These modules are placed in the path of the thermal light field before and after the parametric crystal, respectively. Together with the on/off photodetector in the trigger channel, the crystal forms the photon-addition module.   *(Average of 18 words per sentence)*

Example 6

An international team of researchers says that some key elements of modern behavior were in place by 164,000 years ago. This date pushes back the appearance of complex tools and manipulating symbols by 25,000 to 40,000 years. The date is evidenced by complex stone bladelets and ground red pigment—advances usually seen as hallmarks of modern behavior—coupled with the shells of mussels, abalone, and other invertebrates in a cave in South Africa. These ancient clambakes are the earliest evidence of humans that also include marine resources in their diet.   *(Average of 22 words per sentence)*

Example 7

Attosecond pulses are created when intense laser pulses of femtosecond duration are focused into a gas sample. A process known as high-harmonic generation results to produce light at a range of frequencies that are precisely phased together. Such phasing creates a train of very short, coherent pulses.   *(Average of 15 words per sentence)*

Example 8

Negative global coupling changes pattern formation only qualitatively if the system oscillates. Such coupling can result in the formation of standing waves with wave Number 1 just like for N-NDR-systems. However, for parameter values for which the migration coupling leads to the formation of Turing-like structures, the interaction of both instabilities can yield a complex wave pattern characterized by two wave numbers, $n = 1$ and $n > 1$. *(Average of 21.3 words per sentence)*

Example 9

For a remote reference electrode but close counter electrode (or equi-potential area), the migrational coupling is local, and thus the fronts move at a constant rate. However, the fronts are increasingly accelerated, if the distance between working electrode and counter electrode is larger. *(Average of 21 words per sentence)*

Problem 4-6 Active Verbs

1. Reaction products **were measured** by a mass spectrometer. (**We measured** …)
2. Transplant rejection **increased**.
3. Mitochondria **were removed** by HPLC. (**We removed** …)
4. Amyloid plaques **were measured** twice for each brain. (**We measured** …)
5. WBC count **was not elevated** when aspirin and streptokinase were given.
6. If the symmetric stretch levels **are not corrected**, results **are** misleading.
7. Our results showed that the vaccine **protected** the dogs.
8. PI 3-kinase signaling cascade **regulates** PIP2 levels.
9. Stanozolol **prolonged** appetite.
10. Increasing cluster size **reduces** E_8.
11. Visible light **is emitted** when clusters of Ag or Cu are **agglumerated** in a noble gas matrix.
12. The A-B construct **was sequenced** on an ABI377 automated DNA sequencer using Big Dye Sequencing deoxyterminators (ABI v3.0). (**We sequenced** …)
13. Growth rates did not **differ** significantly between the Shh-treated and control samples during the 1 hr **before protein was added**.
14. We **isolated** plasmids to sequence the A-B insert in the pBluescript (STRATAGENE) vector.
15. Rfp2 overexpression **changed** the expression of many genes that regulate cell cycle.
16. To determine whether cells **migrate**, we dissected mouse brains.
17. COP1 **was inactivated** by light prior to its nuclear depletion.
18. Electrons **were liberated** through the photoelectric effect. OR: The photoelectric effect **liberated** electrons.
19. In all instances, the binding of these soluble molecules to their specific receptors **activates** intracellular signaling cascades. OR: In all instances, when these soluble molecules **bind** to their specific receptors, they **activate** intracellular signaling cascades.
20. NCs **do not interact** with the very small stray fields present in the ferromagnetic column matrix.
21. Buffalo, elephant, and black rhino abundance all **declined** rapidly after 1977.
22. Current **flows** freely in a superconductor because the electrons form pairs that travel unhindered through the material.
23. A powerful methodology developed in our laboratory allows us **to structurally characterize** increasingly complex systems.
24. To accomplish these objectives, solvated gas-phase catalytic intermediates **need to be prepared**.

25. We must understand the central puzzle of how O=O bonds **form**.

26. In the samples containing MMS, replication **slowed**.

27. Water oxidation **produces** biochemical fuel and oxygen.

Problem 4-7 Noun Clusters

1. The results are compared using a **plane-wave basis set of high quality**.

2. The **result group that tested negative** in the penicillin skin test showed no allergic rash.

3. We analyzed a **time series of particle flux**.

4. The **callus tissue of the spinach** culture produced no shoots in 1999.

5. With this unconventional technique, we could easily define the transition temperatures **for the liquid crystal of aromatic hydrocarbons**.

6. In 10 of the 15 patients, we observed a **syndrome of chronic depression**.

7. Previous work suggested that **fish oil diets rich in vitamin A** protect mice against certain mosquitoes.

8. The quantity of TCR signal is directly related to the time for which a T-cell remains in contact with **the MHC-peptide complex that expresses APC**.

9. Our study involved **lymphoblastoid cell lines transformed with the human Epstein–Barr virus** (EBV).

10. In particular, their extraordinary low density (1.00–$0.20\,g/cm^3$) and high surface area (500–$4{,}500\,m^2/g$) make them **ideal candidates for the storage and separation of gases**.

11. Inelastic neutron scattering (INS) and diffuse reflectance infrared spectroscopy do not provide adequate information on **the structural details of the adsorption site**.

12. The **spin susceptibility in samples of different mobilities** behaves critically near the transition.

13. **Abundance distributions in ecological communities** need to be considered in these studies.

Problem 4-8 Pronouns

1. **Bacterial cells become avirulent** when they are cured of viruses.

2. A few microorganisms such as *Mycobacterium tuberculosis* are resistant to phagocytic digestion. **This resistance** is one reason why tuberculosis is difficult to cure.

3. Patients often suffer from infections of *Corynebacterium* and from toxins released by **these bacteria**.

4. An action potential triggers calcium channels to open at the synapse. **This opening/chain of events** causes docked synaptic vesicles to fuse to the membrane and release their neurotransmitter content.

5. When the *E. coli* ompA gene was expressed in *Sodalis*, **Sodalis** displayed a pathogenic phenotype. OR, *Sodalis* displayed a pathogenic phenotype when the *E. coli* ompA gene was expressed by it.

6. Another more costly and labor intensive approach is to perform biochemical analysis of the fats in yeast. **This analysis** has been done for some yeast mutants (47). **The method/This analysis** involves thin layer chromatography, coupled to gas chromatography.

7. Results indicate that binding decreased about 4-fold when the temperature increased from 4 to 17 °C. (Fig. 1). **This temperature increase/ decrease/change** suggests that the binding among the particles may be governed by interactions such as hydrogen bonds or van der Waals forces that weaken when temperature rises.

8. p53 co-immunoprecipitates with ARF-GFP and HDM2 in the presence of E6 (Fig. 1). **This co-immunoprecipitation** can also be seen for total cell lysates (Fig. 1).

9. The identification of this restriction factor helps us to understand the function of this and similar factors. **Its identification/Determining its function** gives us more knowledge of the host cell restriction system.

10. The goal of this study was to identify the molecular mechanisms by which rosiglitazone induces primary adipocytes to increase adiponectin. **To identify this mechanism,** we tested both the mouse 3T3-L1 adipocyte cell line and Ob/Ob mice.

11. Adipose tissue from the same rosiglitazone-treated animals secretes an increased amount of adiponectin. **This increased secretion** indicates that there may be additional factors signaling adiponectin secretion.

12. CTLA-4 inhibits the number of activated T-cells, and this effect is maximal at low TCR strength. **This inhibition** is contradictory to earlier observations where CTLA-4 expression increases when TCR strength is increased.

13. The breakthrough was achieved because new methods and concepts could be applied, which had been developed in nonlinear dynamics to describe the spontaneous formation of order in various disciplines. **This description** in turn was only possible because the underlying laws are universal at a certain abstract level.

14. A current-voltage-characteristic with a negative impedance is a necessary prerequisite for this type of oscillator. However, in the stationary current-voltage-plot, **this impedance** is visible only in a subsystem.

15. Obviously, the disturbance is dampened in the first picture, whereas a spatiotemporal structure forms in the second example. **This formation** can best be described as a superposition of the homogeneous oscillation and a sine-shaped standing wave.

16. The concentration profile perpendicular to the electrode does not adjust spontaneously when the concentration changes at the electrode. **This lack of adjustment** is caused by a high reaction rate.

17. Almost all sulfur atoms are also engaged in lone pair bonding to a phenyl edge (25). Most p-MBAs are linked through chains of such interactions extending from one pole of the nanoparticle to the other (Fig. 4D). **This ordering of p-MBAs** exemplifies the "self-assembly" of a thiol monolayer on a gold surface (26).

18. Orienting C–F bonds antiperiplanar to the lone pairs of the now sp3-hybridized O results in an anomeric nO–σ*CF conjugation with concomitant lengthening of the C–F bonds (35). **This lengthening** reduces the conjugation between O and the aromatic π system and

eliminates the energetic preference of a planar, in-plane conformation.

19. Our previous models tended to underproduce Na compared with the abundances of HE0107–5240. **This problem/underproduction** has been improved in our current presupernova models.

Problem 4-9 Parallelism

1. The pathogenesis observed in other cells, such as circulating monocytes, may differ from **that in** endothelial cells.
2. The stability of these particles appears to be regulated by RNA helicase activity and **by** protein phosphorylation.
3. Dengue hemorrhagic fever is particularly a concern during heterologous secondary dengue virus infections where cross-reactive antibodies are implicated in immune pathology as well as **in** antibody-enhanced infection.
4. MAPs proteins are able to destabilize RNA helices in the presence of ATP **but not DNA helices.**
5. Diabetes can be affected **by both** exercise and diet. OR Diabetes can be affected **both by** exercise and **by** diet.
6. The fat bodies were washed twice with TBS/Tween, incubated in secondary antibody for 2 hr, and **then mounted** onto glass slides.
7. Aptamer complexes could be used as highly sensitive biosensors, detection probes, efficient inhibitors of cancer, **or as powerful tools** for accelerating the process of drug discovery.
8. The pellet was rinsed with 200 μl ethanol, air dried, and **then resuspended** in 10 μl dH$_2$O.
9. These peaks were dependent on kinetic heat flow, not **on** state transition heat flow.
10. Spatial coupling acts on the activator variable and also **on** the inhibitor variable.
11. Other research may indicate that the translating ribosome has the ability to open up downstream mRNA secondary structure and **to overcome** a block in translation initiation in this way.
12. Whether the crisis for liquids would be avoided in some way at lower temperatures than **those at which** measurements have been made is controversial and is of limited relevance to describing the observed behavior.
13. The thiol monolayer is stabilized not only by gold-sulfur bonding but also **by** interactions between p-MBA molecules.
14. Most sulfur atoms, bonded to a phenyl ring and **to** gold atoms in two different shells, are also chiral centers.
15. To keep the time-varying term to a minimum, we study the balance over two long periods, from January to June and **from** July to December.
16. The internal pressure must not only depend on volume but also **on** the rate of filling.
17. We provide a broad overview of the project, including the target population, the project's primary aim, the problem areas the project addresses, and **the location where** the services are being delivered.

18. NbCA1 acts as a negative regulator and **as** a suppresser of SGT1, RAR1, and HSP90.

Problem 4-10 Comparisons

1. The kinetics of the protein G accumulation were similar to **those of** endogenous proteins (19).

2. Thus, the pattern observed for the zot gene probe proved to be much more diverse **than** that of the CT genotype.

3. Optimizations for the MMA-co-MVE copolymerization system were better for the GA simulations **than for the** experimental data.

4. The dendrogram showed that *V. cholerae* O22 is more closely related to *V. cholerae* O139 **than to other strains**.

5. Mayer's hypothesis on the spread of this disease is similar to **that of** our present study.

6. Archaebacteria are more ancient **than eubacteria**.

7. Mutant A showed the highest peak.

8. The male dolphin was **larger than** any of the other animals.

9. The different propagation behavior of electrochemical fronts compared to **fronts in** reaction-diffusion systems is due to a different range of spatial coupling.

10. There are several connections of the Au102 nanoparticle structure with **structures reported in** previous works.

11. The Austro-Tai populations were found to be a unity not only in culture but also in genetic structure, **unlike** other groups in East Asia.

12. VSV vectors induce much stronger CD8 T cell responses **than** other viral vaccine vectors, including poxviruses (25).

13. 'A' stars are less numerous **than** their solar-type equivalents.

14. The cosmic ray anisotropy was more complex **than expected**.

15. High concentrations in previous studies (800 times more active sites than in this study) cause a signal decrease of no more than 12%, even when the concentration of polarized nuclei is 2,600-fold higher **than** that in our setup.

16. The New Haven cohort had the same immunological response **as that found for the** Philadelphia and Miami **ones**.

Problem 4-11 Common Errors

1. Altogether, we measured **XYZ** for 5 days.

2. This structure controls the packaging of the DNA into the phage head and **the release of the DNA** after recognizing the receptor on the host.

3. **Data for Fig. 7 were obtained** during the oxidation of hydrogen at the Pt electrode. OR The effect of oxidation of hydrogen at the Pt electrode can be seen in Fig. 7.

4. **Our data indicate** that more work needs to be done to prove our hypothesis.

5. The liver, gall bladder, and spleen of each rabbit **were** examined.

6. We confirmed that eight genes were overexpressed and one gene **was** down-regulated in HUVECs during D2V infection.

7. Contrast medium was infused at a steady rate into the injection port, and the flow **was** calculated from the observed change in CT number at equilibrium.

8. These complexes form *in situ*.

9. The Tibet Air Shower Array experiment was **conducted** at Yangbajing (90.522 E, 30.102 N; 4,300 m above sea level) in Tibet, China.

10. We used a novel **program** to **analyze** the **tumor**. OR We used a novel **programme** to **analyse** the **tumour**.

11. When HBC graphitic nanotubes **are doped** with oxidants, they become electrically conductive.

12. The **study** of Brownian motion and its **generalizations has** had a **profound** impact on physics, mathematics, chemistry, and biology.

13. Having determined that an anisotropy exists, **we rescanned** the full data set from 1 January 2004 to 31 August 2007.

Problem 4-12 Punctuation

1. Moreover, their results can be explained by the fact that the DNA repair defect of the H2A tail deletion mutant is mainly kinetic.

2. Patients with lnd QT syndrome can experience seizures, arrhythmia, and sudden death by ventricular fibrillation.

3. As can be seen in Fig. 3, the amount of amplified product increases in parallel fashion to the amount of inclusions observed.

4. The overall relative contributions of anthropogenic versus biogenic sources of VOCs have not been clearly established. There is geographic variation in the relative contributions of these broad source groups.

5. In conclusion, our results suggest that the pathways that regulate glucose uptake have even more components than previously thought.

6. Arabidopsis transformation will be done using Agrobacterium-mediated T-DNA transformation, thereby allowing *in vivo* expression of our transgene.

7. The goal was to select the best, most efficient deaminases possible.

8. The book, *Handbook of Scientific Writing*, is very helpful.

9. The density matrix of a thermal state, the most classical state of light, is diagonal in the Fock state basis and, accordingly, exhibits no phase dependence.

10. The accumulation of oxidizing equivalents was analyzed by gas-phase ion chemistry, laser spectroscopy, and computational methods.

11. Two key roadblocks had to be overcome: i) we needed to store the energy of four photons; and ii) we needed to design efficient catalysts.

12. Thus, this system of pathways is termed "pioneer metabolism."

13. 'A' stars are less numerous than their solar-type equivalents, the T-Tauri stars, and in general are located farther away from Earth.

14. An insulator has an energy gap separating filled and empty bands of electronic states and thus is electrically inert because a finite energy is required to dislodge an electron.

Chapter 5

Problem 5-1 Prepositions

1. In connection **with**
2. Compared **to/with**
3. In contrast **to**
4. Search **for/of**
5. Correlated **with**
6. A comparison **of** A **with** B.../A comparison **between** A **and** B...
7. Similar **to**
8. To look forward **to**
9. Results shown **in** Fig. 3...
10. Through the decrease **of/in**
11. Analogous **to**
12. Implicit **in**
13. Theorize **about/that**
14. Different **from**
15. Attempt (n.) **at**   attempt (v.) **to**
16. **With/In** respect **to**

Problem 5-2 Articles

1. Oscillatory behavior was reduced in **the** presence of a ring electrode.
2. Sea urchin fertilization is **a/the** model system in developmental biology.
   Here, **a** *would be used if there is more than one such model system in developmental biology, whereas* **the** *would be used if the author considers it the only one or most important one.*
3. Therefore, **the** hypothesis we presented previously is further confirmed.
4. Back to nature.
5. Inorganic chemistry proves to be a very interesting subject.
6. Can you name **an** inorganic molecule?
7. Conversion of **a/the** CK riboside to the base can be regarded as the last step.
   Here, **a** *would be used if there is more than one CK riboside, whereas* **the** *would be used if there is only one CK riboside.*
8. The data for oxidation of hydrogen at the Pt electrode were obtained in **the** absence of chloride ions.
9. **The** theoretical studies cited earlier depict a variety of different patterns in oscillating media.
10. Here we discuss **(the)** physical-chemical mechanisms leading to pattern formation in electrochemical systems.
11. *Title:* Role of Physical Activity on **the** Severity of Diabetes
12. *Title:* Orbital Reconstruction and Covalent Bonding at **an** Oxide Interface
13. *Title:* Correlation of **the** Highest-Energy Cosmic Rays with Nearby Extragalactic Objects
14. *Figure 4:* Effect of temperature on ctx expression.

15. Lichens consist of algae (or cyanobacteria) and fungi, which coexist in symbiosis.
16. Rainfall in **the** Sahara desert occurs rarely.
17. **The** average rainfall in **the** Sahara desert is less than 25 mm per year.
18. Under **the** light microscope, organelles are visible in eukaryotes.
19. At the end of its "life," **an** mRNA molecule is degraded.
20. **The** first probe was chosen upstream of **the** CA1 sequence.
21. **The** NbCA1 promoter was amplified using **the** BD GenomeWalker Universal Kit (BD Bioscience-Clontech, Palo Alto, CA).
22. **The** experimental data show **a** strong interference pattern that qualitatively resembles the pattern induced by a double slit.

Problem 5-3 Verb Forms

1. The analysis was limited to **determining** the change in the population number.
2. We sought to **improve** the yield by adding specific catalysts.
3. Impurities in our samples were barriers to **obtaining** chemically pure products using standard isolation techniques.
4. Ongoing volcanic eruptions limited the amount of time that we could dedicate to **collecting** samples.
5. The reviewers recommend **adding** more information about the bear habitat.
6. As with chemical reactions, the concept of concerted versus consecutive nuclear motions becomes central to **understanding** the elementary steps of the mechanism.
7. In the mid-1990s, Bottomly began **reporting** on immune response to foreign substances.
8. Plants are able **to perceive** pathogen attacks and subsequently induce defense responses.
9. The K616 strain failed **to grow** on calcium-depleted media.
10. We were surprised **to find** so few differences among mutations that determine drug resistance.

Problem 5-4 Verb Forms

1. The conservation of the sequence in similar places in the genomes **suggests** that they play a key role in genome function.
2. Higher-order stimulated Raman scattering (HSRS) in hydrogen or deuterium **has been shown** to exhibit a broad and coherent HSRS spectrum.
3. High-mobility electron systems with tunable density have **led** to prominent advances in science and technology over the past decades.
4. In this study, **it is** clear that our amplifying circuit element is simple, fast, modular, and robust.
5. Frequency and fitness **are** greater for annuals in light gaps and biennials in the understory.
6. *S. multiplicata* females did not **choose** hetero-specific mates regardless of water level.
7. Until now, **there** has been no experimental confirmation of this hypothesis.

8. Microspheres of like charge were **bound** together with nanoparticles of opposite charge to form clusters of two to nine colloids.

9. Each of the experiments **was** performed at two different photon energies.

10. One third of the mice **was/were** infected with *P. chabaudi*.

11. Ultrafast electron microscopy (UEM) and ultrafast diffraction **have** been the methods of choice for studies of molecular and phase transitions.

12. Virus A or virus B **was/is** thought to cause the disease.

Problem 5-5 Adjectives and Adverbs

1. X behaved **differently** than expected.

2. The simulations demonstrate how **sensitively** the formation of the first stars depended on the detailed properties of the still mysterious dark matter.

3. To be highly conserved, genes must play a role so important to survival that evolution keeps them **intact**, weeding out deleterious mutations.

4. Species richness in plants is correlated **biologically** and **geohistorically** and increases ecological opportunities.

5. Diffraction restricts the ability of most electromagnetic devices to image or **selectively** target objects smaller than the wavelength.

6. Mainstream climate science needs to look more **closely** at geoengineering.

7. Hurricane Katrina's impact on U.S. Gulf Coast forests was quantified by linking **ecological** field studies and **empirically** based models.

8. It has been reported that the virus linked to the collapse of honeybee colonies may have arrived in the United States via **recently** imported Australian bees.

9. The flux of cosmic rays at Earth decreases very **rapidly** with energy, from a few particles per square centimeter per second in the low-energy region to less than one particle per square kilometer per century above 10^{20} eV.

10. The analysis offers initial results of an **ambitious** international project that will eventually compare the complete genomes of several dozen TB strains from around the globe.

Problem 5-6 Mixed ESL Errors

1. **It** is not clear if the difference in our results was due to the temperature difference in our measurement.

2. We expected **to observe** a large difference in the output.

3. The environment surrounding invasive breast tumors **exhibits** significant changes.

4. Star formation in the early universe was very **different** from that of the present.

5. The graphene bilayer's band structure is predicted **to give** rise to several unconventional phenomena, such as the Klein paradox and the Vaselago lensing effect.

6. Despite the internationally recognized uniqueness and importance of **the** Peruvian rainforest, the impacts of human activities throughout the region remain poorly understood.

7. At the center is **a/the** Pt atom, surrounded by a 12-atom Pd icosahedron, and **the** second shell is a 42-atom icosahedron with either 3 or 4 Pt atoms; **the** third and fourth shells are high-symmetry structures with 60 and 50 atoms, respectively.

8. We hypothesize that **the** concept of metamaterial-inspired nanoelectronics ("metactronics") can bring the tools and mathematical machinery of the circuit theory into optics, electronics, metamaterials, and nanoscale devices.

9. The production of an individual platelet begins when a hematopoietic stem cell begins to differentiate. **It** is complete when a fragment of a bone marrow megakaryocyte is released into the vasculature and begins to circulate.

10. In support **of** the latter model, platelets formed from proplatelet processes *in vitro* are functional.

11. The fossil's small size suggests that the individual belongs to *H. habilis* or a new species, but the more modern traits, such as long legs and modern body proportions, have **proven** the individual to be in *H. erectus* and have **shown** this species was adapted for long-distance locomotion.

12. Although the other species in the *Gorilla* genus, **the** Eastern Lowland Gorilla, **is** far less numerous than **the** Western Lowland Gorilla, IUCN ranks the former one level lower at "endangered" because it is outside the current area of Ebola outbreaks.

13. Our results indicate that disruption of NbCA1 function **accelerates** HR cell death.

14. Tanzania's Lake Natron is **the** only **known** breeding site for East Africa's lesser flamingos.

15. This is **the** first report on cellular calcium signaling during **the** disease resistance response in plants.

Problem 5-7 Mixed ESL Errors

a)

Dengue fever, transmitted most often by the bite of an infected *Aedes aegypti* mosquito, was often seen as an obscure, only **occasionally** fatal disease of tropical countries, and progress toward a vaccine and drugs to treat it has been slow. Since the disease **became** more virulent and **spread** into new geographic areas, vaccine research has taken on a new urgency. A vaccine can't come a moment too soon. It is estimated that the number of dengue cases tops 50 million **annually**, which is similar **to** the number of malaria cases. The warm, crowded cities of Latin America and Asia provide an ideal habitat for the main vector, *A. aegypti*, which breeds in stagnant water and likes **to feed** on humans in quick succession.

b)

We have discovered **an** imported virus that may be associated **with** the sudden disappearance of honey bees in the United States, **known** as colony collapse disorder (CCD). This syndrome caused **surprising** losses of up to 90% of hives in some apiaries. The suspect is a pathogen called Israel acute paralysis virus (IAPV). Our team **found** the virus in most of the affected colonies they tested but in almost no healthy ones. If **the** virus **proves** to be the cause of CCD, it could have **international** economic implications. Since 2005, U.S. beekeepers, especially those struggling **to keep** up with the insatiable demand for almond pollination in California, have imported several million dollars' worth of bees from Australia. We report that we have found IAPV in imported Australian bees.

c)

The basic anatomy of East Asia's dust storms is **fairly well** established. The common term "sand storms" is incorrect, however, as sand particles are too heavy **to get** lifted high into the atmosphere. Thus, little of **the** dust that afflicts East Asia comes from deserts, where erosion over the millennia has carried away most of the smaller particles. Instead, the dust originates in dry lakebeds and arid lands on desert fringes. In springtime, the crust of the undisturbed soil is broken **up** by plowing and livestock. The dust gets lifted into the air by updrafts created through the temperature difference between a cold atmosphere and a surface warmed by spring sunlight. Subsequently, the airborne dust is carried south and east by winds. It gets sucked into **the** upper atmosphere when it reaches low-pressure pockets at the mountain ranges that ring northern China and Mongolia. Easterly winds transport **the** particulate matter to East Asia, and sometimes across the Pacific Ocean to North America.

d)

A typical laser produces coherent radiation and emits light in a narrow, low-divergence beam and with a well-defined wavelength. In contrast, a light source such as **the** incandescent lightbulb emits over a wide spectrum of wavelength. Lasers have **become** a multibillion dollar industry. The most widespread use of lasers is in optical storage devices such as compact disc and DVD players in which the laser (a few millimeters in size) scans the surface of the disc. However, laser use is not limited to **applying** this technology to optical storage devices. Another common application of lasers **is** as bar code readers or laser pointers. In industry, lasers are used for cutting steel and other metals and for inscribing patterns (such as the letters on computer keyboards). Lasers are also **commonly** used in various fields in science, especially spectroscopy, **typically** because of their well-defined wavelength or short pulse duration. In addition, lasers are **sought** for military and medical applications.

Chapter 6

Problem 6-1 Paragraph Organization

Phytophthora infestans, which precipitated the Irish potato famines in the mid-19th century, remains the most economically important potato pathogen. **Infection by the pathogen involves two phases: a biotrophic phase up to 36 hr postinoculation in which *P. infestans* forms haustoria and requires living plant tissue and an ensuing necrotrophic phase in which infected host tissue becomes necrotic (1).**

The second sentence, which functions as a second topic sentence here, is about the two phases of infection. These phases are subsequently described in parallel form.

Problem 6-2 Paragraph Organization

1Ionotropic glutamate receptors fall into two general categories (7, 8). 1'*These receptors are NMDA receptors or non-NMDA receptors.* 2**NMDA receptors are activated by N-methyl-D-aspartate (NMDA).** 2'**Upon activation,** ion channels are opened, allowing ions to rush into the cell and thus cause an excitatory postsynaptic potential (Fig. 1). 3*In contrast,* non-NMDA receptors like PCP bind to kainate and quisqualate only and prevent ion flow and an excitatory postsynaptic potential, even in the presence of NMDA. 4Thus, PCP acts as a strong noncompetitive antagonist of NMDA.

The topic sentence is about the two categories of glutamate receptors. After reading the topic sentence, the reader expects to find out what the two categories are. Sentence 1' fulfils the expectation created by the topic sentence. Sentence 2 is revised to be parallel to sentence 3. Sentence 2 is also split in two to divide the different ideas into different sentences.

Problem 6-3 Paragraph Organization

Copper Oxide Superconductors

1**Two features characterize copper oxide superconductors.** 2One feature is, of course, their unprecedented high transition temperatures. 3The other feature is that their normal-state properties are not those of ordinary metals; they are not consistent with the traditional Fermi-liquid quasiparticle picture that is a cornerstone of our understanding of the metallic state.

The topic sentence states the message of the paragraph, "two features characterize ... superconductors." The topic is the subject; the action is in the verb.

Problem 6-4 Paragraph Organization

1'**Salmon use different methods to find their way during their homeward journey in the fall.** 1Salmon may use a magnetic or sun compass to orient themselves. 2As described by Brown et al. (15), **they may also use olfactory cues learned as smelts** to find the river and tributary of their birth. 3Salmon may reenter fresh water in spring, summer, or fall, but spawning occurs in the fall, and the life cycle of the salmon begins anew.

The topic sentence states the message of the paragraph, "the different methods salmon use to find their way during their homeward journey in the fall." The topic is the subject in every sentence, resulting in a consistent point of view throughout the paragraph. Sentence 2 is parallel to sentence 1.

Problem 6-5 Paragraph Organization

1Interleukin I (IL-I) is a mediator produced by activated macrophages and many other cell types (1). 2IL-I initiates inflammation and thus stimulates various cells. 3**IL-I is a co-signal in T-cell activation.** 4**Furthermore**, IL-I plays a role in activating B cells and **in stimulating** the production of acute phase proteins, muscle catabolism, bone resorption, and fever.

The point of view is now consistent throughout the paragraph. A transition has been added at the beginning of sentence 4 to indicate the logical relationship of sentence 3 to sentence 4, and parallel ideas have been written in correct parallel form (sentence 4: in stimulating).

Problem 6-6 Paragraph Consistency

1Both GAD-positive cell bodies and processes were found in the **thalamic reticular nucleus** and **ventral lateral posterior nucleus**. **2**Almost all of the neurons in the thalamic reticular nucleus appeared to contain GAD-immunoreactivity. **3However**, only small round cells in the ventral lateral posterior nucleus were GAD positive.

In the revision, consistent order has been kept for thalamic reticular nucleus and ventral lateral posterior nucleus, reversing their order in the first sentence. In addition, a transition at the beginning of sentence 3 indicates the logical relationship between sentence 2 and 3 more clearly.

Problem 6-7 Paragraph Consistency

1Lipopolysaccharide (LPS) endotoxin, a component of the bacterial cell wall, stimulates macrophages to produce pro-inflammatory cytokines. **2**Two such cytokines have been shown to be critical mediators of septic shock (7). **3**The first of these mediators is the tumor necrosis factor α (TNF-α). **4**Its excessive production causes systemic capillary leakage, tissue destruction, and ultimately lethal organ failure and death (1-4, 7). **5**TNF-α possesses a signal peptide and is first synthesized as a membrane-associated protein. **6The membrane-associated form of TNF-α is cleaved by a member of the metalloproteinase family, the TNF-α converting enzyme (TACE), generating soluble TNF-α (8, 9). 6′The second of the mediators is interleukin-1β (IL-1β). 7′IL-1β acts in early immune responses as a chemotactant for lymphocytes and as a signal for eosinophil and basophil degranulation (10). 8′Unlike TNF-α, IL-1β is synthesized as a precursor molecule. 9This precursor is then cleaved and activated by the cysteine protease caspase-1.**

Consistent order has been ensured by adding a sentence before sentence 7, which follows through with naming the second mediator. In addition, sentences 7 and 8 have been rewritten in parallel form and order to sentences 4 to 6, and sentence 6 has been rearranged to display consistent form.

Problem 6-8 Paragraph Consistency

Apart from seawater acidification, the ocean in a high CO_2 world will experience other changes, including <u>higher surface temperatures</u>, **increased** <u>stratification</u>, and decreased mixed-layer depths. **Higher surface temperatures appear to increase the C:N ratio at the organism level, although the direct effect of higher surface temperatures on net community carbon-to-nutrient ratios is unclear.** <u>Increased stratification</u>, in contrast, would decrease the supply of nutrients to the surface layer, reducing overall primary and export production. **Decreased mixed-layer depth tends to increase phytoplankton carbon-to-nutrient ratios, which would augment the direct CO_2 effect on C:N:P stoichiometry observed in this study.** Thus, concurrent with a decrease in the strength of the biological pump caused by a lower nutrient supply, the pump's efficiency is likely to increase at increased CO_2.

The order of the key terms introduced in the topic sentence corresponds with that displayed in the remaining sentences. Key terms are repeated exactly throughout, and parallel form is kept when signaling the subtopics.

Problem 6-9 Paragraph Consistency and Transitions

1The emission properties of a nanoscale optical emitter can be significantly modified by the proximity of a nanowire that supports surface plasmons. 2In principle, three distinct decay channels exist. 3First, direct optical emission into free-space modes is possible with a rate modified from that of an isolated quantum dot owing to the proximity of the metallic surface. **4Second, t**he optical emitter can also be damped nonradiatively owing to Ohmic losses in the conductor. **5Third, and m**ost important, the tight field confinement and reduced velocity of surface plasmons can cause the nanowire to capture the majority of spontaneous radiation into the guided surface plasmon modes, much like a lens with extraordinarily high numerical aperture.

In the revised version, the three decay channels are now consistently introduced, and transitions to sentences 4 and 5 are parallel to the transition used in sentence 3.

Problem 6-10 Key Terms

A model system to study the underlying mechanism that connects Type II diabetes, life span, and **fat storage** is the nematode *C. elegans*. **Nematodes** are a great system to study the connection of diabetes, **life span**, and **fat storage** because they have a well-conserved insulin signaling pathway that affects life span and fat storage, suggesting that there is an underlying molecular mechanism that connects insulin signaling, **life span**, and **fat storage**. In addition to the known insulin signaling pathway, the entire genome of the nematode has been sequenced, and many molecular and genetic tools are available that will allow a comprehensive identification of the genes that affect insulin-like signaling, **life span**, and fat **storage** in *C. elegans.*

Key terms are now repeated exactly throughout the paragraph. To avoid too much repetition, you may consider replacing 'life span' and 'fat storage' with a category term such as 'these two factors' occasionally.

Problem 6-11 Key Terms

Opioid drugs are avidly self-administered by both humans and laboratory animals (2,3). **Two such opioids, heroin and morphine**, mimic the endogenous opioid neurotransmitters, known as endorphins, by binding to one or more of the mu or delta opioid receptors in the brain (4).

The phrase "Two such opioids, heroin and morphine" establishes a link between the key terms in the first and second sentence.

Problem 6-12 Key Terms

1We measured **systolic blood pressure** in 5000 individuals aged ten to 18 years using a random zero sphygmomanometer. **2We found that** mean **systolic blood pressure** in girls was 116 mmHg at age 15. **3 In contrast, mean systolic blood pressure** in fifteen-year-old boys was 128 mmHg.

4Thus, **systolic blood pressure** in boys was about 10% higher **than** that in girls.

"Systolic blood pressure" is introduced in sentence 1 and used consistently and exactly throughout the paragraph. In addition, results are signaled and placed in parallel form. The transition "In contrast" indicates that a contrasting finding is following. The faulty comparison in sentence 4 has been corrected by using "than that in" rather than "compared to."

Problem 6-13 Transitions

1Substantial arteriosclerotic lesions can be produced by a diet rich in cholesterol (1). 2Aside from these lesions, other tissues are also effected (2–7) by high cholesterol levels. **3For example,** the liver undergoes severe fatty degeneration (2). **4In addition, the rabbit eye undergoes pathological changes** (3,5).

Problem 6-14 Transitions

1Marine coastal ecosystems are among the most productive and diverse communities on Earth (*1*) and are of global importance to climate, nutrient budgets, and primary productivity (*2*). **2Yet/However,** the contributions that coastal ecosystems make to these ecological processes are compromised by human-induced stresses including overfishing, habitat destruction, and pollution (*3–5*). 3These stressors particularly impact benthic (bottom-living) invertebrate communities because many species are sedentary and cannot avoid disturbance. 4**Thus,** marine coastal ecosystems are likely to experience a proportionally large change in biodiversity should present trends in human activity continue (*6–8*).

Problem 6-15 Paragraph Construction

A possible way to write this paragraph is the following:

Cancer Cells

Cancer cells, which are malignant tumor cells, differ from normal cells in three ways: (1) they dedifferentiate—for example, ciliated cells in the bronchi lose their cilia; (2) they can metastasize, meaning they can travel to other parts of body; and (3) they divide rapidly, growing new tumors, because they do not stick to each other as firmly as normal cells do.

Problem 6-16 Paragraph Construction

Depending on the author's emphasis, there are various possibilities on writing a paragraph using the provided list. Here is one version:

Reactive Oxygen Species (ROS)

Reactive oxygen species (ROS) form as a natural by-product of the normal metabolism of oxygen and include oxygen ions, free radicals, and peroxides. ROS are highly reactive due to the presence of unpaired valence shell electrons. At low levels, ROS play important roles in cell signaling. At high levels, such as during times of environmental stress, ROS levels can increase dramatically, which can result in significant damage to cell structures and ultimately cumulate in oxidative stress.

Problem 6-17 Paragraph Construction

Depending on the author's emphasis, there are various possibilities on writing a paragraph using the provided list. Here is one version:

Superconductivity
Superconductivity is a quantum mechanical phenomenon. The phenomenon occurs in certain materials, such as in tin and aluminum and in various metallic alloys but does not occur in noble metals like gold and silver, nor in most ferromagnetic metals. Superconductivity occurs at extremely low temperatures. When the material is cooled below its "critical temperature," the electrical resistance of a superconductor drops abruptly to zero. At zero resistance, an electrical current flowing in a loop of superconducting wire can persist indefinitely with no power source.

Problem 6-18 Paragraph Construction

Depending on the author's opinion and emphasis, there are various possibilities on writing a paragraph using the provided list. Here are three versions:

Combination Drug Therapy
a) Combination drug therapy is of important clinical value. It can be used against organisms for which resistance to one drug readily develops, such as in the treatment of tuberculosis. However, combination drug therapy can be expensive. It can also increase the risk of toxicity and superinfection.
b) There are positive and negative factors that should be considered when using combination drug therapy. Combination drug therapy is of important clinical value. For example, combination drug therapy is used against organisms for which resistance to one drug readily develops. It is, for example, applied in the treatment of tuberculosis. On the other hand, combination drug therapy can be expensive, and it can increase the risk of toxicity and superinfection.
c) Combination drug therapy should be avoided wherever possible. It is not only expensive, it can also increase the risk of toxicity of each single drug. Furthermore, it can increase the risk of superinfection. Yet, some clinicians consider combination drug therapy of important clinical value because it can be used against organisms for which resistance to one drug readily develops. For example, it is applied in the treatment of tuberculosis.

Problem 6-19 Paragraph Construction

1Chikungunya is a painful viral disease of the tropics.

2It is rarely fatal but can cause severe fevers, headaches, fatigue,

?

nausea, and muscle and joint pains. 3Medical entomologists have worried about the astonishing ascent of the Asian tiger mosquito. 4Many viral diseases are transmitted by *Aedes albopictus*. 5It was first found in the United States in second hand tires imported from Asia in Houston, Texas, in 1985. 6Plants shipped in water containers, such as the popular Lucky bamboo, often carry the eggs of the mosquito. 7In Italy, almost 200 people were infected by the mosquito this summer. 8Scientists wonder whether *Aedes albopictus* has the potential to touch off much larger outbreaks in Europe and the United States.

Arrows connecting key terms show that sentences 2 and 3, as well as sentences 3 and 4, are not linked. In addition, the link between sentences 4 and 5 is not clear, and that between sentences 5 and 6 comes rather late in the sentence, as do the links between sentences 5 and 7, and 5 and 8.

Revised Version:

Chikungunya is a painful viral disease of the tropics.

It is rarely fatal but can cause severe fevers, headaches, fatigue, nausea, and muscle and joint pains. **The disease is caused by the Asian tiger mosquito *Aedes albopictus*, a daytime biter that displays not only nasty bites but also transmits many viral diseases. The astonishing ascent of this mosquito has worried medical entomologists. It was first found in the United States in Houston, Texas, in 1985 in secondhand tires imported from Asia. The eggs of the mosquito are also often carried in plants shipped in water containers, such as the popular Lucky bamboo. This summer, the mosquito has infected almost 200 people in Italy. Because of its fast and wide spread,** scientists wonder whether

A. albopictus has the potential to touch off much larger outbreaks in

Europe and the United States.

In the revised version of the paragraph, the links between all the sentences are clearly established and occur early in the sentences. This way, the reader can establish a relationship between sentences immediately.

Problem 6-20 Point of View

1When the temperature falls at constant pressure, most pure materials pass from gas to liquid to solid. **2However, some materials called "liquid crystals," which are rod-like molecules, display** a large number of phases (ref.).

OR:... **An exception to these materials are "liquid crystals," which are rod-like molecules that display** a large number of phases (ref.).

Problem 6-21 Transitions/Coherence

1For typical liquid crystals, the high temperature liquid phase is isotropic, meaning that the positions and the orientations of the molecules are scattered about at random. 2At lower temperatures, the substance undergoes a transition into the so-called "nematic" phase in which the molecules tend to orient in the same direction but in which positions are still scattered. **3At still lower temperatures,** the substance passes into the "smectic" phase in which the molecules orient in the same direction, and their positions tend to fall into planes. 4Finally, at even lower temperatures, the molecules freeze into a conventional solid. **5Thus/In general,** lower temperatures show more and more qualitative order.

Problem 6-22 Condensing

1ATP is important for initiation of translation, one of the most energy-consuming processes in living organisms. 2Here we describe an additional and novel function of ATP in translation, which uses an *in vitro* translation system dependent on exogenous mRNA and derived from wild type *S. cerevisiae.*

(47 words)

In the condensed version of the abstract, the important message, the role of ATP, is placed in the first power position in the topic sentence. Unimportant and redundant information has been omitted (thus, "Translation of the genetic message into proteins" has been condensed to just "translation," as the term already implies that the genetic message is translated into proteins), and less important information has been subordinated ("of the most energy-consuming processes in living organisms" and "among other nucleotide-dependent mechanisms"). The noun cluster in sentence 3 has been broken up.

Problem 6-23 Condensing

This passage can be condensed by using a list in addition to removing wordy phrases.

Superconductors are characterized by (a) their unprecedented high transition temperature and (b) their normal state properties. Unlike normal state properties of ordinary metals, those of superconductors are not consistent with the traditional Fermi-liquid quasiparticle picture essential in our understanding of the metallic state.

(43 words)

Problem 6-24 Condensing

One possible way to reduce the abstract is shown here.

Tourette syndrome (TS) tics are commonly treated with habit reversal therapy (HRT) or supportive psychotherapy (SP). Here we compare the efficacy of HRT and SP in reducing tics, improving life-satisfaction, and psychosocial functioning. Our assessment of 30 adult outpatients with TS showed that only HR reduced tic severity, although patients in both groups showed improved life-satisfaction and psychosocial functioning. Tic severity remained reduced and life-satisfaction and psychosocial functioning remained improved at the 6-month follow-up, suggesting that HR reduces specific tics and that SP improves life-satisfaction and psychosocial functioning. These findings may be of value for predicting treatment response to HR.

(100 words)

Problem 6-25 Condensing

The paragraph can be condensed by only stating the most important information, omitting all other irrelevant results.

Our results indicate that only diamond covered with chemically bound hydrogen shows a pronounced conductivity at temperatures between 5 and 25 °C.

(21 words)

Problem 6-26 Condensing

Radionuclide X contamination of ground and surface water is a serious problem in many parts of the world. Its transport in subsurface groundwater systems is strongly affected by radionuclide X adsorption to iron oxides. The solid-solution ratio—the ratio of the mass of adsorbent solid to the volume of solution—as well as carbonate interactions with radionuclide X can substantially affect its adsorption. Here, we examined the effects of solid-solution ratio and total carbonate concentration on radionuclide X adsorption onto a heterogeneous natural subsurface soil in batch adsorption experiments. Our results show that radionuclide X adsorption, even normalized to adsorbent mass, was affected by solid-solution ratio at a fixed total carbonate concentration and that increasing total carbonate concentration could decrease radionuclide X adsorption because aqueous radionuclide X-carbonate complexes are formed.

(132 words)

Chapter 7

Problem 7-1 Outline

General overview

- There is continued climate warming
- Causes for variations in polar ice sheets
  ○ Variations in Earth's orbit around the sun are considered to be a primary external force driving glacial-interglacial cycles
  ○ Anthropogenic climate warming will accelerate the natural process toward reduction in polar ice sheets
- Effects of variations
  ○ Modern polar ice sheets have become unstable within the natural range of interglacial climates
  ○ Sea level is projected to rise between 13 and 94 cm over the next 100 yr
  ○ Increased rates of sea level rise related to polar ice sheet decay is a potential natural hazard

Previous Pleistocene interglacials

- Controversial geologic evidence suggests that current polar ice sheets have been eliminated or greatly reduced during previous Pleistocene interglacials
- Specific example: marine isotope stage 11
  ○ Conflicting evidence for warmer conditions and higher sea level during marine isotope stage 11
  ○ Sea level may have been more than 20 m higher than today during a presumably very warm interglacial about 400 ka during marine isotope stage 11
  ○ Microfossil and isotopic data from marine sediments of the Cariaco Basin support the interpretation that global sea level was 10 to 20 m higher than today during marine isotope stage 11
  ○ The West Antarctic ice sheet and the Greenland ice sheet were absent or greatly reduced during marine isotope stage 11
  ○ Warm marine isotope stage 11 interglacial climate with sea level as high as or above modern sea level lasted for 25 to 30 ky

Problem 7-2 Outline

Overview of current situation

- Current transportation sector of the economy is carbon based
- At present, transportation fuels are primarily produced from crude oil
- Crude oil is not an infinite resource
- Alternative sources such as renewables, coal, and natural gas have only a small market share
- There will be increasing scope for transportation fuel production from other carbon sources

Alternative liquid fuels

- Liquid fuels can be produced from non-crude oil carbon sources by direct means, such as direct coal liquefaction, or indirect means, such as biomass gasification followed by hydrocarbon synthesis

Specific example: Fischer–Tropsch (FT) synthesis of liquid alternative fuels

- General overview of Fischer–Tropsch (FT) synthesis
    - Fischer–Tropsch (FT) synthesis is a key technologies for gas-to-liquid (GTL), coal-to-liquid (CTL), and biomass-to-liquid (BTL) conversion
    - product from Fischer–Tropsch synthesis is a synthetic crude oil, syncrude
    - syncrude has to be further refined to produce transportation fuels
- Problem with Fischer–Tropsch (FT) synthesis
    - little attention has been paid to the refining of Fischer–Tropsch syncrude
    - assumption: Fischer–Tropsch syncrude can be refined similarly as crude oil
    - no refining technologies have been specifically developed for refining of Fischer–Tropsch syncrude
- Refining of Fischer–Tropsch (FT) syncrude
    - There are significant differences between Fischer–Tropsch syncrude and crude oil refining
    - When syncrude is treated as if it is crude oil, refining becomes inefficient, and it violates green chemistry principles
    - Syncrude refining can be more efficient than crude oil refining (2)
    - Responsible syncrude refinery design is necessary from a commercial and an environmental point of view

Problem 7-3 Constructing a Paragraph

The active ingredient in most chemical-based mosquito repellents is DEET (N,N-diethyl-meta-toluamide), which was developed by the U.S. military in the 1940s. DEET is absorbed readily into the skin and should be used with caution. Although recent research suggests that DEET products, used sparingly for brief periods, are relatively safe (2), other studies report diverse side effects. Common side effects to DEET-based products, often due to overapplication, include rash, swelling, itching, and eye-irritation (4), but toxic encephalopathy has also been associated with use of DEET insect repellents (5).

DEET alternatives include eucalyptus oil containing cineol, 1:5 parts diluted garlic juice, soybean oil, neem oil, marigolds, Avon Skin-so-soft bath oil mixed 1:1 with rubbing alcohol, and the juice and extract of Thai lemon grass (7,8). True citronella has not been proven to be a very effective mosquito repellent (3).

Chapter 8

Problem 8-1 Corrected References are in bold:

Bollag RJ, Waldmann AS, Liskay RM (1989) Homologous recombination in mammalian cells. Annu Rev Genet 23:199–225

Capecchi MR (1989a) Altering the genome by homologous recombination. Science 244:1288–1292

Capecchi MR (1989b) The new mouse genetics: altering the genome by gene targeting. Trends Genet 5:70–76

Grimm C, Kohli J (1988) Observations on integrative transformation in Schizosaccharomyces pombe. Mol Gen Genet 215: 87–93

Halfter U, Morris PC, Willmitzer L (1992) Gene targeting in Arabidopsis thaliana. Mol Gen Genet 231:186–193

Kempin SA, Liljegren SJ, Block LM, Rounsley SD, Yanofsky MF (1997) Targeted disruption in Arabidopsis. Nature 389:802–803

Lee KY, Lund P, Lowe K, Dunsmuir P (1990) Homologous recombination in plant cells after Agrobacterium-mediated transformation. Plant Cell 2:415–425

Offringa R, de Groot JA, Haagsman HJ, Does MP, van den Elzen PJM, Hooykaas PJJ (1990) Extrachromosomal recombination and gene targeting in plant cells after Agrobacterium-mediated transformation. EMBO J 9:3077–3084

Offringa R, Franke-van Dijk MEI, de Groot MJA, van den Elzen PJM, Hooykaas PJJ (1993) Non-reciprocal homologous recombination between Agrobacterium transferred DNA and a plant chromosomal locus. Proc Natl Acad Sci USA 90:7346–7350

Paszkowski J, Baur M, Bogucki A, Potrykus I (1988) Gene targeting in plants. EMBO J 7:4021–4026

Puchta H, Hohn B (1996) From centimorgans to base pairs: homologous recombination in plants. Trends Plant Sci 1:340–348

Puchta H, Dujon B, Hohn B (1996) Two different but related mechanisms are used in plants for the repair of genomic double strand breaks by homologous recombination. Proc Natl Acad Sci USA 93:5055–5060

Riele H te, Maandag ER, Berns A (1992) Highly efficient gene targeting in embryonic stem cells through homologous recombination with isogenic DNA constructs. Proc Natl Acad Sci USA 89:5128–5132

Rossant J, Joyner A (1989) Toward a molecular-genetic analysis of mammalian development. Trends Genet 5:277–283

Roth D, Wilson J (1988) Illegitimate recombination in mammalian cells. In: Kucherlapati R, Smith GR (eds) Genetic recombination. American Society for Microbiology, Washington DC, pp 621–653

Schaefer DG, Bisztray G, ZryÈd JP (1994) Genetic transformation of the moss Physcomitrella patens. Biotech Agric For 29:349–364

Struhl K (1983) The new yeast genetics. Nature 305:391–397

Rothstein R (1991) Targeting, disruption, replacement and allele rescue: integrative DNA transformation in yeast. Methods Enzymol. 194:281–301

Timberlake WE, Marshall MA (1989) Genetic engineering in filamentous fungi. Science 244:1313–1317

Problem 8-2

The integration of foreign DNA into eukaryotic genomes occurs mainly at random loci by illegitimate recombination, even when the introduced DNA has homology to endogenous sequences (**Roth and Wilson 1988**; Bollag et al. 1989; Capecchi 1989a; Puchta et **al.** 1994; Puchta and Hohn 1996). Exceptions to this generalization include *Saccharomyces cerevisiae* (**Rothstein 1991**), *Schizosaccharomyces pombe* (**Grimm and Kohli 1988**) and some filamentous fungi (Timberlake and Marshall 1989). Direct integration by homologous recombination (gene targeting) is rare compared with non-homologous integration, but would provide a powerful technique for the molecular genetic study of higher organisms and the determination of gene function. Although gene targeting is now a well established tool for the specific inactivation or modification of genes in yeast and mouse embryonic stem cells (Struhl 1983; Capecchi 1989b; Rossant and Joyner 1989; te Riele et al. 1992), the development of similar techniques for plant systems is still at a very early stage (**Puchta and Hohn 1996**). Some groups have reported genomic integration by homologous recombination in plant cells using DNA introduced by direct gene transfer or *Agrobacterium*-mediated DNA transfer (Paszkowski et al. 1988; Offringa **et al.** 1990, 1993; Halfter et al. 1992). The highest rates of gene disruption so far published for flowering plants are between 0.01 and 0.1% (Lee et al. 1990; Miao and Lam 1995; Kempin et al. 1997), and these are too low to allow routine gene disruption. However, it has recently been shown that integrative transformants resulting from homologous recombination can be obtained following polyethylene glycol (PEG)-mediated direct gene transfer into protoplasts of the moss *Physcomitrella patens*, at an efficiency comparable to that found for *S. cerevisiae* (**Schaefer and ZryÈd 1997**). *(With permission from Springer Science and Business Media)*

Problem 8-3

The carbon ($\delta^{13}C$), nitrogen ($\delta^{15}N$), and sulfur ($\delta^{34}S$) isotope ratios of humans, other animals, and microbes are strongly correlated with the isotope ratios of their dietary inputs (1–5). There are limited differences ($\leq 1\%$) between heterotrophic organisms and their diet in either the $\delta^{13}C$ or $\delta^{34}S$ values (**reference**). Hydrogen ($\delta^{2}H$) and oxygen ($\delta^{18}O$) isotope ratios of organic matter, however, are more useful, because $\delta^{2}H$ and $\delta^{18}O$ values of precipitation and tap waters vary along geographic gradients (**reference**). Although differences in the $\delta^{2}H$ and $\delta^{18}O$ values of scalp hair have been noted in humans (**reference**), less is known about diet–organism patterns of $\delta^{2}H$ and $\delta^{18}O$ values. Four potential sources can be important: dietary organic molecules, dietary waters, drinking waters, and atmospheric diatomic oxygen. Hobson *et al.* provided evidence that $\delta^{2}H$ values of drinking water were incorporated into different proteinaceous tissues of quail (**reference**). Other research showed that the $\delta^{2}H$ values of bird feathers and butterfly wings (both are largely keratin) and water in the region in which the tissue was produced are highly correlated (14, 15).

Kreuzer-Martin *et al.* showed that ~70% of the oxygen and ~30% of the hydrogen atoms in microbial spores (~50% proteinaceous) were derived from the water in the growth medium, whereas the remainder was derived from the organic compounds supplied as substrate (**reference**). *(With permission from National Academy of Sciences, U.S.A.)*

Problem 8-4

Ostracodes are small bivalved Crustacea that form an important component of deep-sea meiobenthic communities along with nematodes and copepods (10). Crustaceans are dense and diverse in the deep sea and are one of the most representative groups of whole deep-sea benthic community **(10, 11)**. Pedersen et al. **(12)** as well as Jackson et al. **(13, 14)** reported that ostracode species have a variety of habitat and ecology preferences (e.g., infaunal, epifaunal, scavenging, and detrital feeders), representing a wide range of deep-sea soft sediment niches. Furthermore, Ostracoda is the only commonly fossilized metazoan group in deep-sea sediments (**reference**). Thus, fossil ostracodes are considered to be generally representative of the broader benthic community. The distribution and abundance of deep-sea ostracode taxa in the North Atlantic Ocean are influenced by several factors, among them, temperature, oxygen, sediment flux, and food supply **(14, 15)**. Several paleoecological studies suggest that these factors influence deep-sea ecosystems over orbital and millennial timescales **(1, 16)**. *(With permission from National Academy of Sciences, U.S.A.)*

Problem 8-5

Droughts can occur in almost all climates, although their onset, severity, and frequencies vary widely. Droughts may be little noticed, if any, in arid areas but result in dramatic loss of crops and water shortage in others. Droughts result when it has not rained enough over an extended period of time, usually a season or more. High temperatures and wind can add to their onset and lengths. Because drought characteristics are much disputed among scientists and politicians, not much progress in drought management has been seen.

Problem 8-6

The deadly H5N1 avian influenza virus could spark a global pandemic that could kill tens of millions once it acquires the ability to spread in humans. Since 2003, H5N1 has been spreading in an ever increasing number of countries, from Asia to Africa and Europe. While outbreaks have surfaced in new countries in 2007, they have been more or less continuous in others, such as in Indonesia and in Nigeria. Although the virus is passed on to humans mainly through domesticated birds, it is unclear what role wild birds play in its ever widening circles. To shed more light on their role in the spread of the virus, lightweight transmitters could be used to track migratory birds by satellite.

Problem 8-7

Three possible versions of paraphrased citations are shown following:
...McCaffrey predicts on average 1–3 Mw ≥ 9 earthquakes per century that can occur at any subduction zone (McCaffrey 2008). He also believes

that the past half century with five M9 events reflects temporal clustering and not the long-term average.

...according to McCaffrey (McCaffrey 2008), one to three magnitude 9 or larger size earthquakes occur 100 years at subduction zones, and the increased frequency of M9 earthquake occurrences in the past half century is a higher than long-term average.

...an average number of one to three magnitude 9 or larger size earthquakes occurs every century at subduction zones, although sometimes, such as in the past 100 years, M9 earthquakes occur with a higher than normal frequency (McCaffrey 2008).

Problem 8-8

Version A would be considered plagiarized. Here, sentences have simply been rearranged, and a few words have been substituted. Overall, Version A is too close to the original. In contrast, Version B is paraphrased. It does not only differ very much from the original, it also quotes the original source.

Chapter 9

Problem 9-1 Figure or Table?

1. Table or figure/map depending if actual numbers are important or trends are meant to be displayed.
2. Table and/or figure/map—Table would provide precise data, but a map may more visually show location of occurrence.
3. Table or graphs. Omit rat photo as it only shows a control rat.
4. Show the electron micrograph—it presents new evidence. The roentgenogram does not present new evidence and therefore should not be shown.
5. Both—a schematic will help to visualize what is being described in the text.
6. Both—a schematic will help to visualize what is being described in the text.

Problem 9-2 Figure or Table

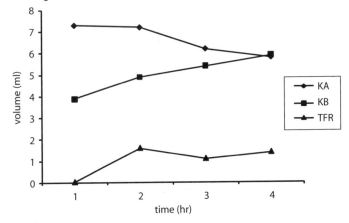

A figure of the average values per time reading would best present the data

Problem 9-3 Figures
- Omit the title – it should go into the figure legend
- Expand the y-axis scale so data points do not fall below x-axis
- Omit the grid lines
- Tick marks should only face outward
- Remove the heavy frame around the graph
- Replace + symbols with closed circles, squares, or triangles

Problem 9-4 Figures
- Different scales for A and B are confusing
- The heavy x-axis bar is distracting
- The key in the middle is confusing
- There are too many tick marks on the y-axis
- It is not immediately obvious why the numbers above the bars are there

As individual values above the graph are apparently important, these data would best be presented in a table.

Chapter 10

Problem 10-1

Despite good signals for the unknown and question, this introduction leaves the reader hanging and wondering what exactly was done and what the results are. The reason for its apparent incompleteness is that it does not state the question precisely nor mention any experimental approach. This introduction also provides a meaningless overview sentence of the discussion to come. Overview sentences such as these should be avoided in research papers, as they do not add anything to introductions of investigative papers.

Astaxanthin is a carotenoid that is found in microalgae, yeast, salmon, trout, krill, shrimp, crayfish, crustaceans, and the feathers of some birds (1,2). Astaxanthin is a natural nutritional component, but it is also used as a food supplement intended for human, animal, and aquaculture consumption (4).

Like many carotenoids, astaxanthin is a colorful, fat/oil-soluble pigment, provising a redish and pink coloration (2). While in certain bird species all adult members display carotenoid containing feathers rich in color, in many gulls and terns an unusual light pink coloring (or flush) to the normally white plumage can be found in highly variable proportions within and across populations (5). It has been suggested that some gulls turn pink because they acquire unusually high amounts of astaxanthin in their diets at the time of feather growth, especially in areas with farm-raised salmon (5). However,

Known

	the exact relationship between astaxanthin and plumage **is not**
Unknown	**fully understood. Here we examine this relationship** in more
Meaningless	**detail and discuss** its implication.
overview sentence	

Question
Purpose—not
stated precisely

Experimental
Approach
missing

Problem 10-2

In this Introduction, the unknown has not been stated, leaving the reader wondering what this study is really about and what it adds to the field.

	Methane clathrate is a solid form of water that contains a large amount of methane within its crystal structure (a clathrate hydrate). Significant deposits of methane clathrate have been found under sediments on the ocean floors.[1] Methane
Known	hydrates are believed to form by migration of gas from depth
Unknown missing	along geological faults, followed by precipitation, or crystallization, on contact of the rising gas stream with cold sea water.
Question/ Purpose	A "Bottom Simulating Reflector" (BSR) was used to detect the presence of methane clathrates along the ocean floor of the Blake Bahama Outer Ridge. Through seismic reflection
Experimental approach	at the sediment to clathrate stability zone interface caused by the unequal densities of normal sediments and those laced with
Main findings	clathrates, we were able to identify several deposits of methane clathrates at depths of 500–1000 m.

Problem 10-3

This Introduction is descriptive and should contain background, discovery statement, description of findings, and implication. The introduction consists largely of background information. The discovery statement is very general and reads like an overview or table of content. Only a partial description of findings is provided ("We also identified …"). No implication is given. Thus, the importance of the findings remains unclear, resulting in a very weak introduction.

| | Ehrlichiae are obligatory intracellular bacteria that infect leukocytes and platelets of a wide variety of mammals and are transmitted by ticks (1). Dogs can be infected by *E. canis*, *E. chaffeensis*, *E. ewingii*, and *Anaplasma phagocytophilum* (*Ehrlichia equi*) (2). Infection with any of these species can cause a severe disease with indistinguishable hematological |
| Known | and clinical anomalies (3). |

E. canis was the first species described in dogs (3–5). It is worldwide distributed, particularly in tropical and sub-tropical regions, and is the causative agent of classical canine monocytic ehrlichiosis which presents three well characterized clinical phases. Dogs treated during the acute phase of the disease normally recover rapidly. However, since this stage of the infection can evolve with moderate or imperceptible clinical signs, infected dogs can develop the subclinical phase and some of them reach the chronic phase (6).

E. chaffeensis is the etiological agent of human monocytic ehrlichiosis. This potentially mortal disease was reported for the first time the year 1987 in USA. E. canis was first thought to be the causative agent due to a cross reaction of sera from patients with an antigen preparation from this species (7). In 1991, the bacterium was isolated and characterized at the molecular level. It was then established that it was a distinct species of Ehrlichia (8). The recognized natural reservoir for this bacterium is the white-tailed deer (9).

Ehrlichiae with tropism for monocytes, lymphocytes, neutrophils and platelets from dogs diagnosed by BCS, have been reported since 1982 (10). The presence of these rickettsiae in monocytes and platelets has been also confirmed using transmission electron microscopy (11,12). In this study we describe primary cultures of monocytes from the blood of a dog with canine monocytic ehrlichiosis. We also identified E. canis and E. chaffeensis using nested PCR with DNA samples extracted from the primary cultures and from dogs with natural and experimental infections.

(Veterinary Clinical Pathology 37(3). pp. 258–265, 2008)

Margin annotations:
- Known
- General discovery statement
- Partial description and Experimental approach
- **Implication missing**

Problem 10-4

This Introduction of an investigative paper contains all required elements (known, unknown, question/purpose, experimental approach). In addition, it also contains main results and describes the overall significance of the paper, rounding up the Introduction nicely.

Septic shock and sepsis syndrome is one of the leading causes of death in hospitalized patients and accounts for 9% of the overall deaths in the United States annually (1-4). While commonly initiated by a bacterial infection, the pathophysiological changes in sepsis are often not due to the infectious organism itself but rather to the uncontrolled production of pro-inflammatory cytokines produced mainly by macrophages. This over-production can result

Margin annotation:
- Known

in an overwhelming systemic inflammatory response that leads to multiple organ failure. Lipopolysaccharide (LPS) endotoxin, a component of the bacterial cell wall, stimulates macrophages to produce pro-inflammatory cytokines such as tumor necrosis factor α (TNF-α) and interleukin-1β (IL-1β), both of which have been shown to be critical mediators of septic shock (7). It is the excessive production of these pro-inflammatory cytokines that causes systemic capillary leakage, tissue destruction, and ultimately lethal organ failure and death (1-4, 7). Thus, the expression of these pro-inflammatory cytokines needs to be tightly regulated during an inflammatory response.

Regulation of mRNA stability is critical for controlling gene expression as the abundance of an mRNA transcript is also regulated by the rate of mRNA degradation (5, 6, 8). The mRNAs encoding most inflammatory cytokines are short-lived, with instability conferred by an AU-rich element (ARE) in their 3' noncoding regions (5, 6). ARE promotes rapid degradation of mRNAs (6) and in some cases translation arrest (9). Furthermore, ARE-mRNAs can be rapidly stabilized upon exposure to certain signals including immune stimulation, UV and ionizing irradiation (5), and heat shock (10). The stability of ARE-mRNAs is controlled by trans-acting AU-rich element binding proteins (6, 16). Hu ARE-binding proteins, such as HuR, stabilize ARE-mRNAs and inhibit mRNA translation (17-19).

In contrast to ARE, AUF1 (or hnRNP-D) (20, 21), TTP (22), BRF1 (23), and KSRP (24) are shown to mediate rapid decay of various cytokine mRNAs. AUF1, an ARE-mRNA destabilizing factor, consists of four isoforms (37, 40, 42, and 45 kD) generated by alternative splicing (20, 21). Increased expression of AUF1 has been correlated with rapid ARE-mRNA degradation in various types of cells (25-27), with p37 AUF1 isoform exhibiting the highest destabilizing activity (25, 26). However,

Known

implication of AUF1 is based largely on correlations with *in vitro* binding to AREs (28, 29) and ectopic overexpression studies in cell lines (25, 26). (25, 26). The role of AUF1 in the regulation of cytokine expression *in vivo* **has not been examined** in a patho-

Unknown

physiological context.

Question/ purpose

To examine the role of AUF1 in promoting inflammatory mRNA degradation *in vivo*, **we generated** AUF1 null mutant mice and studied their response to LPS-induced endotox-

Exp. approach

emia and their expression of pro-inflammatory cytokines. We observed that *AUF1$^{-/-}$* mice were acutely susceptible to endotoxin, showed manifestations typical of endotoxic shock, and had a significantly lower survival rate when challenged by LPS. These phenotypes were associated with over-expression of the pro-inflammatory cytokines TNFα and IL-1β as a result of abnor-

Results

mal stabilization of their mRNAs.

Significance	Our results provide the first *in vivo* evidence implicating AUF1 in regulating inflammatory response. This regulation occurs through targeted degradation of selective cytokine mRNAs, deregulation of which would contribute to the development of endotoxic shock. *(With permission from Cold Spring Harbor Laboratory Press and Robert J. Schneider)*

Problem 10-5

The original introduction is too long and not very focused. The known/background information can be condensed substantially in various ways, two of which are shown following:

Background/ known	The germline is set apart from the somatic tissues early in development. In many species, germ cells contain specialized cytoplasm, known as germ plasm, which contains proteins and RNAs required for germline development. The germ plasm of *C. elegans* includes distinct ribonucleoprotein particles, known as P granules. P granules are initially dispersed in the cytoplasm in oocytes and in the early embryo. Beginning in the P2 blastomere, P granules associate with the nuclear membrane, an association that is maintained in the adult germ cells. A protein with similarity to receptor tyrosine kinases, MES-1, controls the asymmetric partitioning of P granules during the divisions of P2 and P3 (2). The blastomere P4 marks the point of germline restriction, as this cell divide gives rise to a pair of primordial germ cells that proliferate to produce all future germ cells in the animal. The germline blastomeres are transcriptionally silent during the early divisions until after the birth of P4 (3). Therefore, genes that function in the germline establishment in these blastomeres are likely to be maternally deposited proteins or maternally deposited mRNAs under post-transcriptional control. Most proteins present in P granules that have been previously identified contain predicted RNA binding domains, suggesting that regulation of RNAs may be a function of P granules. Some proteins localize exclusively to P granules in both embryonic and adult stages, specifically PGL-1 and PGL-3 (4,5) and the four homologs of *vasa* (-1–4) (6). In contrast, other proteins found in P granules localize to additional cellular locations. For example, PIE-1 is in both P granules and the nucleus, where it represses transcription (7). While many P granule components are required for normal germline development, P granules alone are not sufficient to confer a germline fate. Although more than twenty proteins found in P granules have been identified, a molecular function for this complex has not been elucidated. Additionally, genetic relationships between many P granule components have not yet been reported.

Discovery statement	In this study, we report the identification and characterization of a novel protein, MEG-1, in *C. elegans*. This gene is necessary for the development of the germline of the progeny. We show that loss of *meg-1* results in a severe decrease in germ cell proliferation and abnormalities in the germline blastomeres. Our genetic analysis indicates that *meg-1* functions synergistically with *mes-1* in P granules.
Description	
Significance	

Problem 10-5 Alternate

	The establishment of the germline generation after generation is crucial for the propagation of a species. To maintain the unique ability of the germline to give rise to all tissues of the next generation, the germline must be protected from somatic differentiation signals, which would restrict the fate of the cells. In many species, germ cells contain specialized cytoplasm, known as germ plasm, which contains proteins and RNAs required for germline development. In some cases, this germ plasm includes distinct ribonucleoprotein particles, known as polar granules in *Drosophila* and as P granules in *C. elegans*.
Background/ known	Of the proteins present in P granules that have been previously identified, most contain predicted RNA binding domains, suggesting that regulation of RNAs may be a function of P granules. Some proteins localize exclusively to P granules in both embryonic and adult stages, specifically PGL-1 and PGL-3 (4,5) and the four homologs of *vasa* (-1–4) (6). In contrast, other proteins found in P granules localize to additional cellular locations. For example, PIE-1 is in both P granules and the nucleus, where it represses transcription (7). While many P granule components are required for normal germline development, P granules alone are not sufficient to confer a germline fate. Despite the identification of greater than 20 proteins found in P granules, a molecular function for this complex has not been elucidated. Additionally, genetic relationships between many P granule components have not yet been reported.
Description	
Discovery statement	In this study, we report the identification and characterization of a novel protein in *C. elegans*. This gene is necessary for the development of the germline of the progeny. We have named this gene *egcd-1*. We show that loss of *ecdg-1* results in a severe decrease in germ cell proliferation and abnormalities in the germline blastomeres. Our genetic analysis indicates that *egcd-1* functions synergistically with another gene important in P granules.
Significance	

(With permission from Stefanie W. Leacock, Ph.D.)

Chapter 11

Problem 11-1

1. b. Images of coated specimens were obtained as described by Barnes et al. (23).
   If details are described in a reference, they do not need to be repeated in the text.

2. b. Seedlings were grown in continuous light at 15°C for 21days before being collected.
   Necessary details have been included in b.

3. b. For our study, we selected only healthy males between 60 and 80 years.
   The overview sentence is not needed—it only adds bulk.

4. b. Study subjects were presented with a list of potentially traumatic events, and were asked to use three response categories (*yes, no, unsure*) to indicate if thay had ever experienced them.
   Sentence does not contain the confusing switch in tense found in a.

5. a. The Stress Index Short Form (SI/SF), which is a 36-item question-naire, was used to assess stress after natural disasters (26).
   Verb tense is used correctly here. The statement "which is a 36-item questionnaire" is a statement of general validity and should be present tense.

Problem 11-2

1. The analyses were performed on an Agilent series 1100 HPLC instrument (Agilent, Waldbronn, Germany). *(the additional information is usually not needed and can be omitted)*

2. **Immunoblot analysis was performed** following standard procedure.

3. **The ability to adapt** to changes in light was determined by…

4. PCR products and **oligonucleotides longer than 30bp** (in 50% DMSO, 20 mM) were spotted with SmartArray™ Microarrayer (CapitalBio Corp., Beijing, China).

5. After centrifugation, $10 \times$ buffer **(define the exact content)** was added, and the samples were incubated for 2 min on ice.

Problem 11-3

The purpose of the tests described in this passage is not clear because no such statement has been made. In addition, the last sentence, which explains what types of tests are included in the Wisconsin Card Sorting Test, comes late in the passage, leaving the reader wondering what the test is about throughout the paragraph.

A revised version is shown following. Note that in the revised version, the overall purpose for the test in this particular study is stated (to assess cognitive flexibility), and the background information on the WCST has been moved to the beginning of the paragraph where it helps to provide context.

Wisconsin Card Sorting Test (WCST)

Purpose of test in study is stated, and background information has been placed at the beginning of the passage to provide context

Cognitive flexibility was assessed using the Wisconsin Card Sorting Test (WCST) (Berg, 1948), *in which test subjects have to correctly identify, implement, and remember sorting rules.* To complete this task in our study, participants sorted cards according to color, shape, or number stimuli depicted on the card. Sorting the cards by color was initially verbally reinforced. After a participant responded correctly for 10 consecutive cards in that color category, the participant continued sorting by form and numbers without verbal stimuli or reinforcement. To evaluate our results for cognitive efficiency, the ratio of correct responses to errors was computed, and the numbers of trials, errors, perseverative responses, and perseverative errors were analyzed.

Problem 11-4

Determination of Trypsin Activity

Time not specified

Unspecific temperature and time.

Unspecific— time and temperature?

Unspecific— using what reagents?

Unspecific— how fast, how long, what temperature?

Unspecific— what OD?

Trypsin (TPCK, bovine pancrease) (20ug/µl) dissolved in 0.01 M HCl, was denatured by adding 8M urea, 33 mM Tris, pH 8.0, in a 1:4 (v/v) ratio and then boiling the mixture for 20min. After denaturation, 5µl denatured trypsin was added to 5µl of each crystallization buffer, and the mixture was incubated **at room temperature**. After the first incubation, an equal volume of the crystallization buffer (10µl) was added and the mixture was **incubated again**. Addition of equal volume of buffer together with incubation was repeated twice more (20 and 40µl, respectively). Substrate (0.05g (1%) azocasein in 5ml 10mM Tris, pH 8) (100µl) was added, and the **mixtures incubated**. The mixtures were then **precipitated** with trichloroacetic acid, and **centrifuged**. The supernatants were **read in the spectrophotometer** to determine activity.

(With permission from Elsevier)

Problem 11-5

The first 4 paragraphs of this Materials and Methods section have been well constructed. They contain or repeat key terms and transitions. In addition, word location has been considered, and the topic sentences are easily identified. Transitions and relationships between the paragraphs (except for the last one) could be made clearer through the use of transitions.

Research Question/Purpose: **To construct and test a safe, live, attenuated, oral vaccine candidate, IEM108, immune to CTXΦ infection**

Construction of the candidate IEM108. The 1.15-kb XbaI fragment containing the upstream regulatory and coding regions of ctxB was recovered from pBR (a pUC19-derived plasmid carrying ctxB and rstR, constructed in our laboratory before) and cloned into the XbaI site of pXXB106, containing E. coli-derived thyA (30), resulting in two new constructs, pUTBL1–5 and pUTB1–6. ctxB and thyA have the same transcriptional direction in pUTBL1–5 and opposite directions in pUTBL1–6 (Fig. 1).

The rstR gene and its upstream sequence were amplified from El Tor strain Bin-43 with primers PrstR1 (CCGAATTCACTCACCTTGTATTCG) and PrstR2 (CGGAATTCTCGACATCAAATGGCATG). The amplified fragment was then cloned into the EcoRI site of pUTBL1–5, yielding new construct pUTBL2. Subsequently, an 0.8-kb PvuI fragment of the bla gene in pUTBL2 was deleted to generate pUTBL3. pUTBL3 was then electroporated into IEM101-T to construct IEM108.

Serum vibriocidal antibody assay. Serum vibriocidal antibody titers were measured in a microassay using 96-well plates. The immunized rabbit sera were inactivated at 56°C for 30 min and diluted 1:5 with PBS before use. The prediluted rabbit sera were added into the first well and then serially diluted threefold in PBS. PBS was added to the last well as a negative control. The plates were incubated for 30 min at 37°C with 25 µl of a solution containing 10^2 CFU of *V. cholerae* Bin-43/ml of culture and 20% guinea pig serum as a complement source in PBS. One hundred fifty microliters of 0.01% 2,3,5-trihenyltetrazolium chloride in LB broth was added to each well, and the plates were further incubated for 4 to 6 h at 37°C until the negative-control wells showed a color change. The reciprocal vibriocidal titer is defined as the highest dilution of serum that completely inhibits growth of Bin-43, i.e., no color change.

Rabbit immunization. Eight adult New Zealand White rabbits (2 to 2.5 kg) were divided into naive, IEM101, and IEM108 groups. The naive group consisted of two rabbits that were not immunized. Each immunization group had three rabbits. After fasting for 24 h, the rabbits in both immunization groups were anesthetized with ether. After the abdominal skin was sterilized with an iodine tincture and alcohol, the abdominal cavity was opened by vertical incision (under sterile conditions).

The following are handwritten margin annotations:

Experiment

Construction of IEM108

Experiment

Construction of IEM108
Organization: chronological

Measurement of serum vibriocidal antibodies

Naïve, IEM101, and IEM108 tests

	General exploration was performed to find the ileocecal region. This region was ligated to the inner wall of the abdomen. <u>Then</u> 10^9 CFU of vaccine strain IEM101 or IEM108 were injected into the proximal ileum. <u>Finally,</u> the abdominal cavity was closed. The ligature that tied the ileocecal region to the abdominal wall was removed 2 h later, and the rabbits were given water and feed for 28 days. One rabbit of the IEM101 group died after the operation, probably because of heavy anesthesia. Serum samples were collected from the immunized rabbits prior to the immunization and on days 6, 10, 14, 21, and 28 after the vaccination. The serum titers for the anti-CT antibody and vibriocidal antibody were measured as described above.

Margin labels (top to bottom): Experiment; Consistent order; Naïve, IEM101, and IEM108 tests

	Rabbit ileal loop assay and protection model. <u>To evaluate the protection efficacy *in vivo*,</u> the immunized rabbits were challenged with pure CT and four virulent *V. cholerae* strains (395, 119, Wujiang-2, and Bin-43) of different serotypes and biotypes (Table 1) 28 days after the single-dose immunization. Rabbits were anesthetized and their abdomens were opened as described above. Their intestines were tied into 4- to 5-cm-long loops, and then 10^5 to 10^8 CFU of challenge strains or 1, 2, 3, or 4 μg of pure CT was injected into each loop. Normal saline was used as negative control. At 16 to 18 h post challenge, the rabbits were sacrificed and the accumulated fluid from each loop was collected and measured. The ratio of the volume of accumulated fluid (milliliters) to the length of the loop (centimeters) was calculated for each loop in the challenged rabbits.

Margin labels: Parallel form; Evaluation of protection efficacy

Problem 11-6

1. *Paragraphs 1 and 2 relate directly to the purpose of the study. Paragraph 3 is indirectly related to the purpose.*
2. *Topic sentences have been circled.*
3. *Repeated key terms have been boxed. The passage contains no transitions. Consistent view has been indicated on the margin. Signal of subtopics are the subheadings for such.*
4. *All paragraphs of this Materials and Methods section have been well constructed, although word location has not always been taken into consideration.*

	Method
	Service utilization In collaboration with Child FIRST staff, the evaluation team developed a form that staff used to document the dates, types, and duration (recorded in fifteen-minute increments) of all services provided to children and their families.

Margin label: Parallel form

Examples of services included in-home assessment, classroom assessment and consultation, in-home care coordination, and staff consultation and supervision. Data documented on this form were also used to determine the length of time in the program from entry to discharge.

Family violence and traumatic events The Traumatic Events Screening Inventory–Parent Report Revised, or TESI-PRR (Ghosh-Ippen, Ford, Racusin, Acker, Bosquet, Rogers, et al., 2002), is a twenty-four-item semi-structured interview that determines a history of exposure to traumatic events for children six years old and younger. Parents are presented with a list of traumatic events and asked to use three response categories (yes, no, and unsure) to indicate if the child has ever experienced them.

Twelve of the twenty-four TESI-PRR items were used to screen all children entering the program for a history of family violence and thus determine eligibility for inclusion in the evaluation study: separation from a family member; suicide by someone close to the child; physical assault or physical injury or bruising of child by family member; threat of serious physical harm to child; kidnapping by a family member; witnessing by the child of physical fighting or the use of a gun, knife, or other dangerous weapon by a family member; witnessing of verbal threats to seriously harm by a family member; witnessing of arrest of family member; experiencing inappropriate sexual activity; witnessing inappropriate sexual activity; being yelled at repeatedly in a scary way or told that he or she is no good (verbal abuse); and experiencing neglect. The nonfamily violence items were exposure to serious accident; serious natural disaster; severe illness or injury of someone close; death of someone close; serious medical procedure or life-threatening illness; mugging; animal attack; community violence; direct exposure to war, conflict, or terrorism; exposure to war, conflict, or terrorism via television or radio; or other stressful events. Since the TESI is an inventory-type measurement, it is a reflexive measure in which indicators of internal consistency and other psychometric properties are not suitable.

(With permission from Lyceum Books, Inc.)

Chapter 12

Problem 12-1

1. Disruption of AUF1 did not **significantly alter** the expression level of ARE-binding proteins examined here. (**Chapter 4, Principle 17**)

2. Intravascular coagulation was common in kidney, lung, and liver of *AUF1⁻/⁻* mice after endotoxin challenge, but not **in the organs of** wild-type mice (Figure 2E). (**Chapter 4, Principle 20**)

3. Comparison of the accumulated levels of cytokine secretion after LPS stimulation showed that production of TNFα and IL-1β were **increased 10-fold** in AUF1-deficient macrophages. (**Chapter 2, Principle 2**)

4. In total, 19.7% (542/2,758) of genes located on chromosome 1 and 41.34% (456/1,103) of genes located on chromosome 2 were absent from **strain A**, which seems to indicate higher conservation in chromosome 1. (**Chapter 2, Principle 2**)

5. The number of variant genes in toxigenic strains is much less (missing 0 to 35 genes for each) **than in nontoxigenic strains**, exhibiting the highest conservation as described in previous research (**34**). (**Chapter 4, Principle 20; and missing reference**)

6. The T-test also showed that Cat-315 expression **was decreased significantly** only in the contralateral barrel cortex. (**Chapter 4, Principle 17**)

7. Three of the molecules, CB4, CB6, and CB10, **inhibited enzyme activity** more than 50%. (**Chapter 4, Principle 17**)

Problem 12-2

The paragraph lists only results and conclusions but is missing background/purpose information as well as the experimental approach:

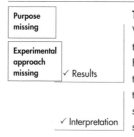

To determine..., we...

We found that the H384A mutant reduced the k_{cat} value more than 3-fold. The apparent K_m values were increased 7-fold for Fru 6-P and 3.5-fold for PPi. The increase of the K_m values and the reduction of the k_{cat} value of the H384A mutant suggest that the imidazole group of His384 is important for the binding stability as well as for catalytic efficiency of Fru 6-P and PPi substrates.

Problem 12-3

The original paragraphs contain many experimental details that should not be placed into the Results section. These details need to be removed as shown:

To evaluate inhibitory effects of the selected molecules, enzymatic reactions were triggered by adding the substrate B and monitoring the absorbance of product. **We found that 6 out of 15 molecules showed appreciable inhibition at 10 μM (Figure 5). Three of those, A3, A6, and A7, exhibited more than 50% inhibition of the enzyme activity and were further diluted to find the minimal inhibitory concentration**

(MIC). Molecules A3 and A6 exhibited 30% and 45% inhibition of the enzyme activity, respectively, even at 1 µM. DMSO was found not to interfere with the enzyme.

Problem 12-4

All the parts of a Results paragraph are there except for an overall interpretation of the results. This section would benefit from such an interpretation.

✓ Background and purpose	Because aggrecan-reactive nets are strongly expressed in the barrel cortex, and the timing of the expression of these PNs coincides with the critical period, we asked if altering whisker sensory input would alter the expression of PNs and aggrecan.
✓ Approach	In our initial study, mice had their whiskers trimmed from the right whisker pad every other day from birth through postnatal day (P) 30. Nissl staining of the barrel cortex illustrated that the development of the barrels was not altered by this manipulation (Fig. 4A,D). Our studies revealed decreased Cat-315 expression in the barrel cortex contralateral to the manipulated whisker pad (Fig. 4B,C) and normal Cat-315 expression in the ipsilateral barrel cortex (Fig. 4E,F) ($n = 11$) compared to controls ($n = 10$).
✓ Results	The number of Cat-315 positive perineuronal nets decreased 22 % as a result of sensory deprivation.
✓ More results	In addition, the repeated measures ANOVA showed a significant interaction between the hemisphere (left vs. right) and treatment (trimmed vs. not trimmed) ($p < 0.05$) (Fig. 4G). The T-test also showed a significant decrease in Cat-315 expression only in the contralateral barrel cortex of trimmed animals compared to both the left and right barrel cortices of the control animals ($p = 0.0015$, 0.0034, respectively). The total number of cells in the barrel cortex of trimmed animals did not differ from that of the controls. **Thus, our results indicate that . . .**
Interpretation of results missing	

(With permission from The Journal of Neuroscience)

Problem 12-5

The main part of the experimental approach is missing in this Results section.

✓ Background	. . . Previous studies have shown that, in biological networks, hubs tend to be essential [7,9], and betweenness of a node is correlated with its degree [20]. We found that degree and betweenness are indeed highly correlated quantities in the networks we analyzed.
Experimental approach missing	
✓ Results	Therefore, we further investigate which one of these two quantities is a better predictor of protein essentiality in both regulatory and interaction networks.
✓ Purpose	

✓ Purpose Partial experimental approach ✓ Results ✓ Interpretation	To disentangle the effects of betweenness and degree, we divided all proteins in a certain network into four categories: (1) nonhub–nonbottlenecks; (2) hub–nonbottlenecks; (3) nonhub–bottlenecks; and (4) hub–bottlenecks (see Figure 1). Even though the two quantities are highly correlated, the number of hub–nonbottlenecks and nonhub–bottlenecks is enough for reliable statistics (see Table S1). This is in agreement with the previous observation by Huang and his colleagues, who found that proteins with high betweenness but low degree (i.e., nonhub–bottlenecks) are abundant in the yeast protein interaction network [18]. *(With permission from PLoS Comput. Biology 3(4), 2007)*

Problem 12-6

All the parts of a Results paragraph are there except for an overall interpretation of the results. This section would benefit from such an interpretation.

✓ Purpose Experimental approach ✓ Results ✓ Interpretation ✓ Experimental approach ✓ Results	To explore the forces behind the strong tendency of husbands to decrease their share of housework, and the low tendency to increase their share of housework in the course of marriage, we looked at the findings from the multivariate event-history models. We first explored the role of economic resources in changing couple's division of housework in the course of marriage (Table 1). We investigate the impact of the spouse's relative economic resources on the likelihood of dividing housework either more equally or less equally in the course of marriage. We find that husbands who work a similar number of hours (Husband = Wife), or lower number of hours (Husband < Wife) than their wives, are less likely to decrease their share in household labor, compared to husbands who work longer hours than their wives (Husband > Wife). Equal earning levels between the spouses also seem to reduce the likelihood for husbands to decrease their share of housework in the course of marriage (model 3b). For couples with an "atypical" female provider earnings ratio (Husband < Wife) the effect is not significant, however. It appears that a winning margin in economic resources does more for the husband than for the wife when housework is redistributed. We also looked at the effects of family formation on the gender division of household tasks to assess how shifts in economic resources play out in this context (Table 2). We found a pronounced and significant effect for both directions of change. During the first year after childbirth, fathers seem to be about twice as likely to decrease their contribution to housework. In the same period, fathers' likelihood to increase their share in housework is reduced by almost 50 percent, compared to childless men.

This push towards a more traditional division of housework seems to come to a halt when the youngest child reaches age two. We found no indication, however, that parents readjust back to a more egalitarian division of housework when kids grow older and mothers return to their previous jobs. The time-varying economic indicators (models 2ab and 3ab) do not seem to explain these processes at all. Here, we need to be cautious not to interpret lack of significance in the economic indicators as lack of relevance, especially since the share of parents in the "non-traditional" resource categories is low. The number of mothers out-earning or working longer hours than their husbands is small in this sample, and the share of continuously working mothers is low.

✓ Results

Interpretation missing

(With permission from Daniela Grunow)

Chapter 13

Problem 13-1

1. *The way the statement is written is very unprofessional. It is better to present contradicting data in a respectful way:*
   Therefore, Ebb's data differs from ours because of variations in…
   OR: Ebb did not take X into consideration when evaluating the obtained data.
   OR: Our results do not agree with those of previous studies (Ebb, 2007).

2. *Use simple words—"necessitate" is not simple.*
   A comprehensive pharmacological evaluation of different rhubarbs **may be needed** for a more rational and accurate clinical application of this important herbal medicine.
   OR: **Different rhubarbs may have to be pharmacologically evaluated in a comprehensive way** for a more rational and accurate clinical application of this important herbal medicine.

3. *Authors should never assume that a reader can interpret a figure or table as well as the authors. Authors need to present their data and explain it clearly and not just simply point readers to a figure, letting them figure out the meaning of it themselves.*
   Counter ions influence X by… (Figure 3B).

4. *Omit or rephrase—do not stake out territory for yourself.*
   Certain single point mutations may affect the conformation of the protein and thus its interaction with the receptor.

5. *The statement sounds almost apologetic. The author does not seem to be assertive. Avoid such statements.*
   Rephrase: The theoretical model presented here provides a new tool to delineate the complex ABC system.

Problem 13-2

Version B is a better first paragraph. It signals the answer, which is followed by supporting evidence. Version A, in contrast, summarizes and repeats results without interpreting them.

Problem 13-3

Version A is a better concluding paragraph. The end is signaled clearly, the key findings are well summarized and interpreted, and the significance is indicated. Version B is simply listing limitations of the study, making for a very weak ending.

Problem 13-4

This is a good final paragraph, as it clearly signals the end of the discussion, summarizes the key findings, and interprets them. The importance of the study is also indicated by the advice given.

Problem 13-5

Signal of the conclusion	Conclusions
Overall key findings	Habitat heterogeneity, as estimated by an advanced land cover classification, provides a stronger prediction of butterfly species richness in Canada than any previously measured factor.
Specific key findings	At large spatial scales, virtually all spatial variability (90%) in butterfly richness patterns is explained by habitat heterogeneity with secondary but significant contributions from climate (especially PET) and topography. Patterns of species turnover
Interpretation of specific key findings	across the best sampled southern region of Canada are strongly related to differences in habitat composition, supporting species turnover as the mechanism through which land cover diversity
Specific key findings	may influence butterfly richness. Differences in climate are unrelated to butterfly community similarity at this scale, suggesting
Interpretation of specific key findings	that the influences of energy on richness may be indirect or limited to within-habitat diversity. These results have significant conservation implications and indicate that the role of habitat heterogeneity may be considerably more important in deter-
Statement of significance	mining large-scale species-richness patterns than previously assumed.

Problem 13-6

The first paragraph of this discussion starts with the answer(s) to the question(s) and contains similar general interpretations as are made in the concluding paragraph. Specific results are added as supporting evidence. The first paragraph is rather long, however, and could be improved by omitting some of the interfering additional details or by moving them into subsequent paragraphs as shown.

The concluding paragraph of this Discussion is well written. It clearly signals the end and summarizes all important findings and their interpretation. In addition, it lends significance to the paper by providing overall significance to its application in the field.

DISCUSSION

Our findings emphasize the need to continue to promote personal protection measures to reduce the risk of Lyme disease infection. **We have** identified three reasonable personal measures that may be protective against Lyme disease when practiced: tick checks, bathing, and insect repellents. Performing tick checks within 36 hours after spending time in the yard may reduce one's risk by as much as 46 percent. In addition, bathing may reduce one's risk by up to 57 percent, and the use of insect repellent may be protective against the disease up to 75% percent.

Studies have suggested it takes more than 24 hours for blacklegged ticks to transmit the etiologic agent[22, 23], **and therefore** prompt removal of ticks found attached to the body is a logical method of Lyme disease prevention. The effectiveness of performing tick checks has been suggested previously [8, 13,] however this is the first time it has been demonstrated in a peridomestic setting.

We also found that controls were more likely than cases to shower or bathe within two hours after spending time in the yard, reducing the risk of the disease. Frequent bathing is not currently among the commonly recommended Lyme disease prevention measures. Although it is unlikely that bathing will remove ticks that have already attached to the body, taking a shower or bath soon after spending time outside may help to remove ticks that are yet unattached, or may create an opportunity to find ticks attached to the body. In addition, the act of bathing may indirectly prevent tick bites in that it necessitates the removal of clothing that may have blacklegged ticks upon them.

Our data also suggest that the use of insect repellent may be protective against disease. This finding refers to all types of repellents, including those that contain N,N-diethyl-3-methyltoluamide (DEET), but not including permethrin insecticide. Responses varied greatly regarding the active ingredients of repellents used, and were often inconsistent with the brands and products named by the respondents. The effectiveness of insect repellents has been shown previously[8, 16], but not in the peridomestic environment. We found that wearing clothing treated with permethrin insecticide was rarely practiced. It may be that residents are unaware of permethrin products, or do not want to spend the time treating outdoor clothing (permethrin must be applied to clothing and allowed to dry for several hours before wearing). Clothing that is pre-treated with permethrin and sold in specialty stores may be too expensive for daily use in one's yard.

...

In summary, this study sought to evaluate the effectiveness of peridomestic Lyme disease prevention behaviors. Though studies suggest that Lyme disease risk in the Northeastern United States is largely peridomestic, members of the study population could have been exposed to ticks outside of their own yards. The analysis attempted to evaluate and control for this non-peridomestic risk by asking participants about their recreational, occupational, and travel exposures to ticks, although it is difficult to determine when and where people are exposed to ticks. Our findings emphasize the need to continue to promote personal protection measures to reduce the risk of Lyme disease infection. In the absence of a vaccine against Lyme disease, it is encouraging that the protective measures identified in our study—tick checks, showering or bathing after spending time in the yard, and use of repellent—are measures that anyone can take at limited expense to reduce their risk of infection. Clinicians and public health practitioners in Lyme disease endemic areas should continue to educate the public about these simple, practical methods for reducing Lyme disease risk after spending time outdoors.

Chapter 14

Problem 14-1

The abstract contains all the important parts and corresponds to the title. All parts are clearly signaled.

	Structural basis for the interaction of chloramphenicol, clindamycin, and macrolides with the peptidyl transferase center in eubacteria
Background	Ribosomes, the site of protein synthesis, are a major target for natural and synthetic antibiotics. Detailed knowledge of antibiotic binding sites is the key to understand the mechanisms of drug action. Conversely, drugs are excellent tools for studying the ribosome function. **To elucidate** the structural basis of ribosome-antibiotic interactions, **we determined** the high-resolution X-ray structures of the 50S ribosomal subunit of the eubacterium Deinococcus radiodurans complexed with the clinically relevant antibiotics chloramphenicol, clindamycin, and the three macrolides: erythromycin, clarithromycin, and roxithromycin. **We found** that antibiotic binding sites are composed exclusively of segments of 23S rRNA at the peptidyl transferase cavity and do not involve any interaction of the drugs with ribosomal proteins. **Here we report** the details of antibiotic interactions with the components of their binding sites. Our results also show the importance of Mg ions for the binding of some drugs. This structural analysis **should facilitate** rational drug design.
Question/ purpose	
Experimental approach	
Results	
Conclusion/ answer	
Significance	*(With permission from Macmillan Publishers Ltd.)*

Problem 14-2

In this Abstract not all required parts are present. The experimental approach, the results, and the conclusion are missing. The Abstract is an investigative abstract (background) but contains an overview sentence at the end, which should be omitted or, better yet, replaced. A suggested addition incorporating the missing portions is shown in bold.

Background	The cyanobacterial circadian pacemaker is an enzymatic oscillator, which orchestrates the metabolism of the bacteria to fit the day and night alternations of this planet. Interactions among KaiA, KaiB and KaiC, the three components of this oscillator,

Background

Question/
purpose

result in many oscillatory properties *in vitro*, including an overall phosphorylation level of KaiC, an apparent segregation between synchronized phosphorylation and dephosphorylation reactions, and the size and the composition of this oscillator. To explain these properties, we propose here a molecular mechanism for this pacemaker within the framework of a cyclic catalysis scheme....

Experimental
Approach
and Results/
description
and
conclusion/
implication
is missing

Added
Experimental
Approach

Added Results/
description and
conclusion/
implication

...as determined mathematically. **The mechanism includes the regulation of KaiC's enzymatic activity by KaiA and KaiB. In this system, high phosphorylation states exhibit high affinities for aggregates of KaiA dimers and KaiB dimers, leading to the formation of their higher oligomeric structures. This model deepens our understanding of the molecular mechanism by which a biological clock operates.**

(With permission from Jimin Wang)

Problem 14-3

Abstract

Background

Discovery
statement

Experimental
approach
and
Description/
Results

Significance

Many insects possess a sexual communication system that is vulnerable to chemical espionage by parasitic wasps. **We recently discovered** that a hitch-hiking (H) egg parasitoid exploits the antiaphrodisiac pheromone benzyl cyanide (BC) of the Large Cabbage White butterfly *Pieris brassicae*. This pheromone is passed from male butterflies to females during mating to render them less attractive to conspecific males. When the tiny parasitic wasp *Trichogramma brassicae* detects the antiaphrodisiac, it rides on a mated female butterfly to a host plant and then parasitizes her freshly laid eggs. **The present study demonstrates** that a closely related generalist wasp, *Trichogramma evanescens*, exploits BC in a similar way, but only after learning. Interestingly, the wasp learns to associate an H response to the odors of a mated female *P. brassicae* butterfly with reinforcement by parasitizing freshly laid butterfly eggs. **Behavioral assays,** before which we specifically inhibited long-term memory (LTM) formation with a translation inhibitor, **reveal** that the wasp has formed protein synthesis-dependent LTM at 24 h after learning. **To our knowledge,** the combination of associatively learning to exploit the sexual communication system of a host and the formation of protein synthesis-dependent LTM after a single learning event **has not been documented before**.

	We expect it to be widespread in nature, because it is highly adaptive in many species of egg parasitoids. Our finding of the exploitation of an antiaphrodisiac by multiple species of parasitic wasps suggests its use by *Pieris* butterflies to be under strong selective pressure.
Significance	

(With permission from the National Academy of Sciences, U.S.A.)

Chapter 15

Problem 15-1

1. Variation of fossil density **in** Triassic sedimentary deposits—*"with" is imprecise*
2. Antibacterial, anti-inflammatory and antimalarial activities of **the African plants *Momordica balsamina* and *Asystasia gangetica*—** *"some" is not specific*
3. Classification of Fowl Adenovirus Serotypes **through/by** genome mapping and sequence analysis of the hexon gene—*"with" is imprecise*
4. Hemolymph-dependent and -independent responses **in** Drosophila immune tissue—*"with" is imprecise*
5. Temperature dependence of **carbon dioxide sequestration by fir trees—***the noun cluster "fir tree carbon dioxide sequestration" is unclear*
6. Stable, immunogenic, and nasal-specific formulation of **Norovirus** vaccine using **virus-like-particles** and adjuvant components—*abbreviations NoV and VLP are unclear*
7. Drug-susceptibility assay for the diagnosis of TB **through** microscopic-observation—*the noun cluster "Microscopic observation drug susceptibility assay" is unclear*

Problem 15-2

1. Differences **in terpenes** of old-world and new-world species of hazelnuts
2. Single nanocrystals of platinum prepared **by partial dissolution of Au-Pt nanoalloys**
3. Widespread increase of bat mortality rates **in the Eastern United States due to White-Nose Syndrome**
4. Transmission of coccidioidomycosis **to a human via a cat bite**

Problem 15-3

1. **Adaptive, cued seed dispersal in the cactus *Mammillaria pectinifera***
2. **Groundwater flow in the South Wales coalfield: historical data informing 3D modeling**
3. **Effect** of negative mood on persistence in problem solving
4. **Hurricane intensities linked** to increased atmospheric dust from the Sahara desert
5. **Optimizing flu virus vaccination strategies: cost effectiveness analysis**

6. Quantification of *antioxidant* activity using a novel cellular *antioxidant* activity assay
   OR: **Next generation quantification of *antioxidant* activity: cellular *antioxidant* activity assays**

7. **Temperature and light requirements** for seed germination and seedling growth of *Sequoiadendron giganteum*

Problem 15-4

The best title is "4. Milk consumption prevents antioxidant effect of blueberries." It includes all important facts and is clear, complete, and succinct.

Problem 15-5

1. Transcoronary transplantation after myocardial infarction
2. Surface characteristics and mineralization of feline teeth
3. Inflammatory mechanisms in pulmonary disease
4. Carbon and Hydrogen in Palladium-Catalyzed Alkyne Hydrogenation
5. Optical Manipulation of Electron Spin in Quantum Dot
6. Structural analysis of *E. coli* hsp90
7. Decline and extinction of Brazilian tree frog
8. Determination of chlorophyll density for corn from spectral reflectance data
9. Rare Structural Variants in Schizophrenia
10. Otoferlin essential at the auditory ribbon synapse
11. Carbon and Hydrogen in Alkyne Hydrogenation
    OR: Catalytic Hydrogenation in Microreactors
12. Multi-Octave Optical-Frequency Combs

Chapter 20

Problem 20-1

Background	As the concentration of CO_2 in our atmosphere increases, so does the amount of CO_2 in the ocean's surface waters. Although the oceans are able to remove some of the increasing carbon from the atmosphere, just how much they are able to remove has
Problem	vital implications in determining the potential severity of global
Objective	warming. To understand the global carbon cycle, we propose to measure the CO_2 content of seawater over a 5-year period
Strategy	using high-precision methods for measuring total dissolved inorganic carbon (DIC), total alkalinity, and carbon and oxygen isotopes of CO_2. Results from this study will provide insight into
Significance	CO_2 absorption by sea water and ultimately increase our understanding of global warming.

Problem 20-2

Abstract

Cardiovirus is a common cause of gastroenteritis. For routine vaccination of Chinese infants, a new cardiovirus vaccine has been recommended.

Here, we propose to evaluate the impact and cost-effectiveness of the Chinese cardiovirus vaccine program using a dynamic model of cardiovirus transmission. Findings from our analysis can be used to inform policy makers to prevent rotavirus infection.

Problem 20-3

This sample abstract contains all necessary elements.

	Abstract
Background	The role that soils play in mediating global biogeochemical processes is a significant area of uncertainty in ecosystem ecology. One of the main reasons for this uncertainty is that we
Problem	have a limited understanding of belowground microbial community structure and how this structure is linked to soil processes.
Hypothesis	Building upon established theory in soil microbial ecology and ecosystem ecology, we predict that the structure of below-
Overall objective	ground microbial communities will be a key driver of carbon and nutrient dynamics in terrestrial ecosystems. We propose to
Strategy	test and develop the established theories by combining state-of-the-art DNA-based techniques for microbial community analysis together with stable isotope tracer techniques. By doing so, we
Significance	expect to advance our conceptual and practical understanding of the fundamental linkages between soil microbial community structure and ecosystem-level carbon and nutrient dynamics.
	(Mark Bradford and Noah Frierer, proposal to private foundation)

Problem 20-4

This Impact/Significance statement focuses on the scientist's needs and wishes rather than on the funders or on society. Most funders will not be interested in the number of papers that will be published. They would like to see how your efforts will advance the field and their cause.

Possible revision:
Funding from the Foundation will provide for a postdoctoral fellow and will lead to more insight into the fundamental principles of X, ultimately leading to better therapy and treatment options for Y.

Problem 20-5

Example:
The proposed project will be led by [name, title]. [name] is an expert in X and will oversee all personnel and research activities A and B.

Chapter 21

Problem 21-1

Background	Plants are our oldest source of medicines. Yet much of Earth's rich plant life remains unexplored. In recent decades, natural drug discovery has concentrated on tropical plants due to their
Problem	great diversity. However, there is equally much diversity for the plants of our oceans, and this plant life has remained untapped.
Objective	The overall goal of this proposal is to identify and purify natural chemicals of oceanic plants and to test their activity as potential medicines. We will apply new chemical fingerprinting technol-
Strategy	ogy for our screens of plant life in the Florida Keys and assess them for potential medicinal use using microbial techniques.
	Identification of plants with compounds active against import- ant human diseases, and subsequent characterization of such compounds, will lay the foundation for new drug development and lead to novel treatment therapies and better outcomes for
Significance	patients.

Problem 21-2

Habit reversal therapy (HRT), a behavioral treatment for tics, may be effective in treating Tourette syndrome (TS), an inherited neurological disorder characterized by chronic motor and vocal tics. We propose to compare the efficacy of HRT in reducing tics, improving life satisfaction and psychosocial functioning in comparison with supportive psychother- apy (SP) in 100 outpatients with TS. We will determine if HR has specific, sustainable, tic-reducing effects and if SP is effective in improving life sat- isfaction and psychosocial functioning. Assessments of response inhib- ition may help to predict treatment response to HR.   *(88 words)*

Problem 21-3

The Abstract contains all necessary elements.

Background	Global warming is arguably one of the most pressing concerns of our time. However, we lack an effective model to predict
Problem	precisely by how much the temperature will rise as a conse- quence of the increased levels of CO_2 and other factors. The
Background	width of this range is due to several uncertainties in different elements of the climate models including the variability in the Sun's rate of energy output. To gain greater insight into the rela-
Objective	tionship between solar energy output and global temperature, **we propose to launch the internationally led ABC satellite in April 2012**. Our aim is to collect for 2 years data on the solar
Strategy	diameter and shape, oscillations, and photospheric temperature variation. We will assess these data to model solar variability.
Significance	Our findings will dramatically advance our understanding of solar activity and its climate effects.

Problem 21-4

	Abstract
Background	The role of nurses continues to expand and shift in response to high societal demand for health care services. <u>However, it is unclear how family and other social responsibilities affect employment status of the profession.</u> **Our objective is to explore and understand the effect of family and other social responsibilities in diverse medical fields to better accommodate nurse professionals in relation to the health care delivery demands of our society.** <u>We will perform a prospective, blinded descriptive study by collecting and analyzing lifestyle, employment, and demographic data on a sampling of nurses in the UK.</u> A validated survey will be mailed to representative cohorts of nurses to determine gender, age, medical field, work hours per week, total time and reasons for leave of absence from the workforce, and total leave time from the workforce that is planned. Standard descriptive statistics and multivariate analysis will be used to identify the impact of family and social responsibilities on work hours based on age and gender in the setting of different fields of medicine. Family and social responsibilities are expected to affect the number of work hours per week as well as the time and reason for work leaves. These responsibilities are anticipated to differ based on age and gender and may be more prevalent in some medical fields than others. Identifying significant differences of employment patterns will enable policy makers to consider the effects of social responsibilities for the nursing profession.
Problem	
Objective	
Strategy	
Expected Outcome	
Significance	

Problem 21-5

The two statements are well written. In the Intellectual Merit section, the impact of the proposed work is clearly stated ("… with potential applications beyond the scope of this proposal…"), and in the Broader Impact section, the educational component is clearly highlighted, as is the fact that a female student will be educated—both components of importance to the NSF.

Chapter 22

Problem 22-1

In this section, the background and need is stated, but the objective/aim is missing.

Background	Healthy older adults often experience mild decline in some areas of cognition. The most prominent cognitive deficits of normal aging include forgetfulness, vulnerability to distraction

Background	and other types of interference, as well as impairment in multi-tasking and mental flexibility. These cognitive functions are the domain of the most evolved part of the human brain known as the prefrontal cortex, the brain region that is the last to fully mature in children and the first to decline as we age. Indeed, prefrontal cortical cognitive abilities begin to weaken already in middle age and are especially impaired when we are stressed.
Need is stated but no aim	Loss of these organizational abilities is a particular liability in this Information Age when our demanding lives require that we multitask and navigate through endless interferences. Thus, understanding how the prefrontal cortex changes with age is a top priority for rescuing the memory and attention functions we need to survive in our fast-paced, complex world.

Problem 22-2

In this section, the background, unknown and the objective are stated and signaled.

Background	Age-induced emphysema ("senile emphysema") is an under recognized and poorly understood phenomenon that occurs in the lungs of most people over the age of 35 years. It leads to the loss of effective oxygen exchange in the lungs, resulting in progressive shortness of breath and if severe, complete respiratory failure. We postulate that cigarette smoke-induced emphysema
Hypothesis	represents a form of accelerated lung aging, akin to what is observed in the skin and cardiovascular system of smokers. Toll-like receptor (TLR) deficiency predisposes people to emphysema. In mice deficient in TLR4, the canonical receptor for lipopolysaccharide (a component of specific bacteria) is an accelerated form of age-induced lung enlargement that resembles human emphysema both histologically and functionally (1). In people, TLR4 function declines with age (2, 3), which may in part contribute to "senile emphysema" even in the absence of significant exposures. Human studies have also found that chronic smoking leads to decreased TLR4 function in the lung and that the level of TLR4 depression is correlated with the severity of COPD (4). The synergistic or additive effects of age and smoking on TLR4 function in susceptible individuals may explain the pathogenesis
Unknown	and temporal characteristics of smoke-induced emphysema, but
Objective	this synergy has not been explored in detail. We aim to explore how TLR4 is affected by age and cigarette smoke.

(Patty Lee, proposal to private funding agency, modified)

Problem 22-3

In this section, the background, unknown and the objective are stated and signaled.

	One potential therapeutic approach for treating chronic HBV infection is through therapeutic vaccination to disrupt the immunological tolerance and to induce an immune response that is capable of controlling the virus. Despite the promise of therapeutic vaccination for treating chronic HBV, progress in this area has been limited (16, 29, 34, 64, 73, 83). A success-
Background	
Need	ful therapeutic vaccination strategy must accomplish two goals. First, immunological tolerance must be broken, and virus-specific T cells must be generated. Second, these T cells must efficiently perform their effector functions including killing target cells and producing the antiviral cytokines such as IFN-γ and TNF-α, which can noncytopathically inhibit virus replication. We will
Objective	test the hypothesis that recombinant VSV vectors are ideally suited for therapeutic vaccination for chronic HBV infection.

(Michael Robek, proposal to federal agency)

Problem 22-4

In this section, the background, unknown and the objective are stated and signaled.

	The Pliocene climate was significantly different from our con- temporary one. Most important, it was significantly warmer— high-latitude temperatures, for instance, were 4–6°C higher than today (4,5). Surface temperatures in polar regions were in fact so much higher that continental glaciers were absent from the Northern hemisphere, and the sea level was approxi- mately 25 m higher than today. A number of numerical simu- lations have previously been conducted using coupled and atmospheric GCM and the data from the PRISM projects (6).
Problem	However, many of these studies have assumed either that the Pliocene tropical climate was similar to the modern one or that the models themselves produced climate conditions in the trop- ics not far different from the modern one.
	Only recently, a wealth of new evidence has accumulated indicating that not only high latitudes but also the tropics and the subtropics had a very different climate in the early Pliocene. In particular, the tropical Pacific was characterized by what is often referred to as "a permanent El Niño-like state" (7). This phenomenon implies that the mean state of the Pacific had a significantly reduced or absent zonal SST gra- dient along the equator. In addition, a number of other dra- matic climatic changes occurred throughout the tropics and the subtropics, resulting in very different patterns of sea sur-
Background	face temperatures from what we typically observe today.

	Because recently discovered changes in the tropical climate of
Objective	the early Pliocene are so significant, we propose to undertake a
	systematic study to understand the physical mechanisms respon-
	sible for such a different climate state.

(Alexey Federov, proposal to federal agency, modified)

Chapter 23

Problem 23-1

This paragraph contains all necessary elements of a well constructed pre-liminary results section.

	ELISA for secreted HBMS. Because HBMS is a secreted protein
Purpose	(46), we determined whether the HBMS protein produced in
	VSV-infected cells is secreted into the culture media. We detected
	HBMS in the media of VSV-HBMS-infected cells but not in unin-
Experimental approach	fected or recombinant WT VSV-infected cells by both Western
	blot (data not shown) and qualitative ELISA (Table 2). We
Results	found by quantitative ELISA that secreted HBMS levels reached
	35 ng/ml in the media of VSV-HBMS-infected BHK cells by
Interpretation of results	24 h post infection. Therefore, the HBMS produced in VSV-
	infected cells is correctly processed for secretion.

(Michael Robek, proposal to federal agency)

Problem 23-2

In this paragraph, context has not been provided. This omission can be confusing for reviewers and should be avoided.

Context missing	Calculations with a General Circulation Model of the atmos-
Experimental approach	phere (6) indicate that during idealized permanent El Niño-
	like conditions, the warming of the Eastern equatorial Pacific
	reduces the area covered by stratus clouds, thus decreasing the
	albedo of the planet. At the same time, the atmospheric concen-
	tration of the powerful greenhouse gas, water vapor, increases.
	This scenario may have happened during the early Pliocene,
Results	amplifying the warm conditions at that time. Consequently,
	projections of the effect of increasing concentration of green-
	house gases on climate should include a thorough consideration
Interpretation/ conclusion	of the tropical climate conditions.

(Alexey Federov, proposal to federal agency, modified)

Problem 23-3

This paragraph contains all necessary elements of a well constructed Preliminary Results section.

Background/ context	There have been few measurements of root exudation *in situ* in forest ecosystems largely due to methodological challenges. As
Purpose of experiment	a first test to examine the role of CO_2 on exudation in loblolly pine, we grew seedlings for 150 days in controlled chambers
Experimental approach	under conditions designed to approximate the CO_2 and N gradients at the Duke FACTS-1 site. We found that pine seedlings increased mass specific exudation rates in response to elevated CO_2, and that the magnitude of the CO_2 effect was greatest in our lowest N treatment (Figure 5a). In contrast, higher N supply
Results	reduced exudation. Our preliminary data suggest that exudation may be increased by 60% in the elevated CO_2 plots, representing an increased flux of ~10 g C m-2 yr-1 to soil.
	. . .
Interpretation of results	Our preliminary data also suggest that elevated CO_2 may increase the importance of rhizosphere processes by increasing the flux of soluble exudates from roots to soil. In addition, our results indicate that CO_2-induced changes in the chemical composition of the exudates released can affect the magnitude of the microbial response and the quantity of N mineralized in the rhizosphere. Given the paucity of root exudation studies from trees *in situ*, and the absence of studies of the exudation response of field grown trees to elevated CO_2, there is a need to
Problem statement	understand the effects of elevated CO_2 on both the quantity and chemical composition of the exudates released from roots.

(Richard Phillips, proposal to federal agency, modified)

Chapter 24

Problem 24-1

Specific Aim: Optimization of triage and treatment in the context of HIV and XDR TB.

Purpose of Experimental design	In this model, we will improve upon preliminary studies to evaluate more precisely whether factors that can be assessed within
	a few days of a patient's admission can be used to determine which patients are most likely to be infected with drug resistant strains. We will thereby determine an optimal empirical treatment regimen for the interim period until their drug sensitivity results are available. Key clinical factors to be incorporated into the model include TB treatment history, hospitalization history, sputum smear results, lack of response to first-line treatment, chest X-ray deterioration, and rapid rifampin, kanamycin, and
Analysis	ciprofloxacin resistance testing results. Parameterizing our model with the data collected, we will calculate the likelihoods that a patient has non-MDR, non-XDR MDR, or XDR TB, and will deter-
Expected outcome	mine what further studies are required to enhance confidence in the likelihood calculations. The model will also account for

Significance/ Justification	the benefits and disadvantages of different treatment regimens in terms of treatment outcomes, side effects, costs, pill burdens, and the probability of acquired or amplified resistance.

(Alison Galvani, proposal to private foundation, modified)

Problem 24-2

Purpose of Experimental design	**I. Laboratory and Field Behavior Study** <u>Objective:</u> To determine the function of butterfly wing patterns by characterizing the detailed behavioral responses of insect and predators, in response to the Monarch butterfly, *Danaus plexippus* L. Laboratory experiments will incorporate information observed in the field.
Experimental design	<u>Approach and Analysis:</u> Field experiments will include a) tethering of live butterflies to vegetation to observe who predates them; b) observation of interactions of insect predators, such as mantids, with Monarch butterflies; and c) observation of interactions of avian predators, such as Blue Jays, with the same butterflies. For our experiments, artificial habitats will
Analysis	have representative irradiance, background color and complexity based on ecological information we collect in the Mohave
Expected outcome are missing	Desert. Our model insect predator will be *Mantis religiosa*, a common mantid species. Using high speed video, we will examine whether the width and number of stripes affect the target of
Significance/ Alternatives are missing	mantid attacks.

Problem 24-3

Purpose	Here we propose to study stress response behaviors in autistic children using both well-developed laboratory paradigms and
Experimental design	dense array electroencephalographic assessments of maturing cortical coherence in response to inhibitory demands on standard neuropsychological tasks under nonstressful and stressful
Analysis	conditions. We will study children's response to stressors in a well-characterized cohort of boys and girls age 5 to 12 years old. We will follow the children over one year with detailed assessments of their social and scholarly development as well as
Expected outcome are missing	their overall health. Laboratory stress procedures include asking the children to have a conversation with a trained professional and to complete simple math and social tasks in a controlled test
Interpretation Significance	setting. These procedures have been found to reliably induce changes in physiological indices of stress (heart rate, salivary cortisol, blood pressure) and to be a reliable index of individual differences in stress response.

Chapter 28

Problem 28-1

Capturing the meaning of the slide by using as few words as possible is important. Details can be filled in by the presenter.

Value of Clinic

- Welcoming environment

- Communication in native language

- Comprehensive care

- Professional and caring clinical teams

- Respected as human beings

Problem 28-2

The slide is not text heavy and contains a good layout, but the contrast between the background and the text is not good. Some of the text is hard to read, let alone find because it blends into the background. Two better alternative slide designs would be

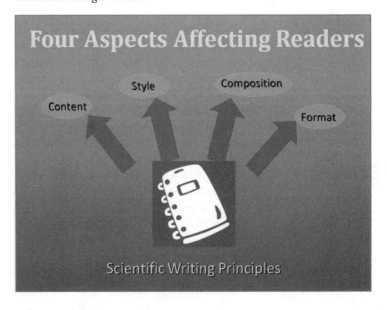

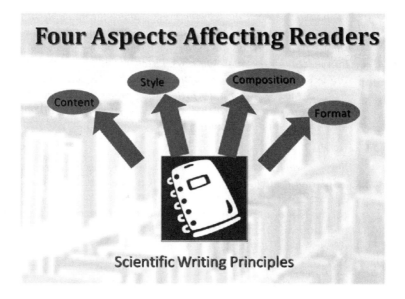

Problem 28-3

Capturing the meaning of the slide by using as few words as possible is important. Details can be filled in by the presenter. Individual sections can be highlighted by an arrow, as shown, or by different colored text.

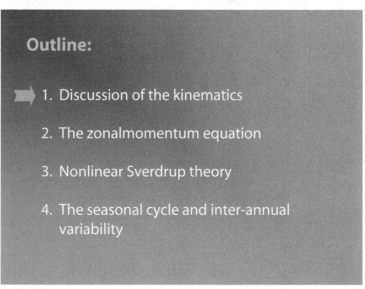

(With permission from Jaclyn Brown, modified)

Problem 28-4

The slide shows a complex table taken directly from a published paper. Avoid using figures and tables directly as published in a paper, as most of the time you end up showing much more information than needed. If it is necessary to show a table, then reconstruct the table anew, and only show relevant portions on a slide. Avoid showing more than four columns and more than seven to eight rows. Listeners will not be able to absorb all this information at once, let alone read the small lettering of a reproduced figure or table. Also, consider highlighting values that are significantly different with a different colored font to draw the viewers' attention to the most important information on a slide.

Problem 28-5

Several problems exist for this slide: (a) the figures have been copied directly from a publication; (b) they lack a title and unnecessarily show a figure legend; (c) the writing is too small to read and the copies look fuzzy; (d) the font of the heading is a serif font—it is better to use a sans serif font, such as Arial; (e) the heading is uninformative; and (f) the slide looks crowded.

Problem 28-6

1. *This ending is very weak and shows that the presenter is not very confident. It is better to end just with a simple "Thank you."*
2. *Asking the audience to focus only on part of a slide and ignore the rest tells the listeners that the presenter has not prepared his or her talk very well and/or that the slide is overcrowded.*
3. *This statement is uninformative and meaningless. When you provide an overview, make sure that the information you give is informative.*
4. *Inexperienced speakers often insert odd phrases such as these subheadings. When you give a talk, you should sound natural, using spoken English and not written English.*
5. *The speaker here is using written English instead of spoken English. This sentence sounds very awkward when used in speech.*
6. *Use of the passive voice and written style is very heavy and boring to listeners. Instead, say, for example, "We determined that...."*

Glossary of English Grammar Terms

Active Voice refers to the form of the verb used in relation to what the subject is doing. In English there are only two voices—passive and active. The *active voice* of a verb is the form of the verb used when the subject is the doer of the action.

Passive Voice: The manuscript was reviewed by the head of the department.

Active Voice: The head of the department reviewed the manuscript.

Adjective a word that modifies a noun or pronoun.

Example: The *young* birds accepted the bait readily.

Adverb a word that modifies a verb, an adjective, or another adverb. Adverbs generally answer one of four questions: how, when, where, or to what extent. Adding the suffix *-ly* to an adjective commonly turns the word into an adverb.

Examples: The reaction was *fast*. (how)

The reaction took place *immediately*. (when)

Appositive a noun, noun phrase, or noun clause that follows a noun or pronoun and renames or describes the noun or pronoun. Appositives are often set off by commas.

Example: S. aureus, *a Gram-positive organism*, can carry many resistance genes

Article a type of adjective which makes a noun specific or indefinite. In English there are three articles: the definite article *the* and the two indefinite articles *a* and *an*. In writing, an *article* is a brief nonfiction composition such as is commonly found in periodicals or journals.

Auxiliary Verb a verb that is used with a main verb. *Be, do,* and *have* are auxiliary verbs. *Can, may, must,* etc. are modal auxiliary verbs.

Clause a group of words containing a subject and verb which forms part of a sentence.

Complement a word that follows a verb and completes the meaning of the sentence or verbal phrase.

Conjunctions words that join words, phrases, or sentence parts.
Examples: and, or, for, but, nor, so, yet, either/or, neither/nor, both/ and, whether/or, not/but and *not only/but also.*

Dangling Modifiers a phrase or clause that says something different from what is meant because words are left out. The meaning of the sentence, therefore, is left "dangling."
Example: Having studied the protocol of Bowen et al., the mice received a diet rich in vitamin B. (Reads like the mice studied the protocol.)

Dependent/Subordinate Clause depends on the rest of the sentence for its meaning. It is usually introduced by a subordinating element such as a subordinating conjunction, transition word, or relative pronoun. It does not express a complete thought, so it does not stand alone. It must always be attached to a main clause that completes the meaning.

Direct Object a noun or pronoun that receives the action of a verb or shows the result of the action. It answers the question "What?" or "Whom?" after an action verb.
Example: We observed *the seals.*

Gerund a verb ending in *-ing* and used as a noun.
Example: We recommend *administering* the drug in combination with X.

Indirect Object tells *to whom* or *for whom* the action of the verb is done and who is receiving the direct object.
Example: The editor returned the manuscript *to us.*

Main or Independent Clause a clause that is not introduced by a subordinating term. It does not modify anything, and it can stand alone as a complete sentence.

Infinitive the simple present form of a verb used as either a noun, adjective, or adverb. The verb of the infinitive is normally preceded by the word *to.* When the infinitive follows some verbs as the direct object, the *to* may be dropped.
Example: Dr. Pacheco helped <u>*to write* the paper.</u>
Dr. Pacheco helped <u>*write* the paper.</u>

Jargon specialized language of a particular trade or group.

Noun a word that signifies a person, place, thing, idea, action, condition, or quality.

Object in the active voice, a noun or its equivalent that receives the action of the verb. In the passive voice, a noun or its equivalent that does the action of the verb.

Participle There are two participles in English: the present participle and the past participle.

The ***present participle*** is formed by adding *-ing* to the base form of a verb. It is used in

i) Continuous or progressive verb forms
    Example: Our competitors were *using* different conditions.
ii) As an adjective
    Example: under *varying* conditions

The ***past participle*** is formed by adding *-ed* to the base form unless it is an irregular verb. It is used

i) As an adjective
    Example: a *known* fact, at the *determined* concentration
ii) With the auxiliary verb "have" to form the past perfect tense
    Example: We have *observed* ...
iii) With the verb "be" to form the passive
    Example: The solution was *mixed*.

Passive Voice the form of the verb used when the subject is being acted on rather than doing something.

Passive Voice: The manuscript was reviewed by the head of the department.

Active Voice: The head of the department reviewed the manuscript.

Person refers to the form of a word as it relates to the subject. In English, there are three persons:

First person refers to the speaker. The pronouns *I, me, myself, my, mine, we, us, ourselves, our,* and *ours* are first person.

Second person refers to the one being spoken to. The pronouns *you, yourself, your,* and *yours* are second person.

Third person refers to the one being spoken about. The pronouns *he, she, it, him, her, himself, herself, himself, his, her, hers, its, they, them, themselves, their,* and *theirs* are third person.

Phrase a group of words not containing a subject and its verb (e.g., *in this experiment, after the eruption*).

Plural *Plural* means "more than one." To show that a noun is plural, an *-s* or *-es* is normally added to the word. There are a few irregular plurals such as *men, children, women, oxen,* and a number of words taken directly from foreign languages such as *alumni* (plural of alumnus) or *media* (plural of medium).

Plural form of pronouns—pronouns that take the place of plural nouns such as *we, you,* and *they.*

Plural form of verbs—verbs that go with a plural subject.

Possessive Case the *possessive case* of a noun or pronoun shows ownership or association. Nearly all nouns and indefinite pronouns show possession by ending with an apostrophe plus an *s.*

Example: the organism's progeny

Prepositions words that relate a noun or pronoun (called the object of the preposition) to another word in the sentence. The preposition and the object of the preposition together with any modifiers of the object are known as a ***prepositional phrase***.

Example: in, of, under, over, for, to, at, with

Prepositional Phrase a phrase beginning with a preposition and ending with a noun or pronoun. The phrase relates the noun or pronoun to the rest of the sentence. The noun or pronoun being related by the preposition is called the ***object of the preposition***.

Pronoun a word takes the place of a noun in a sentence.

Example: this, that, these, both, either

Sentence a group of words that expresses a thought. A sentence conveys a statement, question, exclamation or command and must contain a verb and (usually) a subject. It starts with a capital letter and ends with a period (.), question mark (?), or exclamation mark (!).

Redundant means "needlessly repetitive."

Singular the form of a word representing or associated with one person, place, or thing. The term is normally used in contrast to plural. The singular form of a verb goes with a singular subject.

Subject the main noun (or equivalent) in a sentence about which something is said or who does something.

Tense the tense of a verb shows the time when an action or condition occurred (past, present, or future).

Verb the word or words that express action or say something about the condition of the subject.

References

Alley, Michael. The craft of editing: A guide for managers, scientists and engineers. Springer, New York, 2000.

Alley, Michael. The craft of scientific presentations: Critical steps to succeed and critical errors to avoid. Springer, New York, 2003.

Alley, Michael. The craft of scientific writing. 3rd ed. Springer, New York, 1996.

Altman, Rick. Why most PowerPoint presentations suck. 1st ed. Harvest Books, 2007.

American Medical Association manual of style. 9th ed. Annette Flanigan, et al. Lippincott, Williams & Wilkins, 1997.

American Medical Association manual of style: a guide for authors and editors. 10th ed. Oxford University Press, USA, 2007.

American National Standards Institute, Inc. American national standard for the abbreviation of titles of periodicals. American National Standards Institute, Inc., New York, 1969.

American National Standards Institute, Inc. American national standard for the preparation of scientific papers for written or oral presentation. American National Standards Institute, Inc., New York, 1979.

American National Standards Institute, Inc. American national standard for writing abstracts. American National Standards Institute, Inc., New York, 1979.

Atkinson, Cliff. Beyond bullet points: Using Microsoft® Office PowerPoint® 2007 to create presentations that inform, motivate, and inspire. Microsoft Press, 2007.

Booth, V. Communicating in science: Writing a scientific paper and speaking at scientific meetings, 2nd ed. Cambridge University Press, Cambridge, 1993.

Browner, Warren S. Publishing and presenting clinical research. Lippincott, Williams and Wilkins, 1999.

Browning, Beverly. Perfect phrases for writing grant proposals. 1st ed. McGraw-Hill, 2007.

CBE Scientific Illustration Committee. Illustrating science: Standards for publication. Council of Biology Editors, Bethesda, MD., 1988.

The complete writing guide to NIH behavioral science grants. Editors: Lawrence M. Scheier, William L. Dewey. 1st ed. Oxford University Press, USA, 2007.

Covey, Franklin. Style guide for business and technical communication. Franklin Covey Co., 1997.

Davis, M. and Fry, G. Scientific papers and presentations, 2nd ed. Academic Press, London, 2004.

Davis, Martha. Scientific papers and presentations. Academic Press, 2002.

Day, Robert A. How to write and publish a scientific paper. 5th ed. The Oryx Press, 1998.

Dodd, Janet S. The ACS Style Guide: A manual for authors and editors. American Chemical Society. 2nd ed., 1997.

Ebel, Hans. F., Bliefert, Claus, and Russey, William E. The art of scientific writing: From student report to professional publications in chemistry and related fields. Wiley-VCH, 2nd ed., 2004.

Feibelman, Peter J. A Ph.D. is not enough: A guide to survival in science. Basic Books, 1993.

Foley, Stephen Merriam and Gordon, Joseph Wayne. Conventions and choices: A brief book of style and usage. D.C. Heath and Company, Toronto, 1986.

Fowler, H.W. A dictionary of modern English usage. 2nd ed. Oxford University Press, London, 1965.

Fowler, H.W. A dictionary of modern English usage. Wordsworth Editions Ltd, 1997.

Geever, Jane C. The Foundation Center's guide to proposal writing. 5th ed. Foundation Center, 2007.

Gerin, William. Writing the NIH grant proposal: A step-by-step guide. Sage Publications, Inc., 2006.

Gibaldi, Joseph. MLA Handbook for writers of research pepers. 6th ed. The Modern Language Association of America, 2004.

Glatzer, Jenna. Outwitting Writer's Block and other problems of the pen. The Lyon's Press, 2003.

Gopen, George D. Expectations: Teaching writing from the readers perspective. Pearson Longman 2004.

Gopen, George D. and Swan, Judith A. The science of scientific writing. American Scientist Vol. 78, pp. 550–558, 1990. Also available on the web: https://www.americanscientist.org/issues/id.877,y.0,no.,content.true,page.1, css.print/issue.aspx, last accessed January, 2009.

Greenbaum, Sidney. The Oxford English grammar, Oxford University Press, USA; 1st ed., 1996.

Gustavi, B. How to write and illustrate a scientific paper. Cambridge University Press, Cambridge, 2003.

Hacker, Diana. Rules for writers. 5th ed. Bedford/St Martin's, 2004.

Hall, Mary S. and Howlett, Susan. Getting funded: The complete guide to writing grant proposals, 4th ed. Continuing Education Press, 2003.

Harris, Dianne. The complete guide to writing effective & award-winning grants: Step-by-step instruction. Atlantic Publishing Company, 2008.

Henson, Kenneth. Grant writing in higher education: A step-by-step guide. Allyn & Bacon, 2003.

Huth, E.J. How to write and publish papers in the medical sciences. Williams & Wilkins, 1990.

Huth, E.J. Writing and publishing in medicine. 3rd ed. Lippincott, Williams and Wilkins, 1998.

Iles, Robert L. and Volkland, Debra. Guidebook to better medical writing. Iles Publications, revised ed. 2003.

International Committee of Medical Journal Editors. Uniform requirements for manuscripts submitted to biomedical journals: sample references. http://www.nlm.nih.gov/bsd/uniform_requirements.html, last accessed October 2009.

International Committee of Medical Journal Editors. Uniform requirements for manuscripts submitted to biomedical journals: writing and editing of biomedical publication. http://www.icmje.org/, last accessed October 2009.

Katz, Michael Jay. From research to manuscript. Springer, 2006.

Knowles, Cynthia. The first-time grantwriters guide to success. Corwin Press, 2002.

Korner, Ann M. Guide to publishing a scientific paper. Bioscript Press, 2004.

Lindsay, David. A guide to scientific writing. Longman, 2nd ed., 1995.

Lynch, Jack The English language: A user's guide, Focus Publishing/R. Pullins Company, Newsburyport, MA, 2008.

Malmfors, B., Grossman, M., and Garnsworth, P. Writing and presenting scientific papers. Nottingham University Press. Nottingham, 2004.

Matthews, Janice R., Bowen, John M., and Matthews, Robert W. Successful scientific writing. Cambridge University Press, 1996.

McMillan, Victoria E. Writing papers in the biological sciences. Bedford Books of St. Martin's Press, Boston 1997.

Morgan, Scott and Whitener, Barrett. Speaking about science: A manual for creating clear presentations. 1st ed. Cambridge University Press, 2006.

O'Connor, Maeve. Writing successfully in science. Chapman and Hall, 1991.

Ogden, Thomas E. and Goldberg, Israel A. Research proposals: A guide to success. 3rd ed. Academic Press, 2002.

Peat, Jennifer and Barton, Belinda. Medical Statistics: A guide to data analysis and critical appraisal. 1st ed. BMJ Books, 2005.

Peat, Jennifer, Elliott, Elizabeth, Baur, Louise, and Keena, Victoria. Scientific writing: Easy when you know how. 1st ed. BMJ Books, 2002.

Penrose, Ann M. and Katz, Stephen B. Writing in the sciences: Exploring conventions of scientific discourse. 2nd ed. Longman, 2004.

Perelman, Leslie C., Paradis, James, and Barrett, Edward. The Mayfield handbook of technical and scientific writing. Mayfield Publishing Company, 1997.

Rogers, S.M. Mastering scientific and medical writing: A self-help guide. Springer, Berlin, 2006.

Rubens, Philip. Science and technical writing : A manual of style. Henry Holt and Company, Inc., New York, 1992.

Ryckman, W.G. What do you mean by that? The art of speaking and writing clearly. Dow Jones-Irwin, 1980.

Scientific Style and Format, the CBE Manual for authors, editors and publishers. 6th ed. Council of Biology Editors, 1994.

Silyn-Roberts, Heather. Writing for Science and Engineering. Butterworth-Heinemann, 2000.

Sternberg, Robert. The psychologist's companion: A guide to scientific writing for students and researchers. 4th ed. Cambridge University Press, New York, 2003.

Strunk, W Jr and White, EB. The elements of style. 3rd ed. MacMillan, NY, 1979.

Style Manual Committee, Council of Biology Editors. Scientific style and format: The CBE manual for authors, editors, and publishers. 6th ed. Cambridge University Press, 1994.

Sullivan, K. D. and Eggleston, Merilee. The McGraw-Hill desk reference for editors, writers, and proofreaders. 1st ed. McGraw-Hill, 2006.

Taylor, Robert B. Clinicians Guide to medical writing. Springer, 1. ed., 2004.

Teitel, Martin. "Thank you for submitting your proposal": A foundation director reveals what happens next. Emerson & Church, 2006.

Thurman, Susan. The everything grammar and style book. Adams Media Corporation, Avon, MA, 2002.

University of Chicago Press Staff. The Chicago manual of style. 15th ed. University of Chicago Press, 2003.

Venolia, Jan. Write right!: A desktop digest of punctuation, grammar, and style. 4th ed. Ten Speed Press, 2001.

Venolia, Jan. Rewrite right!: Your guide to perfectly polished prose. 2nd ed. Ten Speed Press, 2000.

Wason, Sara D. Webster's new world grant writing handbook. 1st ed. Webster's New World, 2004.

Williams, Joseph M. Style: Lessons in clarity and grace. 9th ed. Longman, 2006.

Williams, Joseph M. Style: Ten lessons in clarity and grace. Scott, Foresman & Co., 1988.

Yang, Jen Tsi. An outline of scientific writing. For researchers of english as a foreign language. World Scientific Publishing Company, 1995.

Yang, Otto O. Guide to effective grant writing: How to write a successful NIH grant application. 1st ed. Springer, 2005.

Young, Petey. Writing and presenting in English: The Rosetta Stone of science. 1st ed. Elsevier Science, 2006.

Zeiger, Mimi. Essentials of writing biomedical research papers. 2nd ed. The McGraw-Hill Companies, 2000.

Credits

Problem 6-24 Reprinted from *Behaviour Research & Therapy* 44(8), Thilo Deckersbach, Scott Rauch, Ulrike Buhlmann and Sabine Wilhelm, Habit reversal versus supportive psychotherapy in Tourette's disorder: A randomized controlled trial and predictors of treatment response, *1079–1090*, Copyright (2008), with permission from Elsevier.

Example 7-1 Adapted from *Future Microbiol*, 2007 2(6), 571–574 with permission of Future Medicine Ltd.

Example 8-13 *Corrosion Science* Volume 50, Issue 3. 2008: K.E. García, C.A. Barrerob, A.L. Moralesb and J.M. Grenechec Lost iron and iron converted into rust in steels submitted to dry–wet corrosion process. 763–772. Copyright (2008), with permission from Elsevier.

Problem 8-2 *Mol Gen Genet* 261, 1999:92–99. A specific member of the Cab multigene family can be efficiently targeted and disrupted in the moss Physcomitrella patens. A. H. Hofmann, A. C. Codón, C. Ivascu, V. E. A. Russo, C. Knight, D. Cove, D. G. Schaefer, M. Chakhparonian and J.-P. Zrÿd. Copyright Springer (1999). With kind permission of Springer Science and Business Media.

Problem 8-3 *Proceedings of the National Academy of Sciences,* 105(8). 2008: 2788–2793. From the Cover: Hydrogen and oxygen isotope ratios in human hair are related to geography. Ehleringer, J.R., Bowen, G.J., Chesson, L.A., West, A.G., Podlesak, D.W., Cerling, T.E. Copyright (2008) National Academy of Sciences, U.S.A.

Problem 8-4 *Proceedings of the National Academy of Sciences,* 105(5). Abrupt climate change and collapse of deep-sea ecosystems.: 1556–1560. Yasuhara, M., Cronin, T. M., deMenocal, P.B., Okahashi, H. and Linsley, B.K., Copyright (2008) National Academy of Sciences, U.S.A.

Problem 8-7 Used with permission of Geological Society of America, from Global frequency of magnitude 9 earthquakes, McCaffrey, R., Geology 36(3), 2008; permission conveyed through Copyright Clearance Center, Inc.

Problem 8-8 Jefferson, T. A., Stacey, P. J., and Baird, R. W. (2008). A review of Killer Whale interactions with other marine mammals: predation to co-existence. Mammal Review 21(4), 151–180.

Example 9-4 With permission from the author, Rudolf Lurz.

Example 9-5 With permission from the author, Roland Geerken.

Example 9-14b Reprinted with permission from Hanna Richter, Organisation and transcriptional regulation of the polyphenol oxidase (PPO) multigene family of the moss Physcomitrella patens (Hedw.) B.S.G. and functional gene knockout of PpPPO1, Ph.D. Thesis, Universität Hamburg, 2009.

Example 9-23 Reprinted by permission from Macmillan Publishers Ltd: *Heredity*, 93(1), J P Townsend, D M Rand, Mitochondrial genome size variation in New World and Old World populations of Drosophila melanogaster, p. 98–103. Copyright (2004).

Example 9-24 *Journal of Computational Biology* 14(4), Diallo, A.B., Akarenkov, V., and Blanchette, M., Exact and Heuristic Algorithms for the Indel Maximum Likelihood Problem, pp. 446–461. Copyright (2007). The publisher for this copyrighted material is Mary Ann Liebert, Inc. publishers.

Problem 9-4 Used with permission of American Society of Plant Biologists, from Cytokinin Overproducing ove Mutants of Physcomitrella patens Show Increased Riboside to Base Conversion, Peter A. Schulz, Angelika H. Hofmann, Vincenzo E.A. Russo, Elmar Hartmann, Michel Laloue, and Klaus von Schwartzenberg, Plant Physiology, 126; copyright (2001), permission conveyed through Copyright Clearance Center, Inc.

Example 10-8 With permission from the author, Jaclyn Brown.

Example 10-12 From ISPRS Journal of Photogrammety and Remote Sensing (2009), doi: 10.1016/j.isprsjprs.2009.03.001, Geerken, R.A., "An algorithm to classify and monitor seasonal variations in vegetation phenologies and their inter-annual change." Copyright 2009 by Elsevier. Printed with permission.

Example 10-13 Reprinted from Analytical Biochemistry, 230(1), A. Hofmann, M. Tai, W. Wong and C. G. Glabe, A Sparse Matrix Screen to Establish Initial Conditions for Protein Renaturation, 8-15, Copyright (1995), with permission from Elsevier.

Problem 10-3 From Veterinary Clinical Pathology, Volume 37, Number 3. C. N. Gutiérrez, M. Martínez, E. Sánchez, M. De Vera, M. Rojas, J. Ruiz, and F. J. Triana-Alonso. "Cultivation and molecular identification of Ehrlichia canis and Ehrlichia chaffeensis from a naturally co-infected dog in Venezuela," p. 258–265.

Problem 10-4 Adapted from Genes Dev. 20. Jin-Yu Lu, Navid Sadri and Robert J. Schneider, " Endotoxic shock in AUF1 knockout mice mediated by failure to degrade proinflammatory cytokine mRNAs," pp. 3174–3184. Copyright (2006) Cold Spring Harbor Laboratory Press. Adapted with permission.

Problem 10-5 With permission from the author, Stefanie Leacock.

Problem 11-4 Reprinted from *Analytical Biochemistry* 230(1), A. Hofmann, M. Tai, W. Wong, and C. G. Glabe, "A sparse matrix screen to establish initial conditions for protein renaturation," pp. 8–15, Copyright (1995), with permission from Elsevier.

Problem 11-5 From *Infection and Immunity, 71*(10), 2003, pp. 5498–5504. DOI: 10.1128/IAI.71.10.5498-5504.2003. Reproduced with permission from the American Society for Microbiology.

Problem 11-6 From *Best Practices in Mental Health.* Cindy A. Crusto, Darcy I. Lowell, Belinda Paulicin, Jesse Reynolds, Richard Feinn, Stacey R. Friedman, and Joy S. Kaufman. Vol. 4, No. 1, Lyceum Books, Inc. Copyright 2008; Used with permission of Lyceum Books, Inc., permission conveyed through Copyright Clearance Center, Inc.

Example 12-7 Reprinted from *Journal of the American Society for Mass Spectrometry.* 18(1), Ye M. Han J. Chen H. Zheng J. Guo D., "Analysis of phenolic compounds in rhubarbs using liquid chromatography coupled with electrospray ionization mass spectrometry," 82–9, Copyright (2007), with permission from Elsevier.

Example 12-9 Reprinted from *Developmental Biology 156(1)*, Lopez, A. Miraglia, S.J. and Glabe C.G., "Structure/Function Analysis of the Sea Urchin Sperm Adhesive Protein Bindin," pp 24–33, Copyright (1993), with permission from Elsevier.

Example 12-21 With permission from the author, Moshe Herzberg.

Example 12-23 Reprinted by permission from Macmillan Publishers Ltd: Nature 413. Schluenzen, F., Zarivach, R., Harms, J., Bashan, A., Tocilj, A, Albrecht, R., Yonath, A. and Franceschi, F. Structural basis for the interaction of antibiotics with the peptidyl transferase centre in eubacteria, pp. 814–821, Copyright (2001).

Problem 12-4 Reprinted with permission from *The Journal of Neuroscience.* Online by Paulette A. McRae, Mary M. Rocco, Gail Kelly, Joshua C. Brumberg, and Russel T. Matthews. "Sensory Deprivation Alters Aggrecan and Perineuronal Net Expression in the Mouse Barrel Cortex, 27 (20), Copyright 2007 by Society for Neuroscience. Reproduced with permission of Society for Neuroscience in the format Textbook via Copyright Clearance Center.

Problem 12-5 Reprinted from *PLoS Comput Biol 3(4): e59.* Yu H, Kim PM, Sprecher E, Trifonov V, Gerstein M. The Importance of Bottlenecks in Protein Networks: Correlation with Gene Essentiality and Expression Dynamics. 2007.

Problem 12-6 With permission from the author, Daniela Grunow.

Example 13-1 With permission from the author, Neeta Connally.

Example 13-2 Reprinted from *Developmental Biology* 156(1), Lopez, A., Miraglia, S.J. and Glabe, C.G. "Structure/Function Analysis of the Sea Urchin Sperm Adhesive Protein Bindin," pp. 24–33, Copyright (1993), with permission from Elsevier.

Example 13-3 Reprinted from *Current Genetics 29(5),* Backert, S., Lurz, R. and Börner, T. "Electron microscopic investigation of mitochondrial DNA from *Chenopodium album,*" pp. 427–436, Copyright (1996), with permission from Springer.

Example 13-4 Reprinted from *Mol Gen Genet 261,* 1999:92–99. Hofmann, A. H., Codón, A. C. Ivascu, C. Russo, V. E. A. Knight, C. Cove, D. Schaefer, D. G. Chakhparonian, M. and Zrÿd. J.-P. A specific member of the Cab multigene family can be efficiently targeted and disrupted in the moss Physcomitrella patens. Copyright Springer (1999). With kind permission of Springer Science and Business Media.

Example 13-6 Reprinted from *Developmental Biology* 156(1), Angelika Lopez, Sheri J. Miraglia and Charles G. Glabe, Structure/Function Analysis of the

Sea Urchin Sperm Adhesive Protein Bindin, pp. 24–33, Copyright (1993), with permission from Elsevier.

Example 13-9 Reprinted by permission from Macmillan Publishers Ltd: Nature 413. Schluenzen, F., Zarivach, R., Harms, J., Bashan, A., Tocilj, A, Albrecht, R., Yonath, A. and Franceschi, F. Structural basis for the interaction of antibiotics with the peptidyl transferase centre in eubacteria, pp. 814–821, Copyright (2001).

Example 13-12 Reprinted from *Analytical Biochemistry*, 230(1), Hofmann, A. Tai, M. Wong, W. and Glabe, C. G. "A Sparse Matrix Screen to Establish Initial Conditions for Protein Renaturation," pp. 8–15, Copyright (1995), with permission from Elsevier.

Problem 13-4 Reprinted from *PNAS* 102(50), Held, I. M. Delworth, T. L. Lu, J. Findell, K. L. and Knutson, T. R. "Simulation of Sahel drought in the 20th and 21st centuries," pp. 17891–6; Copyright (2005) National Academy of Sciences, U.S.A.

Problem 13-5 Reprinted from *PNAS* 98 (20). Kerr, J. T., Southwood, T. R. E. and Cihlar, J. "Remotely sensed habitat diversity predicts butterfly species richness and community similarity in Canada," p. 11369. Copyright (2001) National Academy of Sciences, U.S.A.

Problem 13-6 With permission from the author, Neeta Connally.

Example 14-2 Reprinted from the *New England Journal Medicine* 336 (19). Luzuriaga, K., Bryson,Y., Krogstad, P., Robinson, J., Stechenberg, B., Lamson, M., Cort, S., and Sullivan, J.L. "Combination treatment with zidovudine, didanosine, and Nevirapine in infants with human immunodeficiency virus type 1 infection," pp. 1343–9. Copyright © 1997 Massachusetts Medical Society. All rights reserved.

Example 14-3a Reprinted from *Science* 2003, 302 (5644). Schiestl, F.P., Peakall, R., Mant, J.G., Ibarra, F., Schulz, C., Franke, S., Francke, W. The Chemistry of Sexual Deception in an Orchid-Wasp Pollination System, p. 437. Reprinted with permission from AAAS.

Example 14-3b Reprinted from *Proc Natl Acad Sci U S A*. 104(13). Zhang, R., Li, G., Fan, J., Wu, D.L., Molina, M.J. Intensification of Pacific storm track linked to Asian pollution. p. 5295. Copyright (2007) National Academy of Sciences, U.S.A.

Example 14-4 Reprinted from *Proc Natl Acad Sci U S A*. 105(48). Sanjay Basu, Gretchen B. Chapman, and Alison P. Galvani. Integrating epidemiology, psychology, and economics to achieve HPV vaccination targets. pp. 19018–19023 Copyright (2008) National Academy of Sciences, U.S.A.

Example 14-6 With permission from the author, Irene Bosch

Problem 14-1 Reprinted by permission from Macmillan Publishers Ltd: Nature 413. Schluenzen, F., Zarivach, R., Harms, J., Bashan, A., Tocilj, A, Albrecht, R., Yonath, A. and Franceschi, F. Structural basis for the interaction of antibiotics with the peptidyl transferase centre in eubacteria, pp. 814–821, Copyright (2001).

Problem 14-2 With permission from the author, Jimin Wang.

Problem 14-3 Reprinted with permission from *PNAS* 106 (3). Martinus E. Huigens, Foteini G. Pashalidou, Ming-Hui Qian, Tibor Bukovinszky, Hans M. Smid, Joop J. A. van Loon, Marcel Dicke and Nina E. Fatouros. Hitch-hiking parasitic wasp learns to exploit butterfly antiaphrodisiac, p. 820. Copyright (2009) National Academy of Sciences, U.S.A.

Example 18-3 From *Science* 5, 321. 2008. Daniel Rosenfeld, Ulrike Lohmann, Graciela B. Raga, Colin D. O'Dowd, Markku Kulmala, Sandro Fuzzi, Anni Reissell, Meinrat O. Andreae. Flood or Drought: How Do Aerosols Affect Precipitation? p. 1309. Reprinted with permission from AAAS.

Example 18-5a Reprinted, with permission, from the *Annual Review of Ecology, Evolution and Systematic,*Volume 38. Aronson, R.B., Thatje, S., Clarke, A., Peck, L.S., Blake, D.B., Wilga, C.D., and Seibel, B.A. Climate Change and Invasibility of the Antarctic Benthos, pp. 129–154. Copyright ©2007 by Annual Reviews www.annualreviews.org.

Example 18-5b Reprinted from *Trends in Plant Science* 2(12), Backert, S., Nietsen, B.L., and Boerner, T. The mystery of the rings: structure and replication of mitochondrial genomes from higher plants, p. 477. Copyright (1997), with permission from Elsevier.

Example 18-6 Reprinted from *Science* 5, 321. 2008. Daniel Rosenfeld, Ulrike Lohmann, Graciela B. Raga, Colin D. O'Dowd, Markku Kulmala, Sandro Fuzzi, Anni Reissell, Meinrat O. Andreae. Flood or Drought: How Do Aerosols Affect Precipitation? p. 1309. Reprinted with permission from AAAS.

Example 18-7 Reprinted, with permission, from the *Annual Review of Ecology, Evolution and Systematic*, Volume 38. Aronson, R.B., Thatje, S., Clarke, A., Peck, L.S., Blake, D.B., Wilga, C.D., and Seibel, B.A. Climate Change and Invasibility of the Antarctic Benthos, pp.129–154. Copyright ©2007 by Annual Reviews www.annualreviews.org.

Example 20-3 With permission from Alison Galvani, Yale University.
Problem 20-3 With permission from Mark Bradford, Yale University.
Example 21-4A With permission from Patty Lee, Yale University.
Example 21-5 With permission from Jun Korenaga, Yale University.
Example 21-13 With permission from Hong Tang, Yale University.
Problem 21-5 With permission from Jun Korenaga, Yale University.
Example 22-3 With permission from Richard Phillips, Indiana University.
Example 22-8 With permission from Hong Tang, Yale University.
Example 22-9 With permission from Michael Robek, Yale University.
Problem 22-2 With permission from Patty Lee.
Problem 22-3 With permission from Michael Robek.
Problem 22-4 With permission from Alexey Federov.
Example 23-3 With permission from Michael Robek, Yale University.
Example 23-11 With permission from Michael Robek, Yale University.
Problem 23-1 With permission from Michael Robek, Yale University.
Problem 23-2 With permission from Alexey Federov, Yale University.
Problem 23-3 With permission from Richard Phillips, Indiana University.
Example 24-2 With permission from Mark Bradford, Yale University.
Example 24-3 With permission from Michael Robek, Yale University.
Example 24-4 With permission from Michael Robek, Yale University.
Problem 24-1 With permission from Alison Galvani, Yale University.
Example 27-2 With permission from Roland Geerken, GTZ.
Example 27-5 With permission from Roland Geerken, GTZ.
Example 27-7 With permission from Roland Geerken, GTZ.

Example 27-10 With permission from Betty Liu, Yale University and currently Keck Foundation.

Example 27-11 With permission from Roland Geerken, GTZ.

Example 27-12 With permission from Alexey Federov, Yale University and Jaclyn Brown, Centre for Australian Weather and Climate Research (CAWCR).

Example 28-2 With permission from Patty Lee, Yale University, and Robert Homer, Yale University.

Example 28-3 With permission from Mark Bradford, Yale University.

Example 28-4 With permission from Betty Liu, Yale University.

Example 28-5 Reprinted from *The Neuroscientist 7(5),* Tadzia Grandpre, Stephen M. Strittmatter. Nogo: A Molecular Determinant of Axonal Growth and Regeneration, p. 10. Copyright (2001) by SAGE Publications, Reprinted by Permission of SAGE Publications.

Problem 28-3 With permission from Jaclyn Brown, Centre for Australian Weather and Climate Research (CAWCR).

Problem 28-4 Table reprinted from *Analytical Biochemistry* 230(1), A. Hofmann, M. Tai, W. Wong and C. G. Glabe. A Sparse Matrix Screen to Establish Initial Conditions for Protein Renaturation, p.8, Copyright (1995), with permission from Elsevier.

Problem 28-5 Figure reprinted from *Developmental Biology 156(1)*, Lopez, A., Miraglia, S.J., and Glabe, C.G. Structure/Function Analysis of the Sea Urchin Sperm Adhesive Protein Bindin, p.10, Copyright (1993), with permission from Elsevier.

Example 29-4 With permission from Betty Liu, Yale University and currently Keck Foundation.

Example 29-6 With permission from Mark Bradford, Yale University.

INDEX